ELECTRIC CIRCUITS FOR TECHNOLOGISTS

Jack L. Waintraub, P.E.
Middlesex County College

Edward Brumgnach, P.E.
Queensborough Community College

WEST PUBLISHING COMPANY
ST. PAUL NEW YORK SAN FRANCISCO LOS ANGELES

50 W. Kellogg Boulevard
P.O. Box 64526
St. Paul, MN 55164-1003

Printed in the United States of America

Library of Congress Cataloging-in-Publication Data

Waintraub, Jack L.
Electric circuits for technologists.
Includes index.
1. Electric circuits. 2. Electronic circuits.
I. Brumgnach, Edward. II. Title.
TK454.W35 1989 621.319′2 88-33769
ISBN 0-314-48149-4

TO OUR FAMILIES—
Betty, Adrienne, and Michael Waintraub
Jean, Ian, and Lara Brumgnach

TO OUR PARENTS—
Harry and Ida Waintraub
Amalia Brumgnach and In Memory of Miro Brumgnach

Contents

ELECTRIC CIRCUITS FOR TECHNOLOGISTS

Preface

The field of electricity and electronics is constantly changing, but the study of electric circuits remains the foundation of electrical/electronics technology. This textbook is geared toward students at 2- and 4-year colleges and technical institutes who are embarking on a career in electrical/electronics technology. It is written to provide a foundation for future electrotechnical studies.

The book is divided into four sections, to be used in traditional two semesters or two quarters of an electric circuits (dc/ac) course. It can also be covered in a one-semester course with the instructor's discretion as to the selection of topics. The sequence of topics was designed to suit most electrical technology programs of study. It is thorough in coverage but does not belabor any particular area.

The objectives of this textbook are:

1. To present the material in a very clear and concise manner best suited for *electrical/electronics engineering technology* programs.
2. To provide a proper balance between *theory and applications*.
3. To maintain a mathematical level that reflects *current admission and first-year* postsecondary requirements of electrical/electronics technology students.
4. To make use of the *computer as a tool* in the solution of electric circuit problems.
5. To provide a pedagogically sound basis for future studies of electrical/electronics technology.

6. To enable the instructor to convey the material in a most efficient and effective manner.

Each chapter includes many illustrative examples and a large number of end-of-chapter problems to reinforce the knowledge gained in the chapter. The reading level is carefully monitored for consistency. Many of the examples and end-of-chapter problems are computer-oriented. Computer analysis methods in addition to conventional analysis methods are shown. Sample computer programs are provided. The computer and the calculator are used as a problem-solving tool.

Chapters 1 through 4 are used to introduce essential electrical concepts and to define the fundamental electrical quantities. Dc resistive circuits are studied in Chapters 5 through 8. Sinusoidal and other time-varying functions are introduced in Chapter 9. Ac circuits are covered in Chapters 10 through 12. Chapters 13 through 15 include such topics as electrical power, resonance and filter circuits, and instrumentation.

SPICE is introduced in Appendix H to familiarize the student with a computer analysis program widely used throughout the industry.

The first half of the book assumes that the student has a basic knowledge of algebra and is concurrently taking a technical mathematics course with trigonometry, which will be utilized in the second half of the book.

This textbook is designed to serve a wide audience in electric circuit courses. It is written as a modern technology text, with the students and instructor in mind.

We thank all the people who helped us make this project a reality. We are thankful to all the reviewers for their valuable and constructive criticism, and especially to Professor A. W. Avtgis, Professor D. B. Beyer for his contribution during the early stages of the process, Mrs. Eileen Schreck for typing the manuscript, Mr. T. Michael Slaughter, the editor who gave us a great deal of support, and to all the editorial staff at West Educational Publishing connected with this project.

Jack L. Waintraub, P.E.
Edward Brumgnach, P.E.

SECTION I

ELECTRIC QUANTITIES AND CIRCUITS

CHAPTER 1

Electricity and Electronics

GLOSSARY

amplifier—an electronic circuit used to magnify electric signals
computer—electronic system used to carry out calculations
electrical engineering—a branch of engineering dealing with electricity and magnetism
electrical machinery—motors and generators
electricity—study of resulting forces when two charged bodies are placed in the vicinity of each other

electronics—branch of electrical engineering dealing with processing and control of electric signals
integrated circuit (IC)—a self-contained complete electronic circuit that occupies a small area
magnetism—the property of being a magnet; having the power to attract
telegraph—communication system used to transmit messages by electric impulses
telephone—a system used for transmitting and receiving speech
transformer—a device used in power distribution
transistor—semiconductor electronics device that controls current flow and exhibits current-gain characteristics
vacuum tube—a device used as the main element in electronic circuits and systems; has been replaced by transistors in most applications
voltaic cell—source of electric energy

INTRODUCTION

Electricity has made a tremendous impact on society. Its practical applications have revolutionized the way we live. Electricity is the study of resulting forces when two charged bodies are placed in the vicinity of each other. Electricity and **magnetism** are closely interrelated. Moving electrical charge creates magnetism.

The study of the history of electricity is important to the student of electrical and electronics technology. It is through the understanding of the development of a field that we acquire a knowledge of the subject matter. The study of electric phenomena and circuits, to which this text is devoted, is based on the inventions and developments outlined in this chapter.

The **electrical engineering** field deals primarily with electricity and magnetism. Through continuous advances made in this field, practical uses of electricity are continually developed to benefit humankind.

Electronics is a branch of electrical engineering that resulted from the discovery of signal processing devices and the development of electrical systems. It encompasses those areas that employ semiconductor devices as primary elements. The general field of electronics is divided into many branches.

1.1 THE ELECTRICAL PHENOMENON

The knowledge of the presence of magnetism dates back to 600 B.C., when Thales of Miletus observed a force of attraction between magnetite and soft iron. Another of Thales' observations was the attraction of lightweight objects to certain materials when those materials were rubbed. It was not until 1551 that the distinction between electric and magnetic materials was made. Jerome Cardan distinguished between them on the basis of the materials becoming electrified by rubbing.

Early theories of electricity were developed using fluid concepts. Jerome Cardan was the first to treat electricity as a fluid. Subsequently, theories of electricity as fluids were proposed by C. F. DuFay in 1733 and by Benjamin Franklin in 1747. The significance of Franklin's theory was the fact that it was a single fluid theory, where an excess of fluid was considered positive electrification and a fluid deficiency, negative electrification. The fluid concept was later dropped, but the notion of positive and negative as applied to charge, and the concept of electricity as a substance, remained.

J. B. Priestly, in 1767, and C. A. Coulomb, in 1785, independently formulated the relationship between charges and the forces acting upon them. This marked the first quantitative development in the electrical field. This relationship came later to be known as *Coulomb's Law,* or as the principal law of electrostatics*.

The development of the **voltaic cell** by A. Volta in 1800 opened the road to applications of electrical theory. The voltaic cell, or the battery cell, is the first known continuous source of electrical energy. The discovery of the energy cell was also a significant electrochemical development.

One of the most significant developments of the nineteenth century was in the area of communication. In 1800 J. Groat introduced the electrical **telegraph,** and in 1840 Samuel F. B. Morse was granted a patent for the electromagnetic telegraph. In 1827 Joseph Henry invented the electromagnet, which was used by Morse in his telegraph.

In 1820 H. C. Oersted observed the presence of a magnetic field due to the flow of electric currents, and Michael Faraday showed in 1831 its dual nature—namely, the development of induced currents due to changing magnetic fields. Faraday further developed the concept of magnetic fields, without ever formulating his ideas mathematically.

The contributions made by Faraday led indirectly to the development of numerous electrical devices, including electrical machinery (motors and generators). The first electrical generator built was the dynamo, and the first central dc generating station in the United States was in operation in New York in 1882. The first ac generating station was supplying power in 1886. Ac-type generators were developed by S. Z. DeFerranti.

In 1880 Thomas Edison developed the carbon filament lamp, which marked the birth of the electrical lighting industry. Direct current generators were used initially to power street lights. Commercial electrical lighting played a significant role in the modern industrialization process. The improvement of facilities and working conditions led to more efficient operations.

It was not until the discovery of the **transformer** by L. Gaulard and J. D. Gibbs in 1883 that ac power distribution became a reality. Together with the ac

* Electrostatics is the area of study that investigates electric charges at rest.

generator, it became possible to generate electricity at high current–low voltage levels, and with the use of a transformer to distribute the power at high voltage–low current levels. This enabled the distribution of power over long distances.

The ac polyphase induction motor was developed by N. Tesla in 1888, and its potential of becoming the industrial workhorse was immediately recognized. Even today, this type of electric motor is in common use. The frequency of generated ac was standardized in 1891 to 60 Hz.

Electromagnetic field theory was mathematically formulated by James Clerk Maxwell in 1865. The famous "Maxwell's Equations" are studied today by all electrical engineering students as the mathematical basis for electromagnetic principles.

The propagation of electromagnetic radiation, or waves, was proven in an experiment by Heinrich Hertz in 1887. The remarkable discovery of the existence of negatively charged particles was demonstrated in an experiment by J. J. Thomson in 1897.

The invention of the **telephone** by Alexander Graham Bell in 1876, is considered to be the origin of modern voice communication. The first unit contained an electromagnetic microphone of the same construction as the earphone, which was later replaced by the carbon microphone developed by Edison.

1.2 THE FORMULATION OF LAWS OF ELECTRICITY

The first law of electrostatics was formulated by C. A. Coulomb in 1885. Coulomb's Law, which states that the force of attraction or repulsion between two charged bodies is proportional to the product of charges but inversely proportional to the square of distance between their centers, is an application of Newton's Inverse Square Law to electricity.

The French scientist Andre Marie Ampere made numerous contributions in many fields of science. He was the founder of electrodynamics and inventor of methods for measuring electricity and also developed the galvanometer. In 1827 he stated the fundamental law of electrodynamics known as *Ampere's Law,* which is a mathematical formulation of the magnetic force between two electric currents. He is best remembered because of the unit of electric current named for him. This honor was bestowed because he was first to recognize the significance of the rate of flow of charge due to a driving force.

In 1833 Georg Simon Ohm made a contribution to the area of practical electricity by formulating the important, fundamental relationship between current, voltage, and resistance. The relationship is known as *Ohm's Law,* and the unit of resistance, the ohm, was subsequently named after him.

The German mathematician Carl Friedrich Gauss, who lived from 1777 to 1855, provided a formulation of electromagnetic field theory by relating a flux of force lines outside a region to the total charge within the region. *Gauss' Law* has been an underlying factor in the study of fields ever since.

Michael Faraday, in addition to his many other discoveries, formulated the famous *Law of Induction* in 1831. As pointed out earlier, this law lays the foundation for the design of electromagnetic devices.

Gustav-Robert Kirchhoff, who was born in 1824, contributed to the understanding and the analysis of electric circuits through his formulation of the *voltage and current laws*.

Of the existing network theorems discussed in this text, perhaps the most powerful of all is *Thevenin's Theorem*. This theorem provides a method for calculating the performance of a device in a circuit from its terminal properties.

1.3 THE AGE OF ELECTRONICS

The birth of the field of electronics is traced back to Thomas A. Edison's experiment in 1883 dealing with electrical conduction in a partial vacuum. His experimental findings—later to become known as the *Edison effect*—became the basis for vacuum tube devices. **Vacuum tubes** were the vehicle in the development and growth of the electronics industry, but today they are used in only a few limited applications.

Although Thomas Edison received a patent for his invention, the explanation for the phenomenon of conduction in a vacuum did not come until Thomson discovered the electron. The importance of the Edison effect was not realized until 1905, when John Embrose Fleming developed the first vacuum tube diode, which was a direct application of the Edison experiment. At that time wireless telegraphy was in its developmental stage, and the diode was the device needed for the detection of wireless signals.

In 1907, one year after the vacuum tube diode was patented, General Dunwoody found that silicon carbide can be made to exhibit the same characteristics. Silicon carbide was used to build a device called a crystal detector, which was widely used in communication circuits. Renamed the solid-state diode, it is still used today.

Radiotelephony, or radio transmission and reception as we know it today, is a natural development of wireless telegraphy or radiotelegraphy. The most commonly used systems utilize amplitude modulation (AM), which was first introduced by R. A. Fessenden in 1906. AM transmission is best suited for long distance, as it is susceptible to impulse noise. E. H. Armstrong invented and built the first superheterodyne receiver in 1918 and introduced the principle of frequency modulation (FM) in 1933. He has successfully shown that FM has excellent noise characteristics.

The advances in radio were possible due to further improvements of the vacuum tube. In 1907 Dr. Lee DeForest developed the triode vacuum tube, which he called the Audion.

The triode exhibits *amplification* properties. Triodes have also been used in a circuit to generate sinusoidal waves of constant amplitude and frequency without the presence of an external signal. The circuit is referred to as an *oscillator*. These two vacuum tube circuits, along with the diode detector, have made possible radio, television, radar, hi-fi amplifiers, computers, and so on.

Television, a complicated system, was developed over a long period of time, with contributions made by many scientists. The advent of the cathode-ray tube, invented by K. F. Braun in 1897 and based on work done by J. J. Thomson and Sir

William Crookes, made television a reality. However, it was not until 1907 that the use of the cathode-ray tube in television was suggested by the Russian scientist Boris Rosing.

In 1925 the British scientist John L. Baird demonstrated before members of the Royal Institution the first television system with receiver and transmitter in separate rooms. His system was based on mechanical scanning. Baird was also responsible for experimental broadcasting in England from 1929 to 1935. Mechanical scanning had one major drawback, namely, poor sensitivity; eventually, it gave way to an all-electronic system.

The missing link in an electronic image pickup was completed when V. K. Zworykin invented the iconoscope camera tube in 1923. Later, he invented the kinescope—the television picture tube as we know it today. The first all-electronic system was introduced by RCA in 1932. Regular television broadcasting in the United States began in 1941.

Early experimental work on color television was done by Zworykin and many others. Baird was again first when he demonstrated a color television system. His system used mechanical scanning principles. H. E. Ives of Bell Telephone Laboratories transmitted color images in 1929 between New York and Washington using a mechanical method. Also in 1929, Frank Gray developed a method for transmitting a number of signals over a single channel, which later proved to be essential for television transmission. Finally, in 1954, color TV broadcasting became a reality in the United States.

1.4 THE DEVELOPMENT OF THE COMPUTER

The search for a machine that will automatically perform routine calculations dates back to 2000 B.C., when the *abacus,* which is still in limited use today, was invented. In 1615 John Napier developed tables of logarithms. The slide rule, constructed on the principles of logarithms, has the ability to perform complicated arithmetic calculations.

Blaise Pascal invented the first adding machine in 1642. Pascal used the principle of gears with a decade advance, still used today. Pascal's adding machine was later expanded by Gottfried Wilhelm Leibniz to multiply, divide, and extract roots.

The first electrical all-purpose **computer** was built by Howard Aiken at Harvard University with support from IBM. Named Mark I, it used punched cards to handle data. This computer, as well as all modern computers, was based on the following:

1. The "analytical engine," invented by Charles Babbage in 1833. This machine was to have a memory, an input and output media, and was programmable.
2. The mathematics of George Boole (Boolean algebra), devised in 1854.
3. Claude E. Shannon's scheme for the implementation of algebraic expressions with switches and relays introduced in 1938.

The individuals credited with the design and development of the first modern electronic computer are J. P. Eckert, Jr., and J. W. Mauchley of the University of Pennsylvania. The ENIAC (Electronic Numerical Integrator and Calculator) was built in 1954 and employed thousands of vacuum tubes. It occupied 15,000 ft^2 of space and could not store a program, but it was able to perform 500 operations per second for a prewired routine.

Modern electronic computers went through a number of evolutionary stages, referred to as *generations*. As new electronic devices and new programming ideas were developed, the size of the computer began to shrink. Its capabilities were expanded, to the point where people in all developed countries are now in some way serviced by computers.

1.5 THE DEVELOPMENT OF SEMICONDUCTOR ELECTRONICS

The invention of the **transistor** resulted in a revolution within the electronics field. The transistor—a semiconductor device—was invented in 1947 by J. Bardeen, W. Brattain, and W. Shockley of Bell Telephone Laboratories (for their invention, they shared the Nobel Prize in physics in 1956). Like the vacuum tube, the transistor is able to amplify signals but in a much more efficient manner. Thus, the vacuum tube has been virtually replaced by the transistor and other semiconductor devices.

The invention of the transistor accelerated the growth of the field of electronics, to the extent that it now has an impact on all types of industries from business to medicine. A prime factor in the development process was the invention of the **integrated circuit** (IC) by Dr. Robert Noyce (working with Fairchild Semiconductor Industries). An IC is a self-contained complete circuit that occupies a small area. For example, an amplifier or a computer circuit can be made on a chip of silicon not much larger in size than a single transistor. The attributes of such a circuit, in addition to its small size, are high reliability and a considerable reduction in cost. Integrated circuits have made possible sophisticated digital computers as well as microprocessors,* advanced communications systems, hand-held calculators, biomedical instruments, and many more electronic systems. An integrated circuit test system is shown in Figure 1.1.

1.6 THE ELECTRICAL INDUSTRY

The electrical industry, which includes many allied fields, is a multibillion-dollar industry that is continually expanding. With the continued increase in demand for electrical energy, we are experiencing an ongoing boom in the power field. In the computer field, there is a phenomenal need for semiconductor IC devices as the computer finds applications in all types of industries. The defense industry is a major consumer of electronic products. There has been a continual rise in sales of electrical and electronic equipment to the general public, as appliances and home

* Computer on a chip.

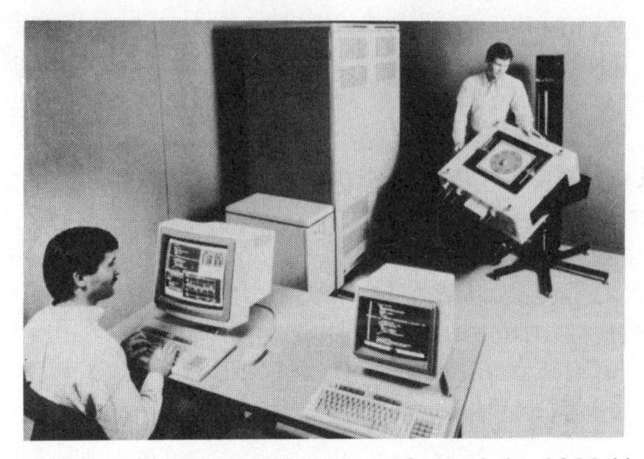

Figure 1.1 Integrated circuit test system. (Copyright 1988 Hewlett-Packard Company. Reproduced with permission.)

entertainment products are improved and expanded. Electronics has become a part of the automobile industry in controls and instrumentation, with a large effort made to develop a practical all-electric car. The reindustrialization process is largely due to the use of electrical and electronic equipment in automated manufacturing processes. A microwave technician utilizing a network analyzer is shown in Figure 1.2.

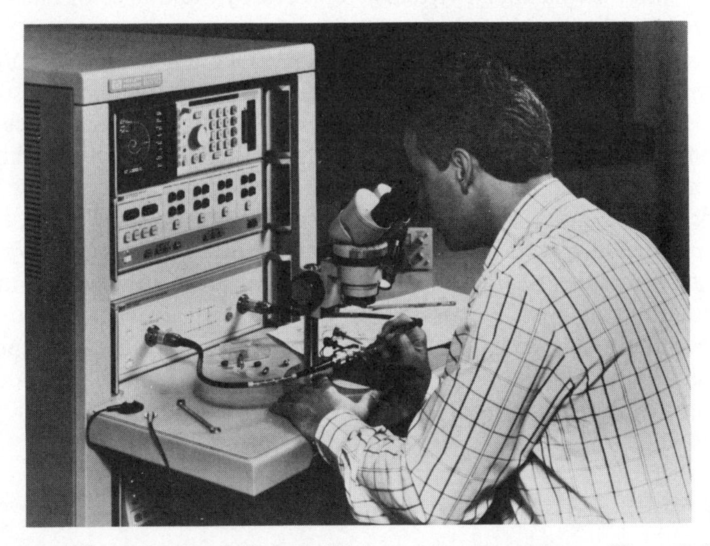

Figure 1.2 Microwave technician utilizing network analyzer. (Copyright 1988 Hewlett-Packard Company. Reproduced with permission.)

1.7 THE ELECTRICAL ENGINEERING FIELD

The electrical engineering field can be divided into two areas: power and electronics, as shown by the block diagram of Figure 1.3. The electrical power area encompasses the generation of electrical energy (by converting it from another

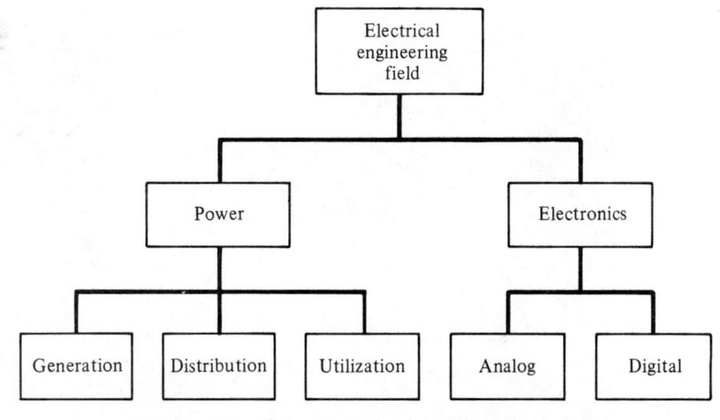

Figure 1.3 The electrical engineering field.

form such as mechanical or chemical), the distribution of the energy, and finally, the utilization of energy. The electronics field concerns itself with the processing of electrical signals.

The electronics branch of electrical engineering deals with processing of information in both analog and digital form. It encompasses a broad range of specialty areas such as electronics, communications, computers, and control of automated processes. Analog electronics is the branch of electronics that processes continuously changing electrical quantities. The word *analog* stems from the fact that the quantities are *analogous,* or represent physical quantities such as temperature, pressure, flow, and so on. For example, an electronic device called an *amplifier* is used to magnify (amplify) electric signals that represent sound or other physical quantities. Digital electronics concerns itself with the processing of discrete electrical quantities that represent numbers. Computers are digital electronics systems.

Digital electronics is the newer of the two electronic fields. Because of its design flexibility and device characteristics, digital electronics is replacing some traditional analog electronics functions in almost all areas of electronics in addition to the new areas created by the digital field.

Engineers, technologists, technicians, and craftsmen work as a team. The engineer is involved in research, design, development, planning, manufacturing, and management of projects. Engineering technologists and technicians work side by side with the engineer to accomplish the practical objectives of the project. An automated board test system is shown in Figure 1.4.

The engineering technologist is concerned with production and operation of manufacturing processes. At times the work of the engineer and the technologist is overlapping. The engineering technician assists the engineers by carrying out the engineering designs. A communication technician using a signal analyzer is shown in Figure 1.5. A technician with temperature-measuring instruments is shown in Figure 1.6.

The engineering technician is usually a two-year college graduate with a technology degree or has a similar technical school education. The education

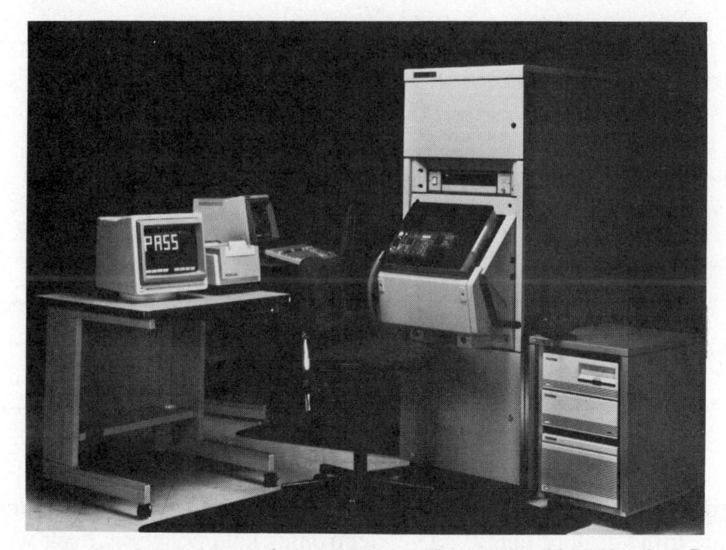

Figure 1.4 Automated board test system. (Copyright 1988 Hewlett-Packard Company. Reproduced with permission.)

requirement for an Engineering Technologist is a baccalaureate degree, commonly obtained after an additional two years of study beyond the associate's degree. The minimum requirement for the engineer is a baccalaureate degree, with many engineers opting for a master's degree and those that chose careers in research and teaching acquire a doctorate.

Figure 1.5 Communication technician using signal analyzer. (Copyright 1988 Hewlett-Packard Company. Reproduced with permission.)

Figure 1.6 Technician with temperature-measuring instruments. (Reproduced with permission from the John Fluke Mfg. Co., Inc.)

Specific jobs performed by engineering technicians and technologists are too numerous to mention. Some of the job titles are:

Electrical/Electronics Laboratory technician

Service technician

Customer representatives

Technical sales technician or sales engineer

Computer repair technician

Manufacturing technician

Electrical designer

Electrical power technician

Instrumentation technician

Biomedical technician

The material in this textbook, which you will use to gain an understanding of electric concepts, is written to serve as a foundation for your future studies in electrical/electronics technology. It is essential then, that you develop an understanding as well as a "feel" for this subject matter.

SUMMARY

This chapter concerns itself with developments that led to the formation of the electrical engineering field. It was shown that magnetism and electricity are closely interrelated. The presence of magnetism and electrostatics was known as early as 600 B.C. However, it was not until the sixteenth century that a distinction was made between electric and magnetic materials. Electric theory was developed during the next two centuries, but applications of the theory were nonexistent until A. Volta developed the voltaic cell in 1800.

The field of electronic communications had its beginnings in the nineteenth century with the development of the electric and the electromagnetic telegraphs. Modern voice communications originated with the invention of the telephone by Alexander Graham Bell.

The field of electronics was born out of an experiment by Thomas A. Edison dealing with electrical conduction in a partial vacuum. This led to the invention of vacuum tube devices, which became the main elements in instrumentation, wireless communications systems, automatic control systems, digital and analog computers, and display devices.

Today, the vacuum tube has been largely replaced by semiconductor devices. Electronic systems are constructed using discrete devices and integrated circuits. Integrated circuits are self-contained complete circuits made generally of a silicon material; they occupy a very small area. All areas of business, industry, and government have been impacted by the developments in the field of electrical engineering.

CHAPTER 2

Physical Basis of Electrical Concepts

GLOSSARY

acceleration—the time rate at which velocity changes
ampere (A)—unit of measurement of electric current
atom—smallest sample of an element that still has all the properties of that element
atomic number—the number of electrons (and therefore of protons) in an atom
charge (*Q*)—electrical property of certain particles of matter
compound—material formed by the chemical combination of elements
conductor—material that has many free electrons at room temperature
coulomb (C)—unit of electrical charge
current—the uniform motion of charged particles
electric current—same as current
electron—atomic particle attributed with negative charge
electronic distribution—the way electrons are distributed in an atom
energy—the ability to do work; also one of the two basic entities involved in the makeup of the physical universe
force—external influence upon a body that tends to effect the motion of the body
insulator—a material with no free electrons at room temperature
joule (J)—unit of energy
kilogram (kg)—unit of mass
law of charges—like charges repel whereas unlike charges attract
mass—quantity of matter
matter—one of the two basic entities making up the physical universe
meter (m)—unit of distance
molecule—smallest sample of a compound that still retains all the properties of that compound
negative charge—electrical property attributed to electrons
neutron—atomic particle with no charge
newton (N)—unit of force
nucleus—the center of the atom, containing protons and neutrons
orbits—paths around the nucleus that groups of electrons follow
positive charge—electrical property attributed to protons
power (*P*)—the rate at which work is done
proton—atomic particle attributed with positive charge
scientific notation—a method of representing numbers as a product of a number from 1 to 10 and a power of ten
second (s)—unit of time
semiconductor—material that at room temperature has very few free electrons
shells—see orbits
speed—rate at which position of a body changes with time
time (*t*)—sequence of events
velocity—the rate at which the position and direction of a body change with time
voltage (V)—entity that will cause electrically charged particles to move uniformly thereby causing a charge; also called electric potential, electromotive force, and electric tension
watt (W)—unit of power

work—the use or the expenditure of energy or the amount of energy used when a force is applied to a mass over a certain distance

INTRODUCTION

The study of electrical concepts commences with the understanding of the vehicle of motion of charge and its mechanical as well as electrical characteristics. This chapter offers the physical background. Atomic structure as well as atomic particles and the concept of charge are introduced and discussed. Qualitative definitions of voltage and current are given. General characteristics of conductors, insulators, and semiconductors are examined.

2.1 MATTER AND ENERGY

Careful observation indicates that there are two basic entities involved in the makeup of the physical world: **matter** and **energy.** Roughly speaking, matter refers to concrete objects such as rocks, trees, and machines. Energy, on the other hand, is the ability to do work. Familiar forms of energy are mechanical, electrical, and thermal. According to the classical law of thermodynamics, also known as the law of conservation of energy:

Energy can be transformed from one form to another but it cannot be created or destroyed.

All matter has the following three basic mechanical properties:

1. Extension into space
2. Mass
3. Existence in time

In order for a material object to be perceptible to the senses or to a measuring device, it has to have extension into space (length, height, and width); its volume must be filled with a certain amount of matter (mass); and the object must exist in time. See Figure 2.1.

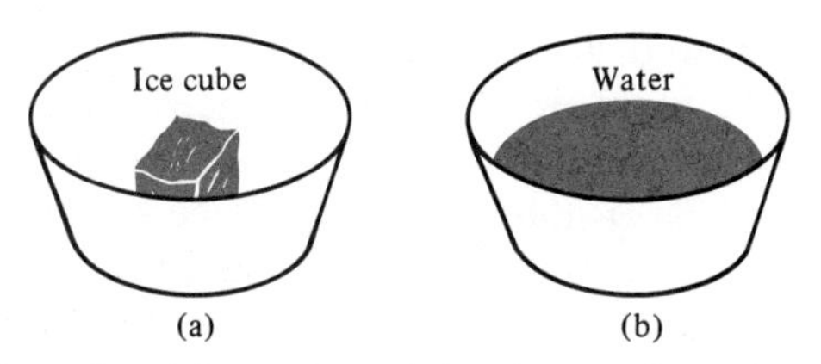

Figure 2.1 Even though matter might change form, it always shows the basic properties of time, mass, and volume.

Some materials also have an electrical property called **charge.** No one knows what charge is, but enough is known about the interaction of charged particles of matter that accurate control of their behavior is possible.

All other mechanical and electrical properties are described in terms of these four: length, mass, time, and charge.

2.2 WORK AND POWER

All physical systems are involved in constant energy transformations and exchanges.

Work *is the use or the expenditure of energy.*

Power *is the rate at which work is done or the rate at which energy is transferred or transformed.*

For example, more power is required to do a certain amount of work in a shorter period of time than in a longer period of time.

The main concern of electrical technology is the *control* of electrical energy. Electrical and electronic systems will do the desired work if they are designed and constructed to deliver the proper amount of energy to the right place at sufficient power.

The technical person must be careful not to misuse the technical definitions of concepts such as energy, work, and power when precise technical definitions are called for. In ordinary speech, for example, the noun *power* may be defined as the ability to act, the ability to exercise control, or any form of energy available for doing work. In a technical sense, the noun power has only one meaning: the time rate at which energy is being transferred or converted into work.

2.3 ELECTRICAL ENERGY SOURCES AND OTHER CIRCUIT ELEMENTS

All electrical and electronic systems need sources of energy in electrical form such as batteries, dc power supplies, and signal energy sources. Once energy has been introduced into the electrical or electronic system, the system is involved in the control of energy transformation and exchanges in two fundamental ways: energy dissipation and/or energy storage.

Practical electrical or electronic devices (resistors, capacitors, inductors, transistors, integrated circuits, batteries, and so on) will simultaneously dissipate and store energy. Even though every device exhibits a complex behavior, the operation of the device may be understood by separating the intertwined energy effects and examining them individually. It is the scope of this book to examine these effects separately.

2.4 MECHANICAL UNITS OF MEASUREMENT

Many systems of units are commonly used to measure the properties of matter. Among these are the International System (SI), the MKS system, the cgs system, and the English system. The more commonly accepted system in scientific and technical fields is the International System. The International System is used almost exclusively throughout this book.

Of the seven basic SI units, three are used here to describe the necessary mechanical and electrical properties of matter needed for our studies.

TABLE 2.1
Basic units used to describe matter

Unit	Entity	Symbol
Meter	Length	m
Kilogram	**Mass**	kg
Second	**Time**	s

*The **speed** at which a body is moving is the rate at which its position changes with time.*

*The **velocity** is the rate at which the position and direction of a body change with time.*

Velocity has two components: speed and direction. If an automobile travels 150 mi in 3 h, its speed is 50 mi/h (see Figure 2.2). If it takes a body 1 s to move a distance of 1 m, then the body's speed is 1 m/s.

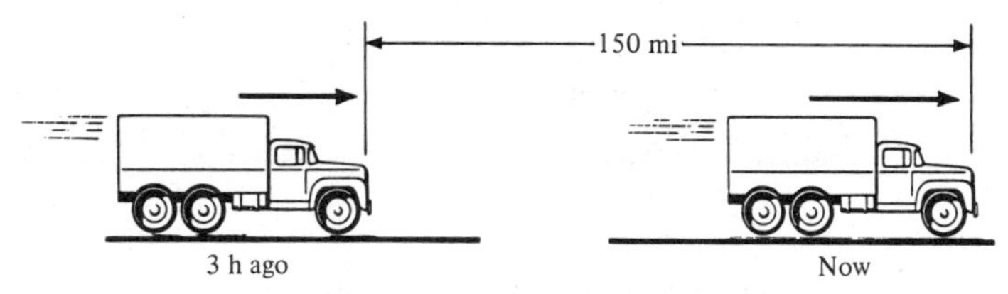

Figure 2.2 Speed is described in terms of distance and time. 150 mi traveled in 3 h gives a speed of 50 mi/h.

*Acceleration is the time rate at which velocity changes. If the direction is not altered, the **acceleration** of a body is the rate at which its speed changes with time.*

If a body is observed to be moving at 2 m/s and 1 s later it is moving at 3 m/s, then it is being subjected to an acceleration of 1 m/s every second. If it takes 1 s to change the speed by 1 m/s, then the acceleration is 1 m/s^2.

***Force** is an external influence upon a body that tends to effect the motion of the body.*

According to Isaac Newton, the force acting on a body is equal to the mass of the body multiplied by the observed acceleration ($F = ma$).

If a 1-kg mass is observed to be moving 1 m/s faster each second, then it is being acted upon by a force of 1 kg · m/s each second, or simply 1 **newton** (N). Since the truck shown in Figure 2.3 was standing still but now begins to move, it must have been acted on by a force. Since after 1 s it is moving at 1 m/s and after 2 s its speed is 2 m/s, the force acting on the truck is 1 N.

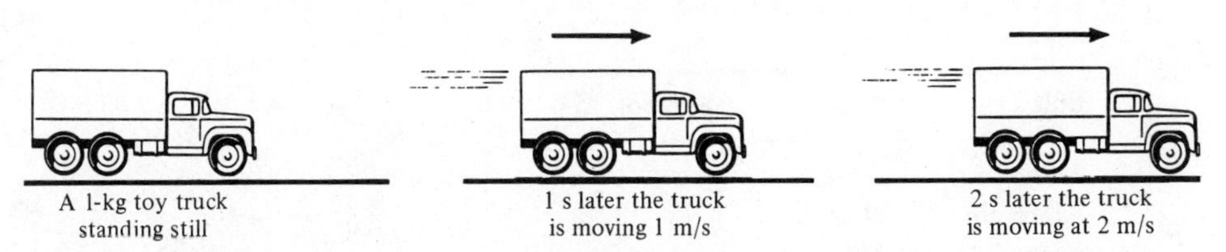

Figure 2.3 Force is an external influence that tends to change the motion of a body. **(a)** A 1-kg toy truck is standing still. **(b)** 1 s later the truck is moving 1 m/s. **(c)** 2 s later the truck is moving at 2 m/s.

***Work** is the amount of energy used when a force is applied to a mass over a certain distance.*

If a force of 1 N is applied to a body for a distance of 1 m then the work done is 1 N · m, or simply 1 **joule** (J). See Figure 2.4.

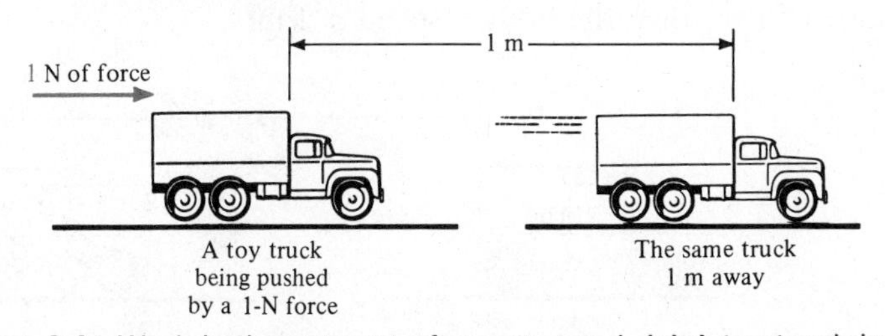

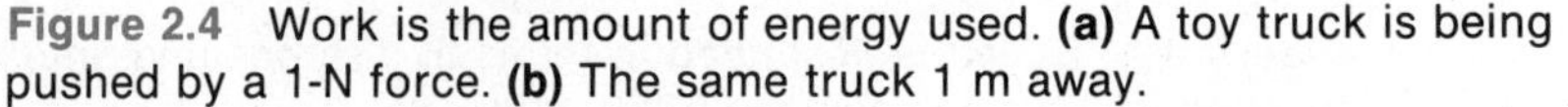

Figure 2.4 Work is the amount of energy used. **(a)** A toy truck is being pushed by a 1-N force. **(b)** The same truck 1 m away.

***Power** is the rate at which work is done.*

If it takes 1 s to do 1 J of work, then the power is 1 J/s, or simply 1 **watt** (W). Figure 2.5 shows that doing the same amount of work in half the time requires twice the power.

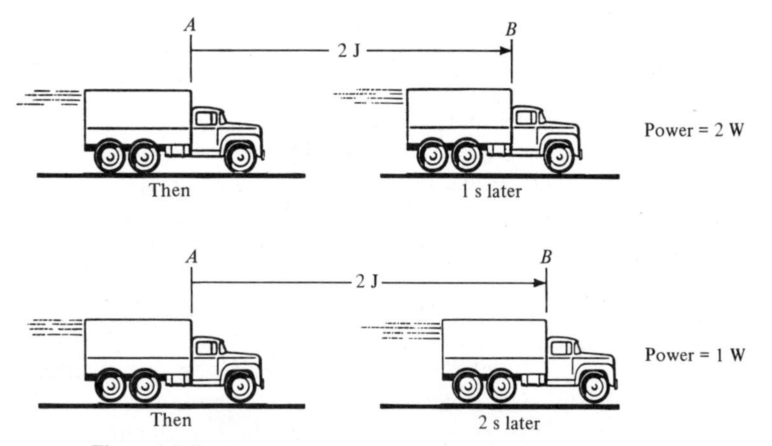

Figure 2.5 Power is the rate at which work is done.

2.5 ELEMENTS AND ATOMS

All matter is made up of approximately 100 known substances called elements. All other substances are composed of these elements. Some common elements are hydrogen, oxygen, carbon, and copper.

Elements combine or react with one another to form ***compounds.***

Hydrogen and oxygen, for example, combine to form water. There are approximately 2 million compounds known today.

A sample of a compound may be subdivided into smaller samples.

The smallest sample of a compound that still retains all the properties of that compound is called a ***molecule.***

A sample of an element may be subdivided into smaller and smaller samples.

The smallest sample of an element that still has all the properties of that element is called an ***atom.***

Atoms of any element may themselves be subdivided. The subdivisions result in particles that no longer have the properties of the element from which they come. Just as a builder uses the same type of bricks, windows, and doors to build two different houses, these particles (having properties common among themselves) are arranged in various combinations to form different elements. Oxygen and copper have completely different characteristics and yet are composed of the same types of particles.

Evidence offered by scientific experiments over the past 200 years suggests that atoms are composed of approximately 200 particles. To facilitate the understanding of the electrical properties of matter this book will consider only **electrons, protons,** and **neutrons.**

The structure of the atom that will be used here was first proposed by the Danish Nobel-prize-winning physicist Niels Bohr. In this model, the number of protons or electrons in the atom of an element is referred to as the **atomic number** of that element. The atomic number of copper is 29, that of silicon is 14, and that of germanium is 32. This means that copper has 29 protons and 29 electrons, whereas silicon has 14 protons and 14 electrons and germanium has 32 protons and 32 electrons.

There is convincing evidence from scientific experiments that protons and neutrons are concentrated in a small volume (called the **nucleus**) in the center of the atom, whereas the electrons exist in **orbits,** or groups circling the nucleus in well-defined paths. The Bohr theory refers to these groups as **shells** and assigns a certain number of electrons to each shell. The way electrons are distributed in an atom is called its electronic distribution. The Bohr model for hydrogen, carbon, and copper is shown in Figure 2.6.

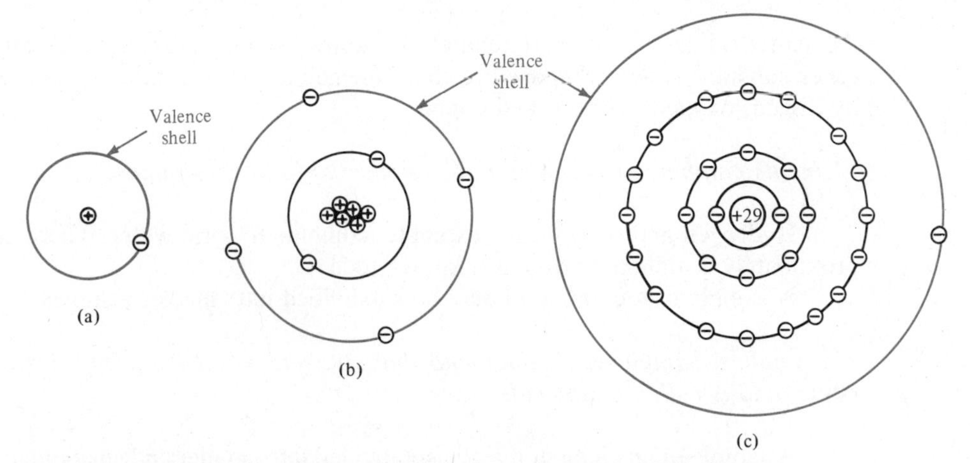

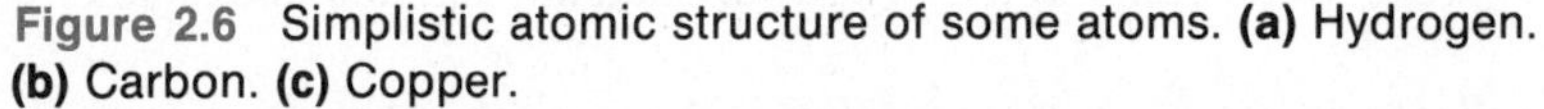
Figure 2.6 Simplistic atomic structure of some atoms. **(a)** Hydrogen. **(b)** Carbon. **(c)** Copper.

As Figure 2.6 shows, the last grouping of electrons in each atom is referred to as the *valence shell.* The number of electrons in this group gives each element its chemical and electrical characteristics. The electrical characteristics of an element depend on how easily electrons in the valence shell can be removed from the parent atom.

Experiments may be performed in which certain atoms lose electrons while other atoms gain electrons. Two materials that both have lost electrons will repel each other. A material that has gained electrons will attract a material that has lost electrons.

The cause of this attraction and repulsion is attributed to the charge of electrons and protons. Arbitrarily, electrons are said to carry **negative charge,** whereas protons are said to carry **positive charge.** Neutrons do not show such properties and are said to be electrically neutral.

The preceding experiments and definition of charge may be summarized by the ***Law of Charges:*** *like charges repel whereas unlike charges attract.*

The Greek word for amber (a yellowish, almost transparent material) is *electron*. The ancient Greeks used the expression "electric force" to refer to the mysterious force of attraction and repulsion exhibited by amber when it was rubbed by a cloth. Today, through knowledge of atomic structure, the source of electric forces, which puzzled the ancient Greeks, is traced to the electrons and the protons. Nevertheless, neither ancient Greek speculation nor modern scientific knowledge has been able to describe the *nature* of the electric charge. The question 'What is electric charge?' remains to this day practically unanswered. Enough however, is known about the behavior of electrically charged particles that their controlled use is one of the most important methods of energy transfer.

2.6 DECIMAL MULTIPLIERS AND POWERS OF TEN

Powers of ten are used to represent very big and very small decimal numbers. Each power of ten has a name and an associated unit symbol. The following list of decimal multipliers and powers of ten is encountered in electrical work.

Number	Power of ten	Unit name	Abbreviation
0.000000000001	10^{-12}	pico	p
0.000000001	10^{-9}	nano	n
0.000001	10^{-6}	micro	μ
0.001	10^{-3}	milli	m
1	10^{0}	unit	
1,000	10^{3}	kilo	k
1,000,000	10^{6}	mega	M
1,000,000,000	10^{9}	giga	G
1,000,000,000,000	10^{12}	tera	T

2.7 SCIENTIFIC NOTATION

Scientific notation is a method of representing numbers as a product of a number from 1 to 10 and a power of ten. The number 378.2 is represented in scientific notation as 3.782×10^2.

Number from 1 to 10 $\times$ power of ten

EXAMPLE 2.1

Write 6,242,000,000,000,000,000 in scientific notation.

Solution

6.242×10^{18}

EXAMPLE 2.2

Write each of the following numbers in scientific notation and represent the power of ten using its abbreviation.

a. $2200 = 2.2 \times 10^3 = 2.2$ kilounits $= 2.2$ k units.

b. $0.005 = 5 \times 10^{-3} = 5$ milliunits $= 5$ m units.

c. $0.000\,0024 = 2.4 \times 10^{-6} = 2.4$ microunits $= 2.4\ \mu$ units.

d. $0.000\,000\,000\,001\,57 = 1.57 \times 10^{-12}$
$= 1.57$ picounits $= 1.57$ p units.

2.8 LAW OF EXPONENTS

The law of exponents is very useful in solving problems involving power-of-ten notation. Recall from algebra that

$$a^m \cdot a^n = a^{m+n}$$

$$\frac{a^m}{a^n} = a^{m-n}$$

$$(a^m)^n = a^{mn}$$

If $a = 10$:

$$10^m \cdot 10^n = 10^{m+n}$$

$$\frac{10^m}{10^n} = 10^{m-n}$$

$$(10^m)^n = 10^{mn}$$

2.9 COULOMB'S LAW

The basic unit of charge is the **coulomb** (C). One coulomb of charge is defined as the total charge contributed by 6.242×10^{18} electrons (or 6.242 followed by 18 zeros, which is read as 6.242 million million million electrons). A body that has gained this many electrons is said to have 1 C of negative charge. A body that has lost this many electrons is said to have 1 C of positive charge. Just as 100 pennies make up $1, so 6.242×10^{18} electrons make up 1 C. The negative charge of one electron therefore is 1.602×10^{-19} C.

According to the French physicist Charles-Augustin de Coulomb, charged bodies attract one another if they are oppositely charged, but they repel one another if they are similarly charged. The relationship that defines this force of attraction or repulsion is known as *Coulomb's Law:*

$$F = k\frac{Q_1 Q_2}{d^2} \tag{2.1}$$

where:
F => force (N)
k => constant (9×10^9 N · m²/C² in free space)
Q_1 => charge on body 1 (C)
Q_2 => charge on body 2 (C)
d => distance between body centers (m)

EXAMPLE 2.3

A body charged with 5×10^{-8} C of positive charge is located at 0.1 m from a body charged with 1×10^{-6} C of negative charge. Find the force of attraction between the two bodies.

Solution

$$F = (9 \times 10^9)\frac{(5 \times 10^{-8})(10^{-6})}{(0.1)^2}$$

$$= 0.045 \text{ N}$$

EXAMPLE 2.4

Two like-charged bodies, as shown in Figure 2.7, are separated by a distance of 0.02 m.

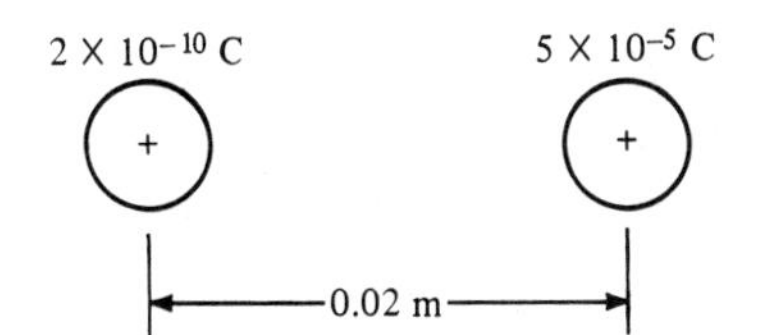

Figure 2.7 Diagram for Example 2.4. Two like-charged bodies.

a. What is the force acting on these bodies?

b. Indicate whether the force is repulsion or attraction.

Solution

a. $F = (9 \times 10^9)\,\dfrac{(2 \times 10^{-10})(5 \times 10^{-5})}{(0.02)^2}$

$= 0.225\ \text{N}$

b. The force is one of repulsion because the bodies are both positively charged.

EXAMPLE 2.5

A body positively charged with 10^{-8} C is placed between two other charged bodies, as shown in Figure 2.8. Compute the force on the body in the middle.

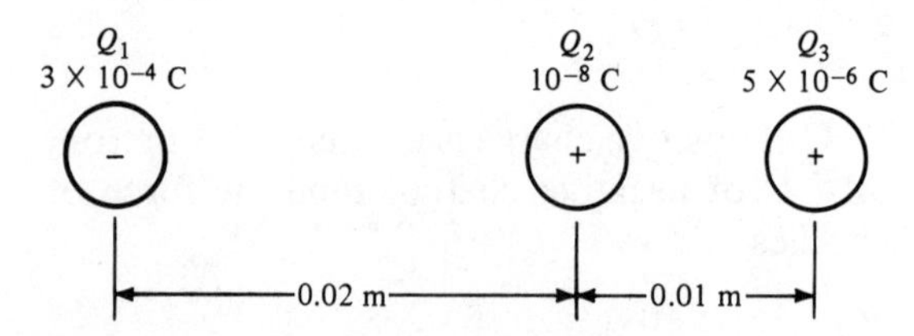

Figure 2.8 Diagram for Example 2.5. Distribution of charged bodies.

Solution

The force on the middle body due to Q_1 is:

$$F = (9 \times 10^9)\,\frac{(3 \times 10^{-4})(10^{-8})}{(0.02)^2}$$

$= 67.5\ \text{N toward } Q_1$

The force on the middle body due to Q_3 is:

$$F = (9 \times 10^9)\ \cup\ \frac{(5 \times 10^{-6})(10^{-8})}{(0.01)^2}$$

$= 4.5\ \text{N toward } Q_1$

Since both forces are in the same direction, the force on Q_2 is:

$F_T = F_1 + F_2 = 67.5 + 4.5$

$= 72\ \text{N toward } Q_1$

A study of electrical phenomena is basically a study of the behavior of electrons under different conditions and in different media. Charge is the electrical property of electrons that results in their behavior.

Experiments show that electrons behave both as particles and as nonmaterial electromagnetic waves at the same time. The treatment of electrons as particles leads us to the concept that charge exists on matter. This makes the electrons themselves subject to mechanical as well as electrical laws. Many "electrical" definitions and units are then necessarily based upon the electromechanical nature of the charged particles.

It should be emphasized that when the word *charge* is used to describe electrical properties, the particles associated with this property must always accompany the charge. Just as one cannot give the yellow color from a tennis ball without giving the ball, one cannot give the charge without giving the charged particle. Modern science is unable to separate the property of "charge" from the particle that has that property.

The sections in this chapter on basic mechanical properties of matter were offered as background for the development and understanding of the definition of basic electrical concepts and units. At this point the student is encouraged to go back and read those sections again.

2.10 ELECTRIC CURRENT

Electric current, or simply ***current,*** *is the uniform motion of charged particles.*

The amplitude of the current (sometimes referred to as its intensity, or its magnitude) is determined by the rate at which electrically charged particles move through a medium (usually a wire). Electric current is measured in amperes (A). Electric current is examined in detail in Chapter 3.

2.11 VOLTAGE (ELECTRIC POTENTIAL)

Voltage *may be described as electrical pressure or as the entity that will cause electrically charged particles to move uniformly, thereby causing a current.*

Voltage is measured in volts (V). A more accurate description of voltage is given in Chapter 3.

2.12 CONDUCTORS, INSULATORS, AND SEMICONDUCTORS

As considered earlier, careful observation shows that all matter is made up of atoms and that these are in turn made up of many particles, three of which are the negatively charged electron, the positively charged proton, and the uncharged neutron. In general, electrons in solids may exist in either of two conditions. They may be tightly bound to the parent atom, or they may be free to move around the structure from atom to atom. The bound electrons are those inside the atomic structures or those involved in bonds between atoms. The free electrons, on the other hand, have by some means been freed from their parent atoms.

*Materials that at room temperature have no free electrons are referred to as **insulators**.*

Under normal conditions, the application of a voltage across an insulator will not result in a current.

*Materials that at room temperature have very few free electrons are referred to as **semiconductors**.*

Under normal conditions, the application of a voltage across a semiconductor causes insignificant currents. Pure semiconductors are practically useless electrically. They become important when they are treated with other materials in order to increase their conduction in a controlled way. Modern electronic industry is so successful in controlling the properties of semiconductors that these materials have revolutionized not only technology but our very way and quality of life.

*Materials that at room temperature have many free electrons are referred to as **conductors**.*

Good conductors such as copper, aluminum, silver, and gold are used to make electrical wires, cables, and contacts. These wires and cables are then used to transfer electrical energy in a controlled way from one place to another. Good insulators are used to cover the conductors in order to isolate the energy carrying conductors from the outside world so that accidental contact is not made with the conductor.

Conductivity is examined in greater detail in Chapter 4. Semiconductors are used to make amplifying and switching devices that control the transfer of electrical energy and the processing of information.

SUMMARY

This chapter presented an overview of the physical basis of electrical concepts. Length, mass, and time were described as the basic mechanical properties of matter, whereas charge was defined as the basic electrical property. Energy was described to be the ability to do work, and, consequently, work was described as the expenditure of energy. Power was defined as the time rate at which work is being done. Velocity was defined as the rate at which position changes with respect to time, and acceleration was defined as the rate at which velocity changes with respect to time. Force was defined as the external influence that tends to change the motion of a body. Atoms were observed to be the smallest components of an element. Electrons, protons, and neutrons were observed to be components of atoms. Electrons were defined as having negative charge, protons positive charge, and neutrons no charge at all.

The Bohr model of the atom was presented and the electronic distribution of atoms was defined as the way the atom's electrons arrange and group themselves around the nucleus of an atom. The last group of electrons was called the valence shell. It was noted that the number of electrons in the valence shell gave the material all its electrical properties (insulator, conductor, or semiconductor).

Coulomb's Law of Charges was discussed both quantitatively and qualitatively. The law of exponents was reviewed and used to solve problems with Coulomb's Law. Electric current was defined as the uniform motion of charged particles, whereas voltage was defined as the "electric pressure" that causes current. Conductors, insulators, and semiconductors were examined in light of the previous treatment of atomic structure.

SUMMARY OF IMPORTANT EQUATIONS

$$F = k\frac{Q_1Q_2}{d^2} \tag{2.1}$$

Table of quantities and their useful units

Quantity	Symbol	SI unit	Symbol	English unit	Symbol
Length	l	meter	m	foot	ft
Mass	m	kilogram	kg	slug	slug
Time	t	second	s	second	s
Charge	q, Q	coulomb	C	coulomb	C
Velocity	v	meters per second	m/s	feet per second	ft/s
Acceleration	a	meters per second per second	m/s^2	feet per second per second	ft/s^2
Force	F	newton	N	pound	lb
Work	W	joule	J	foot-pound	ft-lb
Energy	E	joule	J	foot-pound	ft-lb
Power	P	watt	W	foot-pound per second	ft-lb/s
Current	i, I	ampere	A	ampere	A
Voltage	v, V	volts	V	volts	V

REVIEW QUESTIONS

2.1 State the three basic mechanical properties and one basic electrical property of matter.

2.2 State the units to describe the four properties from Question 2.1 in the SI system of units.

2.3 Define force, velocity, and acceleration. Give their SI units.

2.4 Define energy, work, and power. Give their SI units.

2.5 Name the three basic components of the Bohr atom and give the associated charge with each.

2.6 Explain how the electrical characteristics of a material are a direct result of the number of valence electrons.

2.7 Summarize Coulomb's Law of Charges.

2.8 Define electric current and voltage.

2.9 Discuss the two possible states of electrons in a solid.

2.10 Define conductors, insulators, and semiconductors.

2.11 Why are powers of ten used extensively in scientific and technical work?

2.12 Give the abbreviations for 1,000,000 units and 0.000,001 units.

2.13 Give the abbreviations for 1000 units and 0.001 units.

2.14 Give the abbreviations for 10^{-9} units and 10^{-12} units.

2.15 List three forms of energy.

2.16 Explain the law of conservation of energy and give an example.

PROBLEMS

2.1 A body travels at 10 m/s. How far will it go in 5 s?

2.2 A body travels 20 m in 4 s. Compute the velocity. (ans. 5 ms)

2.3 A force causes a 15-m/s^2 acceleration when applied to a 5-kg mass. Compute the magnitude of the force.

2.4 A 50-N force is applied to a 10-kg mass. Compute the resulting acceleration. (ans. 5 m/s^2)

2.5 An applied force of 25 N causes a body to move 10 m. Compute the amount of work done on the body.

2.6 In moving a body, 30 J of energy are expended in 10 s. Compute the power. (ans. 3 W)

2.7 Convert the following to decimal form.
a. 4.7×10^3
b. 3.54×10^{-4}
c. 8.2×10^5
d. 1.9×10^{-6}

2.8 Express the following numbers in scientific notation.
a. 10,300 (ans. 1.03×10^4)
b. 0.0057 (ans. 5.7×10^{-3})
c. 1,990,000 (ans. 1.99×10^6)
d. 0.000,0042 (ans. 4.2×10^{-6})
e. 0.000,000,000,0075 (ans. 7.5×10^{-12})
f. 0.000,000,00257 (ans. 2.57×10^{-9})

2.9 Convert the units given to the units indicated.
a. 100 milliunits to microunits
b. 37 kilounits to units
c. 2.2 megaunits to kilounits
d. 4.7 picounits to microunits
e. 3.3 microunits to nanounits
f. 4500 units to kilounits
g. 220 milliunits to microunits

2.10 Convert the measurements given to the indicated units.
a. 470 ohms to kilohms (ans. 0.47)
b. 2200 ohms to kilohms (ans. 2.2)
c. 3.3 kilohms to ohms (ans. 3300)
d. 2200 ohms to megohms (ans. 0.0022)
e. 100 kilohms to megohms (ans. 0.1)

2.11 Convert the measurements given to the indicated units.
a. 0.5 amps to milliamps
b. 700 millivolts to volts
c. 1200 microamps to milliamps
d. 10,000 picofarads to microfarads
e. 10 millihenrys to henrys

2.12 Perform the following operations using powers of ten and the law of exponents.
a. (1000)(10,000)/(100,000) (ans. 10^2)
b. (0.001)(10,000) (ans. 10)
c. (10)/(5000) (ans. 2×10^{-3})
d. (20,000)(0.000,001) (ans. 2×10^{-2})

2.13 Compute the force of attraction in free space between the charged bodies shown in Figure 2.9.

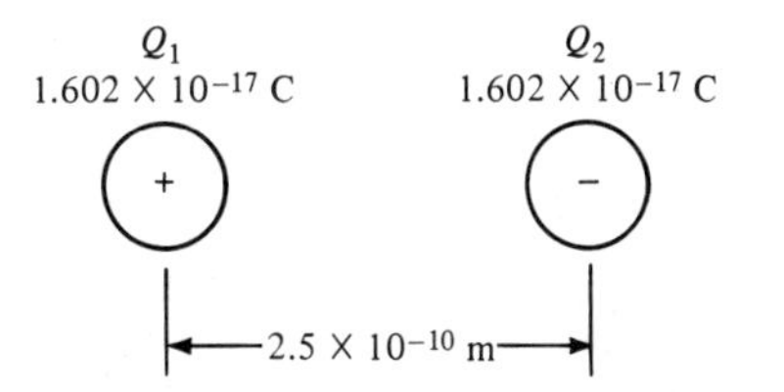

Figure 2.9 Diagram for Problem 2.13.

2.14 Two positively charged bodies in free space are known to have identical charges and are repelling each other with a force of 25 μN. The bodies are 10 m apart. Compute the charge on each body. (ans. 2.78×10^{-19} C)

2.15 Two similarly charged bodies are in free space, as shown in Figure 2.10. The spring scale indicates a 50-mN force. Compute the distance between the two bodies.

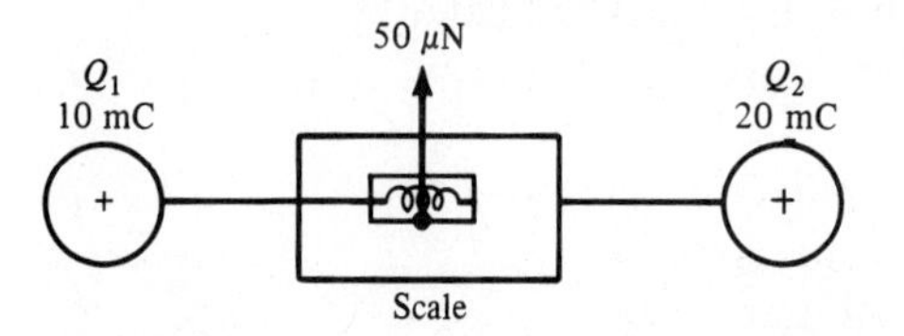

Figure 2.10 Diagram for Problem 2.15.

2.16 Compute the pulling force that a 20-C positively charged cloud exerts on a 30-C negatively charged cloud when they are 200 m apart. How many cars could be lifted by this force if it is assumed that the average car can be lifted by a 1000-N force? (ans. 1.35×10^8 N, 135,000 cars)

2.17 An automobile traveled 24 mi in 40 min. What is the speed of the auto in miles per hour?

2.18 A force of 100 N causes a 25-kg mass to accelerate. Calculate the acceleration. (ans. 4 m/s)

2.19 How many electrons contribute to a charge of 0.2 C?

2.20 Calculate the charge associated with 10^6 electrons. (ans. 1.60×10^{13} C)

CHAPTER 3

Current and Voltage

GLOSSARY

ampere (A)—unit of current
battery—a device made up of many electric cells
constant current source—an electrical device which is capable of supplying electrical energy while maintaining a constant current
constant voltage source—an electrical device which is capable of supplying electrical energy while maintaining a constant voltage
conventional current—the assumed flow of positive charge
current—the uniform motion of charged particles

electric cell—an electrical device capable of maintaining a voltage while providing a current
electric circuit—a connection of electrical components with an energy source, wire conductors, and a load
electric current—same as current
electric potential—same as voltage
electron current—the uniform flow of electrons
EMF—electromotive force; voltage generated by a voltage source entity that causes current
load—a device that transforms electrical energy into other forms such as heat and light
potential difference—same as voltage
voltage—work done per unit charge; measured in volts (V)

INTRODUCTION

There are a few terms and concepts that are essential to the understanding of the field of electrical and electronic technology. In this chapter, the concept of electric current as well as its quantitative definition are presented. The notion of conventional current is also presented.

Constant voltage and constant current sources are investigated. The qualitative and quantitative definition of voltage as well as the idea of potential difference are discussed. The algebraic notation for voltage is also presented.

3.1 ELECTRIC CURRENT

Qualitatively, **electric current** is defined as the uniform motion of charged particles. By virtue of this motion, charged particles involved in an electric current are responsible for the transfer of energy from one place to another in an electrical system.

***Current** is the phenomenon that allows energy to be transferred from one place to another in an electrical system.*

In physical terms, one might say that electric current is a carrier of energy. A familiar analogy for electric current is water flowing in a pipe. The uniformly moving water transfers energy from one place of the water system to another. The reason that water flows is that it is under pressure due to a force. Water flows downhill due to earth's gravitational force and uphill when force is supplied by a water pump.

To define electric current, a wire with cross section A, as shown in Figure 3.1, is used.

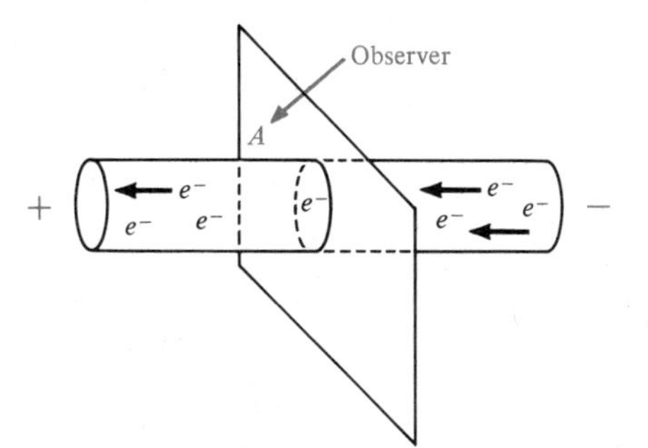

Figure 3.1 Cross section of a wire (e^- = electron).

If no force is exerted on the free electrons in the wire, the electrons are in random motion. An observer at cross section A will note an equal number of electrons moving in each direction, resulting in a net flow of charge equal to zero. Some electrons will recombine with stationary ions (atoms that have previously lost an electron), whereas other electrons will collide with still other atoms to cause other electrons to become free. This is analogous to water in a pipe under no pressure. If a large positive charge is placed on one end of the wire and a large negative charge is placed at the other end, the free electrons in the wire are attracted to the positively charged end and are repelled by the negatively charged end. Since negative charge exists on each electron, uniform charge movement results. Electrons are supplied by the negatively charged region at the right end of the wire, so that a continuous uniform movement of charge is maintained. This is analogous to supplying a force (pressure) to water in a pipe, resulting in a uniform flow of water molecules while maintaining a continuous supply of water.

The observer at point A in Figure 3.1 is able to determine the number of electrons and hence the quantity of charge that passes the cross section in a unit of time.

Quantitatively, ***electric current*** *is defined as the amount of charge flow per unit time.*

This is analogous to the amount of water flowing through a pipe in a unit of time (mass flow per unit time). It should be noted that if the pressure, which is the force applied to the water, is changed the amount of water flow is also changed. In like manner, if the number of positive and negative charges at the respective ends of the wire is changed, the current also changes. For present purposes however, assume that the number of the charges remains constant. In other words, a constant electric force is being used to cause the electrons to move.

The SI unit of electric current is the ***ampere*** *(A).*

Strictly speaking, SI defines the ampere as the current that, if maintained in two straight parallel conductors of negligible cross section and infinite length and spaced 1 m apart in a vacuum, produces between the conductors a force of 2 ×

10^{-7} N per meter of length. The authors propose the much simpler *charge over time* rule as the fundamental definition for electric current. In equation form, this is given by

$$\boxed{I = \frac{Q}{t}} \tag{3.1}$$

where: I => current in amperes (A)
Q => amount of charge in coulombs (C)
t => time in seconds (s)

This relationship is known as *Ampere's Law*. The symbol for electric current is *I*. This is the first letter from the French word *intensité* (current intensity).

If the observer at point *A* counts the number of coulombs of charge that pass through the cross section in 1 s, a measure of current is obtained.

The charge movement of 1 C *in* 1 s *is equal to* 1 A *of current.*

EXAMPLE 3.1

How many electrons are required to pass the cross section of the conductor in Figure 3.1 in 1 s to cause a current of 1 A?

Solution

The charge on one electron is 1.602×10^{-19} C. Since 1 C is required in 1 s, the number of electrons n is:

$$n = \frac{1\text{ C}}{1.602 \times 10^{-19}\text{ C/electron}}$$
$$= 6.242 \times 10^{18}\text{ electrons}$$

EXAMPLE 3.2

If a charge of 14 C passes a cross section of wire in 3.8 s, determine the current.

Solution

By Eq. (3.1):

$$I = \frac{14\text{ C}}{3.8\text{ s}} = 3.68\text{ C/s}$$
$$= 3.68\text{ A}$$

EXAMPLE 3.3

If 100 μC of charge flows in a wire in 20 s, determine the current.

Solution

By Eq. (3.1),

$$I = \frac{100\ \mu\text{C}}{20\ \text{s}} = 5\ \mu\text{A}$$

EXAMPLE 3.4

A current of 14.2 mA is measured in a wire. How much charge will pass in a time of 5 s? Express the answer in millicoulombs (mC).

Solution

Rearranging Eq. (3.1) results in $Q = It$:

$$\begin{aligned} Q &= (14.2 \times 10^{-3}\ \text{A})(5\ \text{s}) \\ &= 71.0 \times 10^{-3}\ \text{C} \\ &= 71\ \text{mC} \end{aligned}$$

3.2 CONVENTIONAL CURRENT

In the early days it was not known that the electrons actually move in a wire. The theorists of the day assumed that positive charges were moving. They guessed incorrectly! In fact, the positive charges (ions) have very little movement; they are essentially stationary. However, we will see shortly that it is not really important whether the positive or the negative particles move. Consider that even though the electrons are in motion, an observer moving with the electrons will conclude that the positive charges are moving.

The assumption that the positive charges are moving became a convention that is still used today.

It is the direction of apparent positive charge movement that defines the direction of ***conventional current.***

Electron current and conventional current are shown in Figure 3.2. The electrons are supplied from a source of energy in order to maintain a continuous flow of charge. The current in a wire is measured directly by an instrument appropriately called an "ampere meter," or simply *ammeter*. This instrument is discussed in a later chapter.

Since current is the uniform motion of charged particles and charges in motion are the mechanism by which energy transfer occurs, it is then advanta-

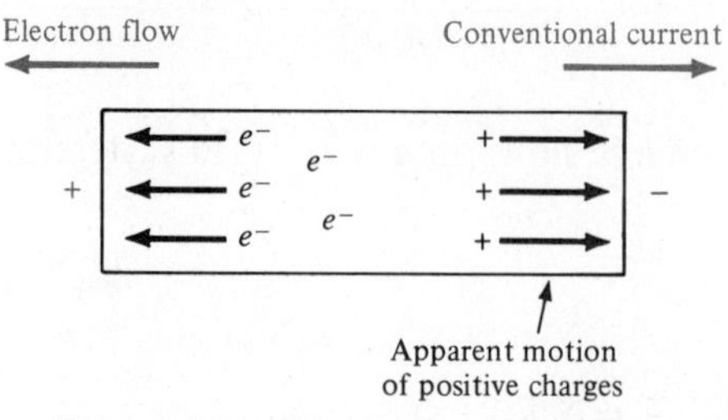

Figure 3.2 Conventional current.

geous to consider electric current as a medium of transferring energy from one point to another. Consider a source of electrical energy connected by wire to an electric device (called a load), as shown in Figure 3.3.

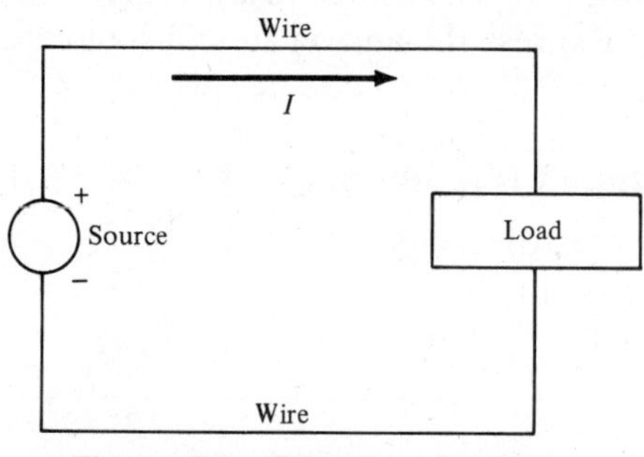

Figure 3.3 Transfer of energy.

A **load** is a device that transforms electrical energy into other forms such as heat or light. A light bulb is an example of a load. The wire allows current to be maintained and the current transfers energy from the source to the load. The energy source generates a force that "pushes" the charged particles around the closed path of wire. This is analogous to a force pushing an object over a distance. If the object is moved along a surface, mechanical energy is transformed into heat (thermal energy) due to the friction between the object and the surface. Similarly, charges moving in the wire have energy imparted to them by the energy source. The current (charges in motion) is the mechanism by which the energy is transferred to the load. If the source remains connected and continues to supply energy, the process will continue. Likewise, the object on the surface will continue to move if the force on the object persists and therefore keeps it moving.

*A connection of electrical components with an energy source, wire conductors, and a load is called an **electric circuit.***

3.3 ENERGY SOURCES

There are two classifications of electrical energy sources, the voltage source and the current source.

*The **constant voltage source** supplies electrical energy while maintaining a constant voltage.*

*The **constant current source** supplies electrical energy while maintaining a constant current.*

The source of electrical energy, which maintains a constant voltage, is called a source of electromotive force, or **EMF,** and is given the general symbol shown in Figure 3.4.

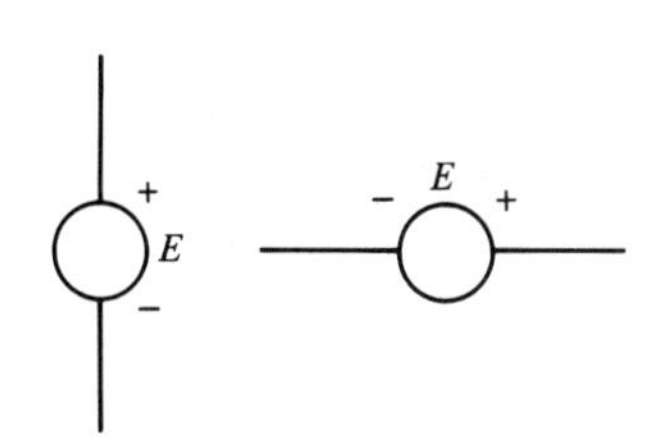

Figure 3.4 Symbol for electromotive force.

The plus (+) and minus (−) signs on the source are used to indicate its polarity.

In the circuit of Figure 3.3, the direction of the current is clockwise because positive charges are assumed to flow out of the positive terminal of the source, around the circuit, and back to the negative terminal. It should be noted that the current is the same everywhere in the circuit of Figure 3.3. The instant that the positive charges enter one side of the source, other positive charges are assumed to leave the other side. Likewise, at any section in the circuit, the same number of charges enter as leave that section.

One specific source of EMF is the **electric cell.** The standard symbol for an electric cell is shown in Figure 3.5a. Several of these cells connected together form a **battery.** The standard symbol for a battery is shown in Figure 3.5b. Note that the longer bar in the battery symbol represents the positive terminal, and the shorter bar represents the negative terminal.

In essence, the source (EMF) supplies energy, which causes the charges to flow (current) in the circuit and the load. The load transforms the electrical energy into other forms such as heat and light. The current will continue to be maintained

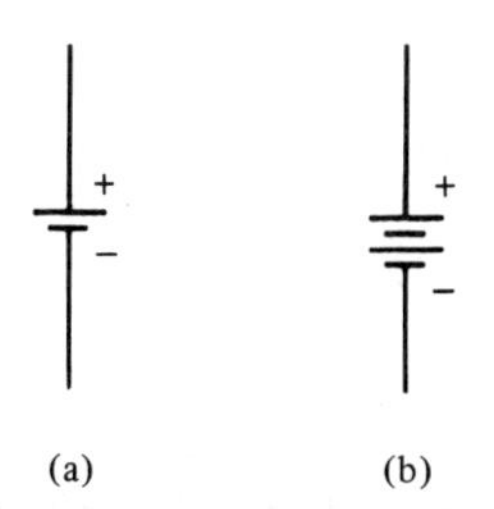

Figure 3.5 Symbols for voltage sources. **(a)** Symbol for a cell. **(b)** Symbol for a battery.

as long as the EMF can keep the charge in motion. The source must continue to supply energy and the circuit must not be interrupted if current is to be maintained.

3.4 VOLTAGE

Whenever a force is used to move something, in our case charged particles, work is being done. Work is done (energy is being expended) when charges are being pushed around a circuit. The measure of the work done on each unit of charge is called *voltage.*

***Voltage** is the work done per unit charge.*

Voltage is sometimes referred to as electric potential. In equation form,

$$V = \frac{W}{Q} \tag{3.2}$$

where: V => voltage (V)
W => energy or work (J)
Q => charge (C)

The SI unit of voltage is the volt (V).

One joule per coulomb is 1 V.

Since the source imparts energy to the charged particles, it is in fact a source of electrical energy. The measuring unit for the EMF is, therefore, the volt. A source of EMF of 6 V, for example, supplies 6 J of energy to each coulomb of charge in a circuit in which current is maintained.

Let us examine a source of EMF with no current. Consider the circuit shown in Figure 3.6, which contains a 6-V source, a switch, and a load.

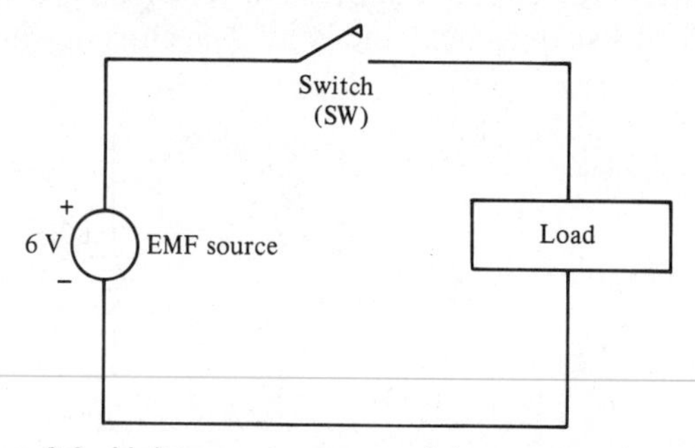

Figure 3.6 Voltage source supplying energy to a load.

When the switch is closed, the source supplies 6 J/C and current is established. The energy given to the circuit by the source is utilized by the load. If the load is a motor, for example, it converts electrical energy into the form of rotational mechanical energy. The process continues as long as the circuit is complete and the current is maintained.

If the switch is opened, the current stops because the circuit is no longer complete. However, the source of EMF has the *potential* to do work. This is obvious, since if the switch is closed, work is done.

The voltage across the source is a measure of its potential to do work.

If 6 J/C are to be maintained, the source must maintain a potential of 6 V when the switch is closed. For the case of the battery, chemical reactions within the battery supply the electrical energy necessary to maintain the source at 6 V. Other types of sources employ different methods of voltage generation. This is discussed in detail in Chapter 6.

An analogy will clarify the concept of voltage and work done by an electrical system. Consider Figure 3.7: a hill with charges on top.

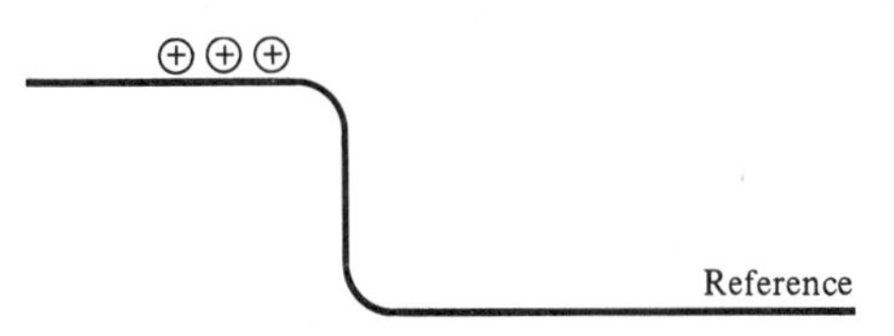

Figure 3.7 An analogy of voltage.

The charged particles, because of their elevation, possess potential energy (analogous to the switch being open in Figure 3.6). Energy has not yet been expended. If the charges are pushed over the edge, kinetic energy of motion will result (analogous to the switch being closed in Figure 3.6). If the charges are continuously supplied and pushed over the edge, the system is analogous to the circuit in Figure 3.6 with the switch closed.

EXAMPLE 3.5

Compute the work done by a source of EMF of 10 V operating on 2 C of charge.

Solution

Rearranging Eq. (3.2) gives $W = VQ$. Hence

$$\begin{aligned} W &= (10 \text{ V})(2 \text{ C}) \\ &= 20 \text{ J} \end{aligned}$$

EXAMPLE 3.6

Compute the voltage required to cause 14 J of work to be done on 4 C of charge.

Solution

By Eq. (3.2),

$$V = \frac{14 \text{ J}}{4 \text{ C}}$$
$$= 3.5 \text{ V}$$

EXAMPLE 3.7

The 12-V battery of a car supplies 80 A for 4 s while starting the car. Compute the work done by the battery.

Solution

By Eq. (3.1),

$$Q = (80 \text{ A})(4 \text{ s})$$
$$= 320 \text{ C}$$

By Eq. (3.2),

$$W = (12 \text{ V})(320 \text{ C})$$
$$= 3840 \text{ J}$$

3.5 POTENTIAL DIFFERENCE

Referring again to the concept of a hilltop, it is clear that an object has more potential energy at the top of the hill than at the bottom. The top of the hill represents a higher potential than the bottom. If an observer stands at the top, the bottom has a smaller potential.

A source of EMF has the potential of doing work on charged particles.

The voltage across the terminals of a source of EMF is known as its ***potential difference.***

Thus a source of 12 V maintains a potential difference of 12 V between its terminals. The unit of volt (V) is thus used to measure both EMF and potential difference. Often the expressions "voltage" and "potential difference" are used interchangeably. Here are some clarifying examples.

EXAMPLE 3.8

Given the source as shown in Figure 3.8:

a. Determine the potential difference maintained by the source.

b. Determine the voltage at point *A* with respect to point *B*.

c. Determine the voltage at point *B* with respect to *A*.

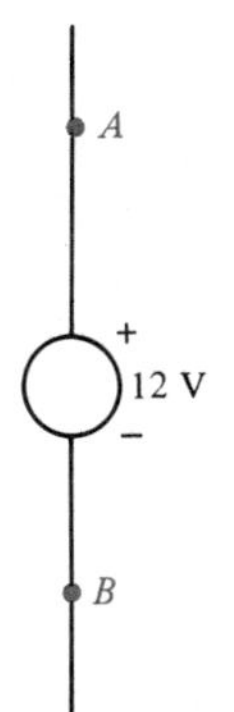

Figure 3.8 Describing a source.

Solution

The positive terminal may be thought of as representing the top of the hill, whereas the negative terminal may be thought to represent the bottom. Point *A* is 12 V higher than point *B* and, conversely, point *B* is 12 V lower than point *A*.

a. The potential difference maintained by the 12-V source is 12 V.

b. The voltage at point *A* with respect to point *B* is +12 V.

c. The voltage at point *B* with respect to point *A* is −12 V.

EXAMPLE 3.9

Determine the potential difference between *A* and *C* in Figure 3.9a.

Solution

The analogy for the source is the combination of hills shown in Figure 3.9b. Point *B* has a potential 12 m higher than point *C*, and point *A* has a potential 10 m higher than point *B*. Therefore, point *A* has a potential 22 m higher than point *C*.

For the voltage sources, the potential difference between *A* and *C* is 10 V + 12 V = 22 V.

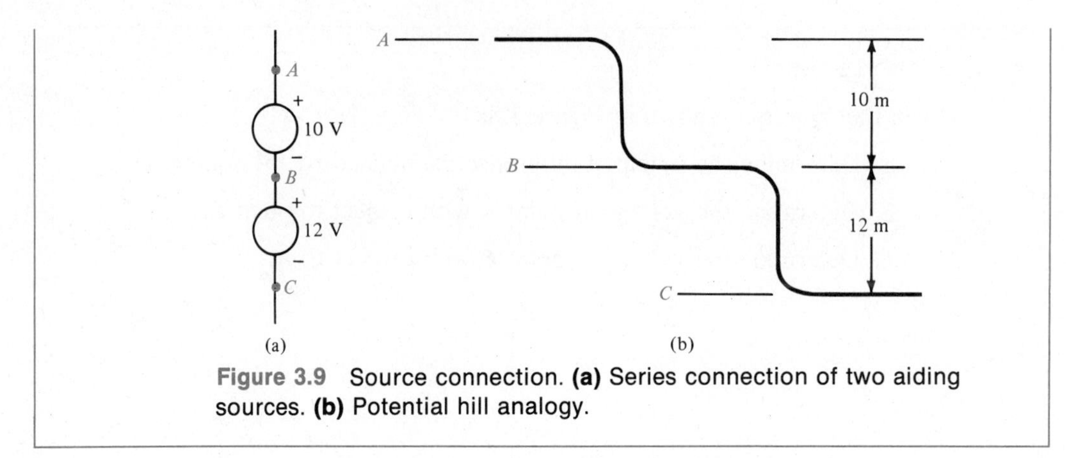

Figure 3.9 Source connection. **(a)** Series connection of two aiding sources. **(b)** Potential hill analogy.

The potential difference between any two points in a circuit is merely the difference in voltage between the two points.

The concept of positive and negative voltage is really a matter of reference. Refer to Figure 3.9 and observe the following (formulate a hill analogy if necessary):

The voltage at point A with respect to point B is $+10$ V.

The voltage at point B with respect to point C is $+12$ V.

The voltage at point A with respect to point C is $+22$ V.

The voltage at point B with respect to point A is -10 V.

The voltage at point C with respect to point B is -12 V.

The voltage at point C with respect to point A is -22 V.

EXAMPLE 3.10

Given the sources shown in Figure 3.10a, determine the potential difference between points A and C.

Solution

Using the diagram in Figure 3.10b, an observer starting at C goes down 5 m to B, then up 7 m to A. The difference of 2 m exists between points A and C with A being higher. Thus, the potential difference between A and C is $+2$ V, and the voltage at A with respect to C is $+2$ V.

In addition, note the following:

The voltage at A with respect to B is $+7$ V.

The voltage at B with respect to A is -7 V.

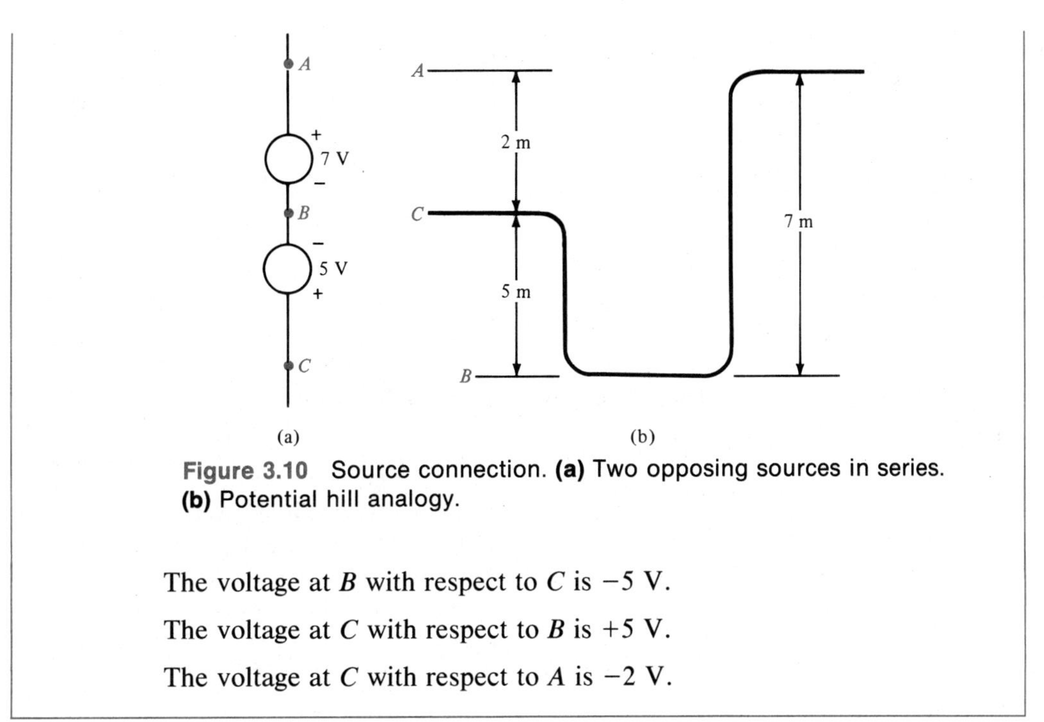

Figure 3.10 Source connection. **(a)** Two opposing sources in series. **(b)** Potential hill analogy.

The voltage at B with respect to C is -5 V.

The voltage at C with respect to B is $+5$ V.

The voltage at C with respect to A is -2 V.

3.6 ALGEBRAIC NOTATION FOR VOLTAGE

A standard notation is normally used to represent the voltage or potential difference between points in a circuit. The letter V is used with subscripts to indicate the points in question with respect to a common reference (usually called a *ground*). The first subscript usually indicates the point in question and the second point indicates the reference. Thus V_{ab} is the voltage at a with respect to point b. Consider the following example.

EXAMPLE 3.11

Given the source as shown in Figure 3.11, determine the voltages V_{ab} and V_{ba}.

Solution

$$V_{ab} = +12 \text{ V}$$

$$V_{ba} = -12 \text{ V}$$

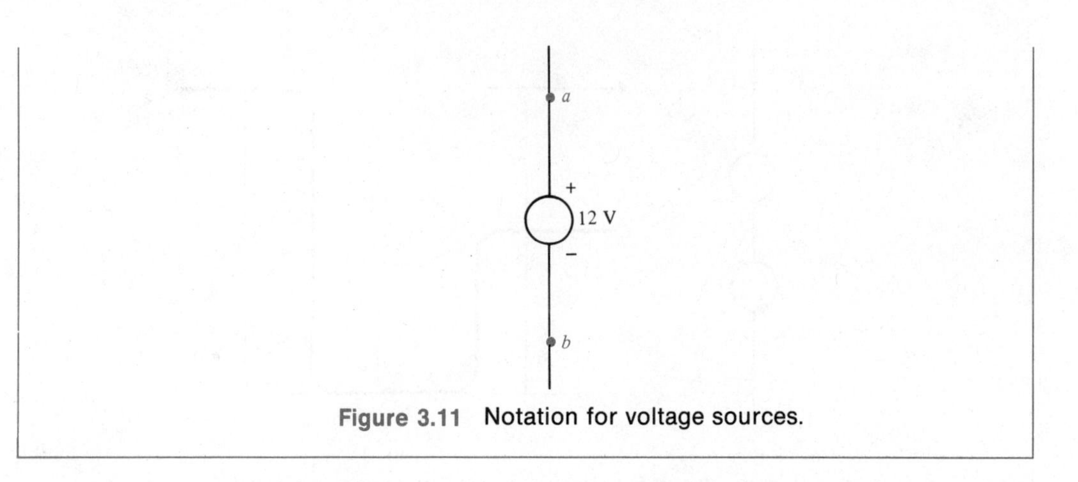

Figure 3.11 Notation for voltage sources.

The use of arrows is another method employed to indicate the voltage, with the arrowhead indicating the point in question and the tail indicating the reference. Figures 3.12, 3.13, and 3.15 show the labeling of a source with arrows.

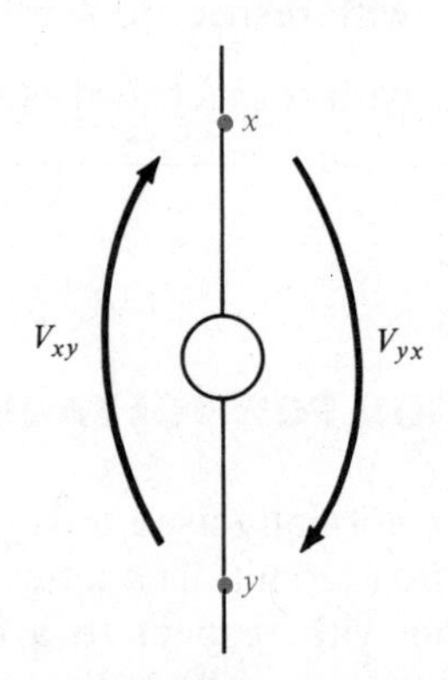

Figure 3.12 Voltage source labeling.

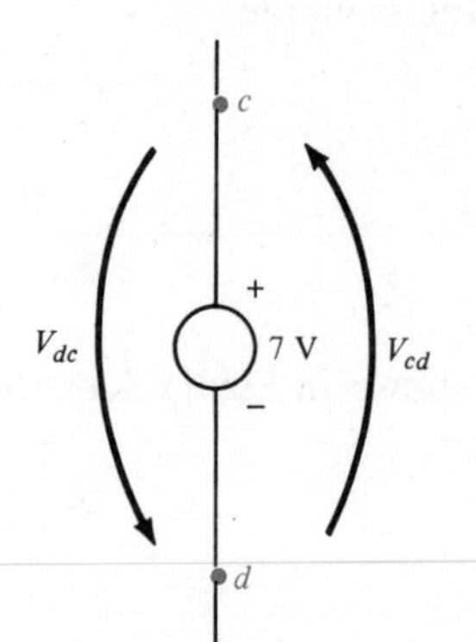

Figure 3.13 Voltage source labeling.

EXAMPLE 3.12

a. Given the circuit in Figure 3.14a, determine the voltages V_{ab}, V_{ac}, V_{bc}, and V_{ba}.

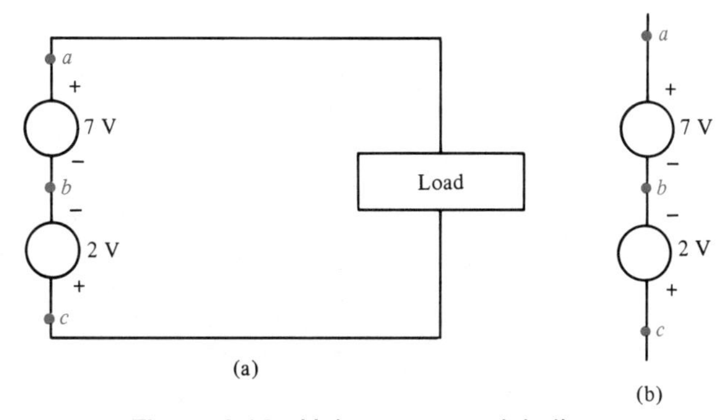

Figure 3.14 Voltage source labeling.

b. Indicate the voltage polarities with arrows.

Solution

a. Referring to Figure 3.14b:

$V_{ab} = 7$ V; $V_{ac} = 5$ V; $V_{bc} = -2$ V; $V_{ba} = -7$ V

b. Figure 3.15 shows the voltage polarities with arrows.

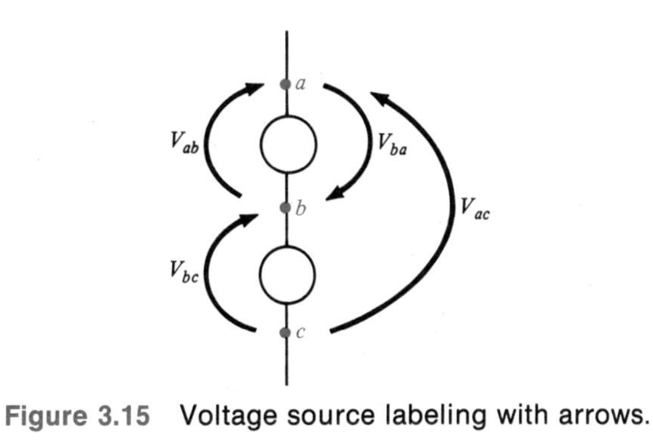

Figure 3.15 Voltage source labeling with arrows.

The use of arrows can be extended to indicate the polarity of a source. Thus, a source of 5 V can be shown as in Figure 3.16a. The notation is equivalent to Figure 3.16b.

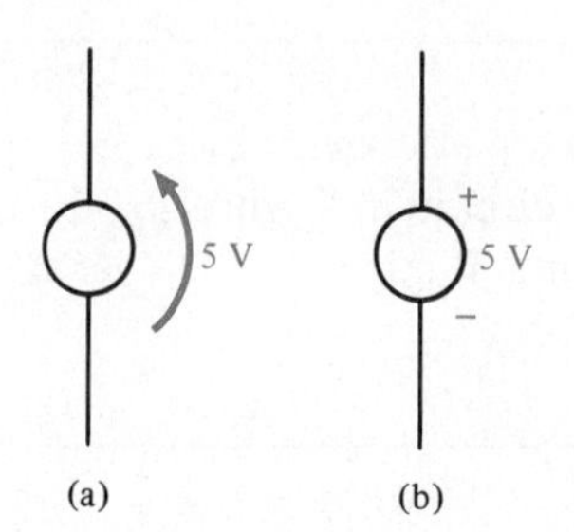

Figure 3.16 Voltage source labeling. **(a)** Polarity with arrows. **(b)** Polarity with + and − signs.

SUMMARY

This chapter has given the definitions of current and voltage and has developed the concept of potential difference. Conventional current has also been defined.

Current as the flow of charged particles per unit time and voltage as the work done per unit charge represent important definitions essential to further study in circuit theory. The use of proper units is important if common understanding is to exist in the discussion of electric circuits. The definition of potential difference as the difference in voltage between two points in a circuit was given. These methods are used throughout the text and indeed throughout the field of electrical, electronic, and computer technology. The definition of a circuit and a discussion of electromotive force together with common analogies are important concepts that are useful in later chapters.

SUMMARY OF IMPORTANT EQUATIONS

Charge and current $$I = \frac{Q}{t} \tag{3.1}$$

Work and voltage $$V = \frac{W}{Q} \tag{3.2}$$

REVIEW QUESTIONS

3.1 State the definition of electric current.

3.2 What is the unit of electric current?

3.3 What is voltage?

3.4 What is the unit of voltage?

3.5 What is meant by a source of EMF?

3.6 What is an electric cell?

3.7 What device is made up of several electric cells?

3.8 What is a constant voltage source?

3.9 What is a constant current source?

3.10 Discuss the terms potential difference, potential drop, and potential rise.

3.11 Give a definition of the term circuit.

3.12 Is it possible to develop circuit theory with the assumption that positive charges are in motion rather than electrons? Explain.

3.13 What is an electrical load?

3.14 In what way is charge related to current?

3.15 Give an explanation of the term potential hill.

3.16 What is the difference between electron current and conventional current?

3.17 How are voltage and current related?

3.18 How is energy transferred from one place to another in an electric system?

3.19 What is Ampere's Law?

3.20 What instrument is used to measure current?

3.21 What instrument is used to measure voltage?

PROBLEMS

3.1 A charge of 10 C passes through a cross section of a wire in 2.5 s. Determine the current in amperes.

3.2 If the work done on the charge in Problem 3.1 is 2 J, determine the voltage required. (ans. 0.2 V)

3.3 A current of 3 mA is maintained in a wire. Determine the amount of charge that passes through the cross section of the wire in 2 s.

3.4 A 12-V source of EMF is causing a 3-mA current; determine the work done in 4 s. (ans. 144 mJ)

3.5 A 3-A current is maintained through a wire. Determine the amount of time the current existed if 180 C passed through the wire.

3.6 A 20-V source is connected by wire to a load. If the load uses 15 J of energy in 10 s, determine the current. (See Figure 3.17 for the circuit.) (ans. 75 mA)

3.7 Consider the circuit shown in Figure 3.18. The source of EMF is given as 25 V and the current is 4 mA. How much work is done by the load in 2 s?

3.8 Determine the time necessary for 28,800 J of energy to be dissipated by a load as heat by a 2-A current at 120 V. (ans. 120 s)

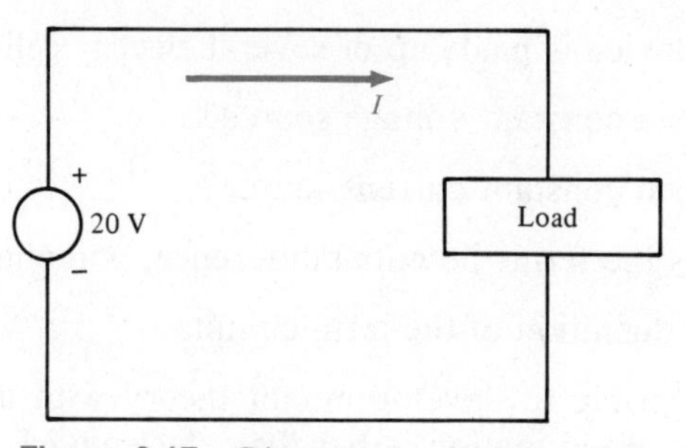

Figure 3.17 Diagram for Problem 3.6.

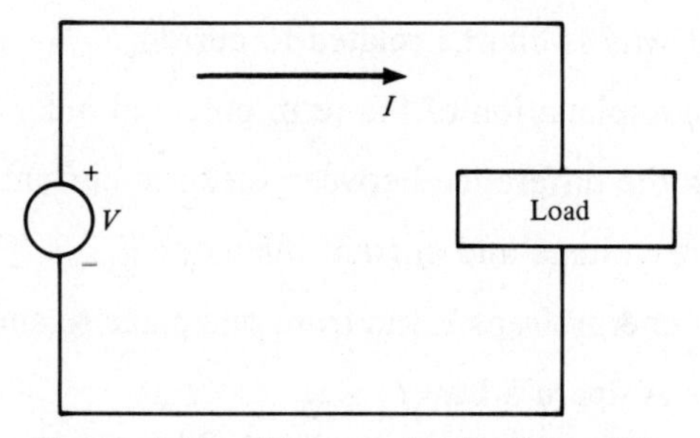

Figure 3.18 Diagram for Problem 3.7.

3.9 Express each of the following in scientific notation (e.g., 0.0015 mA = 1.5×10^{-6} A).
a. 14.72 A
b. 1200 kV
c. 0.000172 C
d. 702 μA

3.10 Given 1.47 μC passing through a cross section of a wire in 3 ms, determine the current in:
a. Amperes (ans. 0.00049 A)
b. Milliamperes (ans. 0.49 mA)
c. Microamperes (ans. 490 μA)

3.11 Five milliamperes are maintained through a cross section of a wire. Determine the time needed for 1.2 C to pass through the wire.

3.12 Determine the number of
a. Milliamps in 1 uA (ans. 0.001 mA)
b. Microamps in 1 mA (ans. 1000 μA)
c. Millivolts in 1 kV (ans. 1,000,000 mV)
d. Kilovolts in 1 μV (ans. 0.000 000 001 kV)
e. Volts in 1 μJ per mC (ans. 0.001 V)

3.13 Given the source as shown in Figure 3.19, determine V_{ab} and V_{ba}.

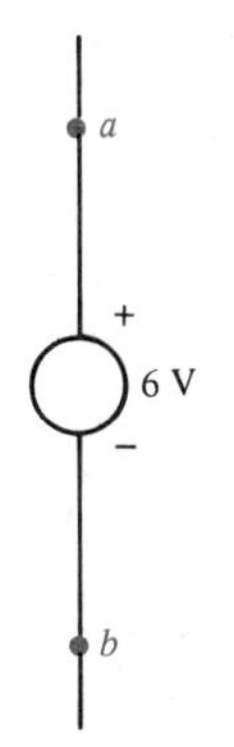

Figure 3.19 Diagram for Problem 3.13.

3.14 Given the sources as shown in Figure 3.20, determine V_{ab}, V_{ac}, V_{bc}, V_{ca}. (ans. +6 V, +8 V, +2 V, −8 V)

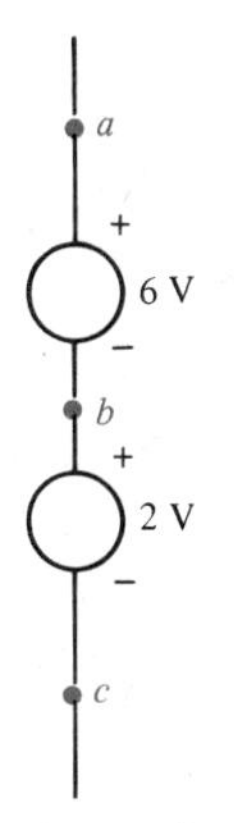

Figure 3.20 Diagram for Problem 3.14.

3.15 An 8-cylinder automobile engine is operating at 1000 rev/min and delivers 12,000 V to each spark plug. Determine the energy delivered by each spark to each spark plug if each spark consists of 0.5 C of charge.

3.16 Given the circuit shown in Figure 3.21, show that $V_{AC} = V_{AB} + V_{BC}$ and that $V_{CA} = V_{BA} + V_{CB}$. Give a familiar analogy as part of the proof. (ans. $V_{AC} = +2 - 8 = -6$ V)

3.17 Given the circuit shown in Figure 3.22, an observer at point A counts 10 μC of charge passing in 1 s. At what rate is energy being supplied to the load?

3.18 An automobile radio operates at 12 V for 10 min. If 72 J of energy were delivered to the radio, determine the current to the radio during the 10 min. (ans. 10 mA)

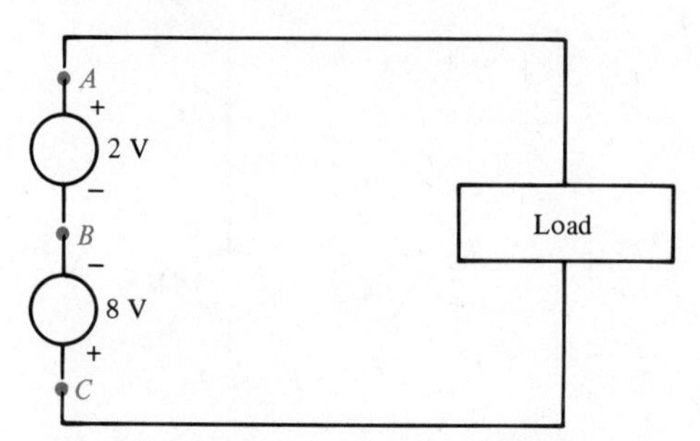

Figure 3.21 Diagram for Problem 3.16.

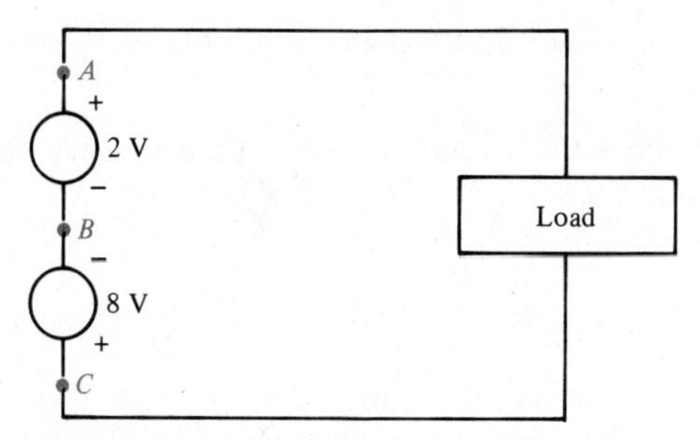

Figure 3.22 Diagram for Problem 3.17.

3.19 A light bulb connected across 120 V operates at 1.67 A. Determine the energy used by the bulb in a 24-h period.

3.20 A motor is converting electrical energy to mechanical energy and is operating at 3 A from a 120-V source. How long must the motor operate to convert 1000 J of electrical energy into mechanical energy? (ans. 2.78 s)

3.21 In a circuit there are three voltage test points: A, B, and C. V_{AB} was measured to be +20 V. V_{AC} was measured to be +50 V. Compute V_{BC}.

3.22 A technician is measuring a voltage from reference to a test point. The schematic diagram indicates that she should read +12 V, but the digital meter is indicating −12 V. What probably happened? (ans. positive test lead was connected to reference, negative test lead to test point)

3.23 Captain Kirk wants to fire 100,000 C of electrons at an attacking starship. How long does he have to maintain a 1000-A current through the electron collector to accumulate the needed electrons?

CHAPTER 4

Ohm's Law and Resistance

GLOSSARY

American Wire Gauge (AWG) Numbers — classification of wires according to their diameters; the higher the number, the smaller the diameter
aspect ratio—ratio of length to width of a material
conductance—ability to conduct charge current and maintain an electric current; measured in siemens
diffused resistor—integrated circuit resistor part of a silicon chip
photoconductor—a two-terminal device whose resistance is a function of illumination
potentiometer—variable resistor
power—work done per unit time
resistance—measure of the opposition to electrical current
resistor—current-opposing device
rheostat—variable resistor with two of its terminals utilized; usually a power variable resistor
semiconductor diode—a device that allows current in one direction only
sheet resistance—resistance of a square of material
thermistor—device whose resistance decreases with increasing temperature

INTRODUCTION

The equation that relates voltages and current is perhaps the single most important relationship of circuit theory. It is upon this relationship that all circuit theory rests. The quantity that relates voltage and current is resistance.

Resistance *is a characteristic of a material and is a measure of the opposition to electric current.*

A mechanical analogy to electric resistance is friction. Friction, or mechanical resistance, develops between a moving object and a surface. The method by which a device opposes current and thus shows resistance from a physical viewpoint before Ohm's Law is discussed.

4.1 RESISTANCE

The nature of resistance can be seen by examining the flow of electrons in a material caused by an external force. "Free" electrons in the slab of material of Figure 4.1 are in motion due to the external EMF. The electrons in motion collide with other electrons; energy is then expended and transferred into heat. The electron flow is diminished in direct proportion to the heat generated in the material. This is a reduction in current and therefore indicates the presence of resistance.

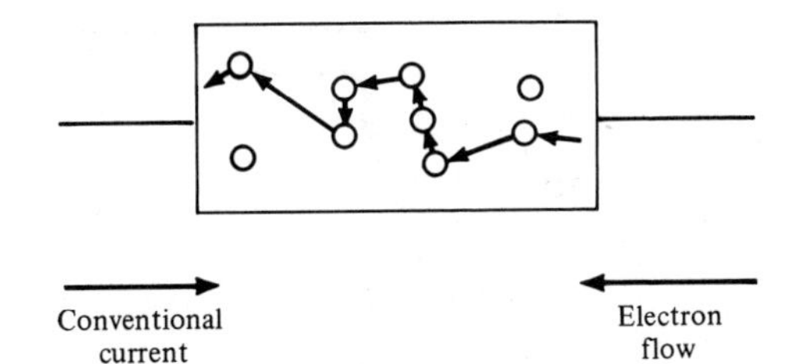

Figure 4.1 Movement of electrons in a slab of material.

All materials exhibit resistance. In the case of metal wires, such as copper, the resistance of small lengths is usually so small as to be negligible. A discussion of wire resistance and types of resistance is presented later in this chapter. The letter R is used to represent resistance. The electrical symbol of a resistor is shown in Figure 4.2.

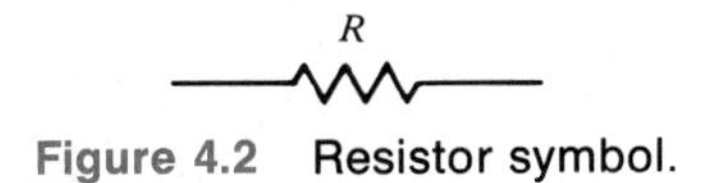

Figure 4.2 Resistor symbol.

4.2 OHM'S LAW

In the mid-1800s Georg Simon Ohm discovered that a circuit consisting of a source of EMF connected to a resistive load—called a **resistor**—as shown in Figure 4.3, resulted in a ratio of voltage to current equal to the resistance of the load. This became known as Ohm's Law. In equation form, Ohm's Law can be stated in any of the following forms:

$$\boxed{R = \frac{V}{I}} \tag{4.1a}$$

$$\boxed{V = IR} \tag{4.1b}$$

$$\boxed{I = \frac{V}{R}} \tag{4.1c}$$

The wire connecting the EMF to the resistor of Figure 4.3 is considered to have negligible resistance.

The MKS (SI) unit of resistance is the ohm. The greek letter omega (Ω) is used to designate ohms.

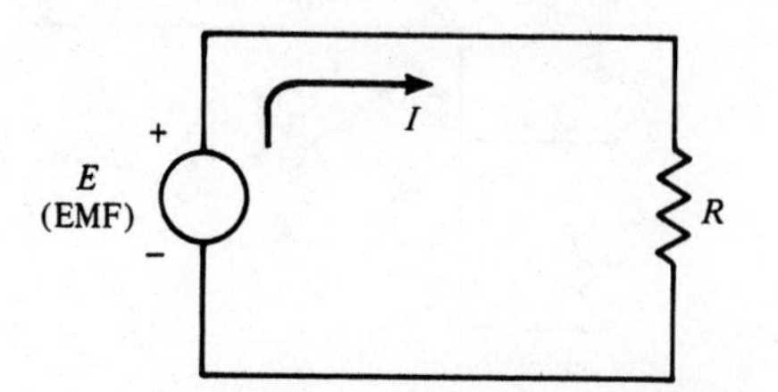

Figure 4.3 Source supplying a resistor.

Thus 10 ohms is written 10 Ω. Ohm's Law can also be represented graphically as shown in Figure 4.4. The slope of the line is the reciprocal of the resistance. The resistance is equal to $\Delta V/\Delta I$ (reciprocal of slope). The resistance value can thus be determined from a graphical representation of Ohm's Law. *The straight line, or linear, graph of Figure 4.4 results from a constant resistance with changes in voltage.*

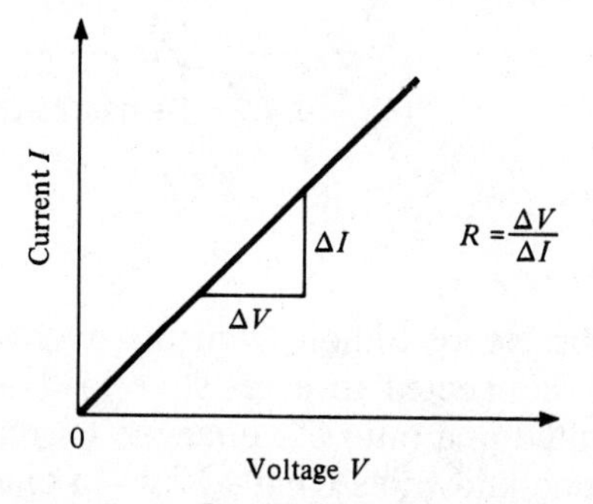

Figure 4.4 Graph of linear resistor.

EXAMPLE 4.1

Determine the value of resistance R if a current of 2 A is maintained through the resistor and a voltage of 6 V is measured across it.

Solution

By (4.1a),

$$R = \frac{V}{I}$$

$$= \frac{6}{2} = 3\ \Omega$$

Ohm's Law is certainly valid for any resistor, linear or otherwise. A device that *does not* exhibit a straight-line characteristic is said to be *nonlinear*. A graphical representation of a nonlinear resistor is shown in Figure 4.5. The resistance

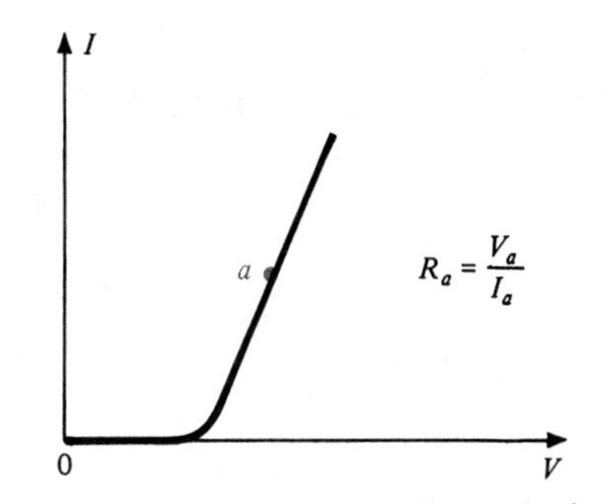

Figure 4.5 Graph of nonlinear resistor.

varies with changes in voltage. The value of the resistance at a particular voltage can be computed from the ratio of voltage to current at that point. A more detailed discussion of the causes of nonlinearity appears later in this chapter.

A resistor is thus able to allow current; the values of the resistance and voltage determine the magnitude of the current. The resistance value in ohms is a measure of the *ability to oppose the current.* Another useful quantity is the **conductance** of a device. The conductance is the reciprocal of resistance and it indicates the ability to allow current. The letter G is used to represent conductance; in equation form,

$$\boxed{G = \frac{1}{R}} \tag{4.2}$$

The SI unit for conductance is the siemens (S).*

Since the wire in Figure 4.3 is considered to have negligible resistance with the direction of the current as shown, a convention of establishing the voltage polarity across a resistor can be made. Consider the circuit in Figure 4.6. All the top portion of the circuit is considered to be one point (A) and the bottom another point (B). The positive terminal of the source is electrically the same as one terminal of the resistor. This being the case, *the terminal in which the current*

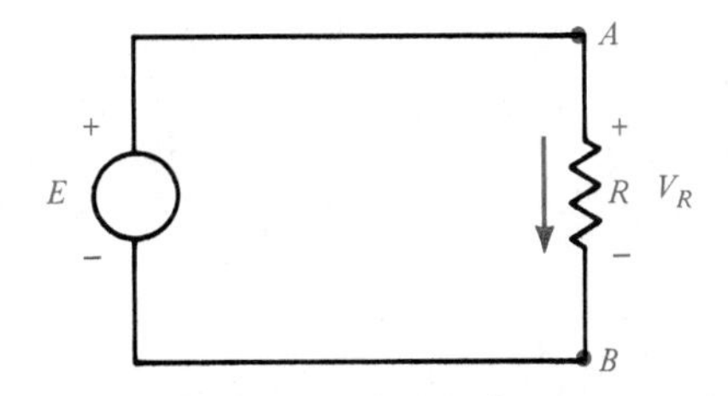

Figure 4.6 Indication of polarity across resistor.

* The unit of conductance was formerly the mho (ohm spelled backwards).

enters the resistor is positive and the terminal from which the current leaves the resistor is negative. Figure 4.6 indicates this polarity. The convention of establishing the polarity of the voltage across a resistor due to a current in a particular direction is essential for electric circuit analysis presented in later chapters. Figure 4.7 offers illustrations of current direction and the resulting voltage polarity.

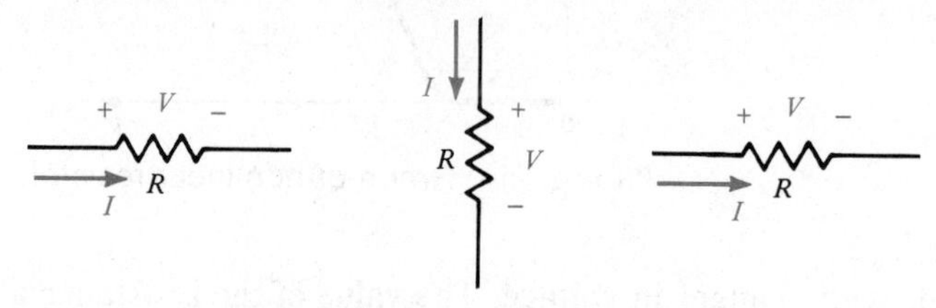

Figure 4.7 Current directions.

EXAMPLE 4.2

Determine the direction of the current and the value of the current for the resistor shown in Figure 4.8.

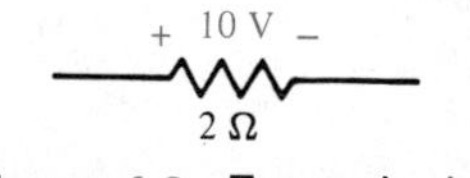

Figure 4.8 Example 4.2.

Solution

The current is to the *right:*

$$I = \frac{V}{R}$$

$$= \frac{10}{2} = 5 \text{ A}$$

The resistor is one type of load, which is a current-opposing device. The voltage source in Figures 4.3 and 4.6 is supplying energy to the circuit and the energy is delivered to the load resistor, where it is converted to heat energy. A motor load, on the other hand, would convert electrical energy to mechanical energy.

4.3 ENERGY AND POWER

Energy was discussed briefly in the previous chapter and is again discussed here as it relates to power.

***Power** is defined as the work done per unit time.*

A mechanical analogy is velocity, which is the distance traveled per unit time. In equation form power, P, is

$$\boxed{P = \frac{W}{t}} \tag{4.3}$$

The SI unit of power is the *watt*. One watt is equal to 1 J/s. Equation (4.3) is usually shown in the form

$$\boxed{W = Pt} \tag{4.4}$$

The homeowner is charged by the utility company for *kilowatt-hours* (kWh); the number of kilowatt-hours is found by multiplying power in kilowatts by time in hours. The use of 100 kWh at 8¢ per kilowatt-hour costs $8.

EXAMPLE 4.3

If energy costs 8¢ per kilowatt-hour, determine the cost of running a 200-W bulb for 3 days.

Solution

$$200\ \text{W} = 0.2\ \text{kW}, \qquad 3\ \text{days} = 72\ \text{h}$$

By Eq. (4.4),

$$W = (0.2)(72) = 14.4\ \text{kWh}$$

$$\begin{aligned}\text{Cost} &= (\text{energy})(\text{cost per unit of energy}) \\ &= (14.4)(0.08) = \$1.15\end{aligned}$$

The power rating of a resistor is important, since it states the maximum rate at which the resistor can dissipate heat energy into the surrounding environment and still operate within its specified value.

For another type of device, such as a motor, the power rating states the rate at which the motor can do mechanical work. The power rating of a resistor indicates the maximum wattage at which the stated value of resistance will remain valid. If the rating is exceeded, the resistance value changes and the device may in fact be destroyed.

Recall from Chapter 3 the relationship of voltage to work:

$$\boxed{V = \frac{W}{Q}} \tag{3.2}$$

or

$$W = VQ$$

since $P = W/t$, then by substitution, $P = VQ/t$; we also recognize that $Q/t = I$ (Ampere's Law). Hence,

$$\boxed{P = VI} \tag{4.5}$$

Equation (4.5) indicates that the voltage across a resistor multiplied by the current through the resistor is the power delivered to the resistor and, therefore, the rate at which the resistor dissipates energy in the form of heat.

EXAMPLE 4.4

Given the circuit of Figure 4.9, determine:

a. the current I through the resistor.

b. the power P to the resistor.

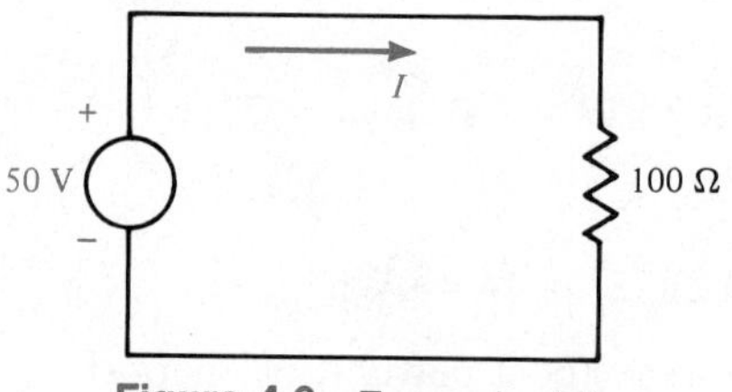

Figure 4.9 Example 4.4.

Solution

a. $I = \dfrac{V}{R}$

$= \dfrac{50}{100} = 0.5 \text{ A}$

b. $P = VI$

$= 50(0.5) = 25 \text{ W}$

The results indicate that the resistor must be able to dissipate, as heat, 25 J/s.

Since $V = IR$ and $P = VI$, by substitution

$$P = (IR)(I) = I^2R$$

and

$$P = V\frac{V}{R} = \frac{V^2}{R}$$

Therefore, these equivalent expressions can be used to calculate power.

$$\boxed{P = VI = I^2R = \frac{V^2}{R}} \tag{4.6}$$

EXAMPLE 4.5

Given the circuit of Figure 4.10, determine the energy dissipated as heat if the circuit operates for 3 s.

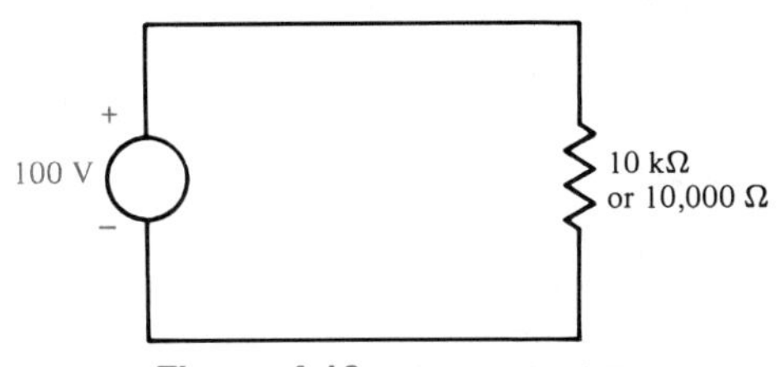

Figure 4.10 Example 4.5.

Solution

From Eq. (4.6),

$$P = \frac{V^2}{R} = \frac{(100)^2}{10{,}000} = 1\ \text{W}$$

1 W = 1 J/s; therefore, in 3 s, $W = (1)(3) = 3$ J of energy is dissipated as heat.

Another approach is to calculate the current first:

$$I = \frac{V}{R}$$

$$= \frac{100}{10{,}000} = 0.01\ \text{A} = 10\ \text{mA}$$

Then

$$P = I^2R = (0.01)^2(10{,}000) = 1\ \text{W}$$

The horsepower (hp) is also used as a unit of power:

1 hp = 746 W

EXAMPLE 4.6

A furnace blower motor is rated at $\frac{1}{2}$ hp. If the motor operates from a 110-V source, determine the current supplied to the motor.

Solution

Converting $\frac{1}{2}$ hp to watts,

$$P = \frac{1}{2} \text{ hp} = \frac{1}{2} \times 746 = 373 \text{ W}$$

From $P = VI$,

$$I = \frac{P}{V} = \frac{373}{110}$$
$$= 3.39 \text{ A}$$

4.4 WIRE RESISTANCE

Although for small lengths of wire, as used in circuits, the resistance is negligible, calculation of wire resistance becomes very important in other areas. The resistance of long sections of wire, such as in power-transmission lines, is significant and must be accounted for in design work. Some resistors are manufactured by winding wire onto a core of insulating material, and thus calculation of wire resistance is important. Modern integrated circuits make use of *film resistors,* for which calculations similar to those of wire resistance are made.

The resistance of a conductor is quite naturally dependent on the physical makeup of that conductor. The shape of the material and the type of material are two physical parameters that affect resistance. Resistance is also effected by variations in temperature.

A length of wire in the form of a cylindrical section is shown in Figure 4.11. It was stated earlier that collisions between electrons in motion (due to EMF) and

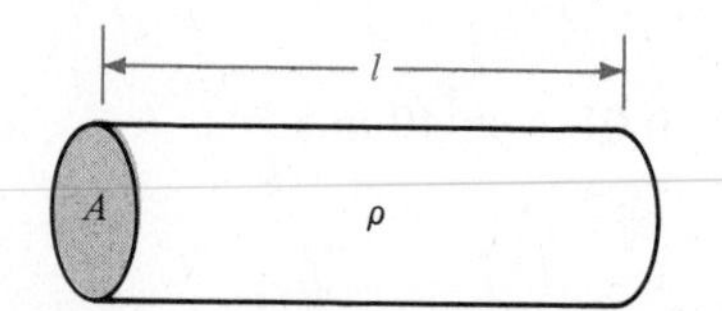

Figure 4.11 Wire configuration.

the atoms of a material are the cause of resistance. Each material has a different atomic structure and, therefore, a different number of collisions take place, causing each material to have a different resistive property. A numerical constant ρ (greek letter rho) is used to specify this property of materials. The larger the value of ρ, the larger is the resistance and thus the lower the conductance. The constant ρ is given the name *resistivity,* and it is a function of temperature.

As the length of wire increases, the resistance increases, since more collisions between electrons and atoms occur; but as the cross-sectional area increases, the resistance decreases. The same effect is apparent in the analogy of a water pipe, as shown in Figure 4.12.

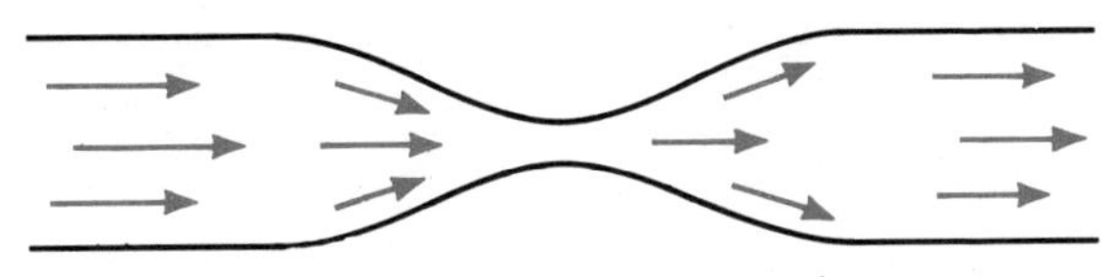

Figure 4.12 Water-pipe analogy.

The resistance of the pipe to water flow is increased in the narrowed section. If this "necking down" were to continue, a point would be reached where only one molecule at a time would fit through the opening. In an electric circuit, to maintain a constant current in spite of increased resistance, the forcing function (voltage) has to be increased. The consequence is that resistance increases with lowered cross-sectional area. The resulting equation for the resistance of a wire is thus

$$\boxed{R = \rho \frac{l}{A}} \tag{4.7}$$

where:
R => resistance (Ω)
l => length (m)
A => area (m^2)
ρ => resistivity (Ω-m)

The SI unit of resistivity is the *ohm-meter* (Ω-m).

The inverse of resistivity is called the conductivity and is symbolized by the greek letter sigma (σ). The unit of conductivity is 1/(ohm-meter), which is siemens per meter (S/m). Table 4.1 shows the resistivities of various materials measured at 20°C.

Materials that have a low resistivity are said to be good **conductors;** those with high resistivity are good **insulators** (poor conductors). Materials that are neither good conductors nor insulators are said to be **semiconductors.**

TABLE 4.1
Resistivity of common electric conductor materials at 20°C

Material	Resistivity in ohm-meter
Silver	1.65×10^{-8}
Copper	1.72×10^{-8}
Aluminum	2.83×10^{-8}
Tungsten	5.49×10^{-8}
Nickel	7.81×10^{-8}
Iron	9.64×10^{-8}
Platinum	9.83×10^{-8}

EXAMPLE 4.7

Determine the resistance of 3 m of copper wire that is circular in cross section and has a diameter of:

a. 0.5 mm

b. 0.25 mm

Solution

a. 0.5 mm = 0.0005 m

$$\text{Cross-sectional area } A = \frac{\pi d^2}{4} = \frac{\pi(0.0005)^2}{4} = 1.96 \times 10^{-7}\ \text{m}^2$$

From Table 4.1, ρ (copper) = $1.72 \times 10^{-8}\ \Omega$-m. Therefore, by Eq. (4.7),

$$R = \frac{(1.72 \times 10^{-8})(3)}{1.96 \times 10^{-7}} = 0.263\ \Omega$$

b. 0.25 mm = 0.00025 m

$$A = \frac{\pi(0.00025)^2}{4} = 0.49 \times 10^{-7}\ \text{m}^2$$

and

$$R = \frac{(1.72 \times 10^{-8})(3)}{0.49 \times 10^{-7}} = 1.025\ \Omega$$

Note: The resistance is increased by a factor of 4 when the diameter is decreased by a factor of 2.

EXAMPLE 4.8

How many meters of copper wire of diameter 1 mm are required to give a resistance of 10 Ω?

Solution

$$1\ \text{mm} = 0.001\ \text{m}$$

$$A = \frac{\pi d^2}{4} = \frac{\pi(0.001)^2}{4} = 7.85 \times 10^{-7}\ \text{m}^2$$

Solving for l in Eq. (4.7),

$$l = \frac{RA}{\rho} = \frac{(10)(7.85 \times 10^{-7})}{1.72 \times 10^{-8}} = 456.6\ \text{m}$$

4.5 AMERICAN WIRE GAUGE (AWG) NUMBERS

Manufacturers have classified wires according to their diameters, with large-diameter wire given small gauge numbers. These are called **American Wire Gauge (AWG) numbers.** Thus number 10 wire has a larger diameter than number 22 wire. Table 4.2 lists several wire gauge numbers and the diameter of each wire. For convenience, the diameters are given in both meters and mils (1 mil = 0.001 in). This is done since many manufacturers continue to use mils as a measurement of wire.

TABLE 4.2
American Wire Gauge numbers

Gauge number	Diameter in mils	Diameter in meters
0000	460	1.168×10^{-2}
000	409.6	1.040×10^{-2}
00	364.8	9.266×10^{-3}
0	324.9	8.252×10^{-3}
1	289.3	7.348×10^{-3}
2	257.6	6.543×10^{-3}
3	229.4	5.827×10^{-3}
4	204.3	5.189×10^{-3}
5	181.9	4.620×10^{-3}
6	162.0	4.115×10^{-3}
7	144.3	3.665×10^{-3}
8	128.5	3.264×10^{-3}
9	114.4	2.906×10^{-3}
10	101.9	2.588×10^{-3}

TABLE 4.2 (continued)

Gauge number	Diameter in mils	Diameter in meters
11	90.74	2.305×10^{-3}
12	80.81	2.053×10^{-3}
13	71.96	1.828×10^{-3}
14	64.08	1.628×10^{-3}
15	57.07	1.450×10^{-3}
16	50.82	1.291×10^{-3}
17	45.26	1.150×10^{-3}
18	40.30	1.024×10^{-3}
19	35.89	9.116×10^{-4}
20	31.96	8.118×10^{-4}
21	28.46	7.229×10^{-4}
22	25.35	6.439×10^{-4}
23	22.57	5.733×10^{-4}
24	20.10	5.105×10^{-4}
25	17.90	4.547×10^{-4}
26	15.94	4.049×10^{-4}
27	14.20	3.607×10^{-4}
28	12.64	3.211×10^{-4}
29	11.26	2.860×10^{-4}
30	10.03	2.548×10^{-4}
31	8.928	2.268×10^{-4}
32	7.950	2.019×10^{-4}
33	7.080	1.798×10^{-4}
34	6.305	1.601×10^{-4}
35	5.615	1.426×10^{-4}
36	5.000	1.270×10^{-4}
37	4.453	1.131×10^{-4}
38	3.965	1.007×10^{-4}
39	3.531	8.969×10^{-5}
40	3.145	7.988×10^{-5}

EXAMPLE 4.9

Determine the resistance of 2 m of number 22 copper wire at 20°C.

Solution

By Eq. (4.7), and recalling that $A = \pi d^2/4$,

$$R = \frac{(1.72 \times 10^{-8})(2)}{\dfrac{\pi(6.439 \times 10^{-4})^2}{4}} = 0.104\ \Omega$$

EXAMPLE 4.10

Determine the length of number 12 aluminum wire required in order to have a resistance of 10 Ω at 20°C.

Solution

The diameter, d, is found in Table 4.2; $d = 2.053 \times 10^{-3}$ m.

$$A = \frac{\pi d^2}{4} = \frac{\pi(2.053 \times 10^{-3})^2}{4}$$

$$= 3.31 \times 10^{-6}\ \text{m}^2$$

Solving for l from Eq. (4.7),

$$l = \frac{(10)(3.31 \times 10^{-6})}{2.83 \times 10^{-8}} = 1169.72\ \text{m}$$

EXAMPLE 4.11

Complete the following table for copper wire.

AWG Number	Ohms per Kilometer at 20°C
0	
5	
10	
15	
20	
25	
30	
35	
40	

Solution

First, the corresponding diameters are found from Table 4.2. Then the resistance values are calculated by Eq. (4.7).

$$R = \frac{\rho l}{A}, \qquad \rho = 1.72 \times 10^{-8}, \qquad l = 1000\ \text{m}$$

AWG	*d* (m)	$A = \pi d^2/4$ (m²)	*R* (Ω)
0	8.252×10^{-3}	5.348×10^{-5}	0.322
5	4.620×10^{-3}	1.676×10^{-5}	1.026
10	2.588×10^{-3}	5.260×10^{-6}	3.270
15	1.450×10^{-3}	1.651×10^{-6}	10.42
20	8.118×10^{-4}	5.176×10^{-7}	33.23
25	4.547×10^{-4}	1.624×10^{-7}	105.91
30	2.548×10^{-4}	5.099×10^{-8}	337.32
35	1.426×10^{-4}	1.597×10^{-8}	1077
40	7.988×10^{-5}	5.011×10^{-9}	3432.45

4.6 TEMPERATURE EFFECTS ON WIRE RESISTANCE

The resistance of wire is affected by temperature. The resistance of a length of wire increases with the increase in temperature and decreases with a decrease in temperature. To develop a quantitative method of determining resistance of wire at various temperatures, it is assumed that resistance changes linearly with temperature. Although this is not quite accurate, it is a very good approximation over most of the range of temperatures encountered for wire. A temperature coefficient, α (alpha), is used to indicate the resistance change of a material to changing temperatures. A tabulation of coefficients is given in Table 4.3.

The method of determining the resistance of a material at a temperature T_2 is given by Eq. (4.8).

$$R_2 = R_1[1 + \alpha(T_2 - 20°\text{C})] \tag{4.8}$$

where R_1 is the resistance at 20°C.

EXAMPLE 4.12

Determine the resistance of 2 m of number 22 copper wire at 100°C.

Solution

From Eq. (4.8) and the value of resistance R_1 given in Example 4.9,

$$\begin{aligned} R_2 &= R_1[1 + \alpha(T_2 - 20°\text{C})] \\ &= 6.8 \times 10^{-5}\,[1 + 0.00393(100 - 20)] \\ &= 8.94 \times 10^{-5}\ \Omega \end{aligned}$$

TABLE 4.3
Temperature coefficients

Material	α in $\Omega \cdot$ °C/Ω at 20°C	
Aluminum	0.0039	
Brass	0.002	
Carbon	−0.005	(Note 2)
Constantan	0.000008	(Note 1)
Copper	0.00393	
Gold	0.0034	
Iron	0.0055	
Lead	0.0039	
Molybdenum-drawn	0.004	
Nickel	0.006	
Platinum	0.003	
Silver	0.0038	
Tantalum	0.0031	
Tin	0.0042	
Tungsten	0.0045	
Zinc	0.0037	

Note 1: Constantan has essentially constant resistance with temperature.

Note 2: Carbon increases resistance with decreasing temperature and is an ingredient in carbon-composition resistors.

Example 4.12 is for the same copper wire as in Example 4.9 at a higher temperature. Notice that the resistance value increased with an increase in temperature.

EXAMPLE 4.13

Repeat Example 4.10 for 100°C.

Solution

Solving Eq. (4.8) for R_1 yields:

$$R_1 = \frac{R_2}{1 + \alpha(T_2 - 20)}$$

$$= \frac{10}{1 + 0.0039(100 - 20)}$$

$$= 7.622\ \Omega$$

Solving Eq. (4.7) for l results in:

$$l = \frac{R_1 A}{\rho}$$
$$= 891.47 \text{ m}$$

4.7 LINEAR AND NONLINEAR RESISTANCE

The graph of a linear resistor is shown in Figure 4.4 and in Figure 4.13a. These graphs show a straight-line function of voltage versus current.

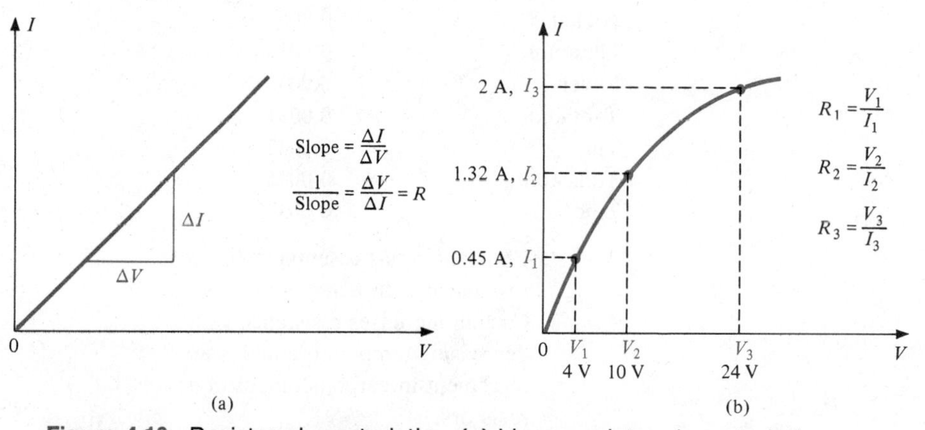

Figure 4.13 Resistor characteristics. **(a)** Linear resistor characteristic. **(b)** Nonlinear resistor (light bulb) characteristic.

A linear resistor is one which has constant resistance with changes in voltage.

At each point on the graph of Figure 4.13a, *the slope—and thus the resistance value—is the same,* even though the voltage and thus temperature of the device is different. *The current through a resistor* (due to a voltage) *causes a heating effect, which changes its temperature.* The graphs of Figures 4.5 and 4.13b indicate a nonlinear resistor. A *nonlinear resistor is one whose resistance changes with a change in some parameter, such as voltage or temperature.** The value of resistance calculated by use of Ohm's Law at several points on the graph produces different values.

Devices are manufactured with this nonlinearity in mind and, depending upon the intended use, are made to be linear or to have a specific nonlinearity. The

* Voltage changes may be the cause of the temperature changes.

incandescent bulb, which has a tungsten filament, is an example of a nonlinear resistor. Notice that the α coefficient of tungsten is quite high compared to carbon.

The voltage-ampere curve of a typical light bulb is shown in Fig. 4.13b. Other examples of nonlinear devices are discussed later in this chapter.

EXAMPLE 4.14

Determine the resistance and the power to the light bulb with 24 V across it. (Use the curves in Figure 4.13b.)

Solution

At 24 V, the current is 2 A. By Ohm's Law:

$$R = \frac{24}{2} = 12\ \Omega$$

The power supplied to the light bulb is:

$$\begin{aligned} P &= VI \\ &= (24)(2) = 48\ \text{W} \end{aligned}$$

4.8 TYPES OF RESISTORS

4.8.1 Discrete Resistors

Discrete resistors can be made from several types of material. Most common resistors are the fixed-carbon composition type. The carbon composition is used for general-purpose, low-voltage applications of resistance tolerance of 5% or greater. Figure 4.14a shows several resistors of this type; the color bands indicate the value of the resistance. The resistance value from the bands is determined by use of Table 4.4 and the position of a particular color band. Figure 4.15 indicates the location of the color bands.

With reference to Figure 4.15, bands 1 and 2 represent the first and second digits of the resistance value; band 3 represents the multiplier; the fourth band represents the tolerance. If the tolerance band is missing, the tolerance is understood to be ±20%. Thus a 2700-Ω, ± 5% resistor would have bands of red, violet, red, and gold.

Red	Violet		Red	Gold
2	7	×	10^2	±5%

A resistor of 51 kΩ ± 20% would have green, brown, and orange bands, with the fourth band missing. A resistor of 7.5 Ω ± 10% would have violet, green, gold,

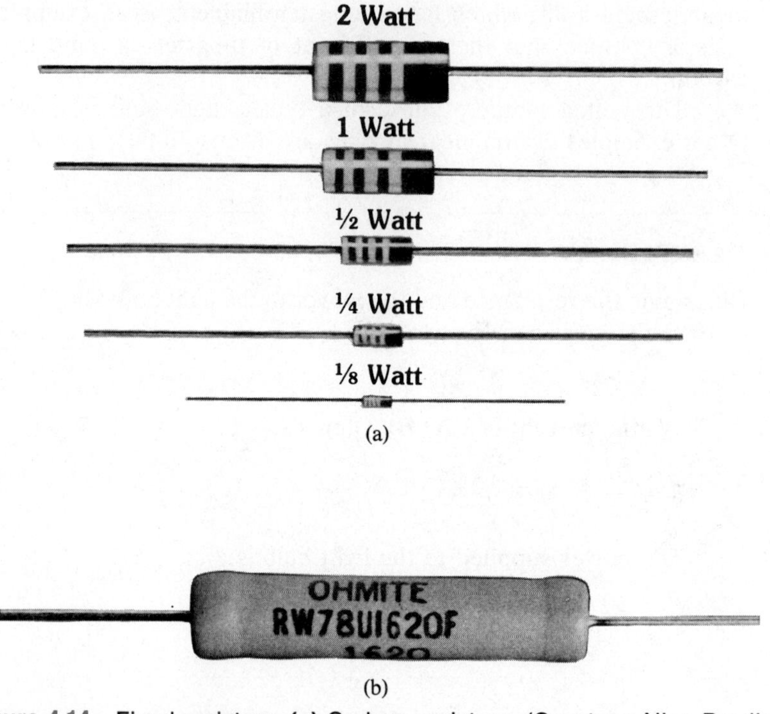

Figure 4.14 Fixed resistors. **(a)** Carbon resistors. (Courtesy Allen Bradley Co.) **(b)** Wirewound resistors. (Courtesy of Ohmite Manufacturing Company, Skokie, IL.)

and silver bands. A fifth band is often included on a resistor; this band represents the reliability of the resistor.

The physical size of the resistor limits the power that can be dissipated by the resistor. Hence, standard carbon-composition resistors are available in a number of sizes to dissipate from $\frac{1}{8}$ to 2 W of power. The larger the body size, the greater the amount of power it can dissipate.

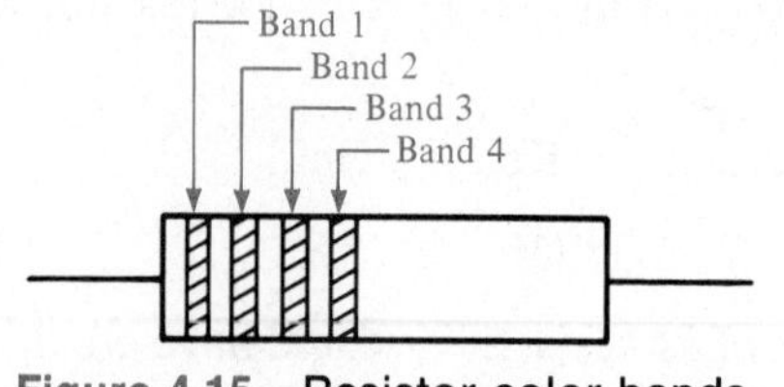

Figure 4.15 Resistor color bands.

TABLE 4.4
Color code

Color band	Value
Black	0
Brown	1
Red	2
Orange	3
Yellow	4
Green	5
Blue	6
Violet	7
Gray	8
White	9
Gold	0.1
Silver	0.01
Tolerances	
Gold	5%
Silver	10%
No 4th band	20%

EXAMPLE 4.15

What is the color code for a resistor of 6800 Ω ± 5% tolerance?

Solution

By Table 4.4, blue, gray, red, gold.

EXAMPLE 4.16

A resistor has a value of 9.1 MΩ ± 20%. What is the color code?

Solution

By Table 4.4, white, brown, green.

Wirewound resistors are used primarily in power applications. Wirewound resistors are made to dissipate from 3 W to hundreds of watts. They are usually constructed of a nichrome wire (alloy of nickel and chromium) wound around a ceramic core to handle the heat dissipated in the wire. Wirewound precision resistors are also available. Because of the ability to control the resistance by controlling the length of the wire, these resistors are available with a tolerance of less than 1%. Wirewound resistors are shown in Fig. 4.14b.

EXAMPLE 4.17

What is the value of a resistor with color code yellow, violet, yellow, silver?

Solution

By Table 4.4,

$$470{,}000 \quad \text{or} \quad 470 \text{ k}\Omega \pm 10\%$$

4.8.2 Variable Resistors

A variable resistor is one whose resistance can be varied by rotating a shaft, moving a slide, or by some similar method. These are three terminal devices. The devices are shown in Figure 4.16a. The electrical symbol for a variable resistor is given in Figure 4.16b.

If a voltage is applied to terminals 1 and 2, a variable voltage exists between

Figure 4.16 Variable resistors. **(a)** Picture. (Courtesy of Allen Bradley Co.) **(b)** Symbols.

terminals 1 and 3 or between 2 and 3. Rotation of the shaft or movement of the slide would change the voltage. If used in this manner, the device is called a **potentiometer,** or "pot." Other connections that are used to obtain a variable resistance are shown in Figure 4.17. Only two terminals of the device are utilized. When connected in this manner, the device is referred to as a **rheostat.**

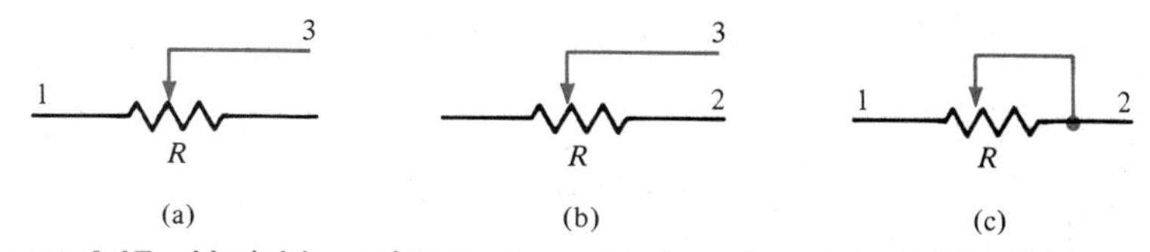

Figure 4.17 Variable resistors connected as rheostats. **(a)** Variable resistance between points 1 and 3. **(b)** Variable resistance between points 2 and 3. **(c)** Variable resistance between points 1 and 2.

Variable resistors of the rotating-shaft type are made either single-turn or multiple-turn. Table 4.5 gives the characteristics of these devices. If several potentiometers share the same shaft, the device is called a *ganged pot*. Potentiometers are used in power applications and in most electronic circuits. Everyone is familiar with the use of potentiometers as volume controls in radio and television.

TABLE 4.5
Characteristics of variable resistors

Single turn	Wide application 50 Ω to 5 MΩ ± 10%, 2–3 W
Multiple turn	For precise settings 10 turns 50 Ω to 250 kΩ ± 3%, up to 5 W
Trimmer pot	For one-time adjustment, single- or multiple-turn, a few ohms to 5 MΩ ± 10%, 1 W

Materials used in the manufacture of variable resistors are generally chosen from the following: carbon, cermet, conductive plastic, wirewound.

4.9 SPECIAL TYPE RESISTORS—NONLINEAR RESISTORS

It was stated earlier that a nonlinear device is one whose resistance changes as a function of some physical parameter such as direct heating, current, illumination, and so on. *The result is that the volt-ampere characteristic is nonlinear; that is, not a straight-line graph.* Several examples of nonlinear devices follow.

4.9.1 Light Bulbs

The ordinary incandescent bulb is a nonlinear resistor. The tungsten filament is a material whose resistance varies with temperature. The large value of temperature coefficient substantiates this fact. As the voltage increases across the bulb, the increased current causes the filament temperature to change. Typical volt-ampere characteristics for a 60-W bulb are shown in Figure 4.18.

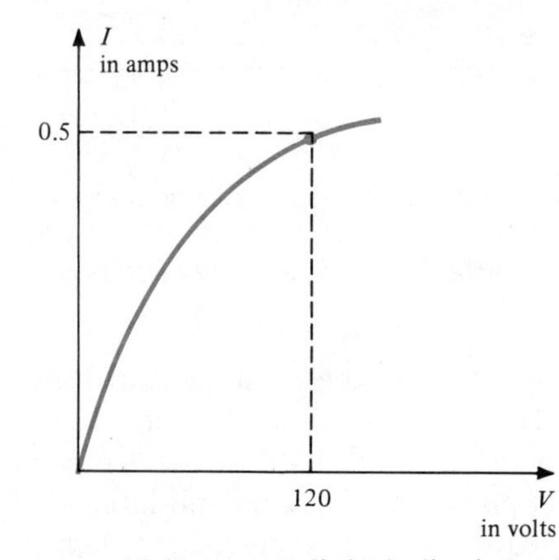

Figure 4.18 Volt-ampere light bulb characteristic.

4.9.2 Photoconductor

The **photoconductor** is a two-terminal device (photocell) whose resistance is a function of illumination; the resistance decreases with increasing intensity of light. These devices are normally made with cadmium sulphide. If a constant voltage is applied across the device and the device is exposed to light, the light photons striking the device cause the current to increase. A typical graph of resistance versus light intensity is shown in Figure 4.19b. The standard unit used for illumination is lumens per square meter (lm/m^2).

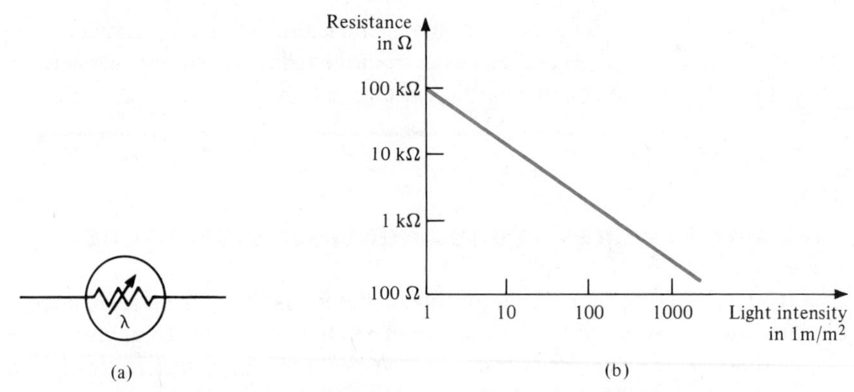

Figure 4.19 Photoconductor. **(a)** Symbol. **(b)** Photoconductor characteristics.

Note that for constant illumination, the volt-ampere characteristics of a photocell are linear. Photocells are used in light sensing, security systems, and counting of objects.

4.9.3 Thermistor

A **thermistor** is a device whose *resistance decreases with increasing temperature;* thus, a negative temperature coefficient (NTC) exists for thermistor material. The primary materials from which thermistors are made are silicon and arsenic.

Thermistors are used to sense temperature and thus are part of temperature-measurement systems. They are also useful in transistor amplifier circuits to aid in the stabilization for temperature changes. Transistor operation is highly sensitive to temperature, and thermistors help neutralize the temperature effect. Figure 4.20 shows how the resistance of a thermistor is affected by temperature.

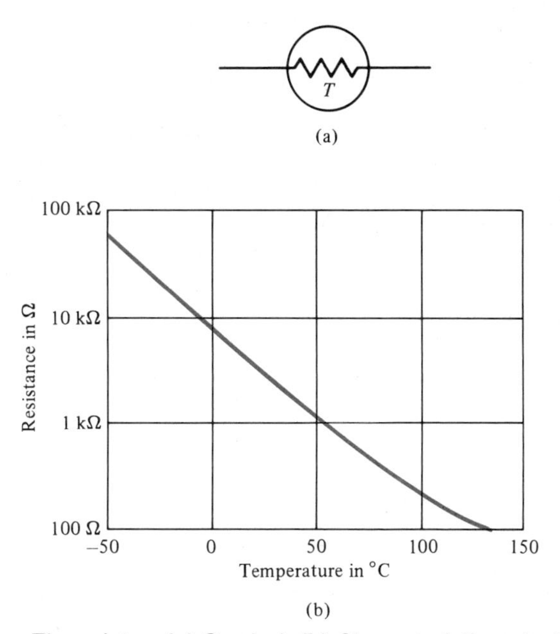

Figure 4.20 Thermistor. **(a)** Symbol. **(b)** Characteristics of a thermistor.

4.9.4 Semiconductor Diode

The **semiconductor diode** is a two-terminal device whose resistance is a function of the polarity and magnitude of applied voltage. Figure 4.21b shows the volt-ampere characteristics for a typical silicon diode. For positive voltage, the resistance of the device is quite low. For negative voltage, the resistance is very high. Diodes have numerous applications in the field of electronics, including computer logic circuits, parts of amplifier circuits, and to many special-application circuits. Ideally, the diode behaves either like an open circuit or a short circuit. Diodes can

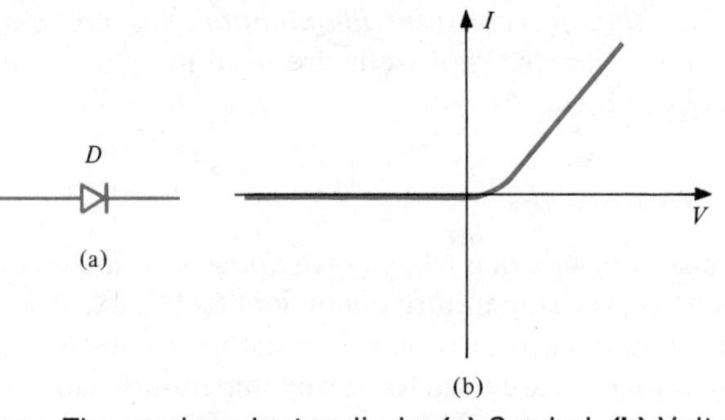

Figure 4.21 The semiconductor diode. **(a)** Symbol. **(b)** Volt-ampere characteristics.

be made effectively to approach the ideal, and this characteristic is used to great advantage.

4.9.5 Integrated Circuit Resistors

Resistors are often manufactured as an integral part of a circuit together with transistors, diodes, and so on. These integrated circuit (IC) resistors are a result of the manufacture of an integrated circuit made on a silicon chip. These resistors are referred to as **diffused resistors.**

The resistance of this material is given by Eq. (4.7), $R = \rho l/A$. Units for diffused resistors are metric but not SI. The resistivity ρ is in ohm-centimeters; length l is in centimeters; cross-sectional area A is in centimeters squared. Resistance R is in ohms. The cross-sectional area equals the product of width w and thickness t; therefore, (4.7) is rewritten as:

$$\boxed{R = \frac{(\rho)(l)}{(t)(w)}} \qquad (4.9)$$

If length l equals width w, a square of material results. A reasonable assumption is that the thickness of a given diffused resistor is constant. Letting $l = w$, Eq. (4.9) gives a result for resistance of ρ/t. This is commonly known as the **sheet resistance, R_s**. Thus,

$$\boxed{R_s = \frac{\rho}{t}} \qquad (4.10)$$

The units of R_s are ohms per square (ohms/square). The concept of sheet resistance is very useful. *Regardless of the dimensions of the square, a given thickness of material results in a constant resistance.* To find the resistance value of a diffused resistor of any length and width, the sheet resistance is multiplied by the length-to-width ratio, l/w:

$$R = R_s\left(\frac{l}{w}\right) \tag{4.11}$$

The ratio l/w is referred to as the **aspect ratio** and is illustrated in Figure 4.22.

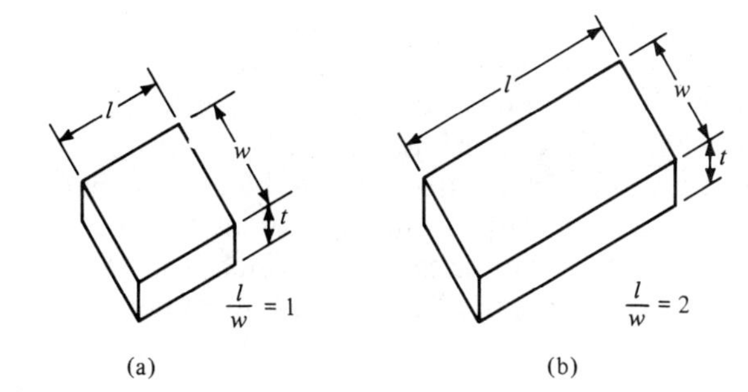

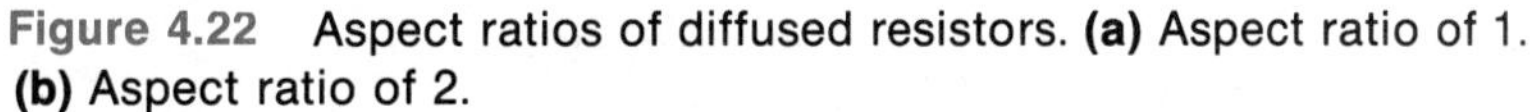

Figure 4.22 Aspect ratios of diffused resistors. **(a)** Aspect ratio of 1. **(b)** Aspect ratio of 2.

EXAMPLE 4.18

It is required to diffuse a resistor of 1500 Ω. The sheet resistance is given as 200 Ω/square.

a. Determine the aspect ratio.

b. If $w = 2\ \mu\text{m}$, determine the length of the resistor in centimeters ($1\ \mu\text{m} = 10^{-6}$ m).

Solution

a. By Eq. (4.11),

$$R = R_s\left(\frac{l}{w}\right)$$

$$1500 = 200\left(\frac{l}{w}\right)$$

Solving for the aspect ratio gives:

$$\frac{l}{w} = 7.5$$

b. Solving for the length l from the aspect ratio,

$$\begin{aligned} l &= 7.5w \\ &= 7.5 \times 2 \times 10^{-6} \\ &= 15 \times 10^{-6}\text{ m, or } 15\ \mu\text{m} \end{aligned}$$

The practical range of diffused resistors is from approximately 20 Ω to 30 kΩ. The absolute tolerance is poor; it can be as high as ±25%. The *tolerance of the ratio* of diffused resistors, however, can be less than ±2%.

4.9.6 Thin- and Thick-Film Resistors

Thin-film technology provides resistors by deposition of a layer of material of approximately 5 Å (ångstroms) to 1 μm (10^{-8} cm to 10^{-4} cm). Materials used for thin-film resistors include nichrome and tantalum. Resistance values range from 10 to 400 Ω/square with a tolerance of better than ±10%. The films are deposited on oxidized silicon, ceramic, or glass substrates in a vacuum.

Thick-film resistors are fabricated by printing a pattern on a ceramic substrate. The patterns range from layers of approximately 0.1 μm upward. The resistance values can range from a few ohms to several megohms, depending on the resistivity of the printing material, which is usually a paste or ink. The tolerance is usually greater than 10% but accuracy of better than 1% can be obtained by trimming.

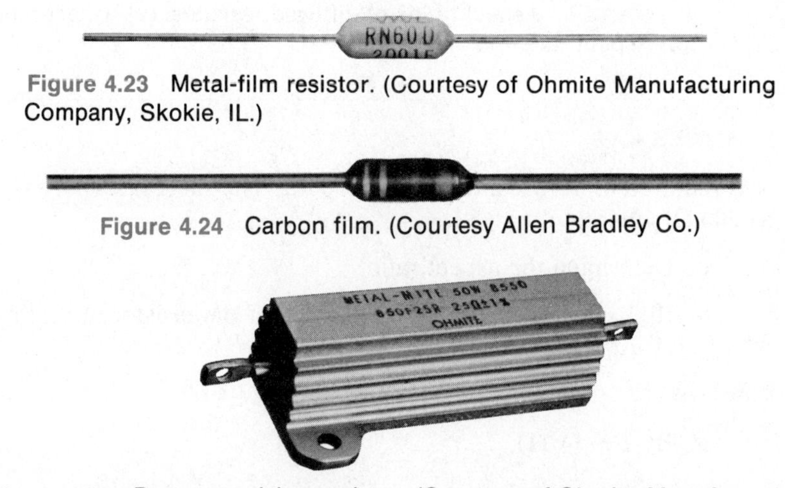

Figure 4.23 Metal-film resistor. (Courtesy of Ohmite Manufacturing Company, Skokie, IL.)

Figure 4.24 Carbon film. (Courtesy Allen Bradley Co.)

Figure 4.25 Power precision resistor. (Courtesy of Ohmite Manufacturing Company, Skokie, IL.)

Figure 4.26 Resistive networks. (Courtesy Beckman Industrial Corp.)

Basic types of fixed resistors are metal film, used in semiprecision applications, with tolerance of 1% to 5%. Carbon-film resistors are used for precision applications of 0.5% to 1%. Wirewound resistors are used in ultraprecision applications of better than 0.5%. Figures 4.23 through 4.26 show these types of resistors.

SUMMARY

Ohm's Law represents the fundamental relationship between the quantities of voltage and current. From Ohm's Law, the relationships of power are developed. Knowledge of Ohm's Law allows for the development of various techniques of circuit analysis.

The concept and relationship for wire resistance were presented in this chapter and a discussion of types of resistors is given. Resistors are available in many values and many forms, both fixed and variable. Many different materials can be used in the manufacture of resistors.

Resistance is a function of temperature, and most materials have a positive temperature coefficient—e.g., increasing resistance with increasing temperature.

Additional types of modern resistors are thin- and thick-film resistors and resistors made as part of an integrated circuit.

SUMMARY OF IMPORTANT EQUATIONS

$$V = IR \tag{4.1b}$$

$$P = \frac{W}{t} \tag{4.3}$$

$$P = VI = I^2R = \frac{V^2}{R} \tag{4.6}$$

$$R = \rho \frac{l}{A} \tag{4.7}$$

$$R_2 = R_1[1 + \alpha(T_2 - 20°\text{C})] \tag{4.8}$$

$$R_s = \frac{\rho}{t} \qquad (t = \text{thickness}) \tag{4.10}$$

$$R = R_s\left(\frac{l}{w}\right) \tag{4.11}$$

REVIEW QUESTIONS

4.1 Define the term resistance.

4.2 Explain the difference between a linear and a nonlinear resistor.

4.3 What is the relationship between resistance and conductance?

4.4 Describe the relation between power and work.

4.5 Explain why utility companies use the kilowatt-hour as the basic billing unit of energy.

4.6 Explain why a larger-diameter aluminum wire must be used to carry the same current as a smaller-diameter copper wire.

4.7 Which material is more desirable as a resistor: one with a low value or one with a larger value of temperature coefficient of resistance?

4.8 Explain why wirewound resistors make very precise resistors.

4.9 Explain how a potentiometer can be connected to function as a rheostat.

4.10 Give an example of where a potentiometer is used.

4.11 What is meant by negative temperature coefficient (NTC) thermistor?

4.12 Explain the difference between a thin-film and a thick-film resistor.

4.13 Explain the function of a photoconductor.

4.14 What is a diffused resistor?

PROBLEMS

4.1 Determine the voltage developed across a 25-Ω resistor with 0.5 A through the resistor.

4.2 What is the conductance of a 25-Ω resistor? (ans. 12.5 V)

4.3 For each of the resistors in Figure 4.27, determine the unknown value and label the direction of the current or the polarity of the voltage.

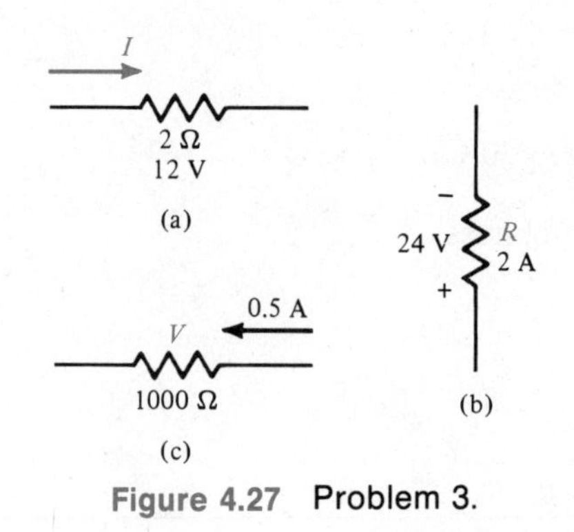

Figure 4.27 Problem 3.

4.4 For each of the figures of Problem 4.3, determine the power to each resistor. (ans. 72 W, 48 W, 250 W)

4.5 If energy costs 10¢ per kilowatt-hour, determine the cost of operation of a 100-W bulb for 1 year.

4.6 What is the cost of using a 2-hp motor for 4 h if the cost of electrical energy is 9¢ per kilowatt-hour? (ans. 53.7¢)

4.7 Determine the time it takes for a 500-W device to use 12 kWh of energy.

4.8 Determine the cost of using each of the following at 9¢ per kilowatt-hour.
a. 75-W bulb for 2 days
b. 1000-W dishwasher for 30 min
c. 6-kW electric stove for 2 h
(ans. 30¢, 4.5¢, $1.08)

4.9 A resistor dissipates 7 J/s with a current of 0.5 A. Determine the value of the resistance.

4.10 How many feet of number 12 aluminum wire are required to have a resistance of 1 Ω? (ans. 383.66 ft)

4.11 Determine the resistance of the copper bar of Figure 4.28.

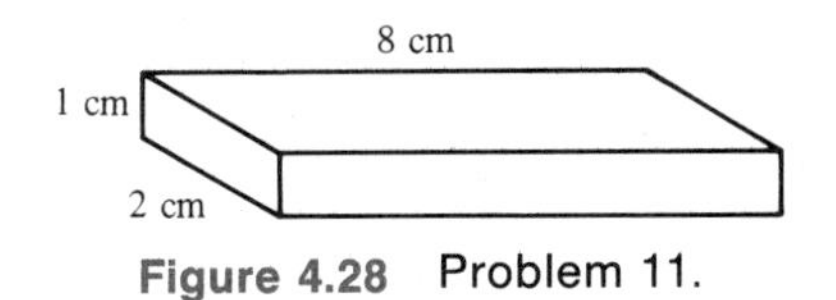

Figure 4.28 Problem 11.

4.12 A resistor has 10 V across it and dissipates 4 W; determine the value of resistance. (ans. 25 W)

4.13 If a 100-Ω resistor dissipates 10 J/s, determine the voltage across the resistor.

4.14 A 2-kΩ resistor has 10 V across it. Determine the energy used in 10 s of operation. (ans. 0.5 J)

4.15 Determine the resistance of 2 m of copper wire that is circular in cross section and has a radius of (a) 0.75 mm; (b) 1 mm.

4.16 How many meters of 2-mm-diameter copper wire are required to give a resistance of 1 Ω. (ans. 182.6 m)

4.17 Determine the resistance of 10 m of number 20 copper wire at 20°C.

4.18 Repeat Problem 4.17 for aluminum. (ans. 0.546 Ω)

4.19 Repeat Problem 4.17 for 100°C.

4.20 Repeat Problem 4.18 for 100°C. (ans. 0.716 Ω)

4.21 Determine the color code for a resistance of 2200 Ω ± 10% tolerance.

4.22 Determine the color code for a resistance of 12 MΩ ± 20%. (ans. brown, red, blue)

4.23 What is the value of a resistor with each color code?
a. Red, red, yellow, gold
b. Blue, gray, yellow, silver

4.24 It is required to diffuse a resistor of 450 Ω. The sheet resistance is 100 Ω/square. Determine the aspect ratio. (ans. 4.5)

CHAPTER 5

Resistive Circuits

GLOSSARY

BASIC—a computer programming language used primarily with microcomputers
computer program—a set of instructions that direct a computer to solve a problem
computer run—the actual running of a computer program, yielding results
Current Divider Rule (CDR)—a rule that allows the direct computation of a particular current when the total current to a parallel combination is known

electric circuit—interconnection of a number of electrical devices or components in one or more closed paths so that current can be maintained to perform a desired function

Kirchhoff's Current Law (KCL)—the sum of the currents entering a node equals the sum of the currents leaving that node

Kirchhoff's Voltage Law (KVL)—in a closed loop or path, the sum of the voltage rises equals the sum of the voltage drops

node—electrical connection point for two or more components

open circuit—circuit with a disconnected or interrupted path

parallel circuit—circuit in which all the components are connected between the same two points

series circuit—circuit in which all the components are connected end to end in string fashion

series-parallel circuit—circuit containing both series and parallel connections of elements

short circuit—a path of zero resistance

Voltage Divider Rule (VDR)—method of calculating the voltage drop across each resistor without first calculating the total current through a series combination

INTRODUCTION

An **electric circuit** is the interconnection of a number of electrical devices or components in one or more closed paths so that current can be maintained to perform a desired function. A **node** is an electrical connection point for two or more components, as shown in Figure 5.1.

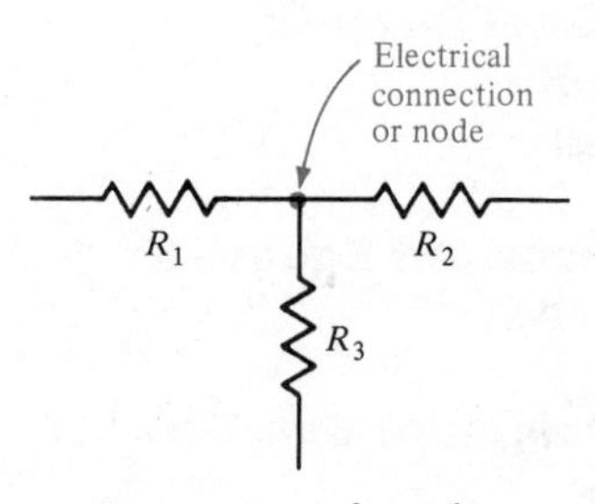

Figure 5.1 A node.

In this chapter, the simple series, the simple parallel, and the more complex series-parallel interconnections are presented. The analysis of circuits is stressed. Both conventional and computer-analysis methods are used. Solution of electric circuits using SPICE is presented in Appendix H.

5.1 SERIES CIRCUITS

*A **series circuit** is a circuit with all the components connected end to end (in string fashion).*

In this way, the same current may be maintained through each component and, therefore, throughout the circuit. Figure 5.2 shows a series circuit made up of four resistors and a battery. The current I in Figure 5.2 is one and the same through each of the four resistors and is also the current maintained by the battery. By definition, then:

In a series circuit the current is the same through all the components.

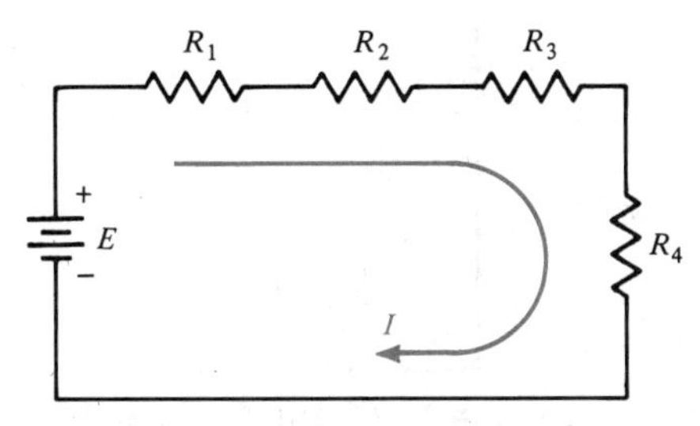

Figure 5.2 A series circuit.

In all circuits, the magnitude of the current maintained by the voltage source depends on the total equivalent resistance that is offered to the voltage source by the circuit. The total equivalent resistance, R_T, of a series circuit is the sum of the individual resistor values.

In equation form,

$$R_T = R_1 + R_2 + R_3 + \cdots + R_N \tag{5.1}$$

where R_N is the value of the Nth resistor (the last resistor in the string).

There are two ways of recognizing resistors in series:

1. Two resistors are in series if they have only one node in common and there is no other component connected to that node.
2. Two resistors are in series if they have one and the same current through them.

In Figure 5.2, R_1 is in series with R_2 because they have one node in common and there is no other component connected to that node. R_1 and R_4 are also in series because even though they have no node in common, they have *one and the same* current through them.

In the following examples, series circuits are analyzed.

EXAMPLE 5.1

Calculate the total resistance and the current through each component in the circuit shown in Figure 5.3.

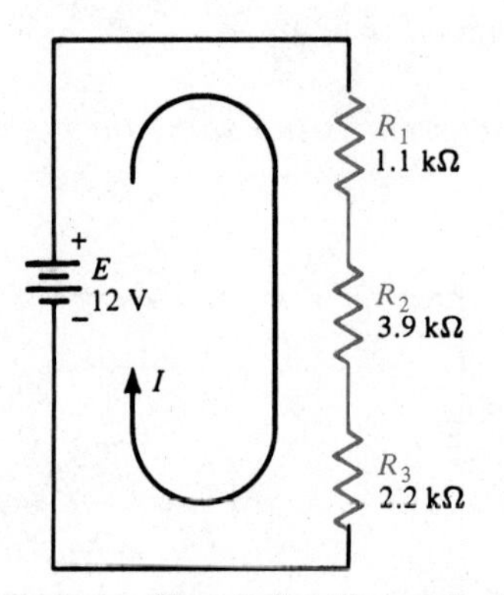

Figure 5.3 Circuit for Example 5.1.

Solution

The total resistance is given by Eq. (5.1).

$$\begin{aligned} R_T &= R_1 + R_2 + R_3 \\ &= 1.1\ \text{k}\Omega + 3.9\ \text{k}\Omega + 2.2\ \text{k}\Omega \\ &= 7.2\ \text{k}\Omega \end{aligned}$$

The total current I is given by Ohm's Law,

$$\begin{aligned} I &= \frac{E}{R} \\ &= \frac{12\ \text{V}}{7.2\ \text{k}\Omega} \\ &= 1.667\ \text{mA} \end{aligned}$$

The current through each resistor is the same (1.667 mA).

EXAMPLE 5.2

Determine the current maintained by the source in Figure 5.4.

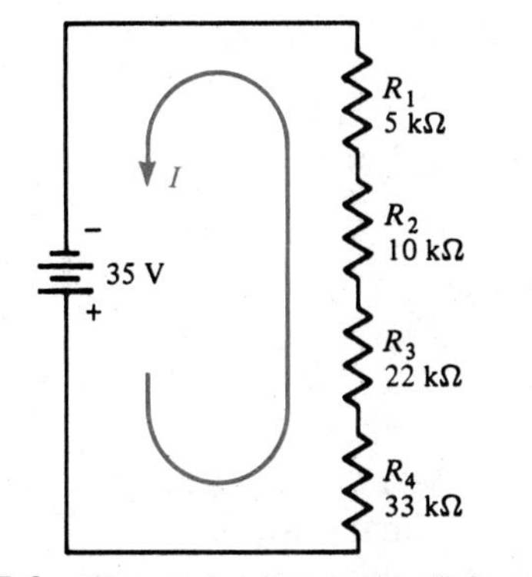

Figure 5.4 Circuit for Examples 5.2 and 5.3.

Solution

The total resistance is given by Eq. (5.1)

$$\begin{aligned} R_T &= 5\text{ k}\Omega + 10\text{ k}\Omega + 22\text{ k}\Omega + 33\text{ k}\Omega \\ &= 70\text{ k}\Omega \end{aligned}$$

The current maintained throughout the circuit is:

$$\begin{aligned} I &= \frac{E}{R_T} \\ &= \frac{35\text{ V}}{70\text{ k}\Omega} \\ &= 0.5\text{ mA} \end{aligned}$$

EXAMPLE 5.3

For the circuit shown in Figure 5.4, determine the power to each resistor and the total power generated by the source.

Solution

The current throughout the circuit, as found in Example 5.2, is 0.5 mA.

As shown previously, power may be calculated by

$$P = (I^2)(R)$$

Therefore, the power to the resistors is:

$$\begin{aligned} P_1 &= (0.5 \times 10^{-3})^2\,(5 \times 10^3) \\ &= 1.25 \times 10^{-3} = 1.25\text{ mW} \end{aligned}$$

$$\begin{aligned} P_2 &= (0.5 \times 10^{-3})^2\,(10 \times 10^3) \\ &= 2.5\text{ mW} \end{aligned}$$

$$P_3 = (0.5 \times 10^{-3})^2 (22 \times 10^3)$$
$$= 5.5 \text{ mW}$$

$$P_4 = (0.5 \times 10^{-3})^2 (33 \times 10^3)$$
$$= 8.25 \text{ mW}$$

The total source power is:

$$P_S = E \times I = (35)(0.5 \times 10^{-3})$$
$$= 17.5 \text{ mW}$$

The total source power in the preceding example may also be calculated by adding the individual powers.

$$P_T = P_1 + P_2 + P_3 + P_4$$
$$= 1.25 \text{ mW} + 2.5 \text{ mW} + 5.5 \text{ mW} + 8.25 \text{ mW}$$
$$= 17.5 \text{ mW}$$

Note that the total power to the resistors is equal to the power generated by the source. This is always true because all the energy delivered to the circuit by the source is dissipated by the resistors. The only components that dissipate energy are resistors.

The following **computer program** in **BASIC** may be written to solve the problem. The results of a **computer run** follow the program.

```
10 CLS
20 REM    *****START OF HEADING AND LEGEND*****
30 REM    SERIES CIRCUIT EXAMPLES 5.2 and 5.3
40 REM    R(N) = Nth RESISTOR
50 REM    NR   = NUMBER OF RESISTORS
60 REM    RS   = TOTAL SERIES RESISTANCE IN OHMS
70 REM    I    = CURRENT IN AMPS
80 REM    RK   = TOTAL RESISTANCE IN K
90 REM    IM   = CURRENT IN MILLIAMPS
100 REM   E    = BATTERY VOLTAGE
110 REM   P(N) = POWER TO RESISTOR R(N)
120 REM   PS   = SOURCE POWER
130 PRINT "COMPUTER SOLUTION FOR EXAMPLES 5.2 and
    5.3"
140 PRINT "TOTAL RESISTANCE, CURRENT AND POWER IN A
    SERIES CIRCUIT"
150 REM   *****END OF HEADING AND LEGEND*****
155 REM
160 REM *****START OF INPUT SECTION*****
170 INPUT"How many resistors are in series"; NR
180 PRINT "Type in the resistor values in ohms"
190 FOR N=1 TO NR
200 PRINT "R(";N;")=";
210 INPUT R(N)
220 NEXT N
230 INPUT "What is the value of the battery
    voltage";E
```

```
240 REM    *****END INPUT SECTION*****
250 REM
260 REM    *****START OF COMPUTATION SECTION*****
270 FOR N=1 TO NR
280 RS= RS + R(N)
290 NEXT N
295 LET I=E/RS
300 LET RK = RS/1000 310 LET IM = I*1000
320 FOR N=1 TO NR
330 LET P(N) = I*I*R(N)
340 NEXT N
350 LET PS = E*I
360 REM    *****END COMPUTATION SECTION*****
370 REM
380 REM    *****START OUTPUT SECTION*****
390 CLS
400 INPUT "FOR HARD COPY TYPE : H";H$ 410 IF H$="h"
    THEN H$="H"
420 CLS
430 PRINT "SOLUTION OF EXAMPLES 5.2 and 5.3"
440 PRINT " THE FOLLOWING ";NR;" RESISTORS ARE IN
    SERIES:"
450 FOR N=1 TO NR
460 PRINT "R(";N;")=";R(N);" OHMS"
470 NEXT N
480 PRINT "RS=";RS;"OHMS OR ";RK;"k"
490 PRINT "I=";I;"AMPS        OR";IM;"MILLIAMPS"
500 FOR N=1 TO NR
510 PRINT "P(";N;")=";P(N);"WATTS"
520 NEXT N
530 PRINT "PS=";PS;"WATTS"
540 IF H$="H" GOTO
550 ELSE END
550 LPRINT "SOLUTION FOR EXAMPLES 5.2 AND 5.3"
560 LPRINT " THE FOLLOWING ";NR;"RESISTORS ARE IN
    SERIES:"
570 FOR N=1 TO NR
580 LPRINT "R(";N;")=";R(N);"OHMS
590 NEXT N
600 LPRINT "RS=";RS;"OHMS     OR";RK;"k"
610 LPRINT "I=";I;"AMPS       OR";IM;"MILLIAMPS"
620 FOR N=1 TO NR
630 LPRINT "P(";N;")=";P(N);"WATTS"
640 NEXT N
650 LPRINT "PS=";PS;"WATTS"
660 REM    *****END OUTPUT SECTION*****
```

Computer Output

```
SOLUTION FOR EXAMPLES 5.2 AND 5.3
 THE FOLLOWING  4 RESISTORS ARE IN SERIES:
R( 1 )= 5000 OHMS
R( 2 )= 10000 OHMS
R( 3 )= 22000 OHMS
R( 4 )= 33000 OHMS
```

```
RS= 70000 OHMS          OR 70 k
I= .0005 AMPS           OR  .5 MILLIAMPS
P( 1 )= .00125 WATTS
P( 2 )= .0025 WATTS
P( 3 )= 5.500001E-03 WATTS
P( 4 )= 8.250001E-03 WATTS
PS= .0175 WATTS
```

5.2 KIRCHHOFF'S VOLTAGE LAW

In his experiments with electric circuits, the German scientist **Gustav Kirchhoff** arrived at a very significant relationship among voltages in a circuit. This relationship is known as **Kirchhoff's Voltage Law** (KVL) and it is stated as follows.

The algebraic sum of the voltages around a closed path in a circuit is zero.

$$\boxed{V_{\text{(around a closed path)}} = 0} \tag{5.2}$$

This can be restated as:

In a closed loop or path, the sum of the voltage rises equals the sum of the voltage drops.

$$\boxed{\Sigma\, V_{\text{rises}} = \Sigma\, V_{\text{drops}}} \tag{5.3}$$

To illustrate this law, consider the circuit of Figure 5.5. The circuit consists of a battery and three resistors connected in series. The entire circuit, then, is a closed loop.

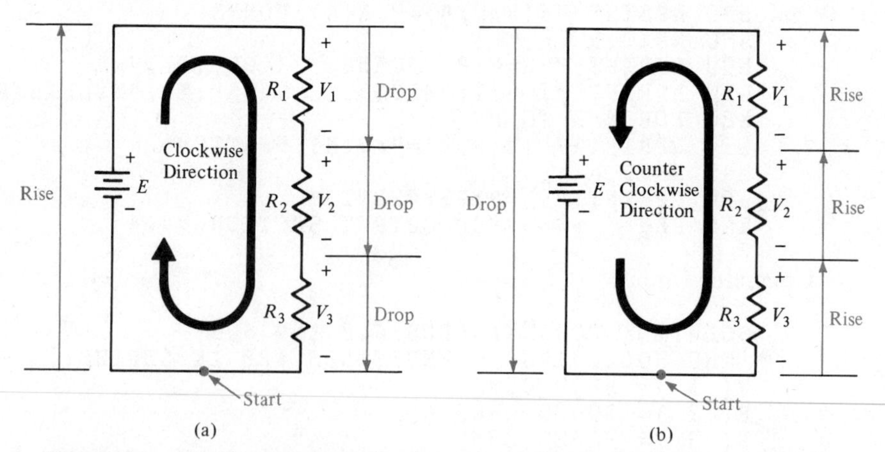

Figure 5.5 **(a)** Clockwise choice for voltage drops. **(b)** Counterclockwise choice for voltage drops.

When a potential difference minus (−) to plus (+) is encountered, it is called a *voltage rise* and is usually considered positive. When a potential difference plus (+) to minus (−) is encountered, it is called a *voltage drop* and is usually considered negative.

The loop can be approached either in a clockwise direction as in Figure 5.5a or in a counterclockwise direction as in Figure 5.5b. The direction does not matter as long as the chosen direction is adhered to throughout the loop. Direction changes in midloop are not allowed.

Choosing a clockwise direction around the loop in Figure 5.5a, there is one voltage rise (E) and three voltage drops, V_1 across resistor R_1, V_2 across resistor R_2, and V_3 across resistor R_3.

Equating the sum of the rises to the sum of the drops, as given by Eq. (5.3), the following relationship is obtained:

$$E = V_1 + V_2 + V_3$$

Using a counterclockwise direction around the loop, as in Figure 5.5b, there are three voltage rises (V_1, V_2, and V_3) and one voltage drop (E). Again, equating the sum of the voltage rises to the sum of the voltage drops gives:

$$V_1 + V_2 + V_3 = E$$

It should be noted that algebraically the two equations are equivalent. The direction chosen for going around the loop does not influence the resulting equation.

The validity of KVL is shown in the following example.

EXAMPLE 5.4

a. In the circuit of Figure 5.6, calculate the voltage drops across the resistors.

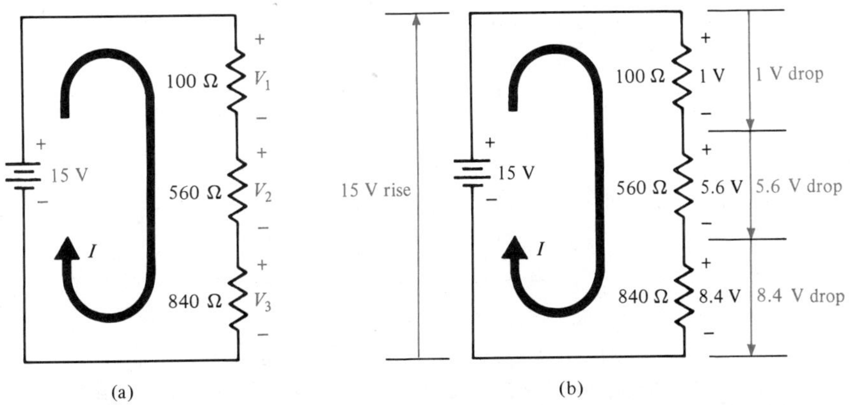

(a) (b)

Figure 5.6 Diagrams for Example 5.4. **(a)** Circuit for Example 5.4. **(b)** Results for Example 5.4.

b. Using a clockwise direction, show that the sum of the voltage rises equals the sum of the voltage drops.

Solution

a. To calculate the voltage drop across each resistor, first solve for the total resistance and the current:

$$R_T = 100 + 560 + 840 = 1.5\ \text{k}\Omega$$

$$I = \frac{E}{R_T} = \frac{15\ \text{V}}{1.5\ \text{k}\Omega} = 10\ \text{mA}\ (=10 \times 10^{-3}\ \text{A})$$

$$V_1 = (I)(R_1)$$
$$= (10 \times 10^{-3})(100) = 1\ \text{V}$$

$$V_2 = (10 \times 10^{-3})(560) = 5.6\ \text{V}$$

$$V_3 = (10 \times 10^{-3})(840) = 8.4\ \text{V}$$

b. The sum of the voltage drops is:

$$1 + 5.6 + 8.4 = 15\ \text{V}$$

The sum of the voltage rises is:

15 V (only one rise)

Since a total of 15 V of drops equals a 15-V rise, KVL is satisfied.

EXAMPLE 5.5

The circuit in Figure 5.7 consists of components connected in series with voltages across them, as shown. Solve for the voltage V_3.

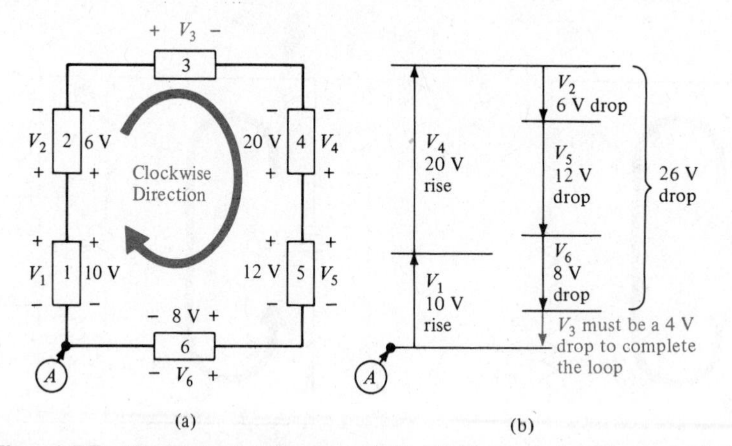

Figure 5.7 Drawings for Example 5.5. **(a)** Diagram for Example 5.5. **(b)** Graphical solution for Example 5.5.

Solution

By KVL around the closed loop, the sum of the voltage rises must equal the sum of the voltage drops.

$$V_1 + V_4 = V_2 + V_3 + V_5 + V_6$$

$$10 + 20 = 6 + V_3 + 12 + 8$$

$$30 = V_3 + 26$$

$$V_3 = 4 \text{ V}$$

V_3 is a 4-V drop in the chosen clockwise direction. Figure 5.7b shows a graphical solution of the problem using arrows. Voltage rises are shown by arrows pointing up, and voltage drops are indicated by arrows pointing down.

5.3 VOLTAGE DIVIDER RULE

It was shown in Example 5.4 that the voltage across each resistor in a series circuit may be calculated by finding the current through the circuit. Ohm's Law can then be applied to solve for the voltage across each resistor. A more convenient method of calculating the voltage drop across each resistor without first calculating the total current through a series combination is sometimes used. This method is called the **Voltage Divider Rule** (VDR).

The Voltage Divider Rule can be stated as follows:

In order to calculate the voltage across a particular resistor in a series circuit, multiply the source voltage by the value of that resistor and divide by the total series resistance of the circuit.

For the general circuit in Figure 5.8, the voltage drop across resistor R_1 is given by:

$$\boxed{V_1 = V_T \frac{R_1}{R_T}} \tag{5.4}$$

Similarly, the voltage drop across any resistor R_X may be found by:

$$\boxed{V_X = V_T \frac{R_X}{R_T}} \tag{5.5}$$

This is illustrated in the next example.

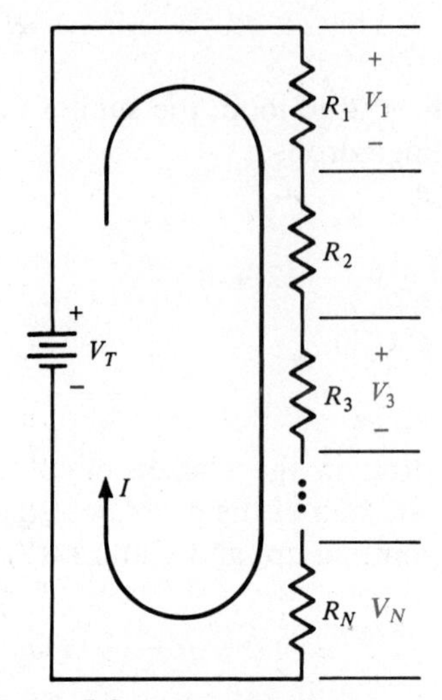

Figure 5.8 General series circuit.

EXAMPLE 5.6

a. Derive the VDR for the two-resistor circuit shown in Figure 5.9a.

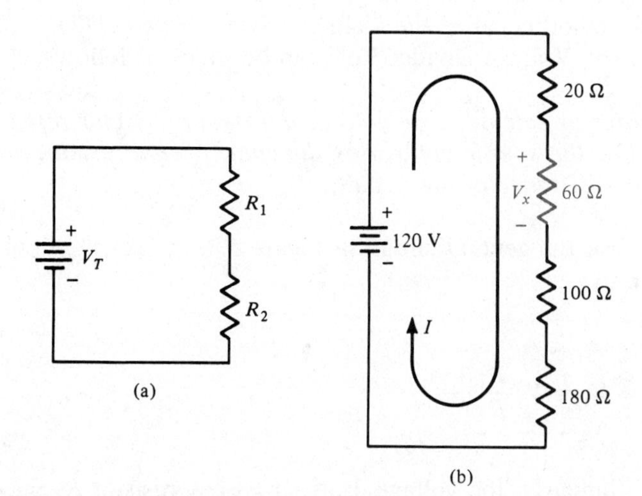

Figure 5.9 Circuits for Example 5.6. **(a)** Circuit for Example 5.6a. **(b)** Circuit for Example 5.6b.

b. Apply the VDR to solve for the voltage across the 60-Ω resistor in the circuit of Figure 5.9b.

Solution

a. The total resistance is:

$$R_T = R_1 + R_2$$

The current in the circuit is:

$$I = \frac{V_T}{R_T}$$

The voltage across resistor R_1 is:

$$V_1 = IR_1$$

Substituting for I:

$$V_1 = V_T \frac{R_1}{R_T}$$

Similarly:

$$V_2 = IR_2$$

$$= V_T \frac{R_2}{R_T}$$

$$= V_T \frac{R_2}{R_1 + R_2}$$

b. By Eq. (5.5),

$$V_{60\Omega} = 120 \frac{60}{20 + 60 + 100 + 180}$$

$$= 120 \frac{60}{360}$$

$$= 20 \text{ V}$$

A computer program in BASIC and the results of a computer run for this example are shown next.

```
10 CLS
20 REM   VOLTAGE DIVIDER EXAMPLE 5.6
30 REM    *****START OF HEADING AND LEGEND*****
40 REM    VT    =TOTAL APPLIED VOLTAGE
50 REM    RX    =RESISTOR IN QUESTION
60 REM    R(N) =Nth RESISTOR
70 REM    RT    =TOTAL SERIES RESISTANCE
80 REM    VX    =VOLTAGE ACROSS RX (TO BE SOLVED FOR)
90 REM    NR    =NUMBER OF RESISTORS IN SERIES
100 PRINT "COMPUTER SOLUTION FOR EXAMPLE 5.6"
110 PRINT "VOLTAGE DIVISION"
```

```
120 REM    *****END OF HEADING AND LEGEND
    SECTION*****
130 REM
140 REM    *****START OF INPUT SECTION*****
150 INPUT "TOTAL VOLTAGE APPLIED IN VOLTS";VT
160 INPUT "HOW MANY RESISTORS ARE IN SERIES";NR
170 PRINT "TYPE IN THE RESISTOR VALUES IN OHMS"
180 FOR N=1 TO NR
190 PRINT "R(";N;")=";
200 INPUT R(N)
210 NEXT N
220 PRINT "TYPE IN THE NUMBER OF THE RESISTOR
    ACROSS WHICH"
230 PRINT "THE VOLTAGE IS DESIRED";
240 INPUT X
250 REM    *****END OF INPUT SECTION*****
260 REM
270 REM    *****START OF COMPUTATION SECTION*****
280 FOR N=1 TO NR
290 LET RT = RT + R(N)
300 NEXT N
310 LET VX = VT * (R(X)/RT)
320 REM    *****END COMPUTATION SECTION*****
330 REM
340 REM    *****START OF OUTPUT SECTION*****
350 INPUT "DO YOU WANT A HARD COPY (Y=YES,
    N=NO)";H$
360 IF H$="yes" THEN H$="YES"
370 IF H$="Y" THEN H$="YES"
380 IF H$="y" THEN H$="YES"
390 PRINT "THE FOLLOWING";N;"RESISTORS ARE IN
    SERIES"
400 FOR N=1 TO NR
410 PRINT "R(";N;")=";R(N)
420 NEXT N
430 PRINT "THE TOTAL APPLIED VOLTAGE
    IS:";VT;"VOLTS"
440 PRINT "THE VOLTAGE ACROSS R(";X;")
    IS";VX;"VOLTS"
450 IF H$="YES" THEN GOTO 460 ELSE END
460 LPRINT "THE FOLLOWING";NR;"RESISTORS ARE IN
    SERIES:"
470 FOR N=1 TO NR
480 LPRINT "R(";N;")=";R(N)
490 NEXT N
500 LPRINT "THE TOTAL APPLIED VOLTAGE
    IS:";VT;"VOLTS"
510 LPRINT "THE VOLTAGE ACROSS R(";X;")
    IS";VX;"VOLTS"
520 REM    *****END OUTPUT SECTION*****
```

Computer Output

```
THE FOLLOWING 4 RESISTORS ARE IN SERIES:
R( 1 )= 20
```

```
R( 2 )= 60
R( 3 )= 100
R( 4 )= 180
THE TOTAL APPLIED VOLTAGE IS: 120 VOLTS
THE VOLTAGE ACROSS R( 2 ) IS 20 VOLTS
```

Sometimes, in a two-resistor voltage divider, one of the two resistor values is unknown. If the total applied voltage, the voltage across the unknown resistor, and the voltage across the second resistor are known, a useful variation of the VDR that solves for the unknown resistor is:

$$R_2 = R_1 \frac{V_2}{V_T - V_2} \tag{5.6}$$

where: R_2 => unknown resistor (Ω)
V_2 => voltage across R_2 (V)
V_T => total applied voltage (V)
R_1 => known resistor (Ω)

EXAMPLE 5.7

A transistor radio that operates from a 9-V battery and requires 30 mA is to operate from a car's electrical system. Design a voltage divider circuit that will assure proper operating conditions for the radio.

Solution

First, a circuit diagram is drawn, as in Figure 5.10. The box labeled R_{RAD} represents the radio with a voltage drop of 9 V across it and a current of 30 mA through it. Resistor R is a series voltage-dropping resistor, across which the extra voltage will be dropped. It is the value of R that is to be computed.

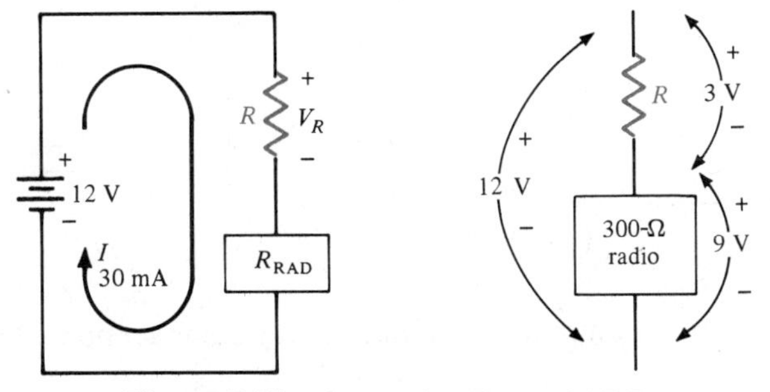

Figure 5.10 Circuit for Example 5.7.

The value of R_{RAD} is calculated by Ohm's Law:

$$R_{RAD} = \frac{9\text{ V}}{30\text{ mA}} = 300\ \Omega$$

The value of R is given by Eq. (5.6):

$$R = R_{RAD}\frac{V_{RAD}}{V_T - V_{RAD}}$$
$$= 300\frac{3}{12 - 3}$$
$$= 100\ \Omega$$

In the preceding example a voltage divider circuit was used to develop a voltage across a particular circuit element. There are numerous applications where voltage divider circuits are used.

5.4 PARALLEL CIRCUITS

*In a **parallel circuit,** all components are connected between the same two points. Conversely, components are in parallel if they are connected between the same two points.*

Examples of components in parallel are shown in Figures 5.11a and 5.11b. In both circuits all the circuit elements are connected between point A and point B. Electrical connection points such as A and B are referred to as nodes.

*A **node** is a point at which two or more circuit components are connected.*

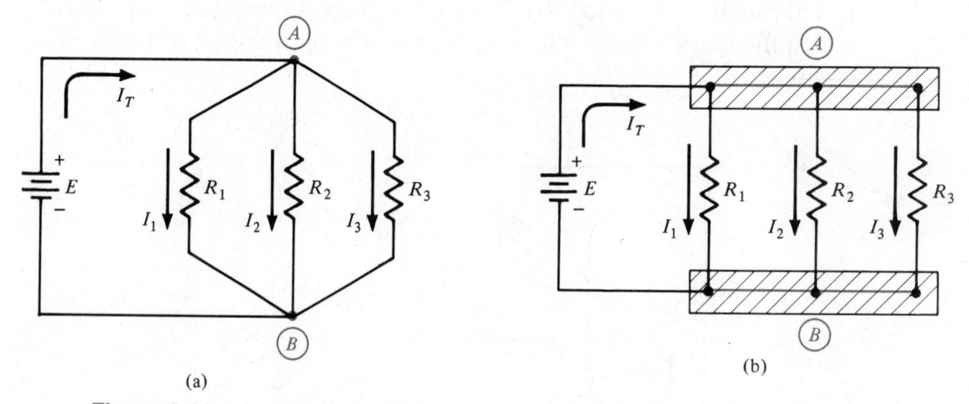

Figure 5.11 Components in parallel. **(a)** Compact node. **(b)** Spread-out node.

Since all the components in the circuit are connected across the same two points or nodes, it follows that

The voltage across all components in parallel is the same.

Hence,

$$\boxed{E = V_1 = V_2 = V_3} \tag{5.7}$$

The total current I_T, as shown in Figures 5.11a and 5.11b, is generated by the source, and it reaches node *A*, where it splits up into three individual resistor currents. At node *B*, the three currents recombine to yield the original total current. This current is then returned to the source. Clearly then, in a parallel connection there is more than one path for current.

The sum of the individual currents must equal the total current because nodes cannot store charge.

This observation was first made by Kirchhoff and is known as **Kirchhoff's Current Law** (KCL). This law is discussed in greater detail later in this chapter.

The current through each of the resistors in Figure 5.11 depends on the particular resistor value. The total current I_T depends on the individual currents; therefore, it depends on the value of the resistors in the circuit. The current through each resistor is calculated by Ohm's Law as follows:

$$I_1 = \frac{E}{R_1}$$

$$I_2 = \frac{E}{R_2}$$

$$I_3 = \frac{E}{R_3}$$

The sum of the individual currents equals the total current:

$$\boxed{I_1 + I_2 + I_3 = I_T} \tag{5.8}$$

This relationship is KCL.

Substituting the voltage over the resistance equation for each current, we get:

$$\frac{E}{R_1} + \frac{E}{R_2} + \frac{E}{R_3} = \frac{E}{R_P}$$

R_P stands for the combined parallel value of all the resistors. Dividing both sides by E we get:

$$\frac{1}{R_1} + \frac{1}{R_2} + \frac{1}{R_3} = \frac{1}{R_P}$$

Therefore, for a circuit that contains N resistors in parallel:

$$\boxed{\frac{1}{R_P} = \frac{1}{R_1} + \frac{1}{R_2} + \cdots + \frac{1}{R_N}} \tag{5.9}$$

Recall from Chapter 4 that the reciprocal of resistance (R) is conductance (G). Therefore, (5.9) becomes:

$$\boxed{G_T = G_1 + G_2 + \cdots + G_N} \tag{5.10}$$

For a two-resistor circuit such as the one shown in Figure 5.12, Eq. (5.9) becomes:

$$\frac{1}{R_P} = \frac{1}{R_1} + \frac{1}{R_2}$$

Solving for R_P, we get:

$$\boxed{R_P = \frac{R_1 R_2}{R_1 + R_2}} \tag{5.11}$$

The preceding is a very useful result known as the *Product over Sum Formula*.

A very popular method for indicating resistors in parallel is the symbol $\|$. For example, $R_1 \| R_2$ is read, "R_1 in parallel with R_2."

EXAMPLE 5.8

In the circuit of Figure 5.12, determine the parallel resistance of the circuit. Use $R_1 = 6\ \text{k}\Omega$ and $R_2 = 12\ \text{k}\Omega$.

Solution

By Eq. (5.11),

$$\begin{aligned} R_P &= R_1 \| R_2 \\ &= \frac{(6\ \text{k}\Omega)(12\ \text{k}\Omega)}{6\ \text{k}\Omega + 12\ \text{k}\Omega} \\ &= 4\ \text{k}\Omega \end{aligned}$$

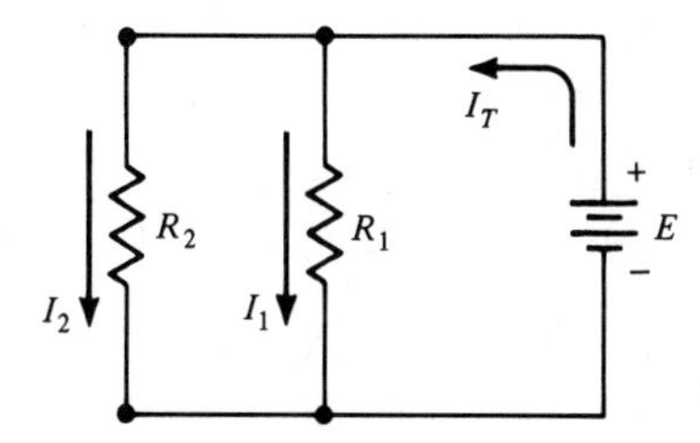

Figure 5.12 A two-resistor parallel circuit.

Alternatively,

$$G_1 = \frac{1}{R_1} = \frac{1}{6\ \text{k}\Omega} = 166.67\ \mu\text{S}$$

$$G_2 = \frac{1}{R_2} = \frac{1}{12\ \text{k}\Omega} = 83.33\ \mu\text{S}$$

$$\begin{aligned} G_T &= G_1 + G_2 \\ &= 166.67\ \mu\text{S} + 83.33\ \mu\text{S} = 250\ \mu\text{S} \end{aligned}$$

$$R_T = \frac{1}{G_T} = \frac{1}{250\ \mu\text{S}} = 4\ \text{k}\Omega$$

The total resistance of several resistors in parallel is always *smaller* than the smallest value of any of the individual resistors. In Example 5.8, the total resistance, 4 kΩ, is smaller than the 6-kΩ or the 12-kΩ component resistors.

If two equal resistors are in parallel, then the total resistance is one-half of the individual resistor value:

$$R_P = \frac{R}{2} \qquad \text{where } R = R_1 = R_2 \tag{5.12}$$

If N resistors of the same value are connected in parallel, then the total resistance is the value of one resistor divided by the number of resistors in parallel (N):

$$R_P = \frac{R}{N} \qquad \text{where } R = R_1 = R_2 = \cdots = R_N \tag{5.13}$$

EXAMPLE 5.9

For the network in Figure 5.13a, calculate the total parallel resistance.

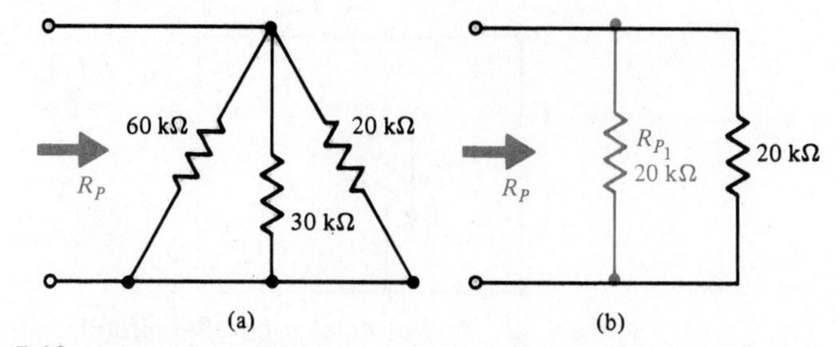

Figure 5.13 Diagrams for Example 5.9. **(a)** Circuit for Example 5.9. **(b)** Alternate solution for Example 5.9.

Solution

The network consists of three parallel resistors. The conductance of each is:

$$G_1 = \frac{1}{60\ \text{k}\Omega} = 16.7\ \mu\text{S}$$

$$G_2 = \frac{1}{30\ \text{k}\Omega} = 33.3\ \mu\text{S}$$

$$G_3 = \frac{1}{20\ \text{k}\Omega} = 50\ \mu\text{S}$$

The total conductance is:

$$\begin{aligned} G_T &= G_1 + G_2 + G_3 \\ &= 16.67\ \mu\text{S} + 33.33\ \mu\text{S} + 50\ \mu\text{S} = 100\ \mu\text{S} \end{aligned}$$

The total resistance is:

$$R_P = \frac{1}{G_T} = \frac{1}{100\ \mu\text{S}} = 10\ \text{k}\Omega$$

The total parallel resistance for the network in Figure 5.13a can also be found by combining two resistors at a time, using the Product over Sum Formula. Finding the parallel resistance of the 60-kΩ and the 30-kΩ by Eq. (5.11):

$$R_{P1} = \frac{(60\ \text{k}\Omega)(30\ \text{k}\Omega)}{60\ \text{k}\Omega + 30\ \text{k}\Omega} = 20\ \text{k}\Omega$$

Next, combining R_{P1} with the remaining 20-kΩ resistor yields the total parallel resistance R_P, as shown in Figure 5.13b:

$$R_P = \frac{(20\ \text{k}\Omega)(20\ \text{k}\Omega)}{20\ \text{k}\Omega + 20\ \text{k}\Omega} = 10\ \text{k}\Omega$$

EXAMPLE 5.10

For the circuit in Figure 5.14, determine the power to each resistor.

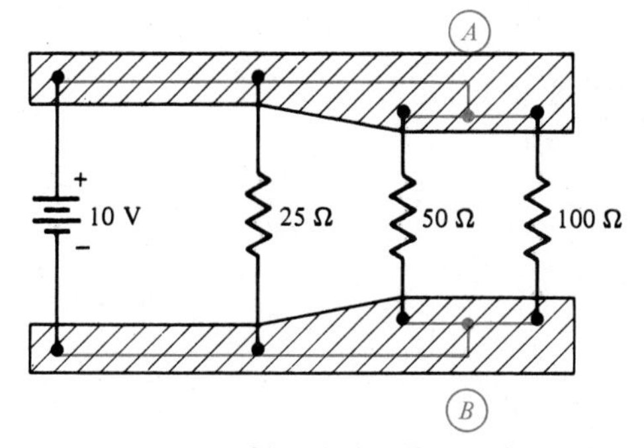

Figure 5.14 Circuit for Example 5.10.

Solution

The three resistors are connected in parallel. The voltage across each is therefore 10 V. The power to each resistor is:

$$P = \frac{V^2}{R}$$

$$P_{25\Omega} = \frac{10^2}{25} = 4\text{ W}$$

$$P_{50\Omega} = \frac{10^2}{50} = 2\text{ W}$$

$$P_{100\Omega} = \frac{10^2}{100} = 1\text{ W}$$

The total power to all three resistors is the sum of the individual resistor powers:

$$\begin{aligned} P_T &= P_{25\Omega} + P_{50\Omega} + P_{100\Omega} \\ &= 4\text{ W} + 2\text{ W} + 1\text{ W} = 7\text{ W} \end{aligned}$$

The total power generated by the source must be equal to the power to the resistors, 7 W.

5.5 KIRCHHOFF'S CURRENT LAW

As mentioned previously, Kirchhoff formulated a relationship for the currents at a node, Kirchhoff's Current Law:

The sum of the currents entering a node equals the sum of the currents leaving that node.

$$\Sigma I_{in} = \Sigma I_{out} \quad (5.14)$$

Consider node A shown in Figure 5.15. The arrows indicate the direction of the current, and by (5.14):

$$I_2 + I_3 + I_5 = I_1 + I_4$$

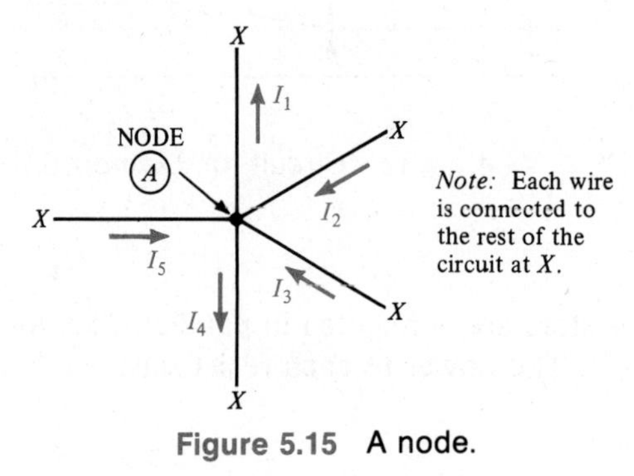

Figure 5.15 A node.

EXAMPLE 5.11

In the circuit of Figure 5.16:

a. Determine all the unknown currents.

b. Calculate the unknown resistor values.

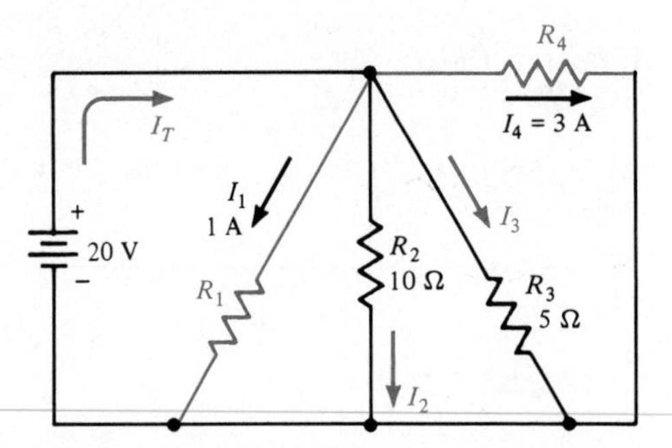

Figure 5.16 Circuit for Example 5.11.

Solution

a. The voltage across each resistor is 20 V. By Ohm's Law:

$$I_2 = \frac{20}{10} = 2\ \text{A}$$

$$I_3 = \frac{20}{5} = 4\ \text{A}$$

By KCL:

$$\begin{aligned} I_T &= I_1 + I_2 + I_3 + I_4 \\ &= 1 + 2 + 4 + 3 = 10\ \text{A} \end{aligned}$$

b. Resistor R_1 is calculated by Ohm's Law:

$$R_1 = \frac{20}{1} = 20\ \Omega$$

Similarly,

$$R_3 = \frac{20}{3} = 6.67\ \Omega$$

EXAMPLE 5.12

Calculate the value of the source voltage and the value of the unknown resistor and unknown current for the circuit in Figure 5.17.

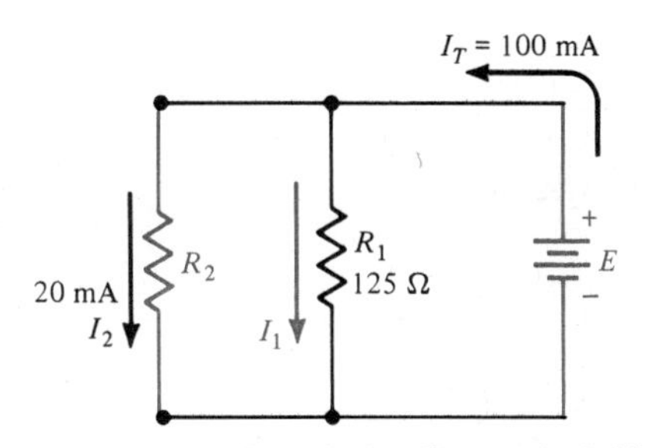

Figure 5.17 Circuit for Example 5.12.

Solution

By KCL:

$$\begin{aligned} I_1 &= I_T - I_2 \\ &= 100\ \text{mA} - 20\ \text{mA} = 80\ \text{mA} \end{aligned}$$

$$E = (I_1)(R_1)$$
$$= (80 \text{ mA})(0.125 \text{ k}\Omega) = 10 \text{ V}$$

By Ohm's Law:

$$R_2 = \frac{E}{I_2}$$
$$= \frac{10 \text{ V}}{20 \text{ mA}} = 0.5 \text{ k}\Omega = 500 \ \Omega$$

EXAMPLE 5.13

Determine the value of conductance G_3, and solve for the currents in the circuit of Figure 5.18.

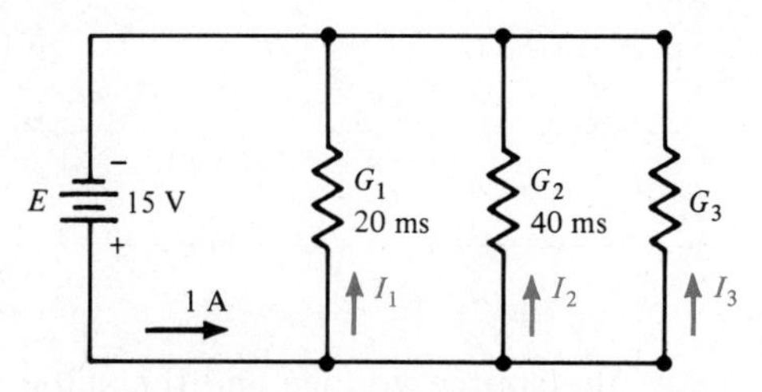

Figure 5.18 Circuit for Example 5.13.

Solution

$$I_1 = (E)(G_1)$$
$$= (15)(20 \text{ mS}) = 300 \text{ mA}$$

$$I_2 = (E)(G_2)$$
$$= (15)(40 \text{ mS}) = 600 \text{ mA}$$

By KCL:

$$I_3 = I_T - I_1 - I_2$$
$$= 1000 \text{ mA} - 300 \text{ mA} - 600 \text{ mA} = 100 \text{ mA}$$

$$G_3 = \frac{I_3}{E}$$
$$= \frac{100 \text{ mA}}{15 \text{ V}} = 6.67 \text{ mS}$$

5.6 CURRENT DIVIDER RULE

It should be apparent from the previous examples that the total current in a parallel combination divides in direct proportion to conductance values (or in inverse proportion to the resistance values). The smaller resistor receives the most current. The **Current Divider Rule** (CDR) allows the direct computation of a particular current when the total current to a parallel combination is known.

The CDR may be stated as:

$$\boxed{I_X = I_T \frac{R_P}{R_X}} \tag{5.15}$$

where: I_X => unknown current through R_X (A)
I_T => total current (A)
R_P => total parallel resistance (Ω)

Consider the circuit in Figure 5.19. This circuit is used to derive the CDR for two resistors.

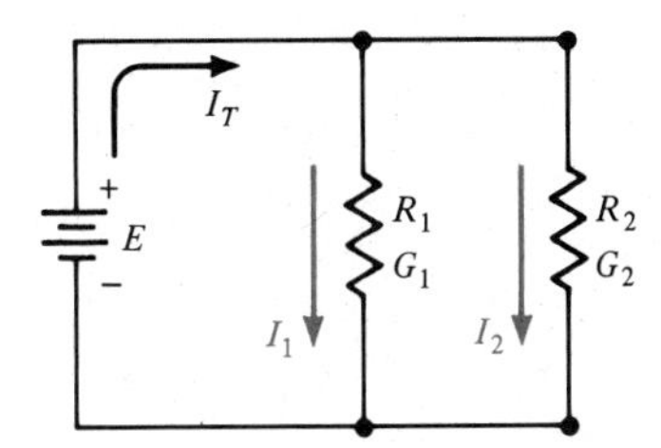

Figure 5.19 Circuit used to derive the Current Divider Rule.

The total parallel resistance is:

$$R_P = R_1 \| R_2$$

The voltage applied is calculated as follows:

$$E = (I_T)(R_P)$$

The current through resistor R_1 is:

$$I_1 = \frac{E}{R_1}$$

$$= \frac{(I_T)(R_P)}{R_1}$$

Rearranging terms,

$$I_1 = I_T \frac{R_P}{R_1}$$

In order to compute the current I_1 when the total current is known, the CDR may be used (for any number of parallel resistors):

$$\boxed{I_1 = I_T \frac{R_P}{R_1}} \tag{5.16a}$$

The CDR also may be stated as follows (for two parallel resistors):

$$\boxed{I_1 = I_T \frac{R_2}{R_1 + R_2}} \tag{5.16b}$$

The CDR also may be expressed in terms of conductances.

$$\boxed{I_1 = I_T \frac{G_1}{G_1 + G_2}} \tag{5.17}$$

$$\boxed{I_2 = I_T \frac{G_2}{G_1 + G_2}} \tag{5.18}$$

For a circuit with N conductors, the current I_X through conductor G_X is:

$$\boxed{I_X = I_T \frac{G_X}{G_T}} \tag{5.19}$$

The following examples illustrate the use of the CDR.

EXAMPLE 5.14

In the circuit of Figure 5.20, solve for the current I_1.

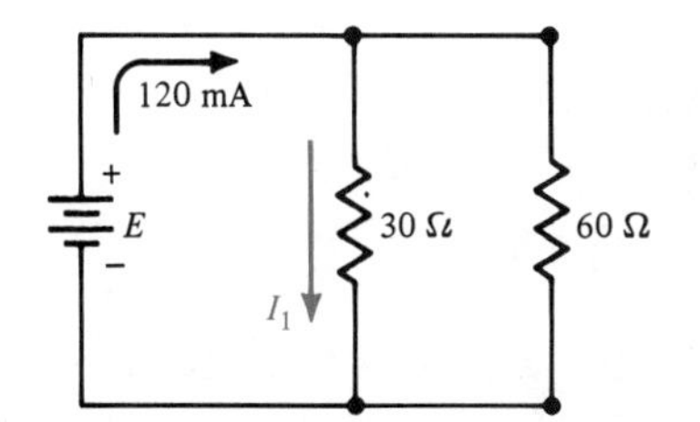

Figure 5.20 Circuit for Example 5.14.

Solution

$$R_P = (30\ \Omega)\|(60\ \Omega)$$
$$= \frac{(30)(60)}{30 + 60} = 20\ \Omega$$

By Eq. (5.15),

$$I_1 = 120\text{ mA}\,\frac{20}{30} = 80\text{ mA}$$

EXAMPLE 5.15

Calculate the current through the 3-kΩ resistor in the circuit of Figure 5.21a.

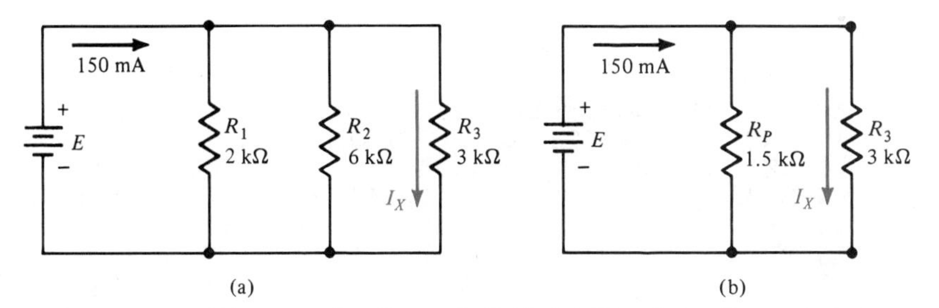

Figure 5.21 Diagrams for Example 5.15. **(a)** Circuit for Example 5.15. **(b)** Partial solution for Example 5.15.

Solution

First, convert the resistance values to conductance values:

$$G_1 = \frac{1}{2\text{ k}\Omega} = 0.5\text{ mS}$$

$$G_2 = \frac{1}{6\ \text{k}\Omega} = 0.167\ \text{mS}$$

$$G_3 = \frac{1}{3\ \text{k}\Omega} = 0.333\ \text{mS}$$

Then add all the conductances to obtain the total conductance:

$$G_T = 0.5\ \text{mS} + 0.167\ \text{mS} + 0.333\ \text{mS}$$
$$= 1.00\ \text{mS}$$

Now apply the CDR, Eq. (5.19):

$$I_X = 150\ \text{mA}\,\frac{0.333}{1.000}$$
$$= 50\ \text{mA}$$

The problem also may be solved by using Eq. (5.15).

$$R_P = \frac{1}{G_T}$$
$$= \frac{1}{1\ \text{mS}} = 1\ \text{k}\Omega$$

$$I_X = 150\ \text{mA}\,\frac{1\ \text{k}\Omega}{3\ \text{k}\Omega}$$
$$= 50\ \text{mA}$$

Following is a computer program in BASIC written to solve this problem. A sample run of the program is also presented.

```
10 CLS
20 REM PARALLEL RESISTOR EXAMPLE 5.15
30 REM    *****START OF HEADING AND LEGEND*****
40 REM        U      = REMEMBERS RESISTOR UNITS
50 REM        E      = TOTAL APPLIED VOLTAGE IN VOLTS
60 REM        IT     = TOTAL CURRENT IN AMPS
70 REM        R(N)   = VALUE OF Nth RESISTOR
80 REM        RP     = TOTAL PARALLEL RESISTANCE IN
                       OHMS
90 REM        G(N)   = CONDUCTANCE OF Nth RESISTOR IN
                       SIEMENS
100 REM       GT     = TOTAL CONDUCTANCE IN SIEMENS
110 REM       I(N)   = CURRENT THROUGH Nth RESISTOR
                       IN AMPS
120 REM       IX     = CURRENT THROUGH DESIRED RESISTOR
                       IN AMPS
130 REM       NR     = NUMBER OF RESISTORS IN PARALLEL
140 REM    *****END OF HEADING AND LEGEND SECTION*****
150 REM
160 REM    *****START OF INPUT SECTION*****
```

```
170 PRINT "COMPUTER SOLUTION FOR CURRENT DIVISION
    EX 5.15"
180 INPUT "HOW MANY RESISTORS ARE IN PARALLEL";NR
190 INPUT "IF RESISTORS ARE IN OHMS TYPE 1, IF THEY
    ARE IN K TYPE 2";U
200 PRINT "TYPE IN THE RESISTOR VALUES:"
210 FOR N=1 TO NR
220 PRINT "R(";N;")=";
230 INPUT R(N)
240 NEXT N
250 PRINT "TYPE IN THE NUMBER OF THE RESISTOR
    THROUGH WHICH"
260 PRINT "THE CURRENT IS DESIRED.";
270 INPUT X
280 INPUT "IF THE TOTAL CURRENT IS IN AMPS TYPE 1,
    IF IN MILLIAMPS TYPE 2";UT
290 INPUT "TYPE IN THE TOTAL CURRENT";IT
300 REM   *****END INPUT SECTION*****
310 REM
320 REM   *****START COMPUTATION SECTION*****
330 FOR N=1 TO NR
340 LET G(N) = 1/R(N)
350 LET GT = GT + G(N)
360 NEXT N
370 LET IX = IT* (G(X)/GT)
380 REM   *****END OF COMPUTATION SECTION*****
390 REM
400 REM   *****START OUTPUT SECTION*****
410 CLS
420 IF UT=2 THEN GOTO 582
430 PRINT " THE FOLLOWING";NR;"RESISTORS ARE IN
    PARALLEL"
440 FOR N=1 TO NR
450 PRINT "R(";N;")=";R(N); "OHMS"
460 NEXT N
470 PRINT "THE TOTAL CURRENT IS:";IT;"AMPS
480 PRINT "THE CURRENT THROUGH THE"; R(X);"OHM
    RESISTOR R(";X;"), IS:";IX;"AMPS"
490 INPUT "FOR HARD COPY TYPE :H";H$
500 IF H$="h" THEN H$="H"
510 IF H$="H" THEN GOTO 520 ELSE END
520 LPRINT " THE FOLLOWING";NR;"RESISTORS ARE IN
    PARALLEL"
530 FOR N=1 TO NR
540 LPRINT "R(";N;")=";R(N);"OHMS"
550 NEXT N
560 LPRINT "THE TOTAL CURRENT IS:";IT;"AMPS"
570 LPRINT "THE CURRENT THROUGH THE";R(X);"OHM
    RESISTOR R(";X;"), IS:";IX;"AMPS"
580 END
582 INPUT "FOR HARD COPY TYPE: H";H$
584 IF H$="h" THEN H$="H"
590 PRINT "THE FOLLOWING";NR;"RESISTORS ARE IN
    PARALLEL"
600 FOR N=1 TO NR
```

```
610 PRINT "R(";N;")=";R(N);"k"
620 NEXT N
630 PRINT "THE TOTAL CURRENT IS:";IT;"MILLIAMPS"
640 PRINT "THE CURRENT THROUGH THE";R(X);"k
    RESISTOR, R(";X;"), IS: ";IX;"MILLIAMPS"
650 IF H$="H" THEN GOTO 660 ELSE END
660 LPRINT "THE FOLLOWING";NR;"RESISTORS ARE IN
    PARALLEL:"
670 FOR N=1 TO NR
680 LPRINT "R(";N;")=";R(N);"k"
690 NEXT N
700 LPRINT "THE TOTAL CURRENT IS";IT;"MILLIAMPS"
710 LPRINT "THE CURRENT THROUGH THE";R(X);"k
    RESISTOR, R(";X;"), IS:";IX;"MILLIAMPS"
720 END
730 REM    *****END OUTPUT SECTION*****
```

Computer Output

```
THE FOLLOWING 3 RESISTORS ARE IN PARALLEL:
R( 1 )= 2 k
R( 2 )= 6 k
R( 3 )= 3 k
THE TOTAL CURRENT IS 150 MILLIAMPS
THE CURRENT THROUGH THE 3 k RESISTOR R( 3 ), IS: 50
MILLIAMPS
```

EXAMPLE 5.16

Determine the value of G_2 in the circuit of Figure 5.22. $I_2 = 2$ mA is known.

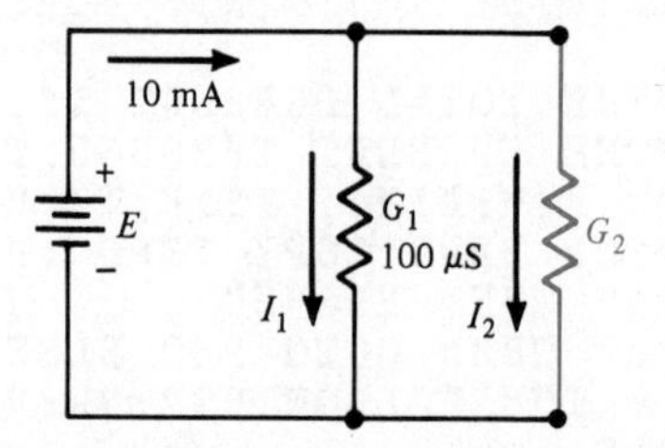

Figure 5.22 Circuit for Example 5.16.

Solution

By Eq. (5.18),

$$2\text{ mA} = 10\text{ mA}\,\frac{G_2}{G_2 + (100\ \mu\text{S})}$$

$$0.2 = \frac{G_2}{G_2 + (100\ \mu\text{S})}$$

$$(0.2)G_2 + 20\ \mu\text{S} = G_2$$

Solving for G_2 we get:

$G_2 = 25\ \mu S$

5.7 SERIES-PARALLEL CIRCUITS

Most practical circuits contain both series and parallel combinations of elements. These are referred to as **series-parallel circuits.** Equipped with Kirchhoff's Voltage Law, Kirchhoff's Current Law, and Ohm's Law, we are now able to solve such circuits. Consider the circuit in Figure 5.23. Resistors R_2 and R_3 are connected in parallel. This parallel combination is, in turn, in series with resistor R_1. The current I_1 is both the current through resistor R_1 and the total source current. The sum of the currents I_2 and I_3 is equal to the total current I_1.

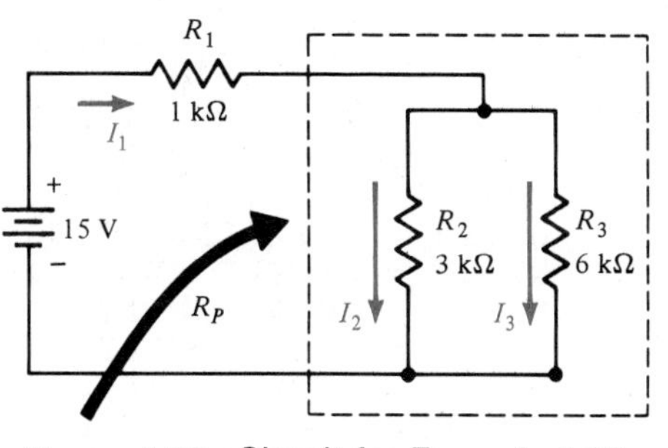

Figure 5.23 Circuit for Example 5.17.

EXAMPLE 5.17

Calculate the current through each resistor in the circuit of Figure 5.23.

Solution

First compute the parallel combination:

$$R_P = \frac{(R_2)(R_3)}{R_2 + R_3}$$

$$= \frac{(3\ k\Omega)(6\ k\Omega)}{3\ k\Omega + 6\ k\Omega} = 2\ k\Omega$$

The total resistance is:

$$R_T = R_1 + R_P$$

$$= 1\ k\Omega + 2\ k\Omega = 3\ k\Omega$$

The total current is computed by Ohm's Law:

$$I_T = I_1 = \frac{E}{R_T}$$

$$= I_1 = \frac{15\ V}{3\ k\Omega} = 5\ mA$$

Using the CDR:

$$I_2 = \frac{R_3}{R_2 + R_3} I_T$$

$$= 5 \text{ mA} \frac{6 \text{ k}\Omega}{6 \text{ k}\Omega + 3 \text{ k}\Omega} = 3.33 \text{ mA}$$

The current I_3 can be calculated by applying the CDR one more time. I_3 can also be calculated by applying KCL:

$$I_3 = I_T - I_2$$
$$= 5 \text{ mA} - 3.33 \text{ mA} = 1.67 \text{ mA}$$

A computer program in BASIC written to solve this problem is presented next, along with a sample computer run.

```
10 CLS
20 REM   SERIES-PARALLEL CIRCUIT EXAMPLE 5.17
30 REM   *****START OF HEADING AND LEGEND
         SECTION*****
40 REM   R2,R3 = PARALLEL RESISTORS
50 REM   R1    = SERIES RESISTOR
60 REM   E     = TOTAL APPLIED VOLTAGE
70 REM   I1    = TOTAL CURRENT
80 REM   I2    = CURRENT THROUGH R2
90 REM   I3    = CURRENT THROUGH R3
100 REM  RP    = R2//R3
110 REM  ALL CURRENTS IN MILLIAMPS
120 REM  ALL VOLTAGES IN VOLTS
130 PRINT "COMPUTER SOLUTION FOR EXAMPLE 5.17"
140 PRINT " TOTAL RESISTANCE AND RESISTOR CURRENTS
    IN A SERIES-PARALLEL CIRCUIT"
150 REM   *****END OF HEADING AND LEGEND
    SECTION*****
160 REM
170 REM   *****START OF INPUT SECTION*****
180 INPUT "TYPE IN THE VALUE OF R1 IN k";R1
190 INPUT "TYPE IN THE VALUE OF R2 IN k";R2
200 INPUT "TYPE IN THE VALUE OF R3 IN k";R3
210 INPUT "TYPE IN THE VALUE OF THE BATTERY
    VOLTAGE";E
220 REM   *****END OF INPUT SECTION*****
230 REM
240 REM   *****START OF COMPUTATION SECTION*****
250 LET RP = (R2*R3)/(R2+R3)
260 LET RT = R1 + RP
270 LET I1 = E/RT
280 LET 12 = I1*R3/(R2+R3)
290 LET I3 = I1*R2/(R2+R3)
300 REM   *****END OF COMPUTATION SECTION*****
310 REM
320 REM   *****START OF OUTPUT SECTION*****
330 CLS
```

```
340 PRINT "SOLUTION TO EXAMPLE 5.17"
350 PRINT "THE SERIES-PARALLEL CIRCUIT OF FIG.
    5-23 HAS THE"
360 PRINT "FOLLOWING COMPONENT VALUES:"
370 PRINT "R1=";R1;"k"
380 PRINT "R2=";R2;"k"
390 PRINT "R3=";R3;"k"
400 PRINT "THE TOTAL RESISTANCE IS:";RT;"k"
410 PRINT "THE TOTAL CURRENT AND THE CURRENT
    THROUGH R1 IS:";I1;"mA"
420 PRINT "I2=";I2;"mA"
430 PRINT "I3=";I3;"mA"
440 INPUT "FOR HARD COPY TYPE:H";H$
450 IF H$="h" THEN H$="H"
460 IF H$="H" THEN GOTO 470 ELSE END
470 LPRINT "SOLUTION TO EXAMPLE 5.17"
480 LPRINT "THE SERIES-PARALLEL CIRCUIT IN FIG.
    5-23 HAS THE"
490 LPRINT "FOLLOWING COMPONENT VALUES:"
500 LPRINT "R1=";R1;"k"
510 LPRINT "R2=";R2;"k"
520 LPRINT "R3=";R3;"k"
530 LPRINT "THE TOTAL RESISTANCE IS:";RT;"k"
540 LPRINT "THE TOTAL CURRENT AND THE CURRENT
    THROUGH R1 IS:";I1;"mA"
550 LPRINT "I2=";I2;"mA"
560 LPRINT "I3=";I3;"mA"
570 END
580 REM   *****END OUTPUT SECTION*****
```

Computer Output

```
SOLUTION TO EXAMPLE 5.17
THE SERIES-PARALLEL CIRCUIT IN FIG. 5-23 HAS THE
FOLLOWING COMPONENT VALUES:
R1= 1 k
R2= 3 k
R3= 6 k
THE TOTAL RESISTANCE IS: 3 k
THE TOTAL CURRENT AND THE CURRENT THROUGH R1 IS: 5
mA
I2= 3.333333 mA
I3= 1.666667 mA
```

EXAMPLE 5.18

Solve for the current through each component and the voltage across each component for the circuit shown in Figure 5.24a.

Solution

The total resistance as "seen" by the source is determined first.

$$R_T = R_{P1} + R_3 + R_{P2}$$

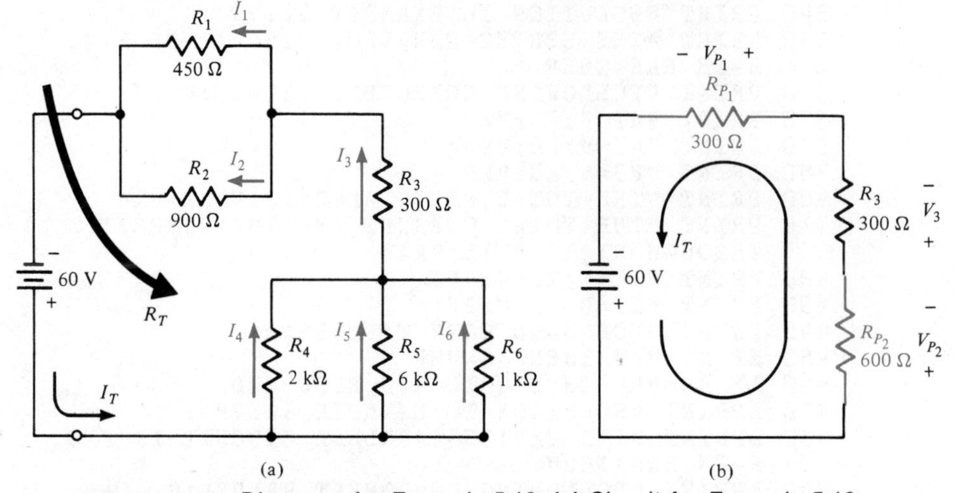

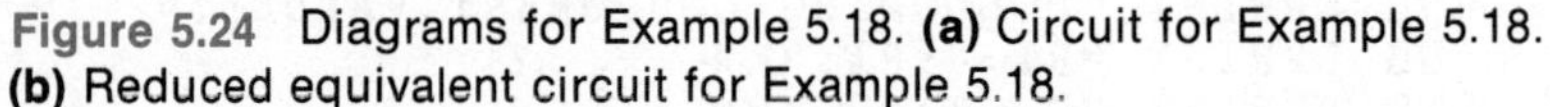

Figure 5.24 Diagrams for Example 5.18. **(a)** Circuit for Example 5.18. **(b)** Reduced equivalent circuit for Example 5.18.

where $R_{P1} = R_1 \| R_2$ and $R_{P2} = R_4 \| R_5 \| R_6$.

$$R_{P1} = \frac{(0.45\ \text{k}\Omega)(0.9\ \text{k}\Omega)}{0.45\ \text{k}\Omega + 0.9\ \text{k}\Omega} = 0.3\ \text{k}\Omega$$

$$G_{P2} = \frac{1}{2\ \text{k}\Omega} + \frac{1}{6\ \text{k}\Omega} + \frac{1}{1\ \text{k}\Omega} = 1.67\ \text{mS}$$

$$R_{P2} = \frac{1}{1.67\ \text{mS}} = 0.6\ \text{k}\Omega$$

$$R_T = 0.3\ \text{k}\Omega + 0.3\ \text{k}\Omega + 0.6\ \text{k}\Omega = 1.2\ \text{k}\Omega$$

A reduced equivalent circuit is shown in Figure 5.24b. The total current, I_T, is given by Ohm's Law:

$$I_T = \frac{E}{R_T}$$

$$= \frac{60\ \text{V}}{1.2\ \text{k}\Omega} = 50\ \text{mA}$$

At this point it is convenient to solve for the voltage drops across the resistors. Note that the voltage drop across the parallel combination of R_1 and R_2 is the same as the voltage drop across each resistor.

$$V_{P1} = (I_T)(R_{P_1})$$

$$= (50\ \text{mA})(0.3\ \text{k}\Omega) = 15\ \text{V}$$

The voltage across R_3 is:

$$V_{R3} = (50\ \text{mA})(0.3\ \text{k}\Omega) = 15\ \text{V}$$

Finally:

$$V_{P2} = V_4 = V_5 = V_6$$
$$= (I_T)(R_{P_2})$$
$$= (50 \text{ mA})(0.6 \text{ k}\Omega) = 30 \text{ V}$$

The currents through all the components could now be found using Ohm's Law.

EXAMPLE 5.19

Calculate currents I_T and I_4 in the circuit of Figure 5.25a.

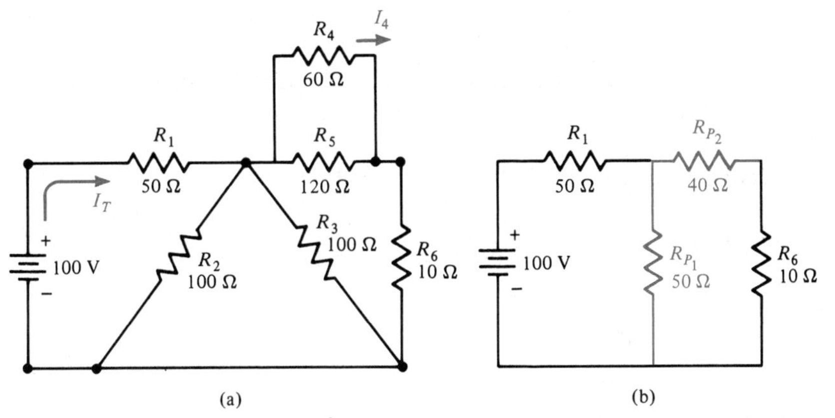

Figure 5.25 Diagrams for Example 5.19. **(a)** Circuit for Example 5.19. **(b)** Reduced equivalent circuit for Example 5.19.

Solution

First combine resistors in parallel and draw the circuit in reduced form.

$$R_{P1} = R_2 \| R_3$$

Since R_2 and R_3 are of the same value,

$$R_{P1} = \frac{100}{2} = 50 \ \Omega$$

Also observe:

$$R_{P2} = R_4 \| R_5$$

$$G_{P2} = \frac{1}{60} + \frac{1}{120} = 0.025 \text{ S}$$

$$R_{P2} = \frac{1}{0.025} = 40 \ \Omega$$

The circuit is now redrawn and shown in Figure 5.25b. Here it is clear that R_{P_2} is now in series with R_6:

$$R_X = 40 + 10 = 50\ \Omega$$

R_X is now in parallel with R_{P_1}:

$$R_Y = R_X \| R_{P_1}$$
$$= 50 \| 50 = 25\ \Omega$$

The total resistance is now the series combination of R_1 and R_Y:

$$R_T = 50 + 25 + 75\ \Omega$$

The total current is found by Ohm's Law:

$$I_T = \frac{100}{75} = 1.33\ \text{A}$$

The current through the parallel combination R_{P2} may be obtained by Eq. (5.15):

$$I_{P2} = 1.33\,\frac{25}{50} = 0.667\ \text{A}$$

Applying CDR to R_4 and R_5:

$$I_4 = 0.667\,\frac{40}{60} = 0.44\ \text{A}$$

5.8 OPEN CIRCUIT

A circuit with a disconnected or interrupted current path is called an ***open circuit.***

In the circuit of Figure 5.26, a battery and a resistor are connected in such a way that current cannot be maintained because the circuit is open at terminals A and B. Since there is no current through resistor R, the voltage drop across it is zero.

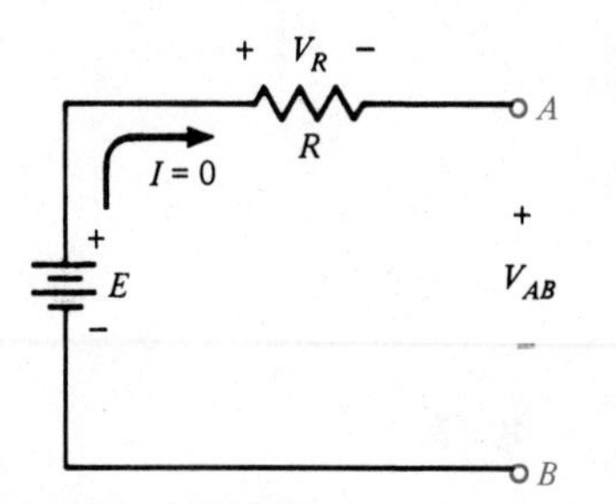

Figure 5.26 Circuit with a battery, a resistor, and an open circuit.

By KVL:

$$E = V_R + V_{AB}$$

$$V_R = 0$$

Clearly:

$$V_{AB} = E$$

Therefore, the open-circuit voltage, V_{AB}, is equal to the source voltage. Consider the following example.

EXAMPLE 5.20

Calculate the open terminal voltage in the circuit of Figure 5.27.

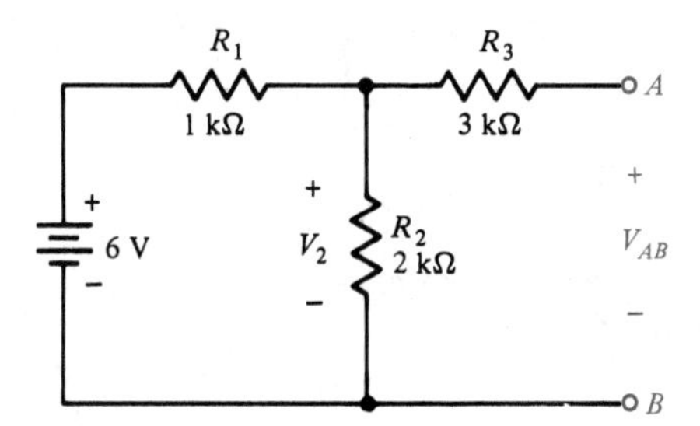

Figure 5.27 Circuit for Example 5.20.

Solution

Since there is no current through the 3-kΩ resistor, the open terminal voltage is:

$$V_{AB} = V_2$$

V_2 may be obtained by applying VDR:

$$V_{AB} = 6\,\frac{2\text{ k}\Omega}{2\text{ k}\Omega + 1\text{ k}\Omega} = 4\text{ V}$$

5.9 SHORT CIRCUIT

A path of zero resistance is called a ***short circuit,*** *or simply* ***short.***

In practice, it is impossible to achieve zero resistance, but it is possible to use very good conductors that have negligible resistance in comparison to other circuit elements. If a good conductor is accidentally placed across a voltage source, a very high current will be established in the conductor. By using current division, it becomes clear that no current will be available to the rest of the circuit because practically all of the current will go through the short circuit.

A circuit component is said to be *shorted out* when a short circuit is placed across it. Almost no current exists in the component.

EXAMPLE 5.21

Calculate the total resistance in the network of Figure 5.28a.

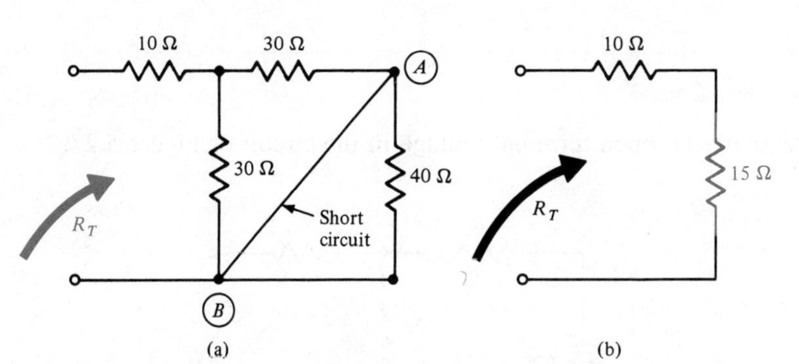

Figure 5.28 Diagrams for Example 5.21. **(a)** Circuit for Example 5.21. **(b)** Reduced circuit for Example 5.21.

Solution

The 40-Ω resistor is shorted out. Points *A* and *B* are therefore electrically the same point. The two 30-Ω resistors are in parallel (30 Ω ‖ 30 Ω = 15 Ω). The total resistance is:

$$R_T = 10 + 15 = 25\ \Omega$$

EXAMPLE 5.22

For the circuit shown in Figure 5.29a, calculate the current I_X.

Solution

Nodes *A* and *B* are shorted together, resulting in no current through the network to the right of the short. Resistors R_4 and R_5 are also shorted out; therefore, they also have zero current. From the reduced circuit of Figure 5.29b:

$$6\ \text{k}\Omega \,\|\, 3\ \text{k}\Omega = 2\ \text{k}\Omega$$

$$R_T = 1\ \text{k}\Omega + 2\ \text{k}\Omega = 3\ \text{k}\Omega$$

$$I_T = \frac{100\ \text{V}}{3\ \text{k}\Omega} = 33.3\ \text{mA}$$

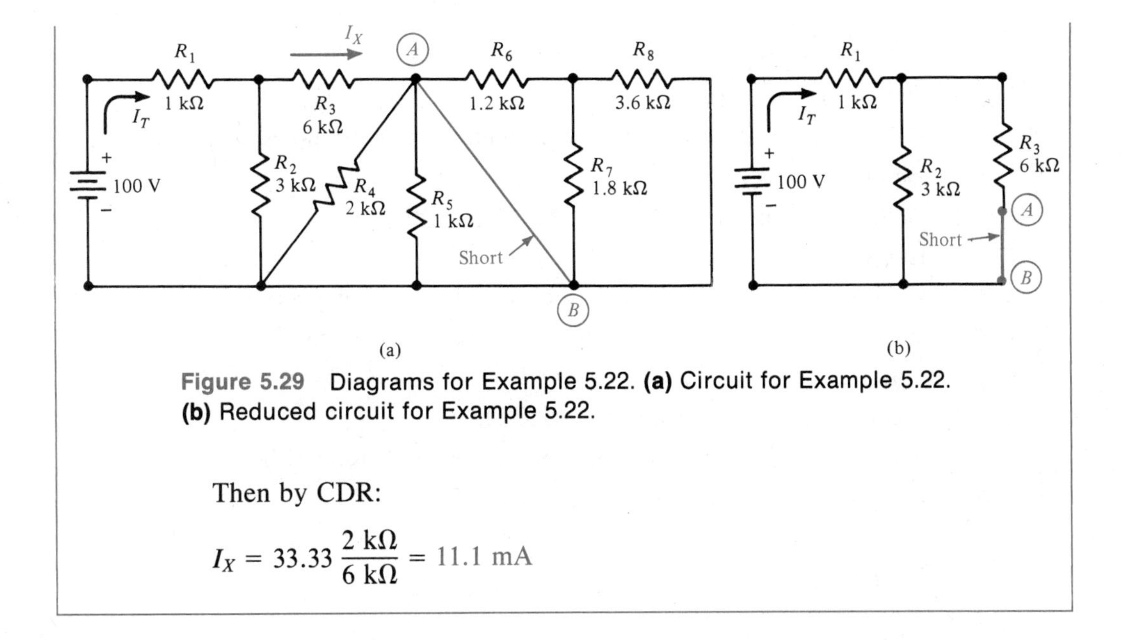

Figure 5.29 Diagrams for Example 5.22. **(a)** Circuit for Example 5.22. **(b)** Reduced circuit for Example 5.22.

Then by CDR:

$$I_X = 33.33 \frac{2\ \text{k}\Omega}{6\ \text{k}\Omega} = 11.1\ \text{mA}$$

SUMMARY

In this chapter, Ohm's Law as well as Kirchhoff's Voltage and Current Laws (KVL and KCL) were studied and applied to the solution of series, parallel, and series-parallel resistive circuits.

The Voltage Divider Rule (VDR) and the Current Divider Rule (CDR) were derived. The concepts of short circuits and open circuits were also examined.

SUMMARY OF IMPORTANT EQUATIONS

Resistors in series:
$$R_T = R_1 + R_2 + R_3 \cdots + R_N \tag{5.1}$$

KVL:
$$\Sigma V_{\text{rises}} = \Sigma V_{\text{drops}} \tag{5.3}$$

VDR:
$$V_X = V_T \frac{R_X}{R_T} \tag{5.5}$$

$$R_2 = R_1 \frac{V_2}{V_T - V_2} \tag{5.6}$$

Resistors in parallel:
$$\frac{1}{R_P} = \frac{1}{R_1} + \frac{1}{R_2} + \cdots + \frac{1}{R_N} \tag{5.9}$$

KCL:
$$\Sigma I_{\text{in}} = \Sigma I_{\text{out}} \tag{5.14}$$

CDR:
$$I_X = I_T \frac{R_P}{R_X} \tag{5.15}$$

$$I_X = I_T \frac{G_X}{G_T} \tag{5.19}$$

REVIEW QUESTIONS

5.1 Define the term electric circuit.

5.2 Define the noun node as it is used in electrical terminology.

5.3 Define a series circuit.

5.4 Define a parallel circuit.

5.5 Define a series-parallel circuit.

5.6 What are the two possible conditions that define two series elements?

5.7 What is the condition that defines two parallel elements?

5.8 State Kirchhoff's Voltage Law (KVL).

5.9 State Kirchhoff's Current Law (KCL).

5.10 Explain in your own words the Voltage Divider Rule (VDR).

5.11 Explain in your own words the Current Divider Rule (CDR).

5.12 Define a short circuit.

5.13 Define an open circuit.

PROBLEMS

5.1 Compute the total resistance R_{AB} of the series networks shown in Figure 5.30.

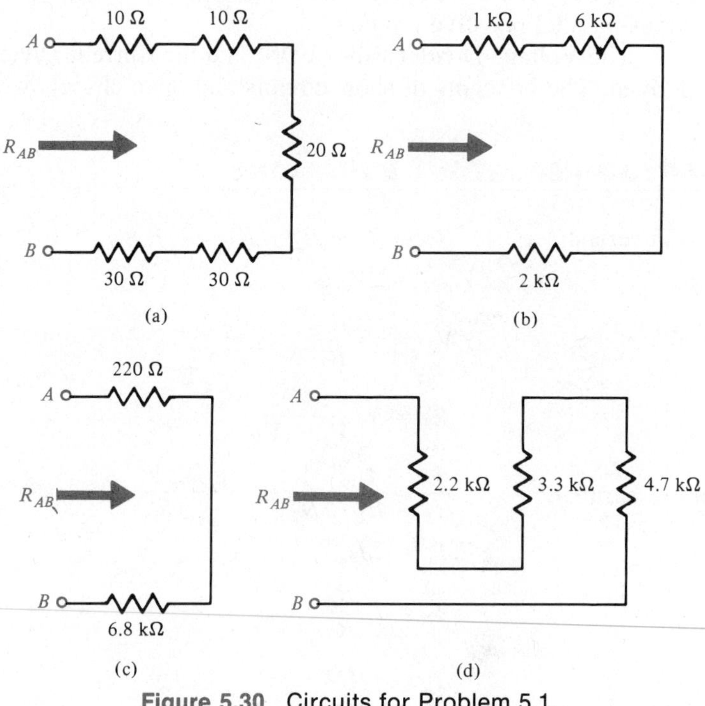

Figure 5.30 Circuits for Problem 5.1.

5.2 Compute the value of resistor R in the networks shown in Figure 5.31. (ans. 20 Ω, 3.3 kΩ, 520 Ω, 10 kΩ)

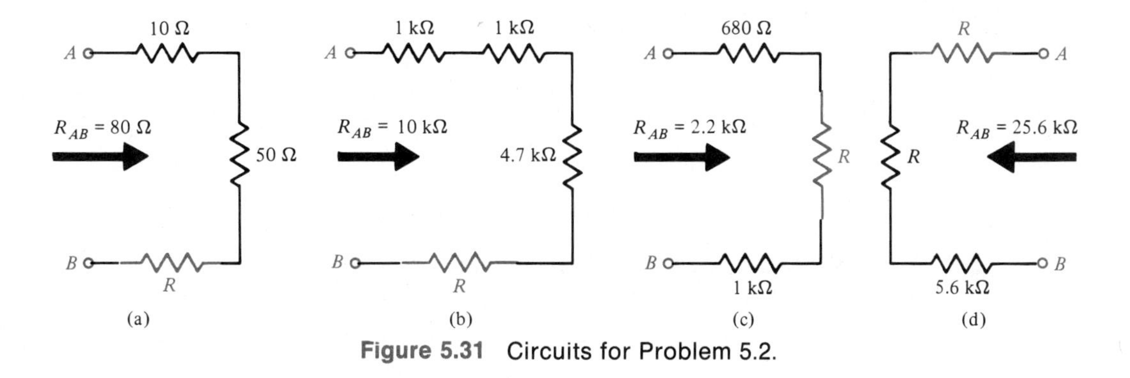

Figure 5.31 Circuits for Problem 5.2.

5.3 Calculate the current I through the resistors in the circuits shown in Figure 5.32.

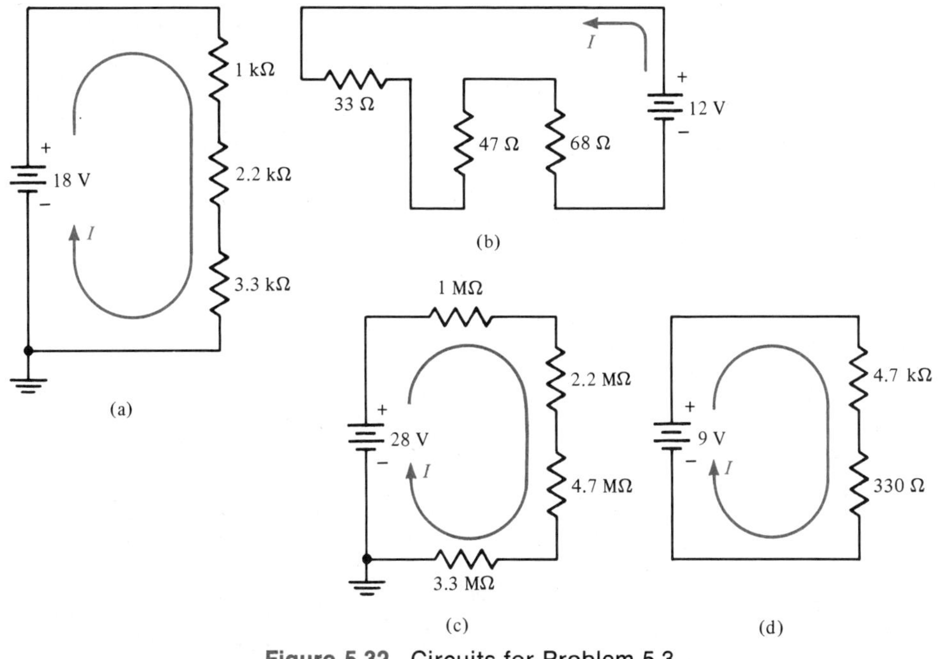

Figure 5.32 Circuits for Problem 5.3.

5.4 For the circuit shown in Figure 5.33, find the source power, the power to the 10-Ω resistor, the power to resistor R, and the value of resistor R. (ans. 240 W, 40 W, 200 W, 50 Ω)

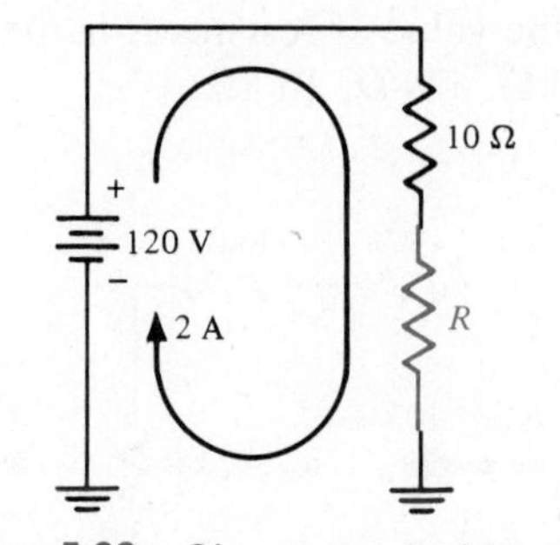

Figure 5.33 Circuit for Problem 5.4.

5.5 Use KVL to compute the unknown voltage V in the circuit shown in Figure 5.34.

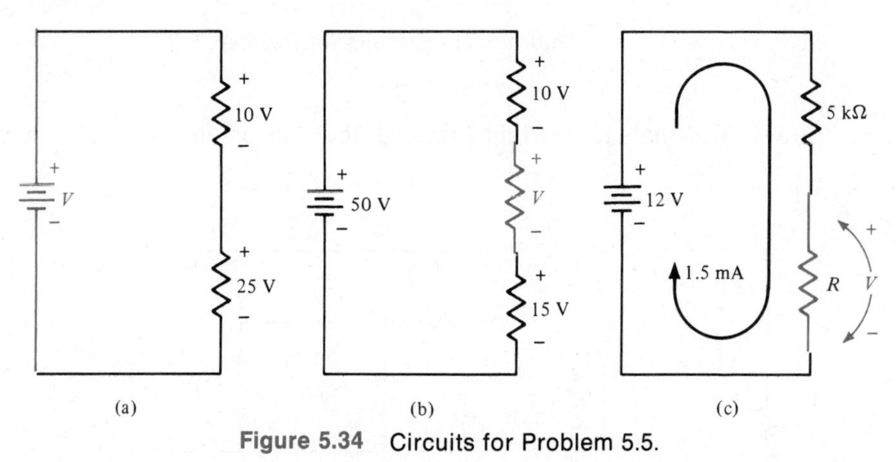

Figure 5.34 Circuits for Problem 5.5.

5.6 Use the VDR to calculate the voltage V in the circuit of Figure 5.35a. (ans. 1.7 V)

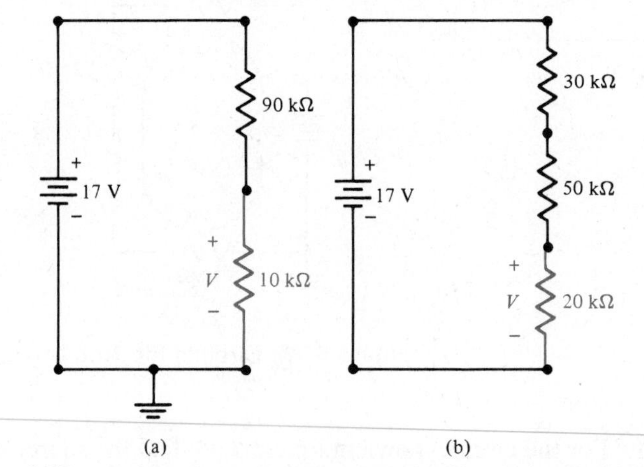

Figure 5.35 **(a)** Circuit for Problem 5.6. **(b)** Circuit for Problem 5.7.

5.7 Use the VDR to calculate the voltage V in the circuit of Figure 5.35b.

5.8 Compute the total resistance R_P and the total conductance G_T for the networks shown in Figure 5.36. (ans. 5 kΩ, 1 Ω, 150 Ω, 569 Ω)

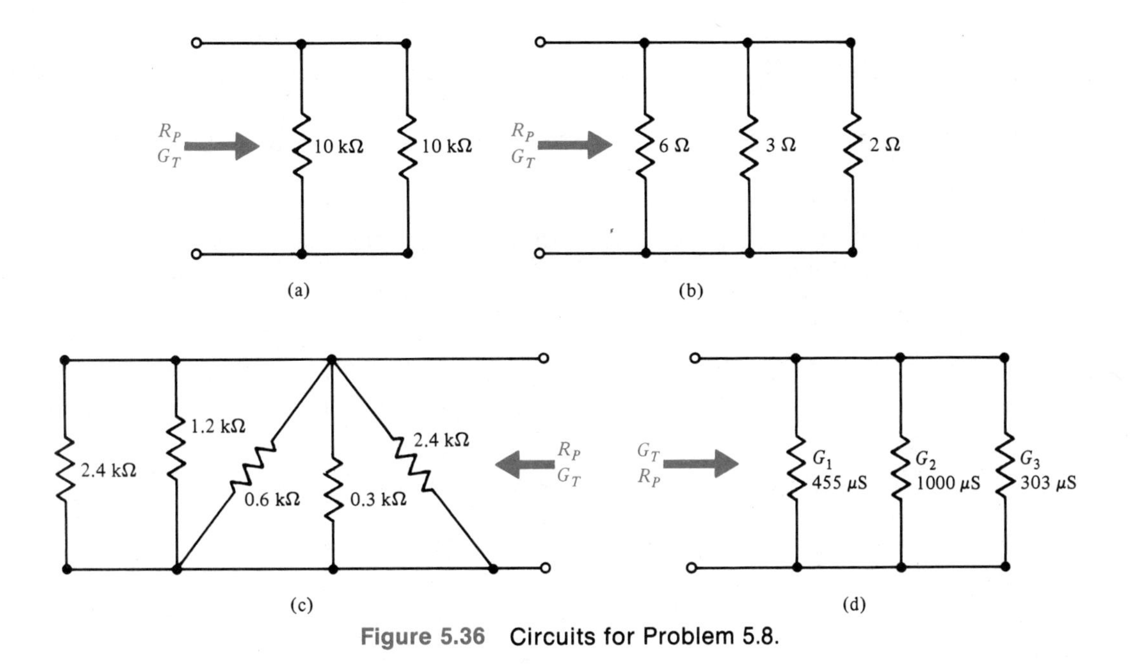

Figure 5.36 Circuits for Problem 5.8.

5.9 Compute the current through each resistor and the battery current for the circuit of Figure 5.37a.

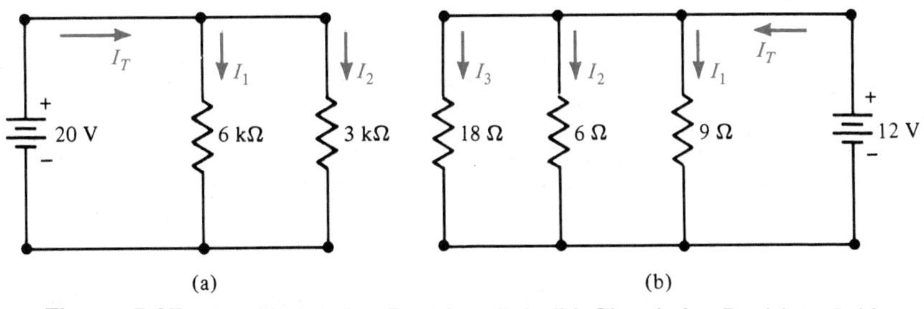

Figure 5.37 **(a)** Circuit for Problem 5.9. **(b)** Circuit for Problem 5.10.

5.10 Compute the current through each resistor and the battery current for the circuit of Figure 5.37b. (ans. 1.33 A, 2 A, 0.67 A, 4 A)

5.11 Use KCL to compute the current I in the circuit of Figure 5.38a.

5.12 Use KCL to compute the current I in the circuit of Figure 5.38b. (ans. 6 A)

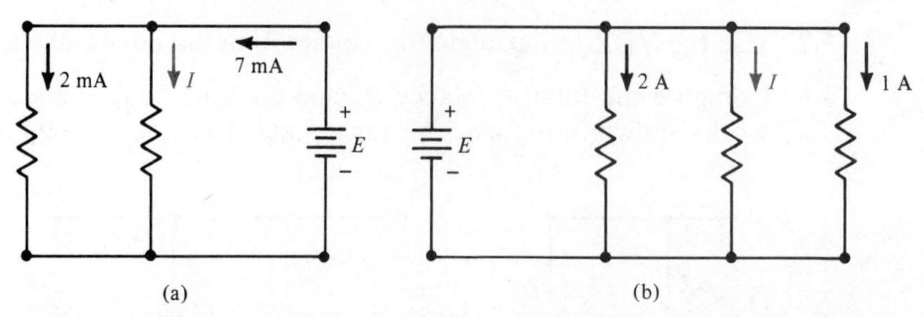

Figure 5.38 **(a)** Circuit for Problem 5.11. **(b)** Circuit for Problem 5.12.

5.13 Use the CDR to solve for the current through each resistor of the circuit of Figure 5.39a.

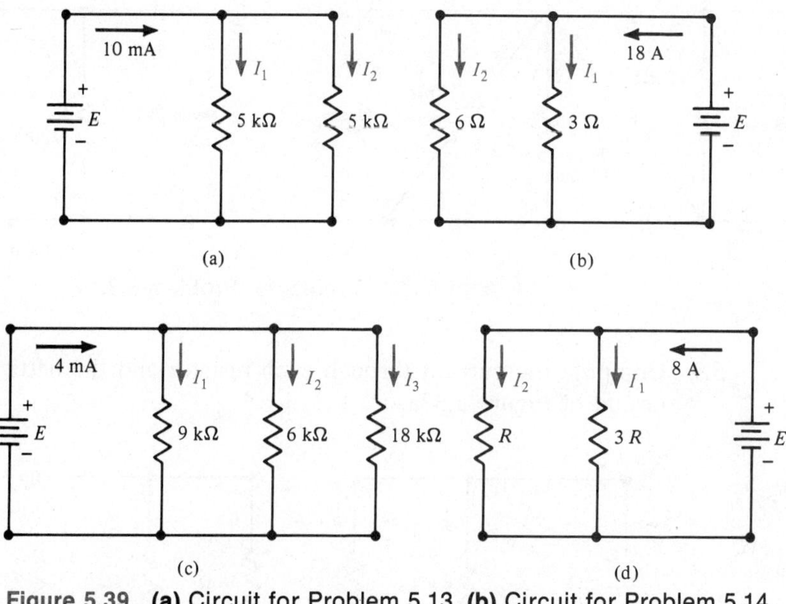

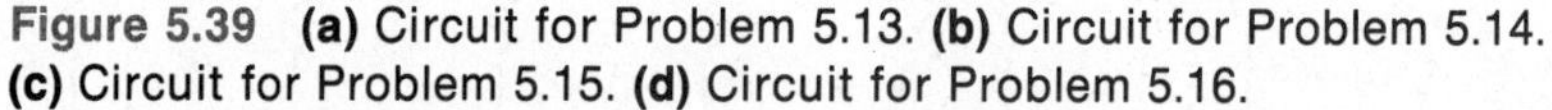

Figure 5.39 **(a)** Circuit for Problem 5.13. **(b)** Circuit for Problem 5.14. **(c)** Circuit for Problem 5.15. **(d)** Circuit for Problem 5.16.

5.14 Use the CDR to solve for the current through each resistor of the circuit of Figure 5.39b. (ans. 12 A, 6 A)

5.15 Use the CDR to solve for the current through each resistor of the circuit of Figure 5.39c.

5.16 Use the CDR to solve for the current through each resistor of the circuit of Figure 5.39d. (ans. 2 A, 6 A)

5.17 Find the total resistance R_{AB} and the total conductance G_{AB} of the series-parallel circuit of Figure 5.40a.

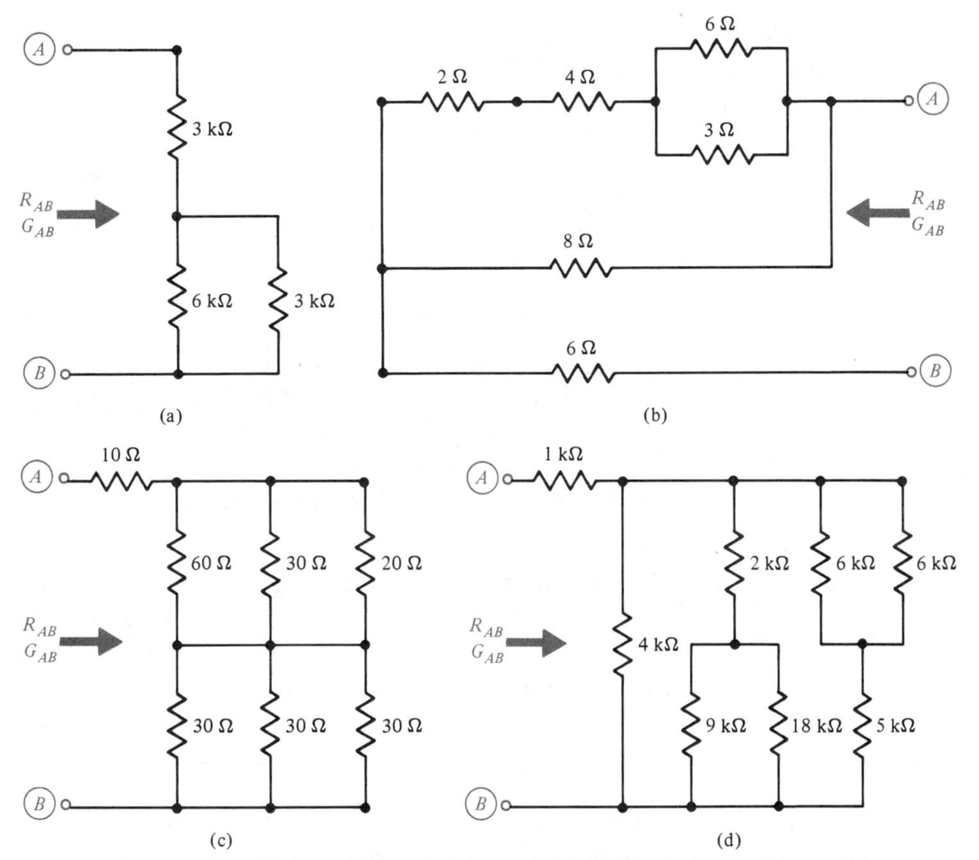

Figure 5.40 **(a)** Circuit for Problem 5.17. **(b)** Circuit for Problem 5.18. **(c)** Circuit for Problem 5.19. **(d)** Circuit for Problem 5.20.

5.18 Find the total resistance R_{AB} and the total conductance G_{AB} of the series-parallel circuit of Figure 5.40b. (ans. 10 Ω, 0.1 S)

5.19 Find the total resistance R_{AB} and the total conductance G_{AB} of the series-parallel circuit of Figure 5.40c.

5.20 Find the total resistance R_{AB} and the total conductance G_{AB} of the series-parallel circuit of Figure 5.40d. (ans. 3 kΩ, 0.33 mS)

5.21 Calculate the value of resistor R in each circuit of Figure 5.41.

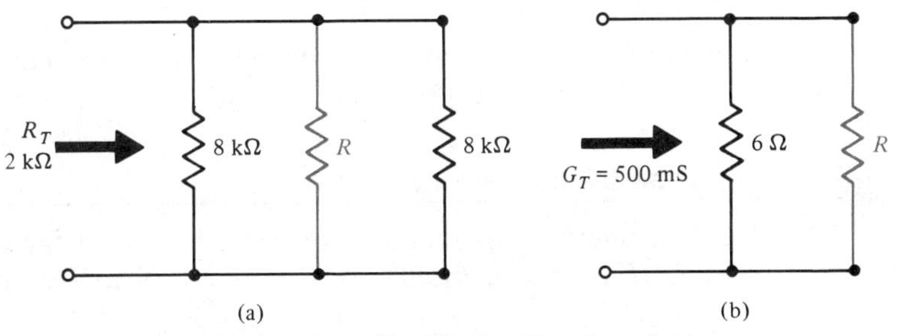

Figure 5.41 Circuits for Problem 5.21.

5.22 For the circuit shown in Figure 5.42, find the current through each resistor. (ans. 3.60 mA, 2.13 mA, 1.47 mA)

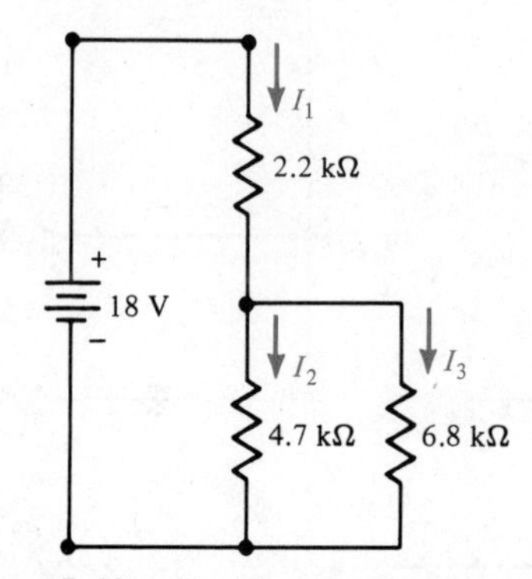

Figure 5.42 Circuits for Problem 5.22.

5.23 Refer to Figure 5.43a and find the value of resistor R so that the voltage across R is 5 V.

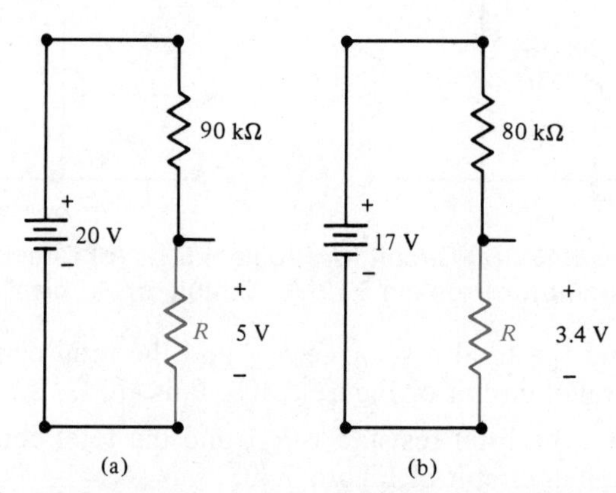

Figure 5.43 **(a)** Circuit for Problem 5.23. **(b)** Circuit for Problem 5.24.

5.24 Refer to Figure 5.43b and find the value of resistor R so that the voltage across R is 3.4 V. (ans. 20 kΩ)

5.25 In Figure 5.44, find the currents I_1 through I_{10}. Also find the voltages V_B and V_{AB}.

Problems for Computer Solution

5.26 Use the computer program developed for Examples 5.2 and 5.3 to solve for the total resistance, current, and for the power to each resistor in Problem 5.3.

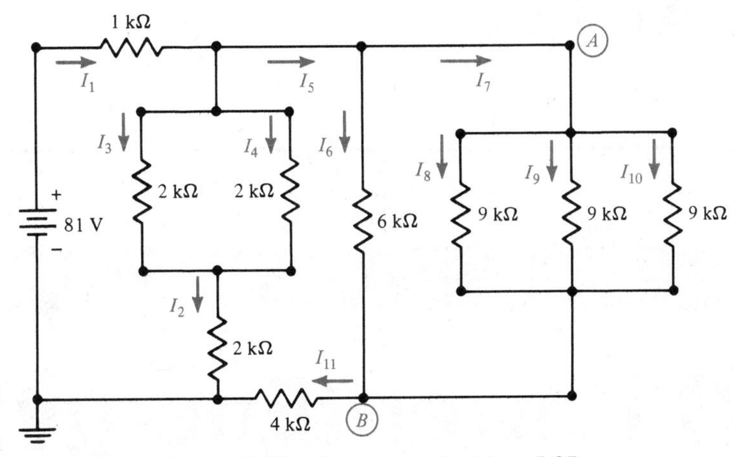

Figure 5.44 Circuit for Problem 5.25.

5.27 Use the computer program developed for Example 5.6 to solve for the voltage across each resistor in Problem 5.6.

5.28 Use the computer program developed for Example 5.6 to solve for the voltage across each resistor in Problem 5.7.

5.29 Use the computer program developed for Example 5.15 to solve for the currents through each resistor in Problem 5.11.

5.30 Use the computer program developed for Example 5.15 to solve for the currents through each resistor in Problem 5.12.

5.31 Use the computer program developed for Example 5.17 to solve Problem 5.22.

CHAPTER 6

DC Power (Energy) Sources

GLOSSARY

dc power supply—device that converts ac to dc power
dependent voltage sources—devices that generate energy as a function of circuit parameters (voltage, current)
electrolyte—salt solution used in battery cells

ideal current source—device that delivers energy to a load at a constant value of current
ideal source—device used to model the behavior of practical sources
ideal voltage source—device that delivers energy to a load, with a constant voltage across its terminals
independent sources—devices that generate energy independent of other circuit components
primary batteries—sources of energy that cannot be recharged
secondary batteries—rechargeable sources of energy
solar cell (photovoltaic cell)—device that converts sunlight to electrical energy
transducer—device used to convert one form of energy to another

INTRODUCTION

The relationship of energy dissipation and power was previously discussed in Chapter 4. Power is the rate at which work is done. It is work done per unit time. Electrical energy is supplied by either voltage or current sources. In this chapter we consider various types of sources, the transfer of energy from a source to a load, and the efficiency of systems.

6.1 IDEAL VOLTAGE AND CURRENT SOURCES

Ideal voltage and current sources are devices used to model the behavior of practical sources.

*An **ideal voltage source** delivers energy to a load, with a constant voltage across its terminals.*

The circuit of an ideal voltage source delivering energy to the load is shown in Figure 6.1. *The terminal voltage, V_T, is constant regardless of the load value and is equal to the source voltage, E_S*. The voltage sources used up until this point have been considered to be ideal.

*An **ideal current source** delivers energy to a load at a constant value of current.*

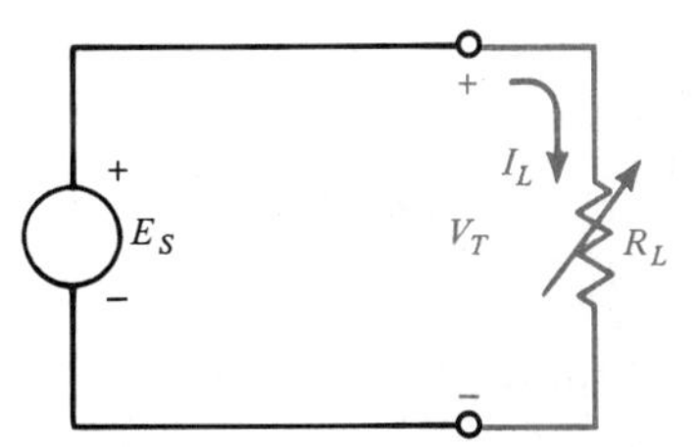

Figure 6.1 Ideal voltage source connected to a varying load resistance.

The circuit in Figure 6.2 depicts a constant current source connected to a variable load. *The load current remains constant with variations in load resistance.*

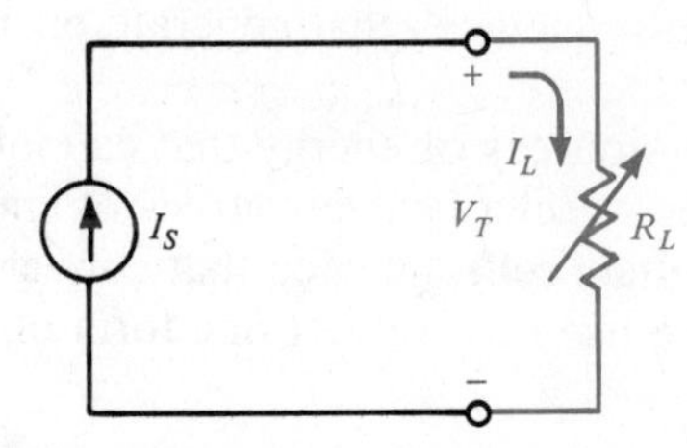

Figure 6.2 Ideal current source and variable load resistance.

EXAMPLE 6.1

What are the values of V_T and I_L in the circuit of Figure 6.1 if R_L varies from 10 to 100 Ω and the source voltage E_S is 10 V?

Solution

From the preceding discussion the terminal voltage remains constant, $V_T = 10$ V. However, the load current varies.

$$I_L = \frac{V_T}{R_L}$$

For $R_L = 10\ \Omega$,

$$I_L = \frac{10}{10} = 1\text{ A}$$

For $R_L = 100\ \Omega$,

$$I_L = \frac{10}{100} = 0.1\text{ A}$$

Hence, the load current varies from 1 A to 0.1 A as the resistance changes from 10 to 100 Ω.

EXAMPLE 6.2

Repeat Example 6.1 for the circuit in Figure 6.2 if the source current I_S is a 2 A current.

Solution

From the preceding discussion, the load current will remain constant:

$I_L = I_S = 2$ A

The terminal voltage varies with load resistance value.

For $R_L = 10\ \Omega$,

$V_T = 10 \times 2 = 20$ V

For $R_L = 100\ \Omega$,

$V_T = 100 \times 2 = 200$ V

The terminal voltage varies from 20 V to 200 V as the load resistance varies from 10 to 100 Ω.

The ideal voltage and current sources under discussion are classified as **independent sources** because the potential of the voltage source is independent of other circuit components. The same is true for the current delivered by the current source; its value does not depend on any other circuit component.

6.2 DEPENDENT SOURCES

Dependent sources are voltage and current sources where the value of the voltage and current of the respective sources is a function of (depends on) voltage or current at other parts of the circuit. The symbols of a dependent voltage and current source are shown in Figures 6.3a and 6.3b, respectively, by the diamond shapes.

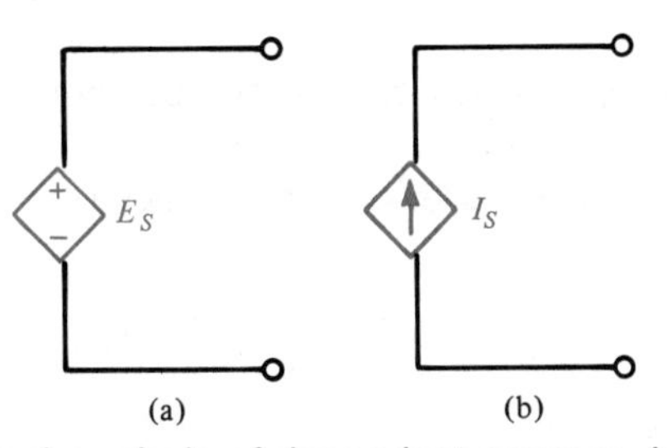

Figure 6.3 Electrical symbols of dependent sources. **(a)** Voltage source. **(b)** Current source.

The circuit shown in Figure 6.4 includes an independent source E_1 and a dependent voltage source E_2. Example 6.3 illustrates the analysis of the circuit.

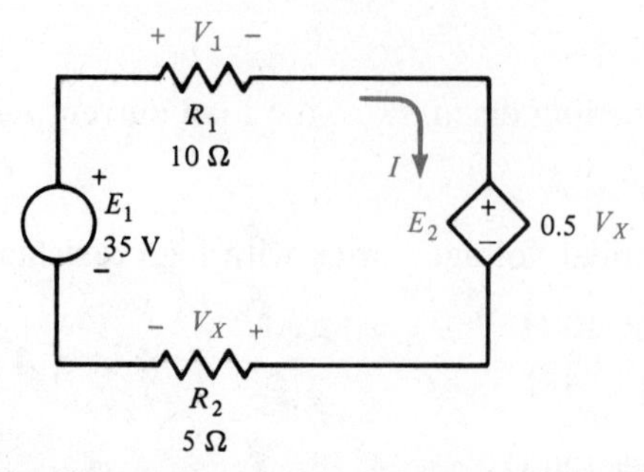

Figure 6.4 Circuit containing dependent and independent voltage sources (Example 6.3).

EXAMPLE 6.3

Determine all the voltages and currents in the circuit of Figure 6.4 and the power delivered by both sources.

Solution

The magnitude of the voltage across the dependent source is a function of the voltage drop across resistance R_2. Kirchhoff's voltage law equation around the loop yields:

$$35 = 10I + (0.5)V_X + 5I$$

where $V_X = 5I$.

Substituting for V_X and solving for I yields:

$$35 = 10I + (0.5)(5I) + 5I$$
$$= 17.5I$$

$$I = \frac{35}{17.5} = 2 \text{ A}$$

Since this is a series circuit, the current in all components is the same, 2 A.

The voltage drop across the resistance is calculated by Ohm's Law:

$$V_1 = 2(10) = 20 \text{ V}$$

and

$$V_2 = V_X = 2(5) = 10 \text{ V}$$

The voltage across the dependent source is:

$$E_2 = 0.5V_X = 0.5(10) = 5 \text{ V}$$

The power delivered by source E_1 is:

$$P_1 = E \times I = 35(2) = 70 \text{ W}$$

Energy is delivered to source E_2 (indicated by the negative sign). Hence,

$$P_2 = -5(2) = -10 \text{ W}$$

Dependent voltage sources can be current-dependent or voltage-dependent, which means that the magnitude of the voltage source can be a function of a current or a voltage in the circuit. Similarly, for dependent current sources the value of the current source can be a function of a voltage across a circuit branch or a current in the circuit.

Dependent sources generally are used to represent certain practical devices in the study of electronic circuits. Such devices as transistors and integrated circuits (ICs) can be represented by electric circuits, which contain dependent sources. Those representations are usually alternating circuit models of the devices. The next example illustrates different circuits that contain dependent sources.

EXAMPLE 6.4

Determine the power to the 1-kΩ resistor in the circuit of Figure 6.5a.

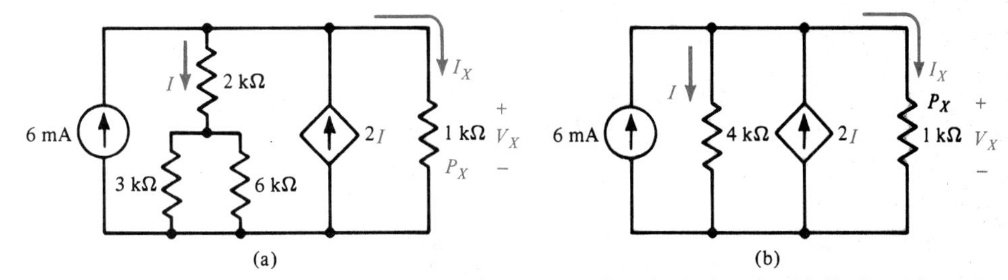

Figure 6.5 Circuit for Example 6.4. **(a)** Original circuit. **(b)** Reduced circuit.

Solution

The equivalent resistance of the 2-kΩ resistor in series with the 3-kΩ and 6-kΩ resistors, which are in parallel, is:

$$R_{eq} = \frac{(3 \text{ k}\Omega)(6 \text{ k}\Omega)}{3 \text{ k}\Omega + 6 \text{ k}\Omega} + 2 \text{ k}\Omega = 4 \text{ k}\Omega$$

Hence, current, I, is maintained through a single equivalent branch of 4 kΩ. The reduced equivalent circuit is shown in Figure 6.5b. By KCL,

$$6 \text{ mA} - I + 2I - I_X = 0$$

The equation above has two unknowns. To solve for the unknowns it is necessary to generate another equation. Since the circuit components are connected in parallel, the voltage across all the components is the same, V_X. The current I can then be expressed as $I = V_X/4\ \text{k}\Omega$; similarly, $I_X = V_X/1\ \text{k}\Omega$. Rewriting the KCL equation in terms of V_X results in:

$$6\ \text{mA} - \frac{V_X}{4\ \text{k}\Omega} + 2\left(\frac{V_X}{4\ \text{k}\Omega}\right) - \frac{V_X}{1\ \text{k}\Omega} = 0$$

Solving for V_X,

$$6 = V_X\left(\frac{1}{4\ \text{k}\Omega} - \frac{2}{4\ \text{k}\Omega} + \frac{1}{1\ \text{k}\Omega}\right)$$
$$= V_X\,(0.75 \times 10^{-3})$$

and

$$V_X = \frac{6 \times 10^{-3}}{0.75 \times 10^{-3}} = 8\ \text{V}$$

The power to the 1-kΩ resistor is:

$$P_X = \frac{V_X^2}{1\ \text{k}\Omega} = \frac{8^2}{10^3} = 64\ \text{mW}$$

EXAMPLE 6.5

The circuit in Figure 6.6 contains a dependent voltage source controlled by a current in the circuit. Calculate the voltage of that source and the voltage V_X.

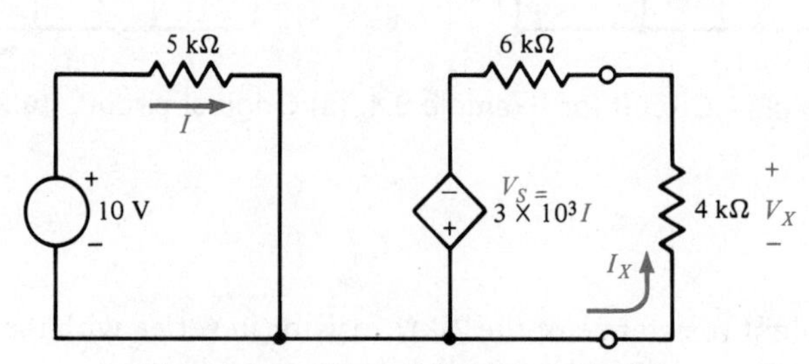

Figure 6.6 Circuit for Example 6.5.

Solution

Notice that the circuit is made up of two separate current loops that are physically isolated. Therefore, the current I is:

$$I = \frac{10\ \text{V}}{5\ \text{k}\Omega} = 2\ \text{mA}$$

and the dependent source has a voltage of:

$$V_S = 3 \times 10^3 I = 3 \times 10^3 \times 2 \times 10^{-3}$$
$$= 6 \text{ V}$$

The voltage V_X is now calculated by using the VDR:

$$V_X = -6 \frac{4 \text{ k}\Omega}{6 \text{ k}\Omega + 4 \text{ k}\Omega} = -2.4 \text{ V}$$

The negative sign is due to the fact that the current of the 4-kΩ resistor will drop a voltage that has opposite polarity to what V_X shows.

6.3 PRACTICAL SOURCES OF ELECTRICAL ENERGY

Practical voltage and current sources are made of materials that have their own characteristics and contribute to the makeup of the electrical parameters of the source.

A practical voltage source, unlike the ideal source examined earlier, does not maintain a constant voltage at its terminals, nor will it be able to deliver an unlimited current to a load connected to the source. Similarly, the practical current source will not be able to deliver a constant current to a load.

The practical voltage source can be represented by an ideal voltage source and an *internal resistance,* R_i, in series with it. The series resistor R_i represents all resistances due to the materials in which current is present when a load is connected to the source. The circuit in Figure 6.7 is that of a practical voltage source and a load, R_L, connected to it. The load resistor R_L represents any load that is connected to the voltage source. Assuming that this source is an automobile battery, R_L can be the starter in the automobile; it can be the light bulb of the flashlight; or the motor of the cassette player. The terminal voltage V_T is what one measures as the voltage of the source. Voltage E_S is the ideal voltage of the source, and R_i is the internal resistance of the source, but both of these (E_S and R_i) cannot be physically separated by the user. The user has access only to the terminals of the source.

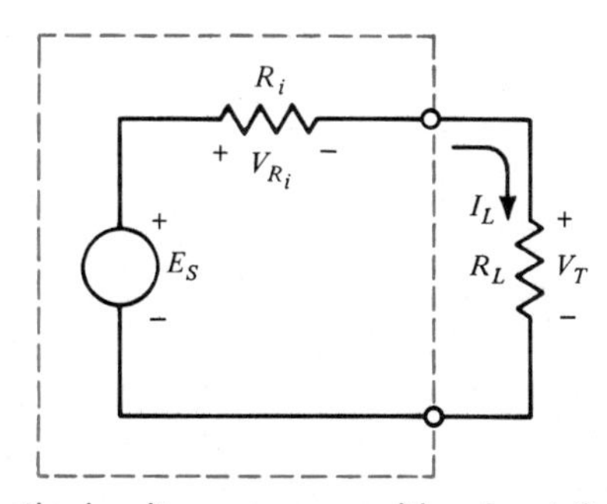

Figure 6.7 Practical voltage source with a load R_L connected to it.

The voltage of the source measured at the terminals is a function of resistor R_L and the internal resistance of the source. Therefore, the larger the value of R_L, the higher the terminal voltage of the source. Conversely, the higher the current drawn by the load, which means a lower value of R_L, the lower the terminal voltage, V_T. *Loading* a source is a term used to specify the current demand on the source. *Heavy* loading refers to a large current drain on the source. *Light* loading is the term used for a small current drain on the source. The lower the value of the internal resistance, the less the terminal voltage varies with changes in loading.

The voltage across the terminals of the source, which is the same as the voltage across the load, is given by

$$V_T = V_L = E_S \frac{R_L}{(R_L + R_i)} \tag{6.1}$$

the load current, I_L, is given by:

$$I_L = \frac{E_S}{R_L + R_i} \tag{6.2}$$

and the power in the load is:

$$P_L = V_L \times I_L = I_L^2 R_L = \frac{V_L^2}{R_L}$$

$$P_L = \left(\frac{E_S}{R_L + R_i}\right)^2 R_L \tag{6.3}$$

By Kirchhoff's voltage law:

$$V_L = E_S - V_{R_i} \tag{6.4}$$

Substituting Eq. (6.2) for V_{R_i},

$$V_L = E_S - \frac{E_S R_i}{R_L + R_i}$$

and

$$V_L = E_S \left(1 - \frac{R_i}{R_L + R_i}\right) \tag{6.5}$$

The circuit in Figure 6.8 shows a typical automobile battery, with an ideal voltage of 13.6 V, an internal resistance of 0.02 Ω, and a variable load.

EXAMPLE 6.6

For the automobile battery in Figure 6.8 calculate the load current, I_L, terminal voltage, V_T, the power lost in the load, P_L, and the power lost by the battery, P_{R_i} for:

a. No load connected, $R_L = \infty$

b. R_L represented by the starter, 0.08 Ω

c. R_L represented by the headlights *on,* 0.5 Ω

d. A short circuit across the terminals

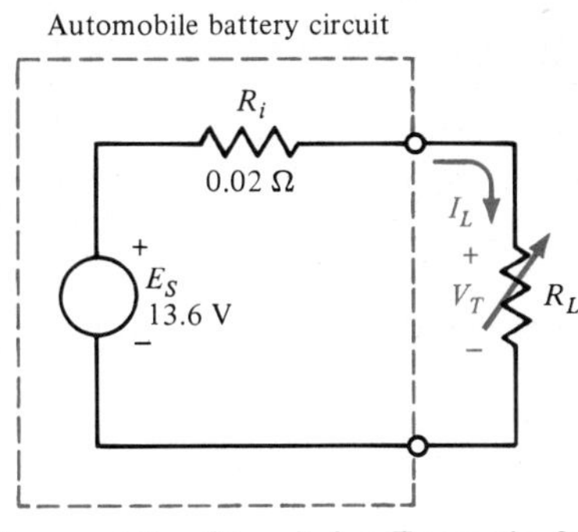

Figure 6.8 Circuit for Example 6.6.

Solution

a. For $R_L = \infty$, there is an open circuit.

$$I_L = 0$$

$$V_T = E_S = 13.6 \text{ V}$$

$$P_L = 0$$

$$P_{R_i} = 0$$

b. For $R_L = 0.08$ Ω, by Eq. (6.2),

$$I_L = \frac{E_S}{R_i + R_L}$$

$$= \frac{13.6}{0.02 + 0.08} = 136 \text{ A}$$

$$V_T = I_L \times R_L$$
$$= 136 \times 0.08 = 10.88 \text{ V}$$

$$P_L = I_L^2 \times R_1$$
$$= 136^2 \times 0.08 = 1479.68 \text{ W}$$

$$P_{R_i} = I_L^2 \times R_i$$
$$= 136^2 \times 0.02 = 369.92 \text{ W}$$

Notice that the power lost in the battery is 369.92 W, or the equivalent of more than 3½ 100-W light bulbs sitting in the battery case.

c. For $R_L = 0.5\ \Omega$,

$$I_L = \frac{13.6}{0.02 + 0.5} = 26.15 \text{ A}$$

$$V_T = 26.15 \times 0.5 = 13.08 \text{ V}$$

$$P_L = (26.15)^2 \times 0.5 = 341.91 \text{ W}$$

$$P_{R_i} = (26.15)^2 \times 0.02 = 13.68 \text{ W}$$

d. For $R_L = 0$,

$$I_L = \frac{13.6}{0.02} = 680 \text{ A}$$

$$V_T = 680 \times 0 = 0$$

$$P_L = 0$$

$$P_{R_i} = (680)^2 \times 0.02$$
$$= 9248 \text{ W or } 9.248 \text{ kW}$$

In the preceding example, with the terminals shorted, the battery lost 9.248 kW. The materials inside the battery heated up considerably. The electrolyte (acid) probably started boiling rather quickly, causing the pressure to increase. This could cause damage to the battery and possibly injure those close to it.

When the terminals are shorted, the current in the circuit is called the short-circuit current, I_{SC}. The short-circuit current is the ratio of E_S/R_i. Conversely, the internal resistance of a voltage source can be calculated by:

$$R_i = E_S/I_{SC} \tag{6.6}$$

the value of E_S can be determined by measuring the voltage across the open terminals, $R_L = \infty$.

A plot of the voltage versus load resistance value R_L is given in Figure 6.9. The voltage drop across the load for the circuit in Figure 6.8 is zero for $R_L = 0$ and

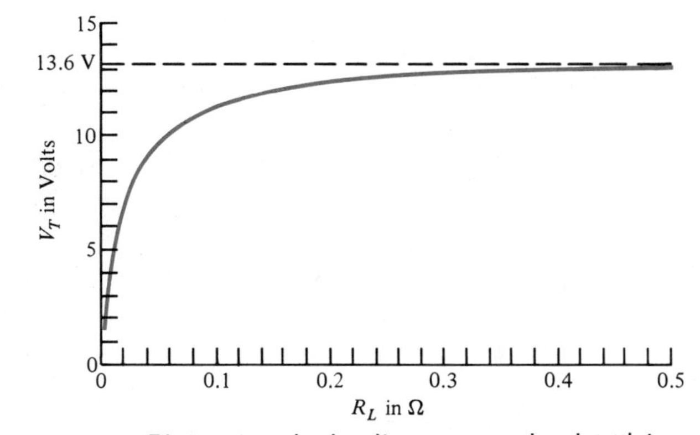

Figure 6.9 Plot of terminal voltage versus load resistance.

approaches the ideal source voltage as R_L becomes large. The plot of load current, I_L versus R_L, is given in Figure 6.10.

Practical current sources are represented by an ideal current source and its internal resistance in parallel, as shown in Figure 6.11. The load current delivered by the source is a function of the value of R_L and R_i and will not be constant with variations in loading as it is with an ideal voltage source. The load current is given by KCL as:

$$\boxed{I_L = I_S \frac{R_i}{R_i + R_L}} \tag{6.7}$$

The terminal voltage is:

$$V_T = I_L \times R_L$$

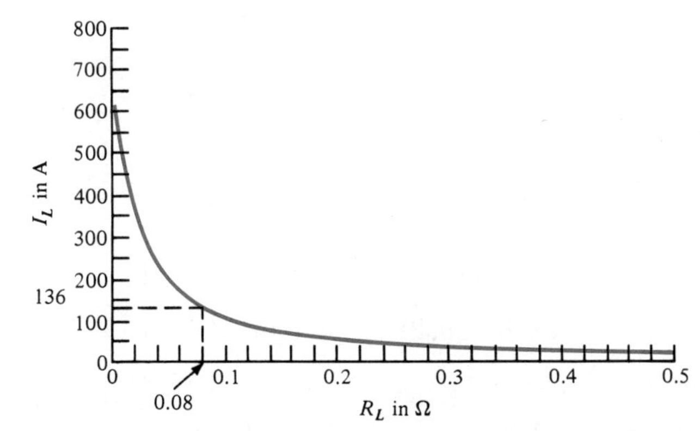

Figure 6.10 Plot of load current versus load resistance of a practical voltage source.

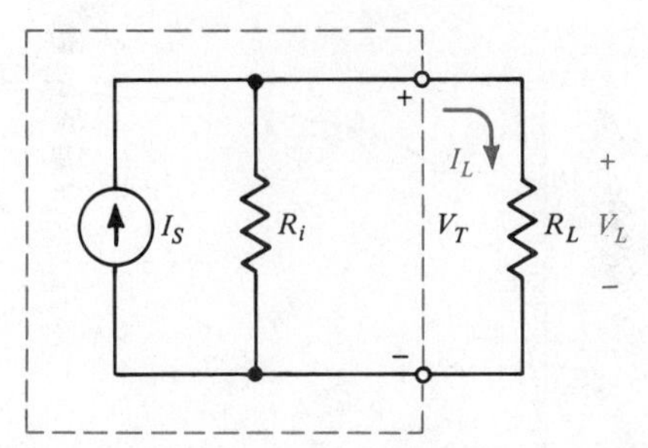

Figure 6.11 Practical current source with load resistor connected across its terminals.

Substituting Eq. (6.7) for I_L,

$$V_T = I_S \frac{R_i \times R_L}{R_i + R_L} \tag{6.8}$$

and the power dissipated by the load is:

$$P_L = I_L^2 \times R_L, \quad \text{or} \quad I_L \times V_L, \quad \text{or} \quad \frac{V_L^2}{R_L}$$

Again, substituting Eq. (6.7) for I_L,

$$P_L = \left(\frac{I_S R_i}{R_i + R_L}\right)^2 R_L \tag{6.9}$$

EXAMPLE 6.7

For the current source and load in Figure 6.11, $I_S = 2$ A, $R_i = 10\ \Omega$, and $R_L = 2\ \Omega$, determine:

a. Load current, I_L

b. Load voltage, V_L

c. Power to the load, P_L

d. Power lost internally in the source, P_{R_i}

Solution

a. By Eq. (6.5),

$$I_L = 2\frac{(10)}{(2 + 10)} = 1.67 \text{ A}$$

b. The load voltage is:

$$V_L = V_T = 1.67 \times 2 = 3.34 \text{ V}$$

c. The power dissipated by the load is:

$$P_L = 3.34 \times 1.67 = 5.58 \text{ W}$$

d. The power dissipated due to R_i is given by:

$$P_{R_i} = \frac{V_T^2}{R_i}$$

$$= \frac{(3.34)^2}{10} = 1.15 \text{ W}$$

6.4 SOURCE EQUIVALENCIES

It was illustrated in the previous section that both current and voltage sources deliver energy to a load. A load was connected to both a voltage and a current source. The quantities measured were load current, load voltage, and power dissipated by the load. Therefore, it is shown next that for every practical voltage source, there is an equivalent practical current source and vice versa.

Comparing Eq. (6.2) with Eq. (6.7), one can conclude that the two are very similar and if $I_S \times R_i$ is substituted for E_S in Eq. (6.2), the result is Eq. (6.7). The same similarities exist with Eqs. (6.1) and (6.8). Substituting $I_S \times R_i$ for E_S results in Eq. (6.8). Therefore, it is concluded that *every practical voltage source can be replaced with an equivalent practical current source and vice versa.* For sources to be equivalent, the internal resistance of both sources must be the same and $E_S = I_S \times R_i$.

EXAMPLE 6.8

For the voltage source shown in Figure 6.12, design the values of the equivalent current source.

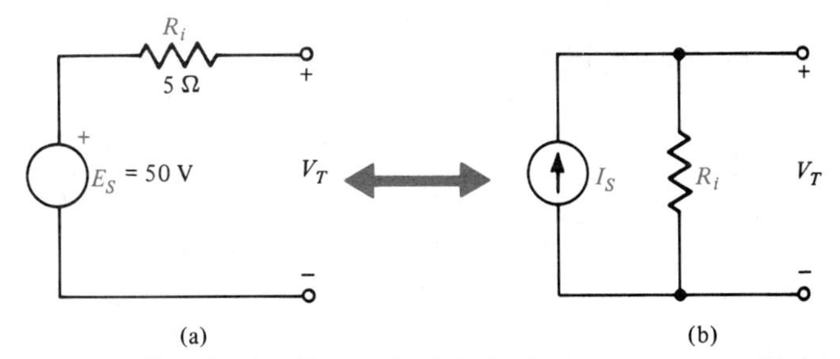

Figure 6.12 Circuits for Example 6.8. **(a)** Voltage source. **(b)** Current source.

Solution

The value of the internal resistance of the current source must be the same as of the voltage source, $R_i = 5\ \Omega$, and $I_S R_i$ must be equivalent to E_S. Therefore,

$$I_S = \frac{E_S}{R_i}$$
$$= \frac{50}{5} = 10 \text{ A}$$

EXAMPLE 6.9

Calculate the power loss by the load and the internal resistance of the two sources in Example 6.8 if a 10-Ω load is connected to each of the sources.

Solution

With a 10-Ω load connected to the *voltage source*, the power to the load is given by Eq. (6.3):

$$P_L = \left(\frac{50}{5 + 10}\right)^2 10 = 111.11 \text{ W}$$

The power lost because of the internal resistance is:

$$P_{R_i} = \left(\frac{50}{5 + 10}\right)^2 5 = 55.55 \text{ W}$$

The power to the load resistor of 10 Ω when connected to the *current source* is given by Eq. (6.9):

$$P_L = \left(\frac{10 \times 5}{5 + 10}\right)^2 10 = 111.11 \text{ W}$$

The power lost because of the internal resistance is calculated by first finding the current through R_i:

$$I_{R_i} = I_S \frac{R_L}{R_i + R_L}$$
$$= 10 \frac{10}{15} = 6.67 \text{ A}$$

and

$$P_{R_i} = I_{R_i}^2 R_i$$
$$= (6.67)^2 \times 5 = 222.2 \text{ W}$$

The preceding results show that the load will dissipate the same amount of power when connected to either the voltage or the current source because the two sources are equivalent. However, the power lost internally in the source is not the same.

6.5 MAXIMUM POWER TRANSFER THEOREM

The *Maximum Power Transfer Theorem* is an important theorem in the study of electricity. It defines the condition for which the power to the load from a particular source is maximum. The theorem applies to both practical current and voltage sources. The theorem states that:

Maximum power is provided to a load by a source if the load resistance is equal to the internal resistance of the source.

$$\boxed{R_i = R_L} \tag{6.10}$$

Under this condition neither the voltage nor the current is maximum. The graph in Figure 6.13 is a plot of power to the load, P_L versus load resistance, R_L. Notice that at $R_L = 0$ and $R_L = \infty$, the power to the load is zero, but at $R_L = R_i$, and only at that value of R_i, the power to the load is maximum. The maximum power to the load is calculated as follows:

$$P_L = I^2 R_L$$

Since $R_i = R_L$, $P_{L_{max}} = (E_S/2R_i)^2\, R_i$, or

$$\boxed{P_{L_{max}} = \frac{E_S^2}{4R_i}} \tag{6.11}$$

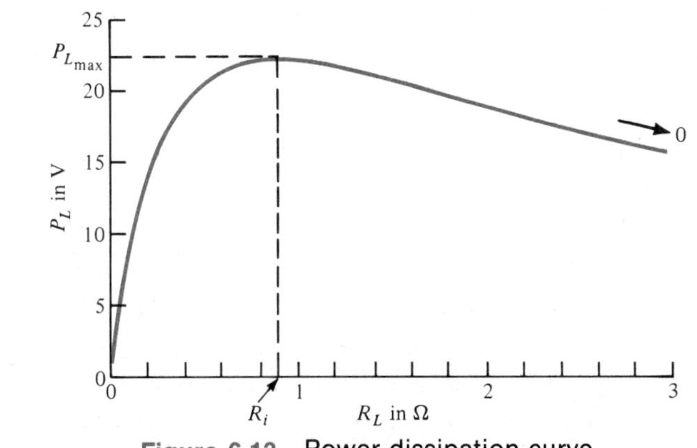

Figure 6.13 Power dissipation curve.

EXAMPLE 6.10

With no load ($R_L = \infty$), the terminal voltage is 9 V. When a 10-Ω load resistor is connected to the source, the terminal voltage is 8.25 V. Determine E_S, the ideal source voltage, R_i, the internal resistance of the source, and the maximum power to the load, and plot the power dissipation curve for a varying load resistance from 0 to 3 Ω, in 0.2-Ω intervals.

Solution

Since the terminal voltage with no load connected is 9 V, then

$$E_S = 9 \text{ V}$$

The current through the load with $R_L = 10\ \Omega$ is

$$I_L = \frac{8.25}{10} = 0.825 \text{ A}$$

and

$$R_i = \frac{9 - 8.25}{0.825}$$

$$= 0.909\ \Omega$$

The maximum power is given by Eq. (6.9).

$$P_{L_{\max}} = \frac{(9)^2}{4(0.909)} = 22.28 \text{ W}$$

The power plot is given in Figure 6.14. The plot was generated on the computer using a standard plotting program and Eq. (6.7) for the values of R_L specified.

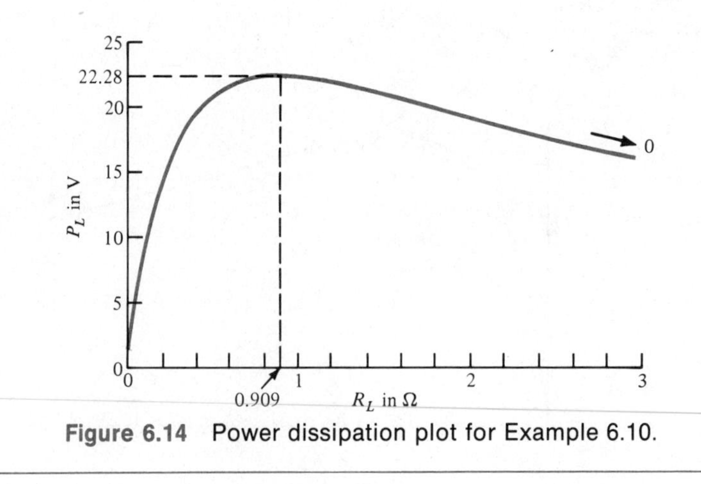

Figure 6.14 Power dissipation plot for Example 6.10.

6.6 DIRECT CURRENT ENERGY SOURCES

Direct current (dc) *energy sources* come in various shapes and sizes. The most familiar dc source is the *battery*. Batteries are *electrochemical* devices used to supply energy to many electronic and electrical systems. Electrical energy is produced by converting chemical energy during battery discharge. Transducers are used to convert one form of energy to another. Dc rotating machine generators are *electromechanical* transducers.

Dc energy sources that convert one form of electrical energy, alternating current (ac), to dc are called *rectifier* types. Since ac energy is readily available in every household outlet and most electronic systems require dc energy for proper operation, rectifier-type converters are used with permanently installed electronic equipment.

6.6.1 Batteries

The electrochemical energy source, the battery, is made up of a number of *voltaic* cells. The Italian scientist Alessandro Volta (1745–1827) invented this source of electrical energy. He showed that two different metallic electrodes, when submerged in a salt solution, develop a potential difference between them due to the chemical reaction of oxidation and reduction that takes place between the metal and the salt solution. His cell consisted of a copper and a zinc plate submerged in a salt solution called an **electrolyte.**

The voltage available across the electrodes of a cell is called the *nominal* voltage. When cells are interconnected to form a battery, only two terminals are made available for connecting a load to the source (battery). The voltage available across the terminals is referred to as the *battery voltage*. A single cell is shown in Figure 6.15a and a battery is illustrated in Figure 6.15b.

Batteries are classified as *primary* or *secondary*. A **primary battery** is used up; it cannot be charged up, and it is thrown away. A **secondary battery** also has a limited life; however, when it cannot deliver current to a load it can be recharged to make it operational again. Primary batteries are used where recharging is not practical and where low cost is essential. Primary batteries are less expensive than their rechargeable counterparts. The most commonly used primary batteries are *carbon-zinc* and *alkaline* dry cells. The most commonly used secondary batteries are the *lead-acid* and *nickel-cadmium* batteries.

Various primary cells and their constructions are shown in Figure 6.16. Rechargeable batteries are shown in Figure 6.17. The lead-acid battery is used in automobiles; it has a nominal terminal voltage of 12 V. It is actually made up of six 2-V cells connected in series. The electrodes are made of lead and the electrolyte is liquid sulfuric acid. Its popularity stems from its ability to deliver very high currents and its excellent recharge characteristics.

Nickel-cadmium batteries are very popular in supplying energy to various electronic devices and equipment. Nickel-cadmium batteries are either of the *vented* or *sealed* type. A vented cell allows the gases to escape to the atmosphere, whereas a sealed battery does not. The sealed-cell batteries are maintenance free

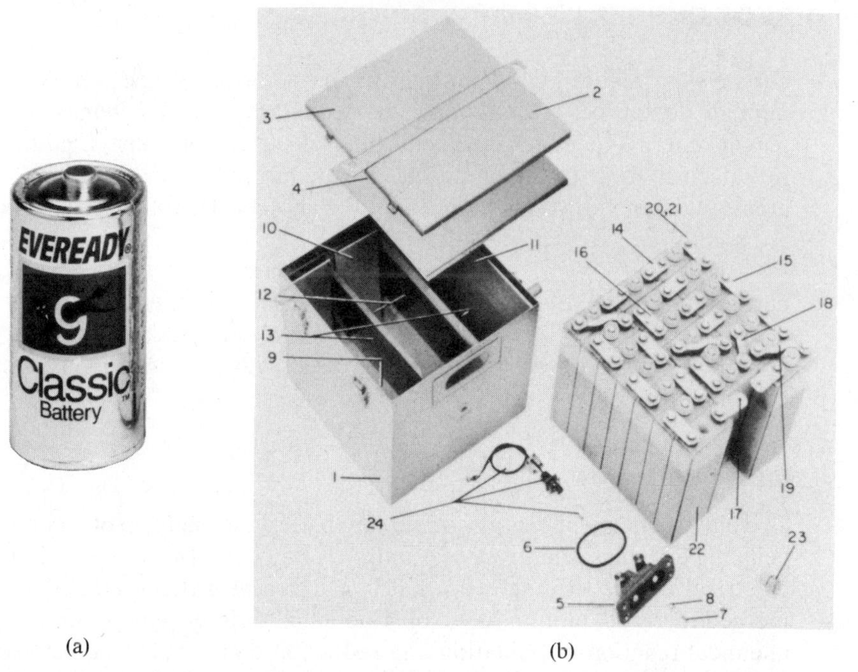

(a) (b)

Figure 6.15 A single cell and a battery. **(a)** Single cell. (Courtesy Eveready Battery Co., Inc.) **(b)** Aircraft battery. (Courtesy Tadiran Electronic Industries Inc.)

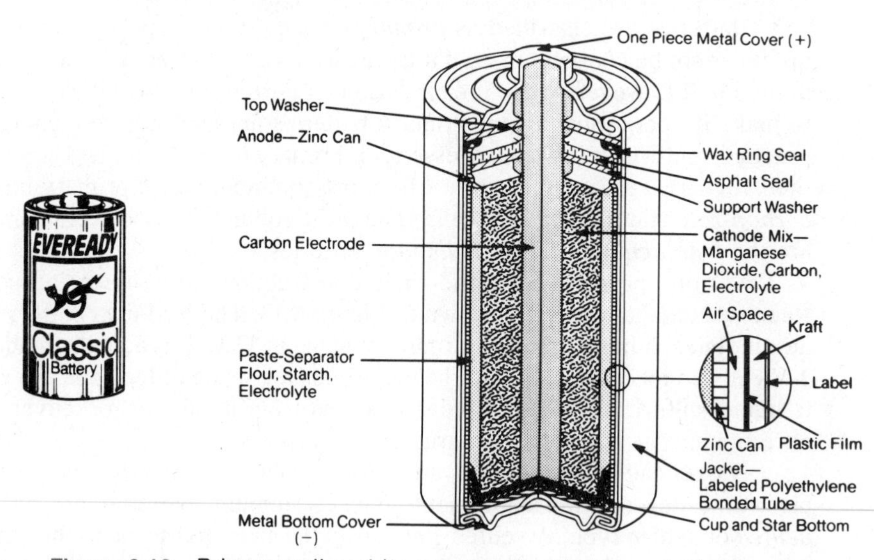

Figure 6.16 Primary cell and its construction (carbon-zinc). (Courtesy Eveready Battery Co., Inc.)

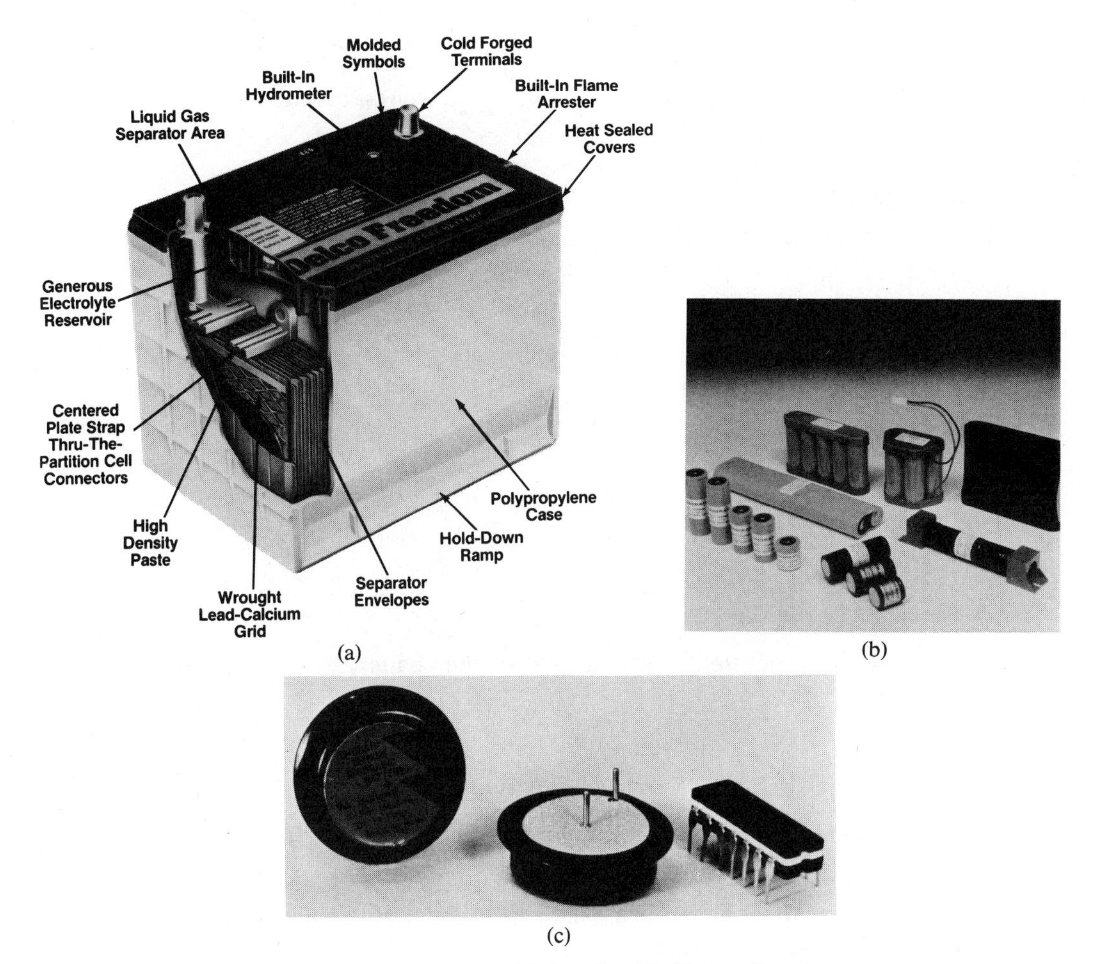

Figure 6.17 Rechargeable cells. **(a)** Lead-acid. (Delco Freedom Maintenance Battery is a Product of General Motors Corporation.) **(b)** Nickel-cadmium (Courtesy Tadiran Electronic Industries Inc.) **(c)** Lithium. (Courtesy Tadiran Electronic Industries Inc.).

and hence are easily adaptable to many applications. Vented cells are made with larger capacity and can be constantly charged. They are used in heavy duty applications such as starting of jet aircraft and other types of engines as well as other military equipment. The vented nickel-cadmium battery is smaller in size and in weight compared with the lead-acid battery. The construction of the vented-cell nickel-cadmium battery is shown in Figure 6.18. The nominal voltage of a nickel-cadmium vented or sealed cell is 1.25 V.

The *lithium* cell is becoming increasingly popular because of several factors:

1. High cell voltage (3.4 V).

2. High energy density, which is defined as *energy in watt-hours per unit weight of the battery*. For example, a Tadiran™ lithium inorganic battery has an energy density of 420 W·h/kg.

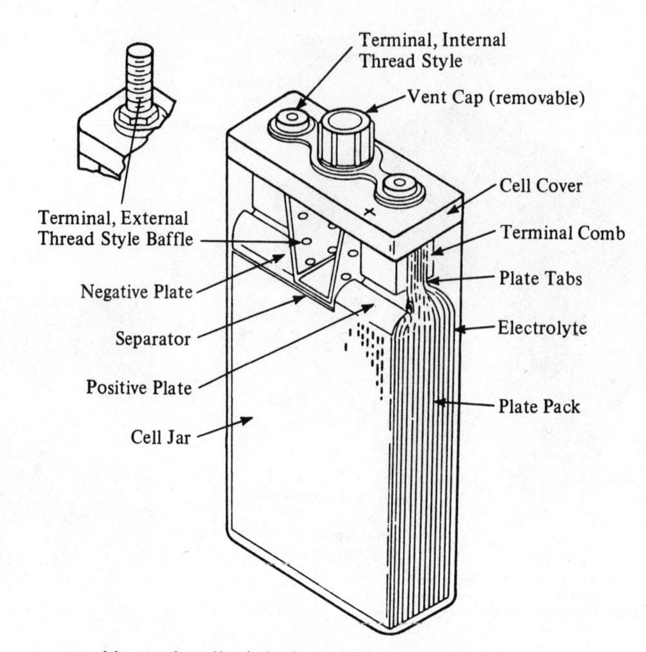

Figure 6.18 Vented-cell nickel-cadmium battery.

3. Long shelf life of up to 10 y is typical.
4. Wide temperature operating range (−55° to +75°C).

These batteries are especially useful in such applications as standby power sources for CMOS memories, pacemakers, portable military equipment, and safety and rescue equipment.

Batteries are rated as to their *capacity C*, which is an indication of how much current can be supplied by the battery for a specified amount of time. It is specified in ampere-hours (A·h). The life of the battery, L, is given by

$$L = \frac{C}{I} \tag{6.12}$$

where L is given in hours, C is the capacity in ampere-hours, and I is the discharge current in amperes. The value of C is also the amount of discharge current that will discharge the battery in 1 h.

Typical discharge characteristics of a nickel-cadmium cell are given in Figure 6.19. The first graph shows the relation between capacity C and temperature. The second graph depicts the change of terminal voltage with change in loading expressed as a percent of nominal capacity. The first graph shows that the capacity of a battery is affected by the ambient temperature.

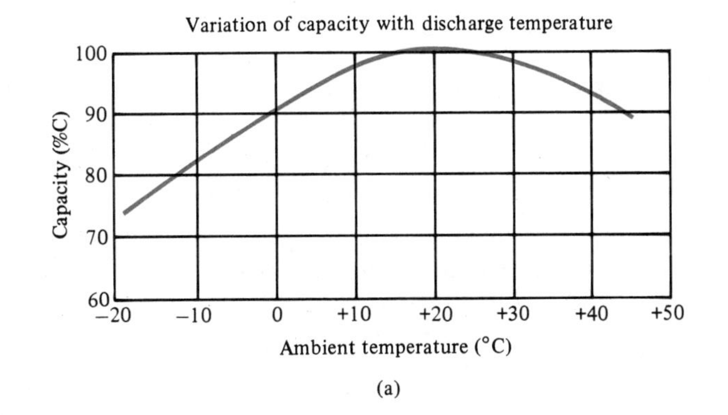

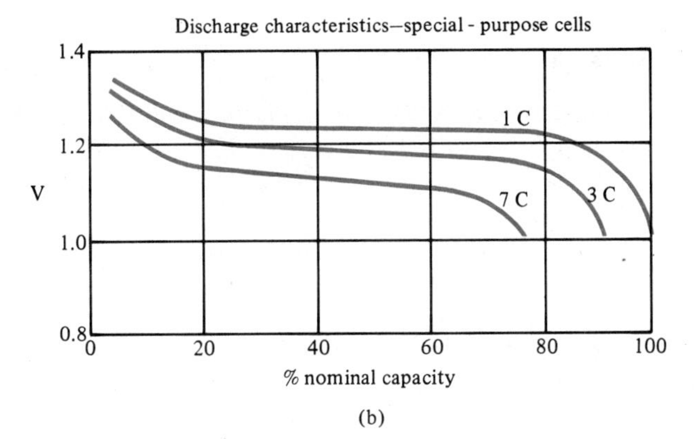

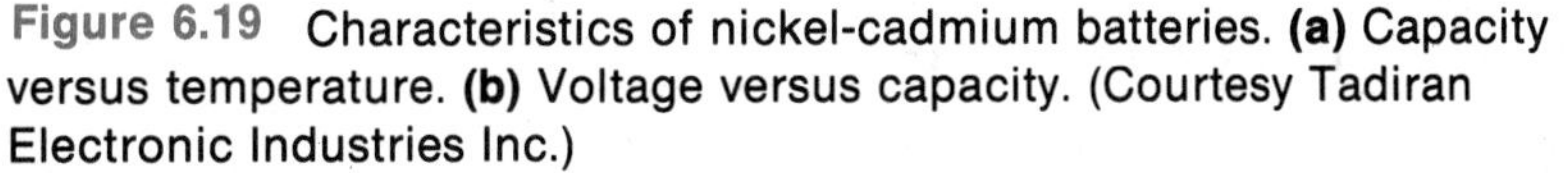

Figure 6.19 Characteristics of nickel-cadmium batteries. **(a)** Capacity versus temperature. **(b)** Voltage versus capacity. (Courtesy Tadiran Electronic Industries Inc.)

EXAMPLE 6.11

A battery has a capacity of C of 20 A·h at 20°C. If it supplies a current of 2 A, what is the life of the cell?

Solution

By Eq. (6.12),

$$L = \frac{20 \text{ A·h}}{2 \text{ A}} = 10 \text{ h}$$

EXAMPLE 6.12

A nickel-cadmium cell has a capacity C of 4.3 A·h at 20°C. Determine from the graph in Figure 6.19a the capacity of the cell at −10°C.

Solution

From the graph, at $-10°C$ the cell is at 82% of its nominal capacity, so

$$4.3 \text{ A·h} \times 0.82 = 3.53 \text{ A·h}$$

The characteristics in Figure 6.19b illustrate the percent of nominal capacity when the cell is discharged at different rates with an end-of-discharge voltage of 1.0 V. Therefore, a cell with a voltage of 1.0 V is considered to be discharged for rating purposes. Note that the battery of Figure 6.19b has a 90% capacity when discharged at a rate of 3C or 3 A of load current if its nominal capacity is 1 A·h.

EXAMPLE 6.13

What is the capacity of a Tadiran™ TDV4.5 cell that is rated at $C = 4.2$ A·h and is being discharged at a rate of 7C?

Solution

From the characteristics in Figure 6.19b, with a 7C discharge rate:

$$\% \text{ nominal capacity} = 75\%$$

Therefore, $C = (4.2)(0.75) = 3.15$ A·h

6.6.2 Nonbattery Energy Sources

As pointed out earlier, rotating machines (generators) are electromechanical sources of electric energy. Generators are used generally to develop large amounts of power and are mechanically driven by gasoline or diesel engines and by wind- and water-powered turbines. Dc generators develop a voltage proportional to the speed of rotation of its shaft. Therefore, small generators are used as speed sensors. When they are used in such applications, they are called *tachometers*.

The **dc power supply** found in most electronic laboratories delivers dc energy by converting ac power from a wall outlet. A dc power supply is shown in Figure 6.20. The process used to convert ac to dc consists of *rectifying* ac and *filtering*

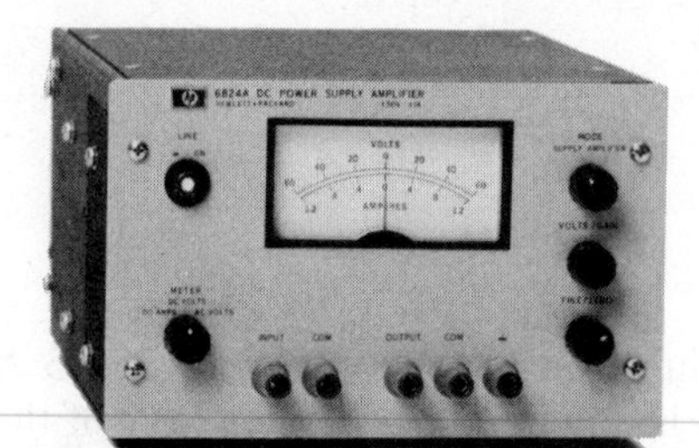

Figure 6.20 Dc power supply. (Courtesy Hewlett-Packard Co.)

out the remaining fluctuations in the signal. This is achieved through the use of electrical/electronic components.

Solar cells are also used to generate electric energy. Solar cells are **photovoltaic** cells used to convert sunlight to electrical energy. A photovoltaic cell is made of semiconductor materials and generates a small voltage (0.5 V) across its two terminals when light is present at its surface. The amount of power generated is proportional to the surface area that is being illuminated. The maximum conversion efficiency is approximately 15%; hence, it is not a very efficient device. The solar-cell battery charger shown in Figure 6.21 can supply up to 3W of power, and it has an area of approximately 189 sq. in.

Figure 6.21 Dc Power supply. (Copyright 1988 Hewlett-Packard Company. Reproduced with permission.)

SUMMARY

Direct current sources of energy are classified as either current or voltage sources. Ideal sources deliver a constant current or voltage to a load regardless of the load value. Practical sources do not deliver constant currents or voltages. The voltage or current delivered to a load is a function of the internal resistance of the source.

Dependent current and voltage sources exhibit terminal currents and voltages that are functions of circuit parameters. Various electronic devices are modeled as dependent sources.

There is an equivalency between practical current and voltage sources. For every practical voltage source there is an equivalent current source and vice versa. Equivalency in this case means that a load connected to either source will experience the same voltage and current.

It was shown that a load connected to a practical source (voltage or current) will dissipate maximum power when it is matched to the source, which *matched* means the resistance value of the load is equal to the internal resistance of the source.

Direct current sources are generally voltage sources. Electrical energy is obtained by converting another form of energy to it. Batteries convert chemical energy to electrical. Rotating machines are electromechanical sources of energy conversion. Solar cells convert light energy to electrical energy, and rectifier-type power supplies convert ac to dc.

SUMMARY OF IMPORTANT EQUATIONS

$$V_T = V_L = E_S \frac{R_L}{(R_L + R_i)} \tag{6.1}$$

$$I_L = \frac{E_S}{R_L + R_i} \tag{6.2}$$

$$P_L = \left(\frac{E_S}{R_L + R_i}\right)^2 R_L \tag{6.3}$$

$$V_L = E_S - V_{R_i} \tag{6.4}$$

$$I_L = I_S \frac{R_i}{R_i + R_L} \tag{6.7}$$

$$V_T = I_S \frac{R_i \times R_L}{R_i + R_L} \tag{6.8}$$

$$P_L = \left(\frac{I_S R_i}{R_i + R_L}\right)^2 R_L \tag{6.9}$$

$$R_i = R_L \quad \text{for maximum power transfer} \tag{6.10}$$

$$P_{L_{\max}} = \frac{E_S^2}{4R_i} \tag{6.11}$$

$$L = \frac{C}{I} \tag{6.12}$$

REVIEW QUESTIONS

6.1 Define the following terms:
  a. Ideal voltage source
  b. Ideal current source

6.2 Define the term dependent source.

6.3 What is the difference between an ideal and a practical voltage source?

6.4 What is the internal resistance of (a) an ideal voltage source and (b) an ideal current source?

6.5 Will the terminal voltage of a current source be nearest to the ideal source voltage if the load resistor is low or high?

6.6 What is the power to the load if the load resistance is ∞?

6.7 What does the term equivalent sources mean?

6.8 Is there an equivalent current source for every voltage source?

6.9 State the Maximum Power Transfer Theorem.

6.10 Is there voltage maximum across a load when it dissipates maximum power?

6.11 What limits the current in a practical voltage source if the terminals are short circuited?

6.12 Describe the function of a transducer.

6.13 Explain the difference between a primary battery cell and a secondary battery cell.

6.14 List several applications where carbon-zinc batteries are used.

6.15 List several uses of nickel-cadmium batteries.

6.16 What is the advantage of a lithium cell over other batteries?

6.17 Why is the capacity of a battery an important rating?

6.18 How does temperature affect the capacity of a battery?

6.19 What is the function of a rectifier-type power supply?

PROBLEMS

6.1 For the circuit in Figure 6.22, calculate the total current, I_T, and voltage drops across each resistor.

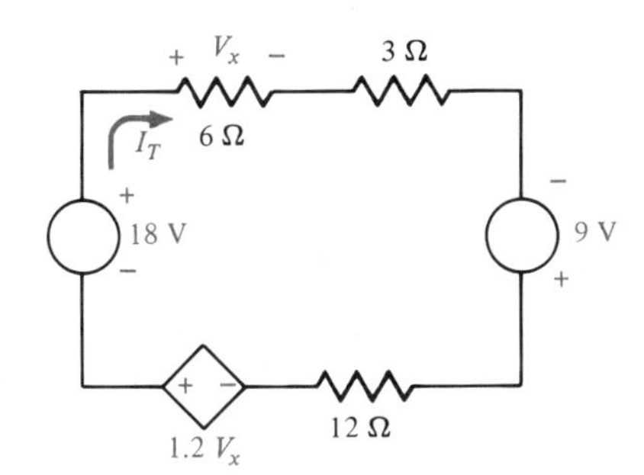

Figure 6.22 Circuit used in Problems 6.1 and 6.2.

6.2 Calculate the power delivered by each source in the circuit of Figure 6.22. (ans. 35.28 W, 17.64 W, 8.64 W)

6.3 Determine the current I in the circuit of Figure 6.23.

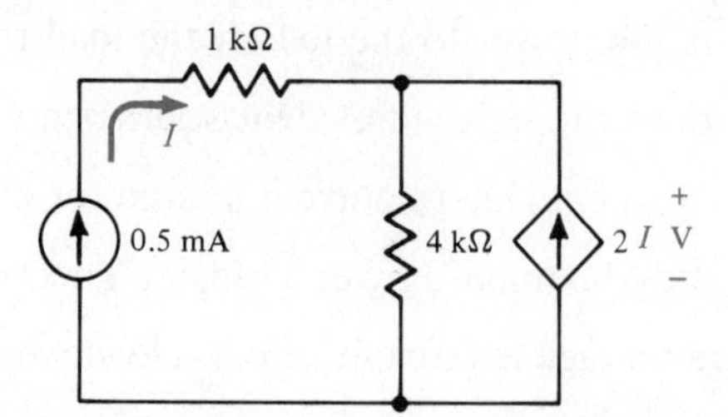

Figure 6.23 Circuit for Problems 6.3 and 6.4.

6.4 Calculate the voltage dropped across the dependent source in Figure 6.23. (ans. 2 V)

6.5 In the circuit of Figure 6.24 calculate the voltage, V, across the parallel branches.

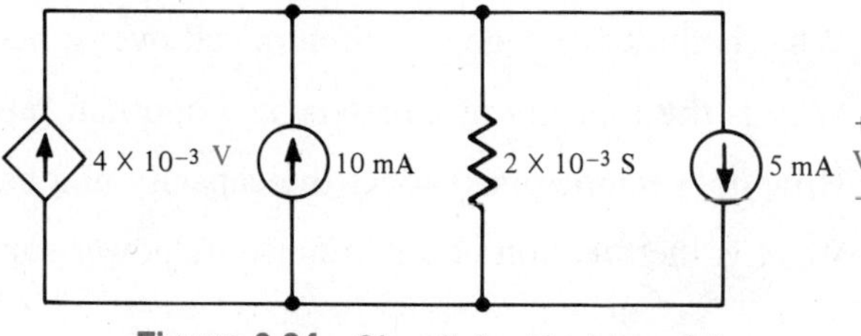

Figure 6.24 Circuit for Problem 6.5.

6.6 For the circuit of Figure 6.25 calculate V_X. (ans. -7.33 V)

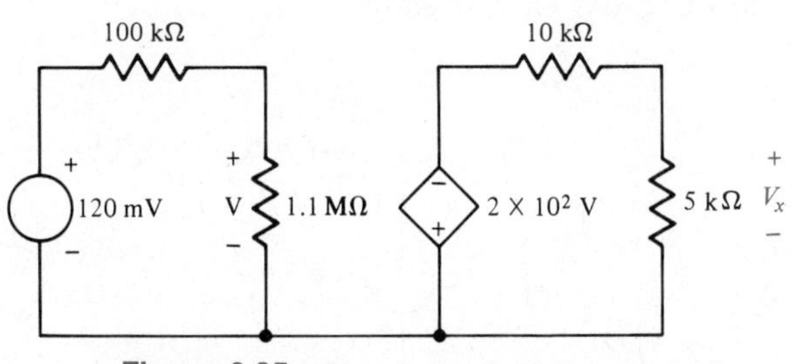

Figure 6.25 Circuit for Problem 6.6.

6.7 In the circuit of Figure 6.26, R_L varies from 1 kΩ to 10 kΩ. Calculate the terminal voltage range as R_L varies.

6.8 Repeat Problem 6.7 for variations in R_L from 100 Ω to 500 Ω. (ans. 60 V–100 V)

6.9 In the circuit in Figure 6.27, calculate the value of I_S for the load voltage, V_L, to be 10 V.

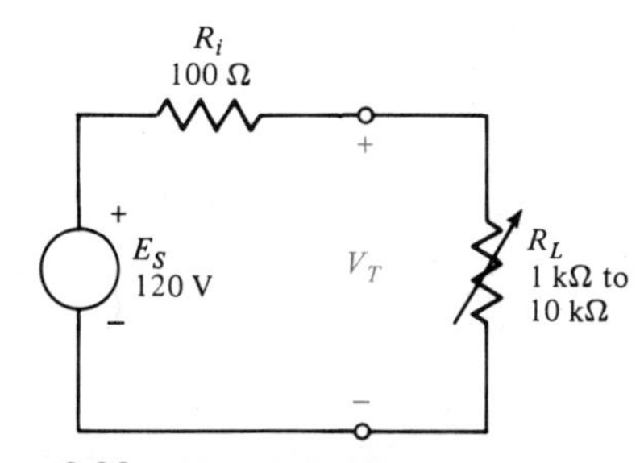

Figure 6.26 Circuit for Problems 6.7 and 6.8.

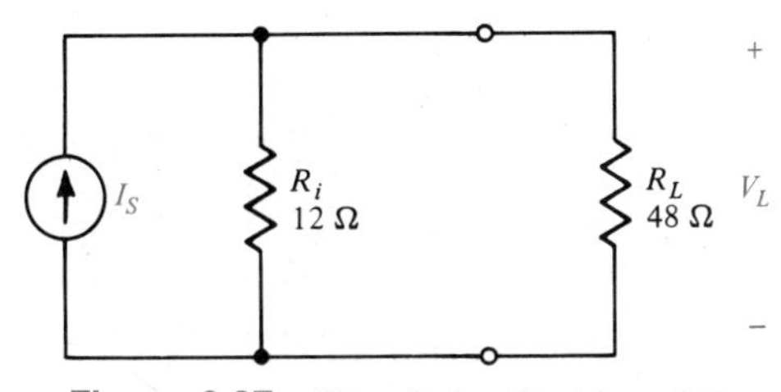

Figure 6.27 Circuit for Problem 6.9.

6.10 Repeat Problem 6.9 for an R_L of 20 Ω. Calculate the power loss in the source resistance R_i. (ans. 1.33 A, 8.27 W)

6.11 A practical voltage source has an internal resistance of 0.5 Ω and an open terminal voltage of 6 V. Design its equivalent current source.

6.12 A practical voltage source has an open terminal voltage of 100 V. What is its internal resistance if the equivalent current source has an ideal current value of 400 mA? (ans. 250 Ω)

6.13 The open terminal voltage of a practical current source is 36 V. The ideal current source is 100 mA. What will be the current through a load resistor connected to its terminals if $R_L = 200$ Ω?

6.14 Calculate the internal resistance of a voltage source with $E_S = 15$ V if a 2-kΩ load connected to the terminals of the source has 14.9 V across it. Calculate the power absorbed by the source resistance. (ans. 13.42 Ω)

6.15 Determine the open terminal source voltage, E_S, and the source resistance if a 150-Ω load connected to the source has 10.5 V across it and a 500-Ω load 10.75 V.

6.16 Determine the voltage, V_T, and the power to the load in the circuit of Figure 6.28. (ans. 14.4 V, 34.56 W)

6.17 For the circuit in Figure 6.28 determine the value of R_X for maximum power transfer to the load. Calculate the power to the load under maximum power transfer conditions.

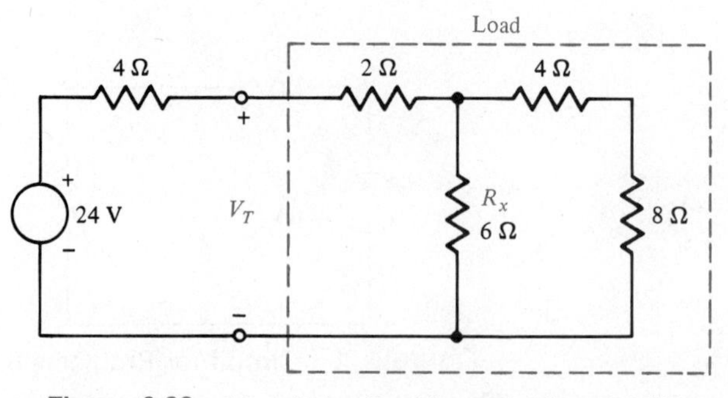

Figure 6.28 Circuit for Problems 6.16 and 6.17.

6.18 A battery with a capacity of 1500 mA·h at 20°C supplies a constant current of 50 mA. Calculate the life of the battery. (ans. 30 h)

6.19 From the graph in Figure 6.19a determine the capacity of the nickel-cadmium cell at −20°C, assuming a nominal capacity of 4.3 A·h.

6.20 From the characteristics of the TDV4.5 cell in Figure 6.19b, determine the capacity of the cell if it is discharged at 3C. (ans. 3.86 A·h)

6.21 For the solar-cell charger shown in Figure 6.21, determine the power an array of such cells mounted on a 2-ft × 2-ft panel can deliver.

SECTION II

DIRECT CURRENT CIRCUIT ANALYSIS

CHAPTER 7

Mesh and Nodal Analysis

GLOSSARY

Cramer's Rule—a method of solving simultaneous equations
determinant—an orderly arrangement of coefficients used in Cramer's Rule
Gauss' Elimination Method—a mathematical technique used to solve simultaneous equations
loop—a closed path in an electric circuit
loop equations—a set of simultaneous equations that results from applying KVL to each loop in an electric circuit
mesh—same as loop
node equations—a set of simultaneous equations that results from applying KCL to each node of an electric circuit

INTRODUCTION

The application of Kirchhoff's Voltage Law around the closed loops of a circuit results in a set of **loop equations.** The application of Kirchhoff's Current Law at the nodes of a circuit results in a set of **node equations.** Loop or node equations are generally written for electric circuits that cannot be solved by combining elements in series or parallel. Either loop or node equations describe the circuit totally. Any circuit property may be found by solving either the circuit loop equations or the circuit node equations. Sometimes a circuit has more loops than nodes; sometimes it has more nodes than loops. If the set of equations is to be solved by hand, the solution is easier if the method with the fewer number of equations is chosen. If a computer solution is considered, then the difference in the number of equations between loops and nodes is really of no great consequence.

This chapter presents *classical* and *standard form* methods of obtaining both loop and node equations. The solution of these equations is then obtained by *Cramer's Rule* and *Gauss' Elimination Method.* Finally, a computer program in BASIC is offered as a method of solving simultaneous equations by computer.

7.1 LOOP (MESH) EQUATIONS

Kirchhoff's Voltage Law (KVL) states that around a closed loop, the sum of the voltage rises always equals the sum of the voltage drops. This very simple yet profound statement allows the circuit analyst to formulate a set of equations that, when solved, gives the currents through all the circuit components.

First the classical way of obtaining loop (mesh) equations by the direct application of KVL around each independent circuit loop is presented. Then the Standard Notation Method of obtaining these same loop equations will be explained.

Once loop equations are obtained, they may be solved by any of several methods. These are presented later in the chapter.

7.2 LOOP EQUATIONS BY KVL

When a current is maintained through a resistor, a voltage drop occurs. The polarity of the voltage drop is such that it is positive where the current enters the resistor and negative where the current leaves the resistor. If two currents are maintained through a resistor, then the two currents as well as their voltage drops superimpose (they either add or subtract).

Figure 7.1a shows a current I_1 being maintained through resistor R and causing a voltage drop V_1 with the polarity indicated. Figure 7.1b shows a current I_2 through resistor R causing a voltage drop V_2 with the polarity indicated. Similarly, Figure 7.1c shows current I_3 through resistor R with the corresponding voltage drop V_3 and its proper polarity. If currents I_1 and I_2 are allowed through the same resistor, then the currents add because they are in the same direction. The corresponding voltage drops also add because they have the same polarity. The polarity of the resulting voltage drop is the same as the individual voltage drops. Figure 7.1d shows both currents, I_1 and I_2, and their corresponding voltage drops and polarities. Figure 7.1e shows the combined, or superimposed, effect of the two currents and corresponding voltage drops. If currents I_1 and I_3 are allowed through the same resistor, then they subtract because they are in opposite directions. The resulting current is in the same direction as the bigger of the two initial currents. The corresponding voltage drops also subtract. The resulting voltage drop has the polarity of the bigger of the two original voltage drops. Assuming that I_1 is bigger than I_3, Figure 7.1f shows both their superposition and the superposition of their voltage drops. Figure 7.1g shows the resulting current and voltage drop.

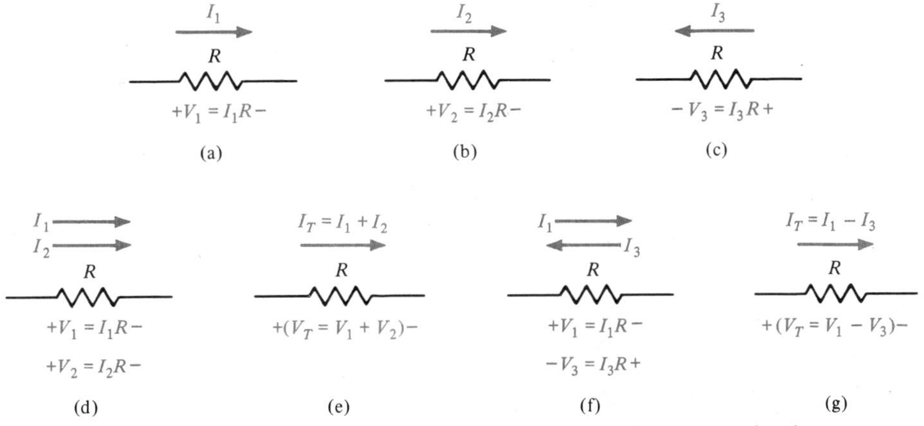

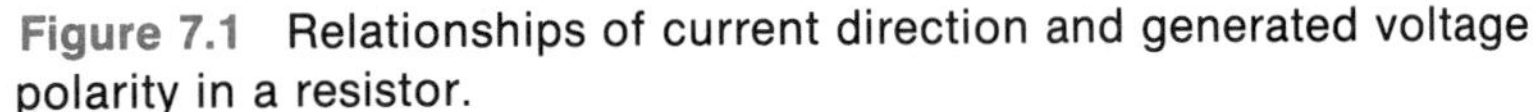

Figure 7.1 Relationships of current direction and generated voltage polarity in a resistor.

A **loop,** or a **mesh,** is a closed path around a circuit.

The rules for obtaining the loop (mesh) equations for a resistive electric circuit with more than one loop follow:

1. Change each current source and its parallel resistor to its corresponding voltage source and its series resistor.
2. Label a loop current in each loop as I_1, I_2, and so on. The direction is arbitrary; however, the equations come out much neater and more symmetrical if all the loop currents are chosen in a clockwise direction.
3. Label a voltage drop across each resistor in the circuit as I_1R_1, I_1R_2, and so on, with the appropriate polarities.
4. Follow the direction of the current in each loop, and for each loop add up all the voltage rises and all the voltage drops. Let the voltage rises equal the voltage drops. Obtain the loop equations by doing the necessary algebra and collecting like terms.

EXAMPLE 7.1

Obtain the loop equations for the circuit shown in Figure 7.2a.

Solution

Figure 7.2b shows the current-to-voltage-source transformation required by Rule 1. Figure 7.2c shows the circuit with the current source and its parallel resistor replaced by its corresponding voltage source and series resistor.

Figure 7.2d shows the circuit with the two necessary loop currents drawn in a clockwise direction and labeled as I_1 and I_2, as required by Rule 2.

Figure 7.2e shows the voltage drop and its proper polarity across each resistor caused by *each* loop current, as required by Rule 3.

Following the direction of I_1 starting at point A there are the following voltage rises around loop 1: 10 V due to the battery and $2I_2$ due to I_2 through the 2-Ω resistor. (The common point A is also called *ground*)

Following the direction of I_1 starting at point A there are the following voltage drops around loop 1: $1I_1$ due to I_1 through the 1-Ω resistor and $2I_1$ due to I_1 through the 2-Ω resistor.

Setting the voltage rises equal to the voltage drops gives

$$10 + 2I_2 = 1I_1 + 2I_1$$

Combining terms, as required by Rule 4:

$$3I_1 - 2I_2 = 10$$

Following the direction of I_2 starting at point A there is the following voltage rise around loop 2: $2I_1$ due to I_1 through the 2-Ω resistor.

Following the direction of I_2 starting at point A there are the follow-

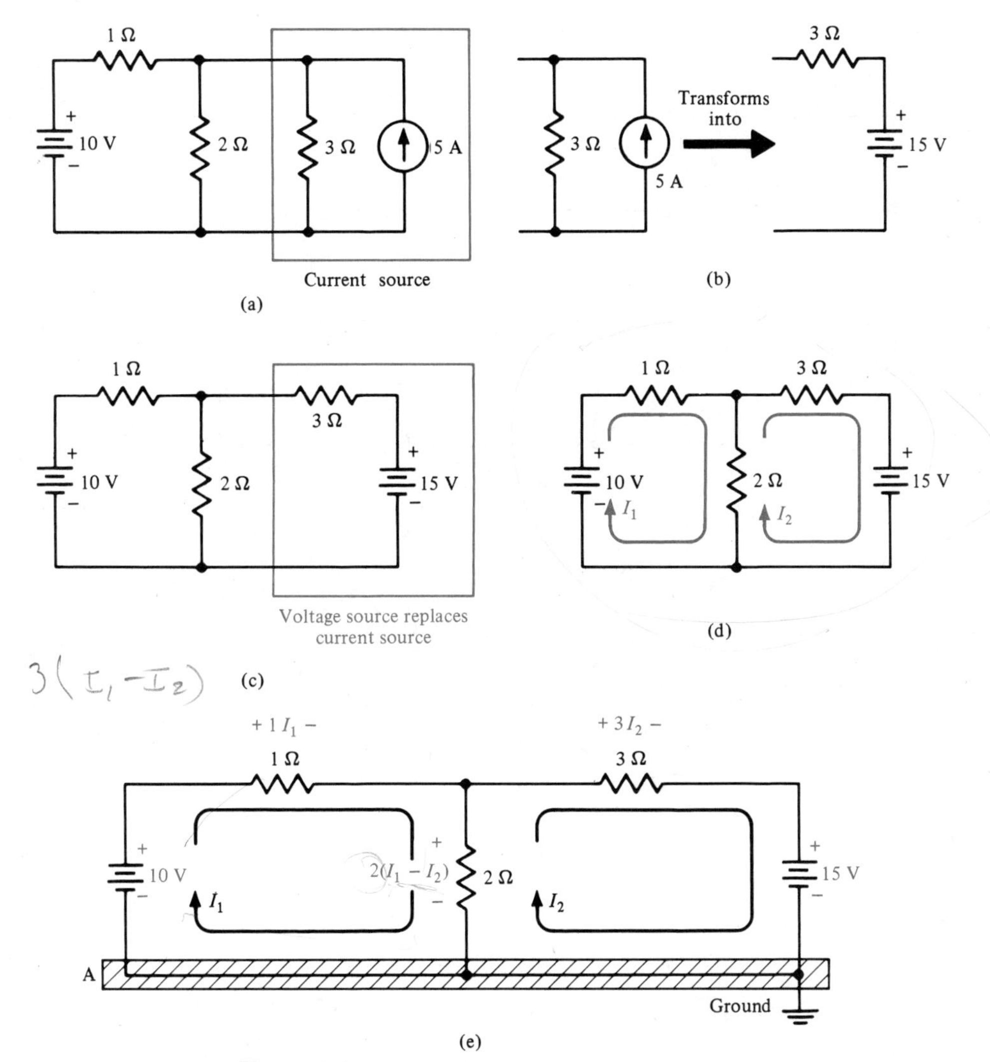

Figure 7.2 Circuits for Examples 7.1 and 7.5.

ing voltage drops around loop 2: $2I_2$ due to I_2 through the 2-Ω resistor, $3I_2$ due to I_2 through the 3-Ω resistor, and 15 V due to the battery.

Setting the voltage rises equal to the voltage drops yields

$$2I_1 = 2I_2 + 3I_2 + 15$$

Combining terms, as required by Rule 4, gives

$$-2I_1 + 5I_2 = -15$$

The two loop equations are, therefore,

$$3I_1 - 2I_2 = 10$$
$$-2I_1 + 5I_2 = -15$$

EXAMPLE 7.2

Obtain the loop equations for the resistive circuit shown in Figure 7.3a.

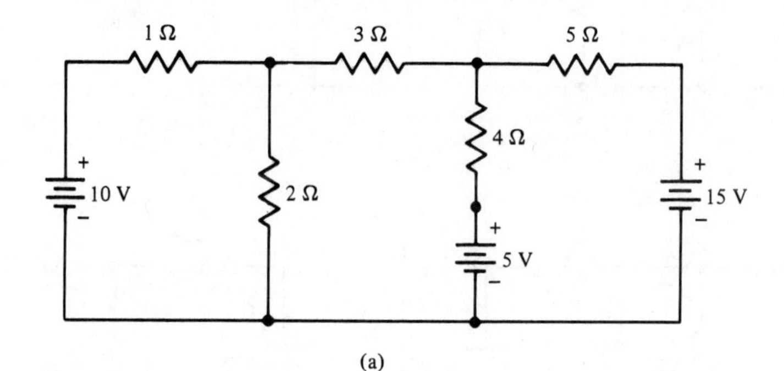

(a)

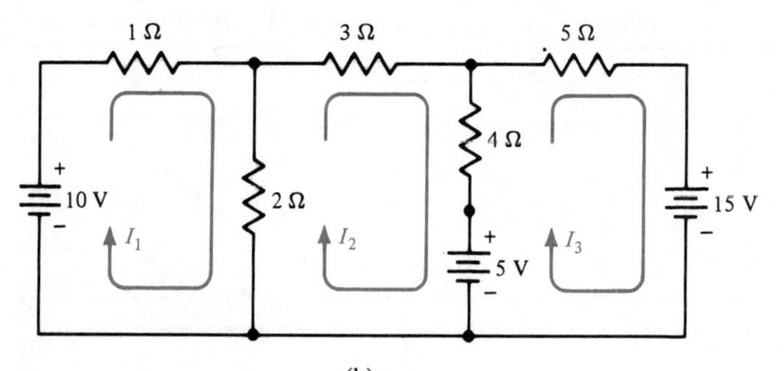

(b)

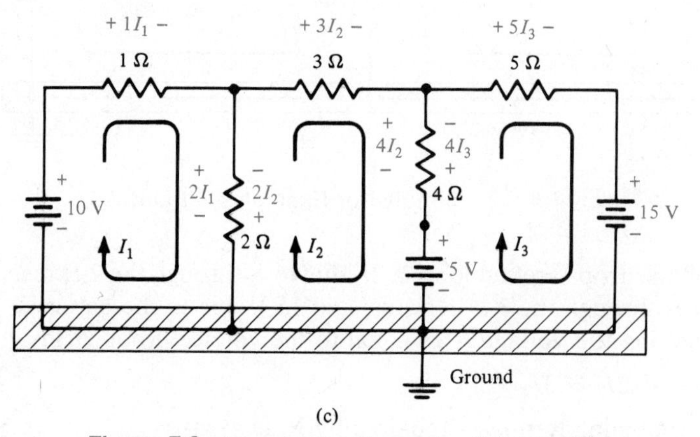

(c)

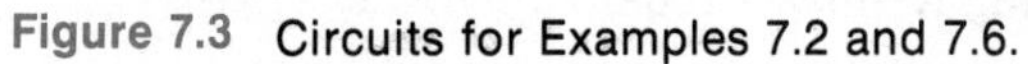

Figure 7.3 Circuits for Examples 7.2 and 7.6.

Solution

Since there are no current sources, there are no conversions necessary.

Figure 7.3b shows the three loop currents drawn clockwise. Figure 7.3c shows the resulting voltage drop due to *each* loop current.

Starting at ground, following the direction of I_1 around loop 1, there are the following voltage rises: 10 V due to the battery and $2I_2$ due to I_2 through the 2-Ω resistor.

Starting at ground, following the direction of I_1 around loop 1, there are the following voltage drops: $1I_1$ due to I_1 through the 1-Ω resistor and $2I_2$ due to I_2 through the 2-Ω resistor.

Setting the voltage rises equal to the voltage drops gives

$$10 + 2I_2 = 1I_1 + 2I_1$$

Collecting terms gives

$$3I_1 - 2I_2 = 10$$

Starting at ground, following the direction of I_2 around loop 2, there are the following voltage rises: $2I_1$ due to I_1 through the 2-Ω resistor and $4I_3$ due to I_3 through the 4-Ω resistor.

Starting at a reference point (usually called ground), following the direction of I_2 around loop 2, there the following voltage drops: $2I_2$ due to I_2 through the 2-Ω resistor, $3I_2$ due to I_2 through the 3-Ω resistor, $4I_2$ due to I_2 through the 4-Ω resistor, and 5 V due to the 5-V battery traversed from + to −.

Setting the voltage rises equal to the voltage drops gives

$$2I_1 + 4I_3 = 2I_2 + 3I_2 + 4I_2 + 5$$

Collecting terms yields

$$-2I_1 + 9I_2 - 4I_3 = -5$$

Starting at ground, following loop current I_3 around loop 3, there are the following voltage rises: 5 V due to the battery and $4I_2$ due to I_2 through the 4-Ω resistor.

Starting at ground, following the direction of I_3 around loop 3, there are the following voltage drops: $4I_3$ due to I_3 through the 4-Ω resistor, $5I_3$ due to I_3 through the 5-Ω resistor, and 15 V due to the battery traversed + to −.

Setting the voltage rises equal to the voltage drops gives

$$5 + 4I_2 = 4I_3 + 5I_3 + 15$$

Collecting terms yields

$$-4I_2 + 9I_3 = -10$$

The three loop equations are:

$$3I_1 - 2I_2 = 10$$

$$-2I_1 + 9I_2 - 4I_3 = -5$$

$$-4I_2 + 9I_3 = -10$$

Inserting the zero terms for the missing currents in the appropriate equations gives

$$3I_1 - 2I_2 + 0I_3 = 10$$
$$-2I_1 + 9I_2 - 4I_3 = -5$$
$$0I_1 - 4I_2 + 9I_3 = -10$$

EXAMPLE 7.3

Obtain the loop equations for the circuit shown in Figure 7.4a.

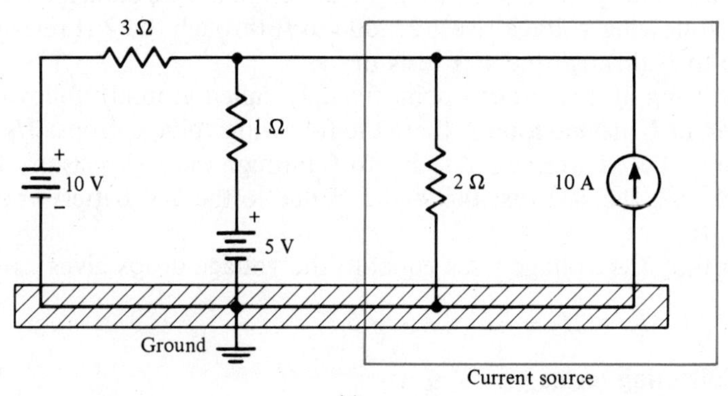

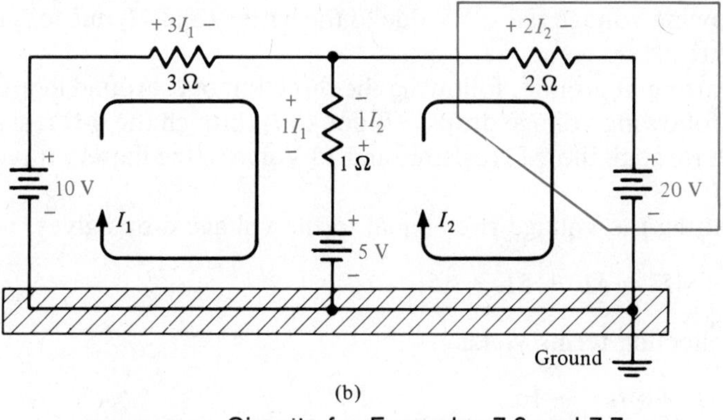

Figure 7.4 Circuits for Examples 7.3 and 7.7.

Solution

Figure 7.4b shows the circuit with the 10-A current source and its parallel 2-Ω resistor replaced by the corresponding 20-V battery and its series 2-Ω resistor. Figure 7.4b also shows loop currents I_1 and I_2 drawn clockwise and the corresponding voltage drop caused by each of the currents through each resistor.

Starting at ground, following the direction of I_1 around loop 1, there are the following voltage rises: 10 V due to the battery and $1I_2$ due to I_2 through the 1-Ω resistor.

Starting at ground, following I_1 around loop 1, there are the following voltage drops: $3I_1$ due to I_1 through the 3-Ω resistor, $1I_1$ due to I_1 through the 1-Ω resistor, and 5 V due to the battery traversed from + to −.

Setting the voltage rises equal to the voltage drops gives

$$10 + 1I_2 = 3I_1 + 1I_1 + 5$$

Collecting terms yields

$$4I_1 - I_2 = 5$$

Starting at ground, following I_2 around loop 2, there are the following voltage rises: 5 V due to the battery traversed from − to + and $1I_1$ due to I_1 through the 1-Ω resistor.

Starting at ground, following I_2 around loop 2, there are the following voltage drops: $1I_2$ due to I_2 through the 1-Ω resistor, $2I_2$ due to I_2 through the 2-ohm resistor, and 20 V due to the battery traversed from + to −.

Setting the voltage rises equal to the voltage drops gives

$$5 + 1I_1 = 1I_2 + 2I_2 + 20$$

Collecting terms yields

$$-1I_1 + 3I_2 = -15$$

The two loop equations are:

$$\begin{aligned} 4I_1 - I_2 &= 5 \\ -1I_1 + 3I_2 &= -15 \end{aligned}$$

EXAMPLE 7.4

Obtain the loop equations for the circuit shown in Figure 7.5a.

Solution

Figure 7.5b shows the conversion of the 10-A, 2-Ω current source to its equivalent 20-V, 2-Ω voltage source. Figure 7.5c shows the circuit with the voltage source and series resistor inserted in the circuit where the current source and parallel resistor were. Figure 7.5c further shows the four loop currents I_1 through I_4 drawn in a clockwise direction. Figure

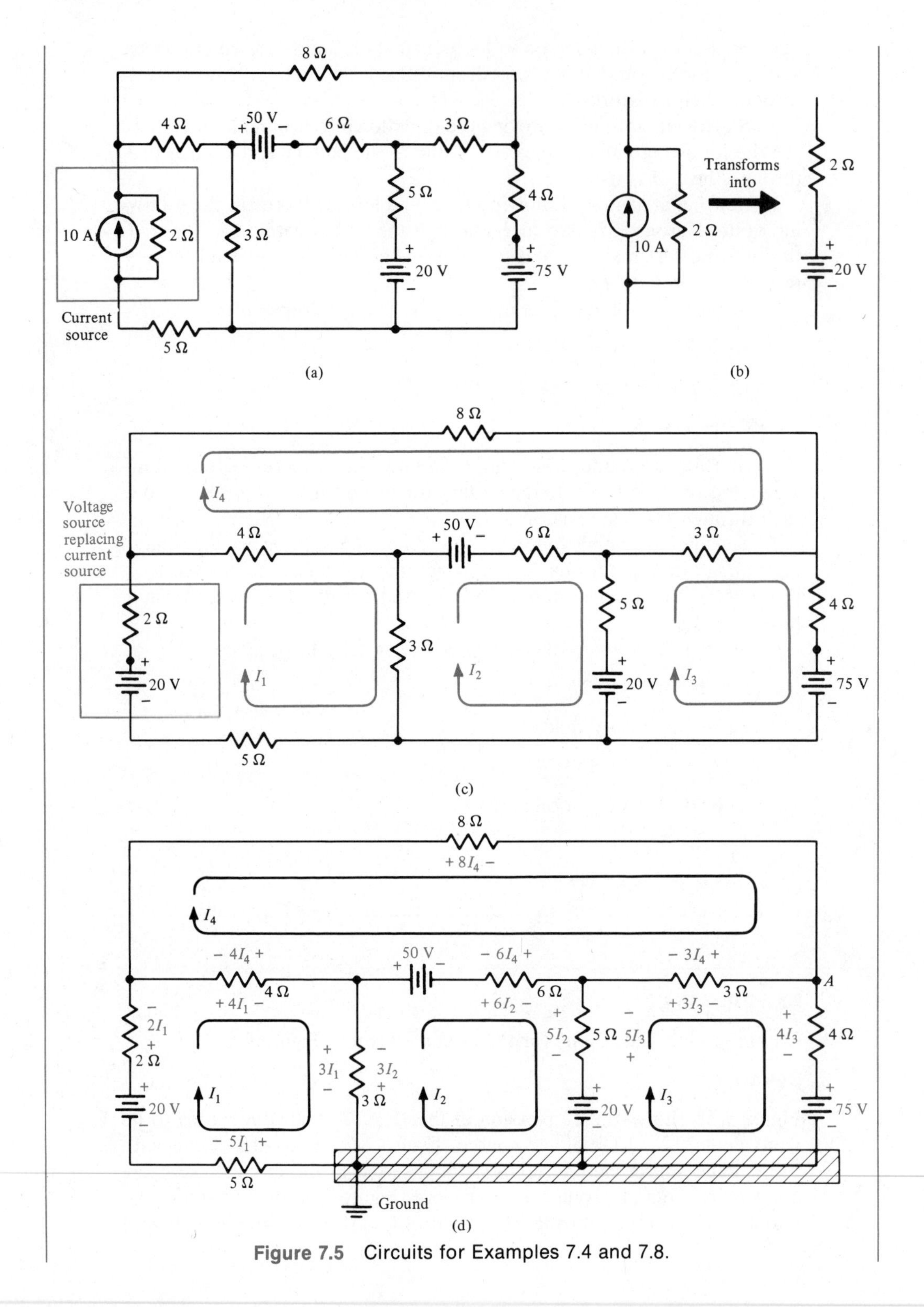

Figure 7.5 Circuits for Examples 7.4 and 7.8.

7.5d shows the circuit with the labeled loop currents and the voltage drops caused by each current through each resistor.

Starting at ground, following I_1 around loop 1, there are the following voltage rises and voltage drops:

Loop 1 voltage rises: 20 V, $4I_4$, and $3I_2$

Loop 1 voltage drops: $5I_1$, $2I_1$, $4I_1$, and $3I_1$

Setting the voltage drops equal to the voltage rises gives

$$20 + 4I_4 + 3I_2 = 5I_1 + 2I_1 + 4I_1 + 3I_1$$

Combining terms yields

$$14I_1 - 3I_2 - 4I_4 = 20$$

Starting at ground, following I_2 around loop 2, there are the following voltage rises and voltage drops:

Loop 2 voltage rises: $3I_1$, $6I_4$, and $5I_3$

Loop 2 voltage drops: $3I_2$, 50 V, $6I_2$, $5I_2$, and 20 V

Setting the voltage rises equal to the voltage drops yields

$$3I_1 + 6I_4 + 5I_3 = 3I_2 + 50 + 6I_2 + 5I_2 + 20$$

Combining terms gives

$$-3I_1 + 14I_2 - 5I_3 - 6I_4 = -70$$

Starting at ground, following I_3 around loop 3, there are the following voltage rises and voltage drops:

Loop 3 voltage rises: 20 V, $5I_2$, and $3I_4$

Loop 3 voltage drops: $5I_3$, $3I_3$, $4I_3$, and 75 V

Setting the voltage rises equal to the voltage drops gives

$$20 + 5I_2 + 3I_4 = 5I_3 + 3I_3 + 4I_3 + 75$$

Combining terms yields

$$-5I_2 + 12I_3 - 3I_4 = -55$$

Starting at point A, following I_4 around loop 4, there are the following voltage rises and voltage drops:

Loop 4 voltage rises: $3I_3$, $6I_2$, 50 V, and $4I_1$

Loop 4 voltage drops: $3I_4$, $6I_4$, $4I_4$, and $8I_4$

Setting the voltage rises equal to the voltage drops yields

$$3I_3 + 6I_2 + 50 + 4I_1 = 3I_4 + 6I_4 + 4I_4 + 8I_4$$

Combining like terms gives

$$-4I_1 - 6I_2 - 3I_3 + 21I_4 = 50$$

The four loop equations, including zero terms for the missing currents, are:

$$14I_1 - 3I_2 - 0I_3 - 4I_4 = 20$$
$$-3I_1 + 14I_2 - 5I_3 - 6I_4 = -70$$
$$0I_1 - 5I_2 + 12I_3 - 3I_4 = -55$$
$$-4I_1 - 6I_2 - 3I_3 + 21I_4 = 50$$

7.3 STANDARD NOTATION FOR OBTAINING LOOP EQUATIONS

The amount of algebra involved just in obtaining the loop equations for an electric circuit can be a cause of error. In order to avoid these errors, it is quite helpful to use the following symbols and systematic method for writing loop equations.

RULE 1. Replace any current sources and their parallel resistors by the equivalent voltage sources and their series resistors.

RULE 2. Label all loop currents in a *clockwise* direction. Label each loop current as I_1, I_2, I_3, and so on.

RULE 3. By inspection write the following set of equations:

$$\begin{aligned} +Z_{11}I_1 - Z_{12}I_2 - Z_{13}I_3 - \cdots - Z_{1N}I_N &= E_1 \\ -Z_{21}I_1 + Z_{22}I_2 - Z_{23}I_3 - \cdots - Z_{2N}I_N &= E_2 \\ &\vdots \\ -Z_{N1}I_1 - Z_{N2}I_2 - Z_{N3}I_3 - \cdots + Z_{NN}I_N &= E_N \end{aligned} \qquad (7.1)$$

This set of simultaneous equations applies to any resistive network with N loops.

DESCRIPTION 1. E_1 is the *total voltage rise* due to batteries in the direction of I_1 in loop 1. Following the direction of the loop current, if the battery is traversed minus to plus ($-$ to $+$), then the battery contributes a voltage rise and the value is positive. Following the direction of the loop current, if the battery is traversed plus to minus ($+$ to $-$), then the battery contributes a voltage drop and the value is negative. E_1 is the algebraic sum of these rises and drops (add the rises, subtract the drops). If there is a voltage source in a branch that is shared by two loops, then it is counted in both loops. The same description applies to voltages E_2 through E_N.

DESCRIPTION 2. Z_{11} is the sum of all the resistors in loop 1. Z_{22} is the sum of all the resistors in loop 2. Z_{NN} is the sum of all the resistors in loop N.

DESCRIPTION 3. Z_{12} is the common resistance between loop 1 and loop 2. Z_{LM} is the resistance common between loop L and loop M. If loops L and M do not touch, then $Z_{LM} = 0$.

DESCRIPTION 4. The set of simultaneous equations is now ready for solution. Use an appropriate method to solve simultaneous equations.

Once the values of the loop currents are obtained, the individual resistor currents are calculated by superimposing the contributing loop currents (magnitude and direction). Notice that the only positive coefficients in the equations are the ones with both subscripts the same (Z_{11} for example).

EXAMPLE 7.5

Use the Standard Notation Method to obtain the loop equations for the circuit shown in Figure 7.2a.

Solution

Apply Rule 1. Figure 7.2b shows how the 5-A current source and its parallel 3-Ω resistor are changed to the corresponding 15-V source and its 3-Ω series resistor. Figure 7.2c shows the circuit with the current source replaced by the voltage source.

Apply Rule 2. Figure 7.2d shows all the loop currents in a clockwise direction.

Apply Description 1:

$$E_1 = +10$$

$$E_2 = -15$$

Apply Description 2:

$$Z_{11} = 1 + 2 = 3$$

$$Z_{22} = 2 + 3 = 5$$

Apply Description 3:

$$Z_{12} = 2$$

$$Z_{21} = 2$$

Apply Rule 3:

$$3I_1 - 2I_2 = 10$$

$$-2I_1 + 5I_2 = -15$$

These are the two loop equations.

EXAMPLE 7.6

Use the Standard Notation Method to obtain the loop equations for the circuit shown in Figure 7.3a.

Solution

Apply Rule 1. Since there are no current sources, the circuit is in the appropriate form for loop analysis.

Apply Rule 2. Figure 7.3b shows the loop currents drawn in a clockwise direction.

Apply Description 1:

$E_1 = 10$

$E_2 = -5$

$E_3 = +5 - 15 = -10$

Apply Description 2:

$Z_{11} = 1 + 2 = 3$

$Z_{22} = 2 + 3 + 4 = 9$

$Z_{33} = 4 + 5 = 9$

Apply Description 3:

$Z_{12} = 2 \quad Z_{21} = 2 \quad Z_{31} = 0$

$Z_{13} = 0 \quad Z_{23} = 4 \quad Z_{32} = 4$

Apply Rule 3:

$$3I_1 - 2I_2 + 0I_3 = 10$$
$$-2I_1 + 9I_2 - 4I_3 = -5$$
$$0I_1 - 4I_2 + 9I_3 = -10$$

These are the required loop equations.

EXAMPLE 7.7

Use the Standard Notation Method to obtain the loop equations for the circuit shown in Figure 7.4a.

Solution

Apply Rule 1. Convert the 10-A, 2-Ω current source to its equivalent 20-V, 2-Ω voltage source, as shown in Figure 7.4b.

Apply Rule 2. Label all loop currents clockwise as shown in Figure 7.4c.

Apply Description 1:

$E_1 = 10 - 5 = 5$

$E_2 = 5 - 20 = -15$

Apply Description 2:

$Z_{11} = 3 + 1 = 4$

$Z_{22} = 1 + 2 = 3$

Apply Description 3:

$Z_{12} = 1$

$Z_{21} = 1$

Apply Rule 3:

$$4I_1 - 1I_2 = 5$$
$$-1I_1 + 3I_2 = -15$$

These are the required loop equations.

EXAMPLE 7.8

Use the Standard Notation Method to obtain the loop equations for the circuit shown in Figure 7.5a.

Solution

Apply Rule 1. Figure 7.5b shows the necessary current-to-voltage-source conversion.

Apply Rule 2. Figure 7.5c shows the circuit with the current source and its parallel resistor replaced by the voltage source and its series resistor. Figure 7.5c also shows the loop currents, all labeled clockwise.

Apply Description 1:

$E_1 = 20$

$E_2 = -50 - 20 = -70$

$E_3 = 20 - 75 = -55$

$E_4 = 50$

Apply Description 2:

$Z_{11} = 5 + 2 + 4 + 3 = 14$

$Z_{22} = 3 + 6 + 5 = 14$

$Z_{33} = 5 + 3 + 4 = 12$

$Z_{44} = 3 + 6 + 4 + 8 = 21$

Apply Description 3:

$Z_{12} = 3 \quad Z_{21} = 3 \quad Z_{31} = 0 \quad Z_{41} = 4$

$Z_{13} = 0 \quad Z_{23} = 5 \quad Z_{32} = 5 \quad Z_{42} = 6$

$Z_{14} = 4 \quad Z_{24} = 6 \quad Z_{34} = 3 \quad Z_{43} = 3$

Apply Rule 3:

$$14I_1 - 3I_2 - 0I_3 - 4I_4 = 20$$

$$-3I_1 + 14I_2 - 5I_3 - 6I_4 = -70$$

$$-0I_1 - 5I_2 + 12I_3 - 3I_4 = -55$$

$$-4I_1 - 6I_2 - 3I_3 + 21I_4 = 50$$

These are the required loop equations.

7.4 CLASSICAL METHOD OF OBTAINING NODE EQUATIONS

The concept of loop currents was used in writing loop equations. KVL was used exclusively because loop currents add up to zero at every node or junction. It was, therefore, not necessary even to consider Kirchhoff's Current Law (KCL).

In the node equation method, a reference node is set up and then all other node voltages are measured with respect to that node. Usually the reference node is called a *ground*. Currents are then assumed at each node and KCL is applied at each node. This results in a set of *node equations*.

Once the node equations are obtained, they may be solved by any of several methods. These are presented later in the chapter.

7.5 NODE EQUATIONS BY KCL

The rules for obtaining the node equations for a resistive electric circuit are as follows:

1. Change each voltage source and its series resistor to its corresponding current source and its parallel resistor.
2. Label one of the nodes as reference (ground) and the rest of the nodes in consecutive order (1, 2, 3, . . .).
3. Change the value of each resistor to its conductance equivalent (change all the ohms to siemens).
4. At each node label currents entering and leaving the node as necessary.
5. At each node apply KCL by setting the currents entering the node equal to the currents leaving the node.

6. Each current may be replaced by its voltage times conductance equivalent. This is obtained by subtracting the voltage of the node to where the current is going from the voltage of the node from which the current is coming and multiplying by the conductance value that the current is traversing.

7. The node equations are obtained by doing the appropriate algebra and collecting terms.

EXAMPLE 7.9

Obtain the node equations for the circuit shown in Figure 7.6a.

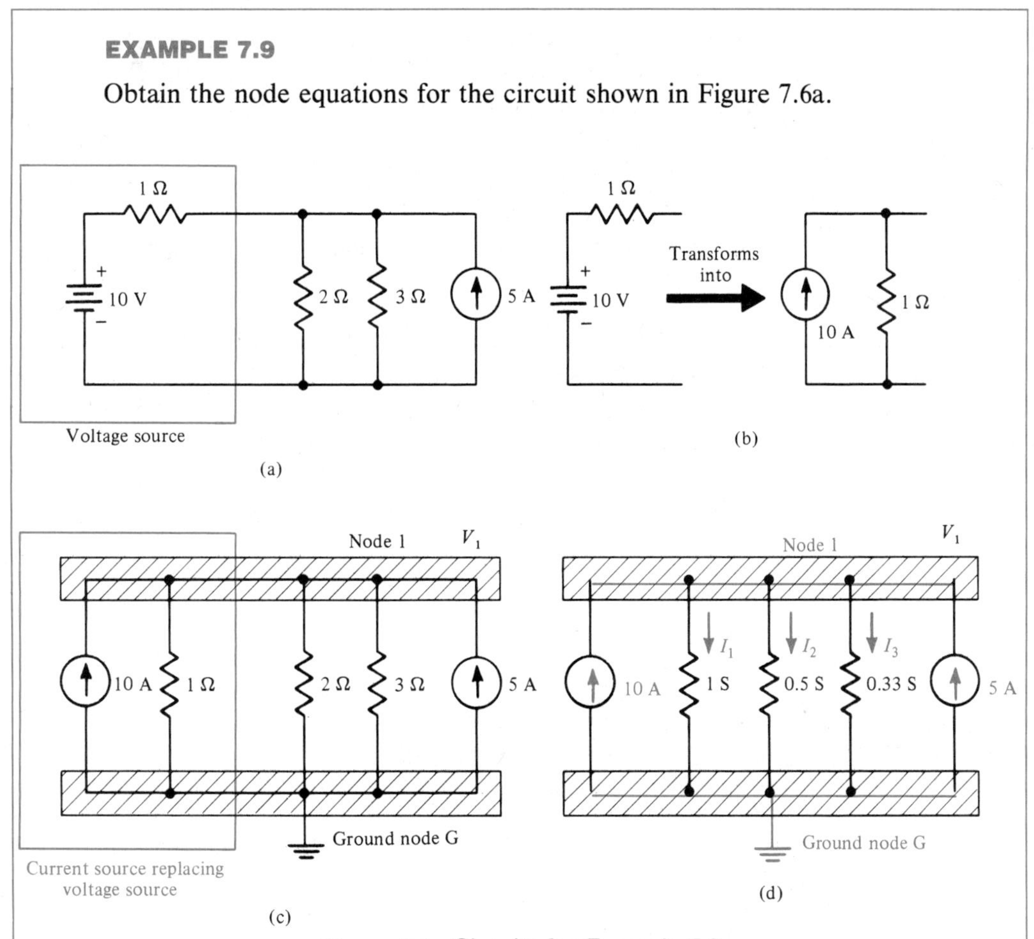

Figure 7.6 Circuits for Example 7.9.

Solution

Apply Rule 1. Figure 7.6b shows the 10-V battery and its 1-Ω series resistor converted to its equivalent 10-A current source and its parallel 1-Ω resistor.

Apply Rule 2. Figure 7.6c shows the circuit with the battery and series resistor combination replaced by the current source parallel resistor combination. Figure 7.6c also shows the reference or ground node labeled as *G* and the ground symbol. Figure 7.6c further shows the only other node (node 1), labeled with the appropriate voltage V_1.

Apply Rule 3. Figure 7.6d shows the circuit with the resistor values replaced by their equivalent conductance values.

Apply Rule 4. Figure 7.6d also shows the currents labeled at node 1. Note that the only currents that had to be labeled were I_1, I_2, and I_3.

Apply Rule 5. The currents entering node 1 are: 10 A due to the current source on the left and 5 A due to the current source on the right. The currents leaving node 1 are: I_1, I_2, and I_3. Setting the entering currents equal to the leaving currents gives

$$10 + 5 = I_1 + I_2 + I_3$$

$$15 = I_1 + I_2 + I_3$$

Apply Rule 6. I_1 is maintained between V_1 and ground through the 1-S conductance. Therefore,

$$I_1 = 1V_1$$

I_2 is maintained between V_1 and ground through the 0.5-S conductance. Therefore,

$$I_2 = 0.5V_1$$

I_3 is maintained between V_1 and ground through the 0.33-S conductance. Therefore,

$$I_3 = 0.33V_1$$

These values may now be substituted into the node equation, with the following result:

$$15 = 1V_1 + 0.5V_1 + 0.33V_1$$

Apply Rule 7. Collecting terms results in:

$$15 = 1.83V_1$$

This is the required node equation. Since there is only one equation, it may be solved directly:

$$V_1 = 8.197 \text{ V}$$

EXAMPLE 7.10

Obtain the node equations for the circuit shown in Figure 7.7a.

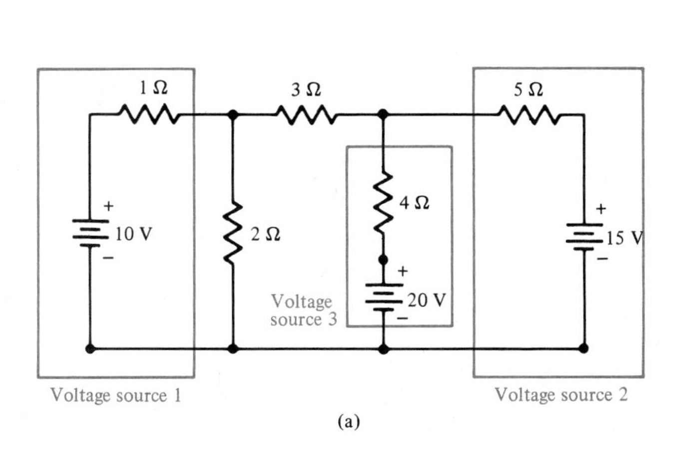

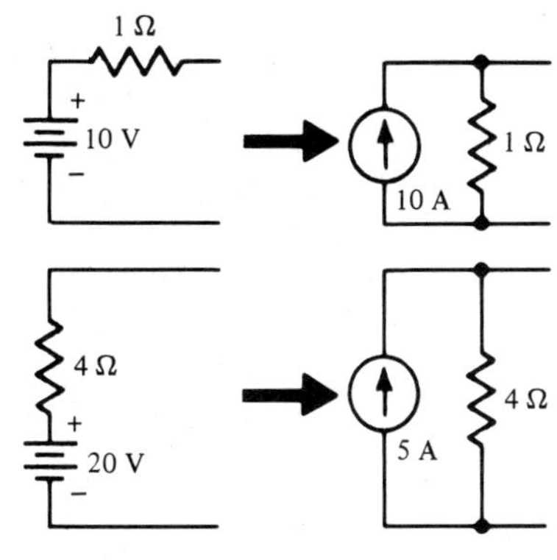

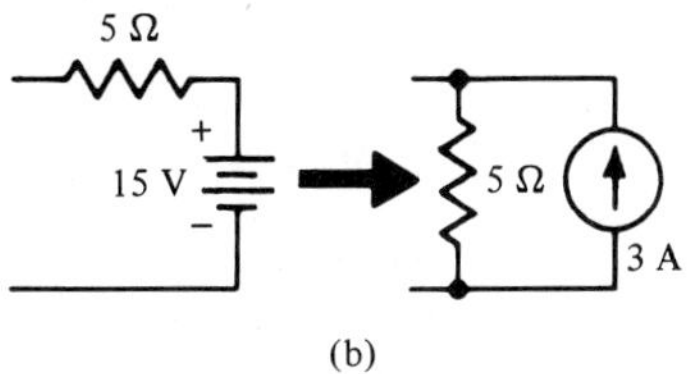

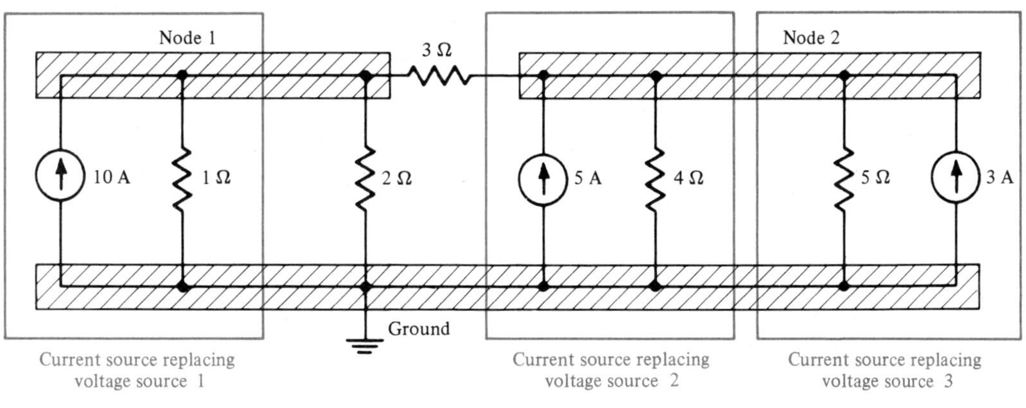

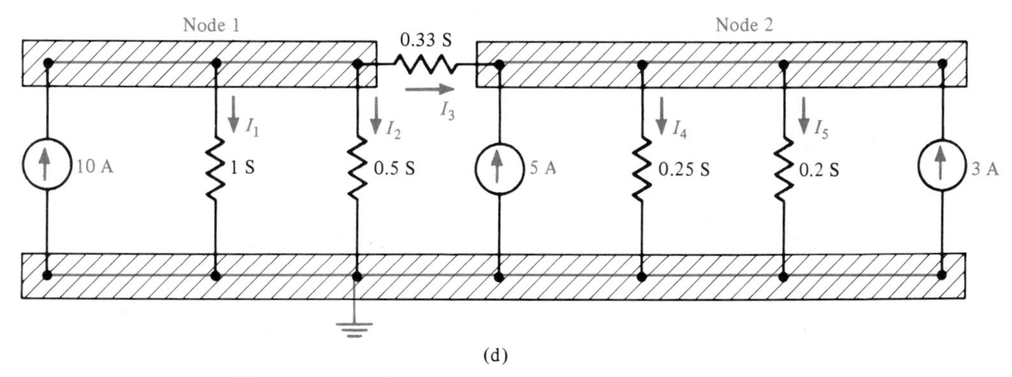

Figure 7.7 Circuits for Examples 7.10 and 7.12.

Solution

Apply Rule 1. Figure 7.7b shows the voltage-to-current-source conversions.

Apply Rule 2. Figure 7.7c shows the circuit with the current sources replacing the voltage sources. Figure 7.7c also shows the ground node labeled with the reference symbol and the other two nodes labeled as node 1 and node 2.

Apply Rules 3 and 4. Figure 7.7d shows the circuit with each resistor represented by its conductance value and the appropriate currents drawn at each node.

Apply Rule 5. At node 1 there is a 10-A current entering the node and I_1, I_2, and I_3 leaving the node. Setting the entering currents equal to the leaving currents gives

$$10 = I_1 + I_2 + I_3$$

At node 2 there are a 5-A current, a 3-A current and I_3 entering the node. I_4 and I_5 are leaving the node. Setting the entering currents equal to the leaving currents gives

$$5 + 3 + I_3 = I_4 + I_5$$

$$8 + I_3 = I_4 + I_5$$

Apply Rule 6. $I_1 = 1V_1$, since it is maintained between V_1 and ground through a 1-S conductance.

$I_2 = 0.5V_1$, since it is maintained between V_1 and ground through a 0.5-S conductance.

$I_3 = 0.33(V_1 - V_2)$, since it is maintained between V_1 and V_2 through a 0.33-S conductance.

$I_4 = 0.25V_2$, since it is maintained between V_2 and ground through a 0.25-S conductance.

$I_5 = 0.2V_2$, since it is maintained between V_2 and ground through a 0.2-S conductance.

Substituting these into the original equations gives

$$10 = 1V_1 + 0.5V_1 + 0.33(V_1 - V_2)$$

$$5 + 0.33(V_1 - V_2) = 0.25V_2 + 0.2V_2$$

Apply Rule 7. Carrying out the algebra and collecting terms results in the following node equations:

$$1.83V_1 - 0.33V_2 = 10$$

$$-0.33V_1 + 0.78V_2 = 8$$

EXAMPLE 7.11

Obtain the node equations for the circuit shown in Figure 7.8a.

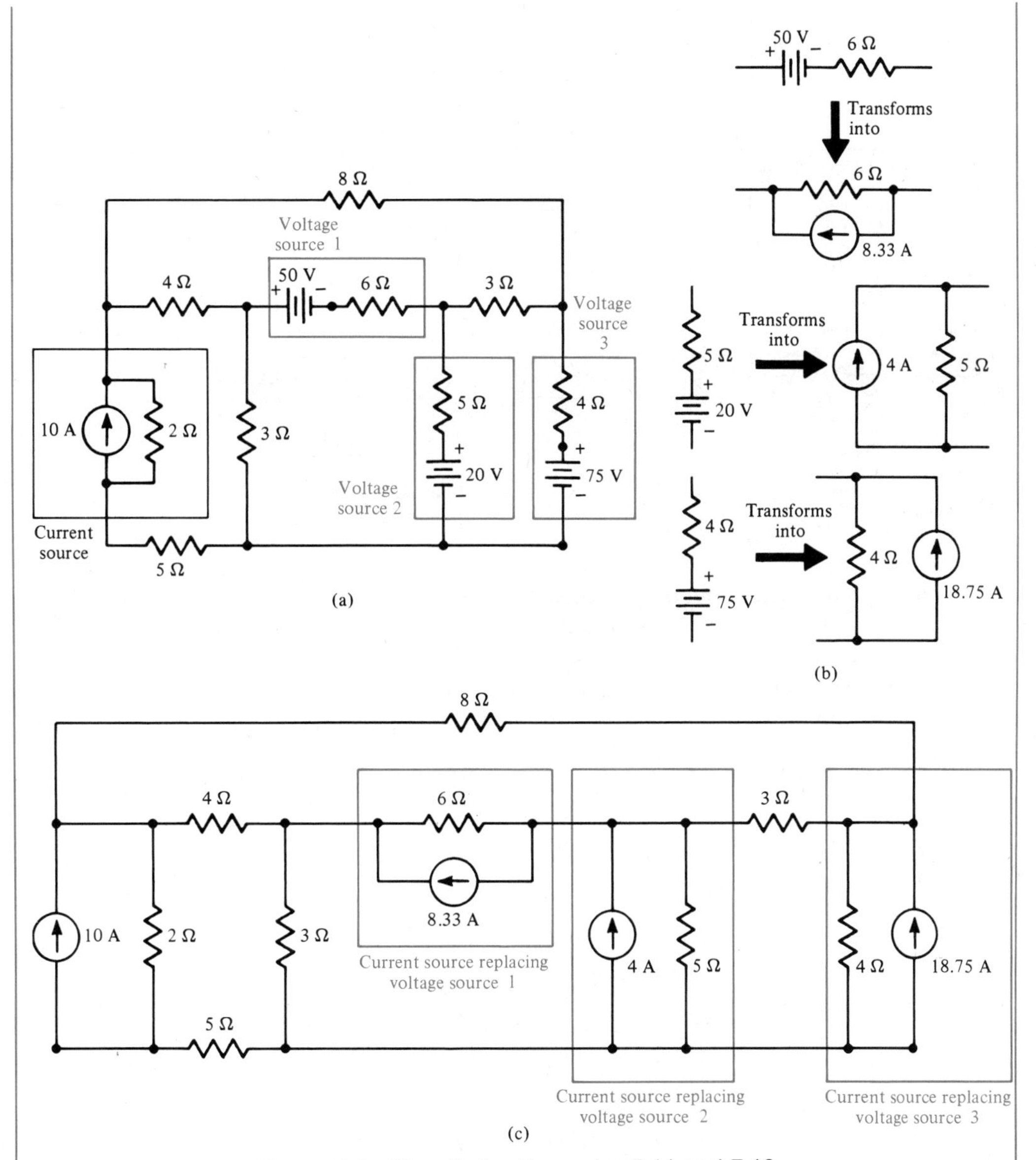

Figure 7.8 Circuits for Examples 7.11 and 7.12.

Solution

Apply Rule 1. Figure 7.8b shows the transformation of each voltage source and its series resistor to the equivalent current source and its parallel resistor.

Apply Rule 2. Figure 7.8c shows the circuit with the voltage sources and series resistors replaced by the equivalent current sources and parallel resistors. Figure 7.8 shows the node that was selected as reference

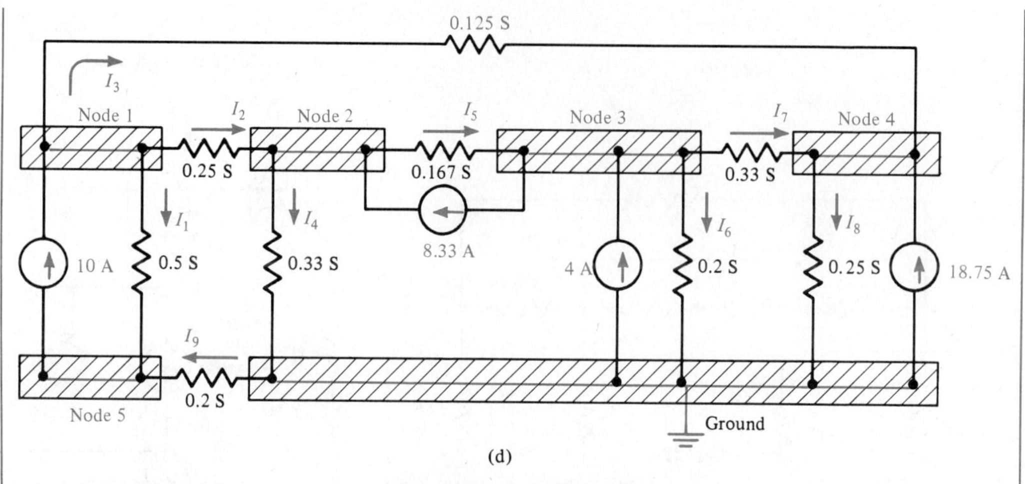

(d)

Figure 7.8 (continued)

labeled with the ground symbol and the other five nodes, labeled node 1, node 2, node 3, node 4, and node 5.

Apply Rules 3 and 4. Figure 7.8d shows the circuit with each resistor value replaced by its equivalent conductance value. Figure 7.8d further shows the currents chosen entering and leaving each node.

At node 1:	*currents entering*	10 A
	currents leaving	I_1, I_2, and I_3
At node 2:	*currents entering*	8.33 A and I_2
	currents leaving	I_4 and I_5
At node 3:	*currents entering*	4 A and I_5
	currents leaving	8.33 A, I_6, and I_7
At node 4:	*currents entering*	18.75 A, I_3, and I_7
	currents leaving	I_8
At node 5:	*currents entering*	I_1 and I_9
	currents leaving	10 A

Setting the entering currents equal to the leaving currents at each node gives the following.

At node 1: $$10 = I_1 + I_2 + I_3$$

At node 2: $$8.33 + I_2 = I_4 + I_5$$

At node 3: $$4 + I_5 = 8.33 + I_6 + I_7$$

At node 4: $$18.75 + I_3 + I_7 = I_8$$

At node 5: $I_1 + I_9 = 10$ A

Apply Rule 6:

$I_1 = 0.5(V_1 - V_5)$

$I_2 = 0.25(V_1 - V_2)$

$I_3 = 0.125(V_1 - V_4)$

$I_4 = 0.33V_2$

$I_5 = 0.167(V_2 - V_3)$

$I_6 = 0.2V_3$

$I_7 = 0.33(V_3 - V_4)$

$I_8 = 0.25V_4$

$I_9 = 0.2(0 - V_5) = -0.2V_5$

Apply Rule 7. Substituting these into the original equations gives:

At node 1:

$10 = 0.5(V_1 - V_5) + 0.25(V_1 - V_2) + 0.125(V_1 - V_5)$

At node 2:

$8.33 + 0.25(V_1 - V_2) = 0.333V_2 + 0.167(V_2 - V_3)$

At node 3:

$4 + 0.167(V_2 - V_3) = 8.33 + 0.2V_3 + 0.333(V_3 - V_4)$

At node 4:

$18.75 + 0.125(V_1 - V_4) + 0.333(V_3 - V_4) = 0.25V_4$

At node 5:

$0.125(V_1 - V_4) - 0.2V_5 = 10$

Carrying out the proper algebra and collecting terms, the required equations are:

$$0.875V_1 - 0.25V_2 - 0V_3 - 0.125V_4 - 0.5V_5 = 10$$
$$-0.25V_1 + 0.75V_2 - 0.167V_3 - 0V_4 - 0V_5 = 8.33$$
$$0V_1 - 0.167V_2 + 0.7V_3 - 0.333V_4 - 0V_5 = -4.33$$
$$-0.125V_1 - 0V_2 - 0.333V_3 + 0.708V_4 - 0V_5 = 18.75$$
$$-0.5V_1 - 0V_2 - 0V_3 - 0V_4 + 0.7V_5 = -10$$

These are the required node equations.

7.6 STANDARD NOTATION FOR OBTAINING NODE EQUATIONS

The amount of algebra involved in obtaining the node equations for an electric circuit can be a cause of errors. It is quite helpful to use the following symbols and systematic method for writing node equations.

RULE 1. Replace each voltage source and its series resistor with the equivalent current source and its parallel resistor. Replace all resistor values by the equivalent conductance values ($G = 1/R$).

RULE 2. Choose a reference node and label it with the ground symbol. This node will have 0 voltage, and all the other node voltages will be measured with respect to this node. Label all the other nodes as 1, 2, 3, and so on. The reference node is usually the one with the most connections.

RULE 3. By inspection write the following set of node equations:

$$
\begin{aligned}
Y_{11}V_1 - Y_{12}V_2 - \cdots - Y_{1N}V_N &= I_1 \\
-Y_{21}V_1 + Y_{22}V_2 - \cdots - Y_{2N}V_N &= I_2 \\
&\vdots \\
-Y_{N1}V_1 - Y_{N2}V_2 - \cdots + Y_{NN}V_N &= I_N
\end{aligned}
\tag{7.2}
$$

This set of simultaneous node equations applies to any resistive network with one reference node and N additional nodes.

DESCRIPTION 1. I_1 is the algebraic sum of the current sources connected to node 1. Current sources supplying current to the node are positive. Current sources withdrawing current from the node are negative. I_2 is the algebraic sum of the current sources connected to node 2. I_N is the algebraic sum of the current sources connected to node N.

DESCRIPTION 2. Y_{11} is the sum of all the conductances connected to node 1. Y_{22} is the sum of all the conductances connected to node 2. Y_{NN} is the sum of all the conductances connected to node N.

DESCRIPTION 3. Y_{12} is the conductance between node 1 and node 2. Y_{N2} is the conductance between node N and node 2. If there is no conductance element between two nodes (say A and B), then $Y_{AB} = 0$.

RULE 4. The set of node equations is now ready for solution. The results yield the voltages at the respective nodes with respect to the reference node.

EXAMPLE 7.12

Using the Standard Notation Method, obtain the node equations for the circuit shown in Figure 7.7a.

Solution

Apply Rule 1. The circuit in Figure 7.7b shows the conversion of each voltage source and its series resistor to the equivalent current source and its parallel resistor.

Apply Rule 2. The circuit in Figure 7.7c shows the current sources replacing the voltage sources. Figure 7.7c also shows the other two nodes labeled as node 1 and node 2. Figure 7.7d shows the circuit with all the nodes labeled and each resistive resistor value replaced by the equivalent conductance value.

Apply Description 1:

$I_1 = 10$ due to the 10-A current source entering node 1

$I_2 = 5 + 3 = 8$ due to the 5-A and the 3-A current sources both entering node 2

Apply Description 2:

$Y_{11} = 1 + 0.5 + 0.33 = 1.83$ sum of all the conductances connected to node 1

$Y_{22} = 0.33 + 0.25 + 0.2 = 0.78$ sum of all the conductances connected to node 2

Apply Description 3:

$Y_{12} = 0.33$ the conductance connected between node 1 and node 2

$Y_{21} = 0.33$ the conductance connected between node 2 and node 1

Apply Rule 3:

$$1.83V_1 - 0.33V_2 = 10$$

$$-0.33V_1 + 0.78V_2 = 8$$

These are the required node equations.

EXAMPLE 7.13

Using the Standard Notation Method, obtain the node equations for the circuit shown in Figure 7.8a.

Solution

Apply Rule 1. The circuits in Figure 7.8b show the conversion of each voltage source and its series resistor to the equivalent current source and its parallel resistor. Figure 7.8c shows the circuit with the current sources replacing the original voltage sources.

Apply Rule 2. Figure 7.8d shows the circuit with each resistance value replaced by the equivalent conductance value. Figure 7.8d also shows the reference node labeled as *G* and the ground symbol. Figure 7.8d further shows the nodes 1 through 5 labeled as such.

Apply Description 1.

$I_1 = 10$ due to the 10-A current source into node 1

$I_2 = 8.33$ due to the 8.33-A current source into node 2

$I_3 = 4 - 8.33 = -4.33$ due to the 4-A current source into node 3 and the 8.33-A current source out of node 3

$I_4 = 18.75$ due to the 18.75-A current source into node 4

$I_5 = -10$ due to the 10-A source out of node 5

Apply Description 2.

$Y_{11} = 0.5 + 0.25 + 0.125 = 0.875$ sum of all the conductances connected to node 1

$Y_{22} = 0.25 + 0.333 + 0.167 = 0.750$ sum of all the conductances connected to node 2

$Y_{33} = 0.167 + 0.2 + 0.333 = 0.7$ sum of all the conductances connected to node 3

$Y_{44} = 0.125 + 0.333 + 0.25 = 0.708$ sum of all the conductances connected to node 4

$Y_{55} = 0.5 + 0.2 = 0.7$ sum of all the conductances connected to node 5

Apply Description 3.

$Y_{12} = 0.25$ $Y_{21} = 0.25$ $Y_{31} = 0$ $Y_{41} = 0.125$ $Y_{51} = 0.5$

$Y_{13} = 0$ $Y_{23} = 0.167$ $Y_{32} = 0.167$ $Y_{42} = 0$ $Y_{52} = 0$

$Y_{14} = 0.125$ $Y_{24} = 0$ $Y_{34} = 0.333$ $Y_{43} = 0.333$ $Y_{53} = 0$

$Y_{15} = 0.5$ $Y_{25} = 0$ $Y_{35} = 0$ $Y_{45} = 0$ $Y_{54} = 0$

Apply Rule 3.

$$0.875V_1 - 0.25V_2 - 0V_3 - 0.125V_4 - 0.5V_5 = 10$$

$$-0.25V_1 + 0.75V_2 - 0.167V_3 - 0V_4 - 0V_5 = 8.33$$

$$0V_1 - 0.167V_2 + 0.7V_3 - 0.333V_4 - 0V_5 = -4.33$$

$$-0.125V_1 - 0V_2 - 0.333V_3 + 0.708V_4 - 0V_5 = 18.75$$

$$-0.5V_1 - 0V_2 - 0V_3 - 0V_4 + 0.7V_5 = -10$$

These are the required node equations. They may be solved to yield the respective node voltages with reference to ground.

7.7 SOLUTION OF SIMULTANEOUS EQUATIONS

The loop and node equations that arise from the analysis of resistive electrical circuits with all component values known are usually solvable, because they contain as many equations as they have unknowns. There are three popular methods of solving these types of equations. These are (1) Cramer's Rule, (2) Gauss' Elimination Method, and (3) matrix manipulation methods. In this book, the first two methods are considered. Both hand solutions and computer solutions will be offered.

When the solution is attempted using pencil and paper or pocket calculator, a system of two or three equations may be readily solved by Cramer's Rule. If the number of equations increases to four or more, then evaluation by Cramer's Rule becomes very lengthy and tedious. Many mathematicians have tried to develop simpler and less time-consuming methods. One of the most popular of these methods is Gauss' Elimination Method. This particular method adapts itself best to computer solution. Both Cramer's Rule and Gauss' Elimination Method are presented as pencil-and-paper computation methods. Gauss' Elimination Method is used for the computer solution.

7.8 2 × 2 DETERMINANTS AND CRAMER'S RULE

Two simultaneous equations with two variables have the form:

$$a_{11}x_1 + a_{12}x_2 = y_1$$

$$a_{21}x_1 + a_{22}x_2 = y_2$$

The system **determinant** is called D and is defined as

$$D = \begin{vmatrix} a_{11} & a_{12} \\ a_{21} & a_{22} \end{vmatrix}$$

The value of such a 2 × 2 determinant is:

$$D = a_{11}a_{22} - a_{12}a_{21}$$

The source column is called S and is defined as

$$S = \begin{vmatrix} y_1 \\ y_2 \end{vmatrix}$$

The particular determinant D_1 is obtained by substituting the source column for the first column in the system determinant:

$$D_1 = \begin{vmatrix} y_1 & a_{12} \\ y_2 & a_{22} \end{vmatrix}$$

The value of D_1 is:

$$D_1 = y_1a_{22} - y_2a_{12}$$

The particular determinant D_2 is obtained by substituting the source column for the second column in the system determinant:

$$D_2 = \begin{vmatrix} a_{11} & y_1 \\ a_{21} & y_2 \end{vmatrix}$$

The value of D_2 is:

$$D_2 = a_{11}y_2 - a_{21}y_1$$

Cramer's Rule then gives

$$x_1 = \frac{D_1}{D} \qquad x_2 = \frac{D_2}{D}$$

EXAMPLE 7.14

a. Use Cramer's Rule to solve the equations obtained in Example 7.1 for I_1 and I_2.
b. Determine the voltage drop across each resistor.

Solution

a. The equations are:

$$3I_1 - 2I_2 = 10$$
$$-2I_1 + 5I_2 = -15$$

$$D = \begin{vmatrix} 3 & -2 \\ -2 & 5 \end{vmatrix} \qquad \begin{aligned} D &= (3)(5) - (-2)(-2) \\ &= 15 - 4 \\ &= 11 \end{aligned}$$

$$D_1 = \begin{vmatrix} 10 & -2 \\ -15 & 5 \end{vmatrix} \qquad \begin{aligned} D_1 &= (10)(5) - (-15)(-2) \\ &= 50 - 30 \\ &= 20 \end{aligned}$$

$$D_2 = \begin{vmatrix} 3 & 10 \\ -2 & -15 \end{vmatrix} \qquad \begin{aligned} D_2 &= (3)(-15) - (-2)(10) \\ &= -45 + 20 \\ &= -25 \end{aligned}$$

$$I_1 = \frac{D_1}{D} = \frac{20}{11} = 1.82 \text{ A}, \qquad I_2 = \frac{D_2}{D} = \frac{-25}{11} = -2.27 \text{ A}$$

The positive value for I_1 indicates that the direction originally assumed for the current is correct.

The negative value of I_2 indicates that the direction originally assumed for I_2 is not correct. The current direction must therefore be reversed.

Figure 7.9a shows the circuit with the original currents assumed to be clockwise. Figure 7.9b shows the circuit with the resulting currents included. Figure 7.9c shows current I_2 drawn

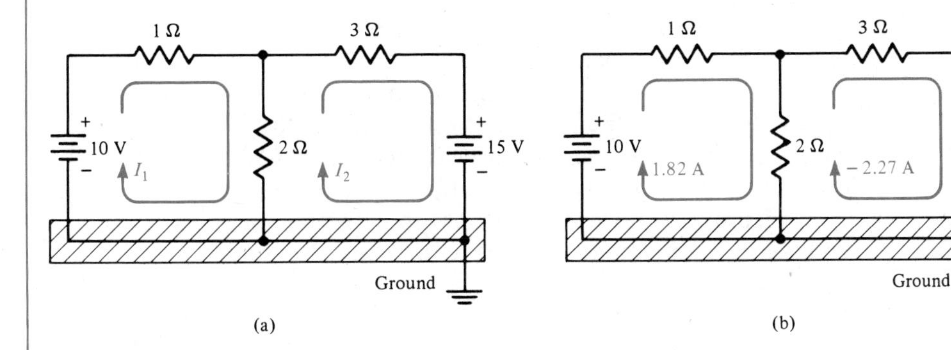

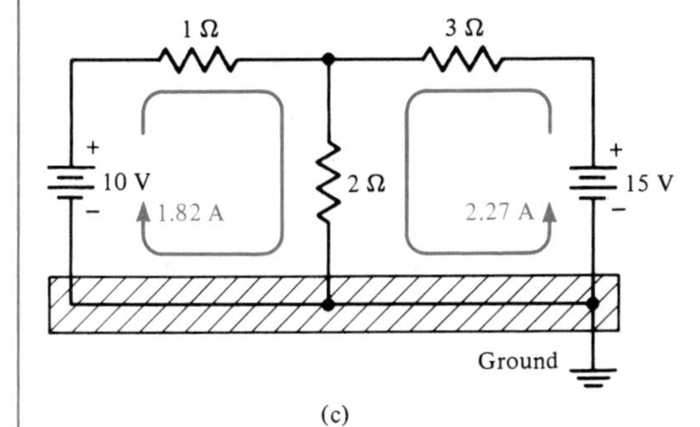

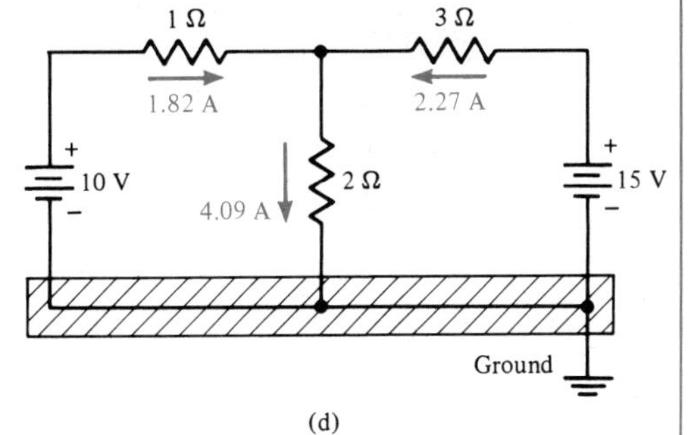

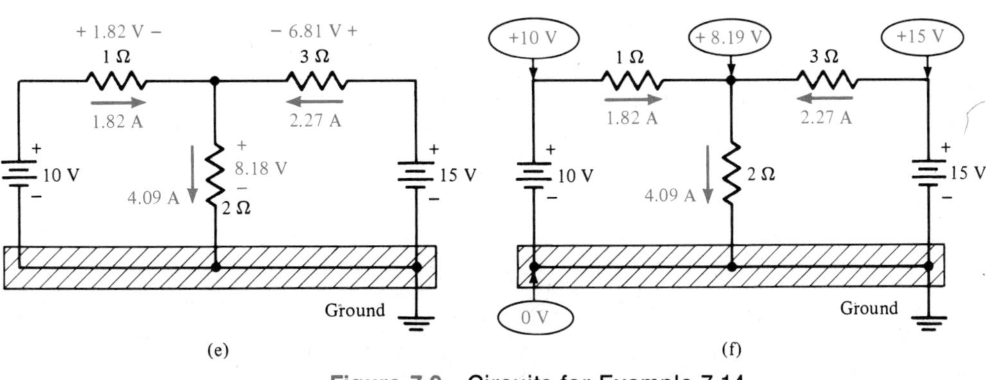

Figure 7.9 Circuits for Example 7.14.

in the opposite direction to the original assumption because the answer came out negative.

b. The current through the individual resistors may now be computed. The 1-Ω resistor has a 1.82-A current through it from right to left. The 3-Ω resistor has 2.27 A through it from right to left. The 2-Ω resistor has a 1.82-A current and a 2.27-A current down through it. Since they are in the same direction, these two currents add, yielding a 4.09-A current down through the 2-Ω resistor. Figure 7.9d shows the circuit with the individual resistor currents.

Each one of these currents through its resistor causes a voltage drop across that resistor. 1.82 A through the 1-Ω resistor causes a 1.82-V drop. 2.27 A through the 3-Ω resistor causes a 6.81-V drop across the resistor. 4.09 A through the 2-Ω resistor causes a 8.18-V drop across the resistor. Figure 7.9e shows the circuit with these voltage drops and their polarities.

Figure 7.9f shows the circuit with currents, voltage drops, and the individual node voltages with respect to ground.

7.9 3 × 3 DETERMINANTS AND CRAMER'S RULE

Three simultaneous equations with three variables have the form

$$a_{11}x_1 + a_{12}x_2 + a_{13}x_3 = y_1$$

$$a_{21}x_1 + a_{22}x_2 + a_{23}x_3 = y_2$$

$$a_{31}x_1 + a_{32}x_2 + a_{33}x_3 = y_3$$

The system determinant is called D and is defined as

$$D = \begin{vmatrix} a_{11} & a_{12} & a_{13} \\ a_{21} & a_{22} & a_{23} \\ a_{31} & a_{32} & a_{33} \end{vmatrix}$$

The easiest way to solve a 3 × 3 determinant is as follows:

1. Reproduce the first two columns and draw the indicated arrows.

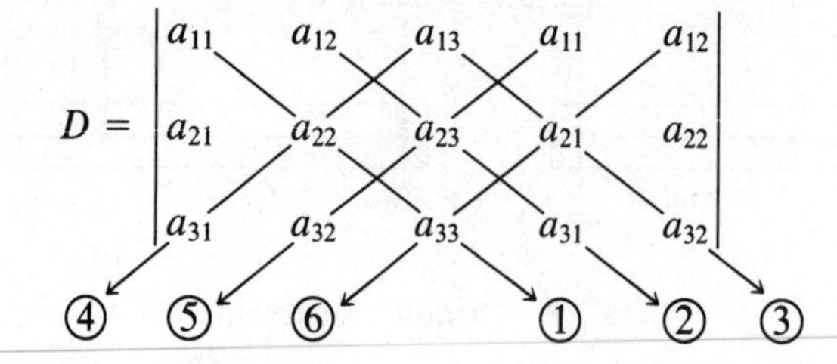

2. Multiply the elements that each arrow goes through:

Arrow 1 = $a_{11}a_{22}a_{33}$

Arrow 2 = $a_{12}a_{23}a_{31}$

Arrow 3 = $a_{13}a_{21}a_{32}$

Arrow 4 = $a_{13}a_{32}a_{31}$

Arrow 5 = $a_{11}a_{23}a_{32}$

Arrow 6 = $a_{12}a_{21}a_{33}$

3. Add the results of arrows 1, 2, and 3 and call it A:

A = arrow 1 + arrow 2 + arrow 3

Add the results of arrows 4, 5, and 6 and call it B:

B = arrow 4 + arrow 5 + arrow 6

4. The value of the determinant then is:

$D = A - B$

The source column is called S and is defined as

$$S = \begin{vmatrix} y_1 \\ y_2 \\ y_3 \end{vmatrix}$$

The particular determinant D_1, is obtained by substituting the source column for column 1 in the system determinant:

$$D_1 = \begin{vmatrix} y_1 & a_{12} & a_{13} \\ y_2 & a_{22} & a_{32} \\ y_3 & a_{32} & a_{33} \end{vmatrix}$$

D_2 is obtained by substituting the source column for column 2:

$$D_2 = \begin{vmatrix} a_{11} & y_1 & a_{13} \\ a_{21} & y_2 & a_{23} \\ a_{31} & y_3 & a_{33} \end{vmatrix}$$

D_3 is obtained by substituting the source column for the third column:

$$D_3 = \begin{vmatrix} a_{11} & a_{12} & y_1 \\ a_{21} & a_{22} & y_2 \\ a_{31} & a_{32} & y_3 \end{vmatrix}$$

Each of these 3 × 3 determinants may be solved as described earlier. Cramer's Rule then gives the solution:

$$I_1 = \frac{D_1}{D} \qquad I_2 = \frac{D_2}{D} \qquad I_3 = \frac{D_3}{D}$$

EXAMPLE 7.15

Use Cramer's Rule to solve the loop equations obtained in Examples 7.2 and 7.6.

Solution

The equations are:

$$3I_1 - 2I_2 - 0I_3 = 10$$

$$-2I_1 + 9I_2 - 4I_3 = -5$$

$$0I_1 - 4I_2 + 9I_3 = -10$$

The system determinant is:

$$D = \begin{vmatrix} 3 & -2 & 0 \\ -2 & 9 & -4 \\ 0 & -4 & 9 \end{vmatrix} \qquad S = \begin{vmatrix} 10 \\ -5 \\ -10 \end{vmatrix}$$

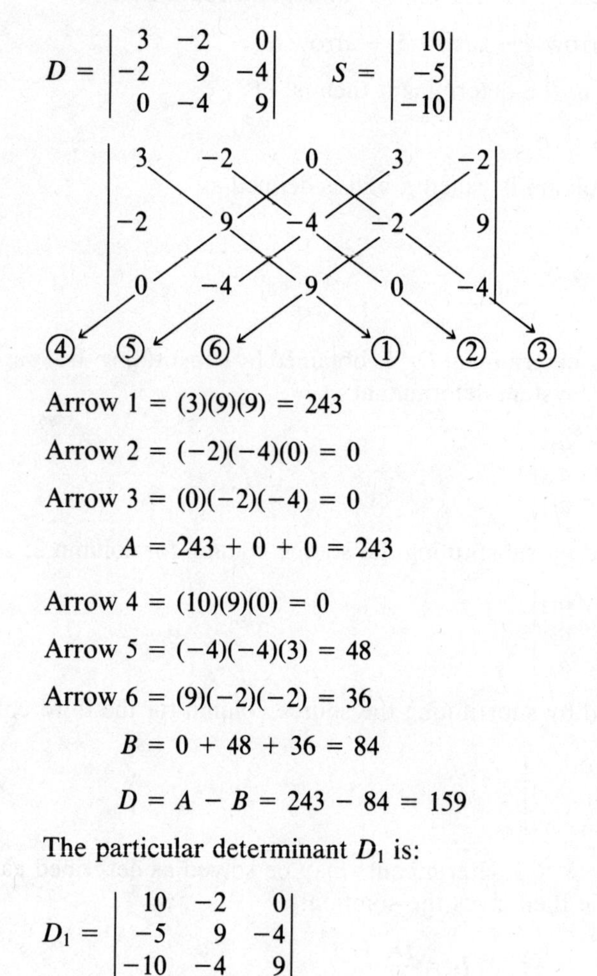

Arrow 1 = (3)(9)(9) = 243

Arrow 2 = (−2)(−4)(0) = 0

Arrow 3 = (0)(−2)(−4) = 0

$$A = 243 + 0 + 0 = 243$$

Arrow 4 = (10)(9)(0) = 0

Arrow 5 = (−4)(−4)(3) = 48

Arrow 6 = (9)(−2)(−2) = 36

$$B = 0 + 48 + 36 = 84$$

$$D = A - B = 243 - 84 = 159$$

The particular determinant D_1 is:

$$D_1 = \begin{vmatrix} 10 & -2 & 0 \\ -5 & 9 & -4 \\ -10 & -4 & 9 \end{vmatrix}$$

Its solution is:

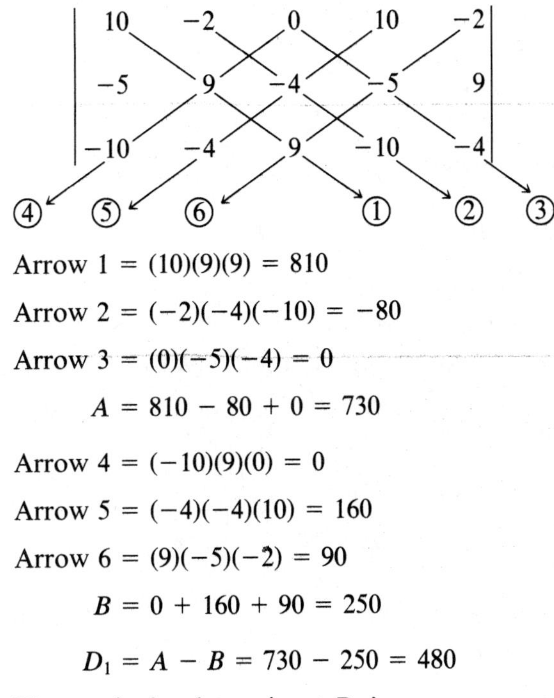

Arrow 1 = (10)(9)(9) = 810

Arrow 2 = (−2)(−4)(−10) = −80

Arrow 3 = (0)(−5)(−4) = 0

$A = 810 - 80 + 0 = 730$

Arrow 4 = (−10)(9)(0) = 0

Arrow 5 = (−4)(−4)(10) = 160

Arrow 6 = (9)(−5)(−2) = 90

$B = 0 + 160 + 90 = 250$

$D_1 = A - B = 730 - 250 = 480$

The particular determinant D_2 is:

$$D_2 = \begin{vmatrix} 3 & 10 & 0 \\ -2 & -5 & -4 \\ 0 & -10 & 9 \end{vmatrix}$$

The solution for D_2 is:

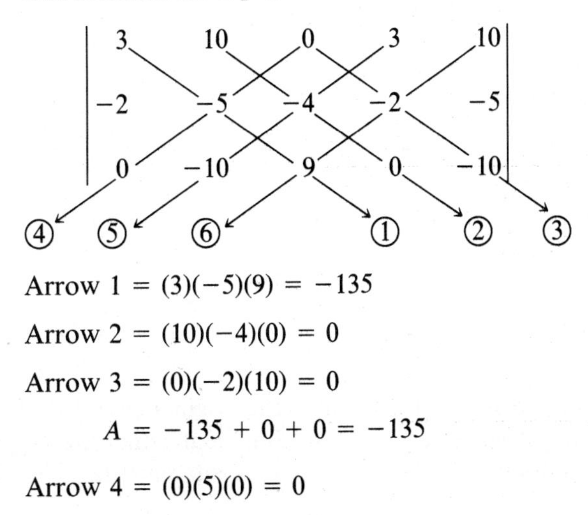

Arrow 1 = (3)(−5)(9) = −135

Arrow 2 = (10)(−4)(0) = 0

Arrow 3 = (0)(−2)(10) = 0

$A = -135 + 0 + 0 = -135$

Arrow 4 = (0)(5)(0) = 0

Arrow 5 = $(-10)(-4)(3) = 120$

Arrow 6 = $(9)(-2)(10) = -180$

$$B = 0 + 120 - 180 = -60$$

$$D_2 = A - B = -135 - (-60) = -135 + 60 = -75$$

The particular determinant D_3 is:

$$D_3 = \begin{vmatrix} 3 & -2 & 10 \\ -2 & 9 & -5 \\ 0 & -4 & -10 \end{vmatrix}$$

Its solution is:

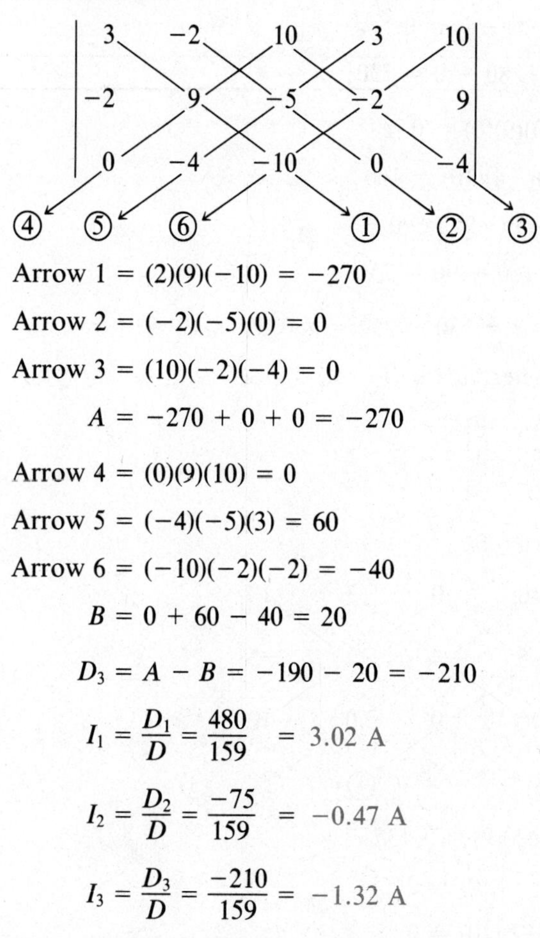

Arrow 1 = $(2)(9)(-10) = -270$

Arrow 2 = $(-2)(-5)(0) = 0$

Arrow 3 = $(10)(-2)(-4) = 0$

$$A = -270 + 0 + 0 = -270$$

Arrow 4 = $(0)(9)(10) = 0$

Arrow 5 = $(-4)(-5)(3) = 60$

Arrow 6 = $(-10)(-2)(-2) = -40$

$$B = 0 + 60 - 40 = 20$$

$$D_3 = A - B = -190 - 20 = -210$$

$$I_1 = \frac{D_1}{D} = \frac{480}{159} = 3.02 \text{ A}$$

$$I_2 = \frac{D_2}{D} = \frac{-75}{159} = -0.47 \text{ A}$$

$$I_3 = \frac{D_3}{D} = \frac{-210}{159} = -1.32 \text{ A}$$

The fact that I_1 is positive indicates that the original choice of direction is correct. The negative result for I_2 and I_3 indicates that the original choice for these currents is incorrect and the currents must be reversed.

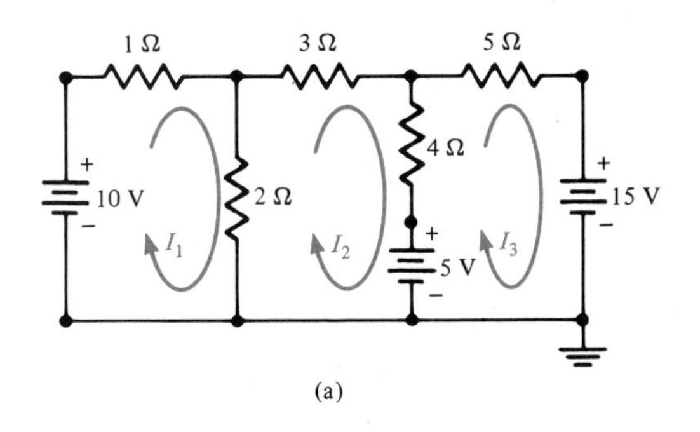

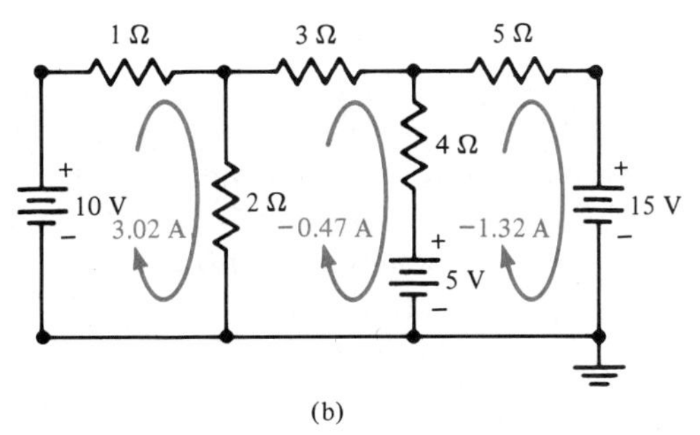

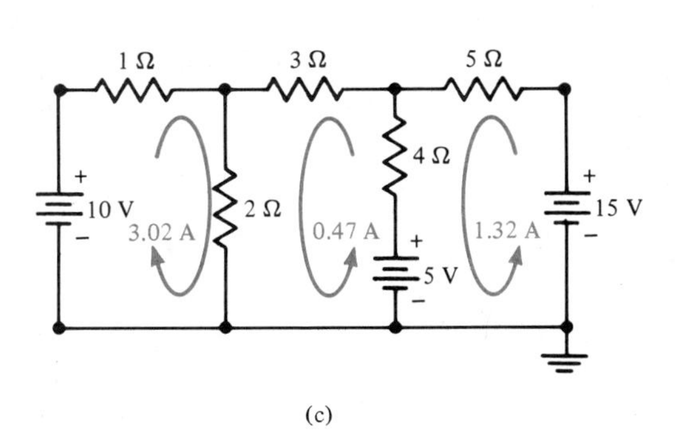

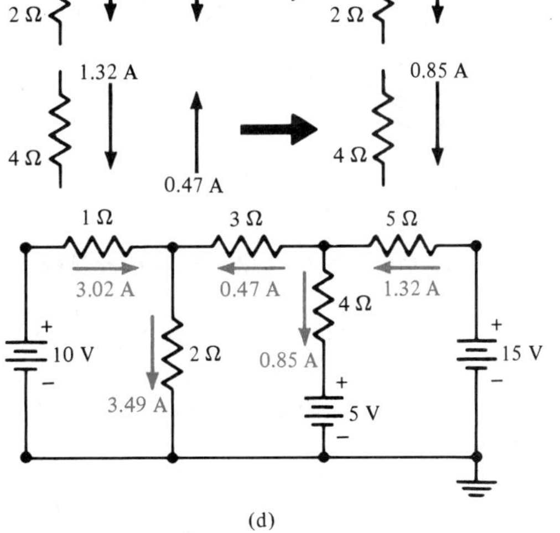

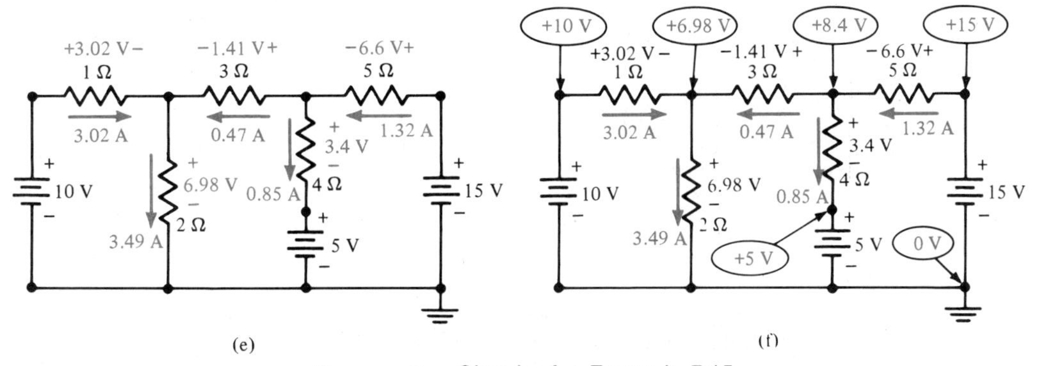

Figure 7.10 Circuits for Example 7.15.

The originally assumed currents are shown in Figure 7.10a. Figure 7.10b shows the current values replacing the corresponding symbols. Note that I_2 and I_3 have negative values. Figure 7.10c shows the direction of I_2 and I_3 as demanded by the negative answers. Figure 7.10d shows how I_1 (3.02 A) and I_2 (0.47 A) combine into a downward current of 3.49 A. The figure also shows how I_3 (0.47 A up) through the 4-Ω resistor subtracts from I_2 (1.32 A down) through the 4-Ω resistor. The 1-Ω resistor has only I_1 (3.02 A left to right) through it. The 3-Ω resistor has only I_2 through it (0.47 A right to left). The 5-Ω resistor has only I_3 through it (1.32 A right to left).

Each one of these currents through the particular resistor causes an individual voltage drop. The 1-Ω resistor has a 3.02-V drop (since $V = IR$, $V = 3.02\text{ A} \times 1\ \Omega = 3.02\text{ V}$). The 3-Ω resistor has a 1.41-V drop across it ($V = 0.47\text{ A} \times 3\ \Omega = 1.41\text{ V}$). The other voltage drops are computed similarly. Figure 7.10e shows the drop across each resistor.

The voltage at each node with respect to ground is shown in Figure 7.10f.

7.10 GAUSS' ELIMINATION METHOD

Gauss' Elimination Method is another mathematical technique used to solve simultaneous equations. It is best explained by example. The system of three equations and three unknowns obtained in Examples 7.2 and 7.6 and solved by Cramer's Rule in Example 7.14 is used to illustrate Gauss' Elimination Method.

EXAMPLE 7.16

Use Gauss' Elimination Method to solve the following system of equations:

$$3I_1 - 2I_2 - 0I_3 = 10$$

$$-2I_1 + 9I_2 - 4I_3 = -5$$

$$0I_1 - 4I_2 + 9I_3 = -10$$

Solution

The column containing the coefficients of I_1 is referred to as the first column, the column containing the coefficients of I_2 is called the second column, and so on. The main diagonal is the diagonal from the upper right to the lower left.

STEP 1. Compare the coefficients of the first variable. Since the largest coefficient of I_1 is in the first equation, there is no need to interchange equations. If the biggest coefficient of I_1 was not in the first equation, then the first equation would have to be interchanged with the one that had the largest I_1 coefficient.

STEP 2. Eliminate the first variable from all but the first equation. The equations are handled so that the top of the first column becomes unity. The rest of the column becomes zero. This is done as follows.

Each coefficient in the first equation is divided by the first coefficient in the equation. This gives unity in the first diagonal position. The first equation then looks like this:

$$\frac{3}{3}I_1 - \frac{2}{3}I_2 + \frac{0}{3}I_3 = \frac{10}{3}$$

$$1I_1 - 0.667I_2 + 0I_3 = 3.333$$

The system of equations looks like this:

$$1I_1 - 0.667I_2 + 0I_3 = 3.333$$

$$-2I_1 + 9I_2 - 4I_3 = -5$$

$$0I_1 - 4I_2 + 9I_3 = -10$$

The first unknown (I_1) is eliminated from the second equation by the following method: The new first equation is multiplied by the first coefficient of the second equation and then subtracted from the second equation.

Since the coefficient of I_1 in the second equation is -2, the first equation is multiplied by -2. This gives

$$-2(1I_1 - 0.667I_2 + 0I_3 = 3.333)$$

$$-2I_1 + 1.334I_2 - 0I_3 = -6.666$$

This is now subtracted from the second equation as follows:

$$-2I_1 + 9I_2 - 4I_3 = -5$$

$$-(-2I_1 + 1.334I_2 - 0I_3 = -6.666)$$

Carrying out the subtraction yields

$$\begin{array}{l} -2I_1 + 9I_2 - 4I_3 = -5 \\ \underline{+2I_1 - 1.334I_2 + 0I_3 = 6.666} \\ 0I_1 + 7.666I_2 + 0I_3 = 1.666 \end{array}$$

Then the system of equations looks like this:

$$1I_1 - 0.667I_2 - 0I_3 = 3.333$$

$$0I_1 + 7.666I_2 - 4I_3 = 1.666$$

$$0I_1 - 4I_2 + 9I_3 = -10$$

Similarly, I_1 is eliminated from the third equation. Since its coefficient is zero already, nothing has to be done. Next, if necessary, exchange the equation with the biggest I_2 coefficient with the second equation. In this problem this is not necessary.

STEP 3. Force the coefficient of I_2 in the second equation to become unity. Then the second equation is divided by the coefficient of I_2. The second equation looks like:

$$0I_1 + \frac{7.666}{7.666} I_2 - \frac{4}{7.666} I_3 = \frac{1.666}{7.666}$$

$$0I_1 + \quad 1I_2 - 0.522I_3 = 0.217$$

The system of equations looks like this:

$$1I_1 - 0.667I_2 - \quad 0I_3 = 3.333$$

$$0I_1 + \quad 1I_2 - 0.522I_3 = 0.217$$

$$0I_1 - \quad 4I_2 + \quad 9I_3 = -10$$

I_2 is now eliminated from the third equation by multiplying the second equation by -4 and then subtracting it from the third equation.

First multiply the second equation by -4:

$$-4(0I_1 + 1I_2 - 0.522I_3 = 0.217)$$

$$0I_1 - 4I_2 + 2.088I_3 = -0.868$$

Then subtract it from the third equation:

$$0I_1 - 4I_2 + \quad 9I_3 = -10$$

$$-(0I_1 - 4I_2 + 2.088I_3 = -0.868)$$

Carrying out the subtraction:

$$\begin{array}{l} 0I_1 - 4I_2 + \quad 9I_3 = -10 \\ 0I_1 + 4I_2 - 2.088I_3 = 0.868 \\ \hline 0I_1 + 0I_2 + 6.912I_3 = -9.132 \end{array}$$

The new system of equations now is:

$$1I_1 - 0.667I_2 - \quad 0I_3 = 3.333$$

$$0I_1 + \quad 1I_2 - 0.522I_3 = 0.217$$

$$0I_1 + \quad 0I_2 + 6.912I_3 = -9.312$$

STEP 4. The third equation has only I_3 with a coefficient other than zero. Solving for I_3:

$$I_3 = \frac{-9.132}{6.912} = -1.321 \text{ A}$$

Substituting this value of I_3 into the second equation, the value of I_2 is obtained:

$$1I_2 - 0.522I_3 = 0.217$$
$$1I_2 - 0.522(-1.321) = 0.217$$
$$1I_2 + 0.690 = 0.217$$
$$I_2 = 0.217 - 0.690$$
$$I_2 = -0.473 \text{ A}$$

Substituting these two values into the first equation, the value of I_1 is obtained:

$$I_1 - 0.667I_2 - 0I_3 = 3.333$$
$$I_1 - 0.667(-0.473) - 0(-1.321) = 3.333$$
$$I_1 + 0.315 - 0 = 3.333$$
$$I_1 = 3.333 - 0.315$$
$$I_1 = 3.108 \text{ A}$$

Therefore, the solutions are:

$I_1 = 3.018$ A $\quad I_2 = -0.473$ A $\quad I_3 = -1.321$ A

These compare favorably with the solutions obtained by Cramer's Rule in Example 7.15:

$I_1 = 3.02$ A $\quad I_2 = -0.47$ A $\quad I_3 = -1.32$ A

7.11 COMPUTER SOLUTION OF GAUSS' ELIMINATION METHOD

It should be obvious by now that the solution of simultaneous equations is tedious at best. The following BASIC computer program uses Gauss' Elimination Method to solve up to eight simultaneous equations. In the unlikely case that more equations have to be solved, the value of MN in line 290 should be changed to equal the number of equations.

```
10    CLS
20    REM *****GAUSS ELIMINATION  by Prof. Edward
      Brumgnach *****
30    REM   *****SUMMARY OF SYMBOLS*****
40    REM      F(R)......ORIGINAL CONSTANTS COLUMN
50    REM      FD(R).....WORKING COEFFICIENT COLUMN
60    REM      A(R,C)....ORIGINAL COEFFICIENT ARRAY
70    REM      B(R,C)....WORKING COEFFICIENT ARRAY
```

```
80     REM     R.........EQUATION NUMBER OR ROW
       NUMBER
90     REM     C.........UNKNOWN NUMBER OR COLUMN
       NUMBER
100    REM     EC........EQUATION COUNTER
110    REM     MN........MAXIMUM NUMBER OF EQUATIONS
120    REM     N.........NUMBER OF EQUATIONS
130    REM     T.........TEMPORARY STORAGE
140    REM     S(R)......SOLUTION ARRAY
150    REM     BG........BIGGEST COEFFICIENT IN
       COLUMN
160    REM     ER........ERROR
170    REM   *****END SUMMARY OF SYMBOLS*****
180    REM *****SOLUTION OF SIMULTANEOUS EQUATIONS
       BY GAUSS ELIMINATION*****
190 GOSUB 230              'GO TO INPUT MODULE
    SUBROUTINE
200 GOSUB 510              'GO TO SOLUTION MODULE
210 GOSUB 1060             'GO TO OUTPUT MODULE
220 END
230    REM      *****INPUT MODULE*****
240 INPUT "FOR EQUATIONS TYPE: 'LE', FOR NODE
    EQUATIONS TYPE:'NE'";ET$
245 IF ET$="le" THEN ET$="LE"
247 IF ET$="ne" THEN ET$="NE"
250 IF ET$="LE" THEN X$="I" : GOTO 280
260 IF ET$="NE" THEN X$="V" : GOTO 280
270 GOTO 240
280 INPUT "HOW MANY EQUATIONS";N
290 MN=8: IF N>MN THEN 270
300 IF N<2 THEN END
310 PRINT "THE COEFFICIENT OF EACH VARIABLE IN EACH
    EQUATION WILL BE ENTERED ONE AT A TIME."
320 FOR R=1 TO N
330 PRINT "EQUATION";R
340 FOR C=1 TO N
350 PRINT "A(";R;",";C;")=";
360 INPUT A(R,C)
370 NEXT C
380 PRINT "F(";R;")=";
390 INPUT F(R)
400 FOR C=1 TO N
410 IF A(R,C)>=0 THEN PRINT
        "+";A(R,C);"(";X$;C;")";
    ELSE PRINT A(R,C);"(";X$;C;")";
420 NEXT C
430 PRINT "=";F(R)
440 INPUT "IS THIS THE RIGHT EQUATION (YES OR
    NO)";Q$
445 IF Q$="yes" THEN Q$="YES"
447 IF Q$="no" THEN Q$="NO"
450 IF Q$="NO" GOTO 330
460 IF Q$="YES" GOTO 470 ELSE 440
470 CLS
480 NEXT R
```

```
485 INPUT "IF YOU WANT A HARD COPY OF THE ANSWERS
    TYPE: H";P$
487 IF P$="h" THEN P$="H"
490 RETURN
500   REM      *****END INPUT MODULE*****
510   REM      *****SOLUTION BY GAUSS
      ELIMINATION*****
520   REM   ***CREATE WORKING COEFFICIENT ARRAYS
      "B" AND "FD"***
530 PRINT "COMPUTING"
540 FOR R=1 TO N
550 FOR C=1 TO N
560 B(R,C)=A(R,C)
570 NEXT C
580 FD(R)=F(R)
590 NEXT R
600   REM        ***END COEFFICIENT CREATION***
610   REM        ***FIND BIGGEST COEFFICIENT IN THIS
      COLUMN***
620 ER=0
630 FOR R=1 TO N-1
640 BG=ABS(B(R,R))
650 EC=R
660 R1=R+1
670 FOR RN=R1 TO N
680 IF ABS(B(RN,R))<BG THEN 710
690 BG=ABS(B(RN,R))
700 EC=RN
710 NEXT RN
720 IF BG=0 THEN 1020
730 IF EC=R THEN 830
740   REM        ***BRING BIGGEST COEFFICIENT TO THE
    TOP***
750 FOR C=1 TO N
760 T=B(EC,C)
770 B(EC,C)=B(R,C)
780 B(R,C)=T
790 NEXT C
800 T=FD(EC)
810 FD(EC)=FD(R)
820 FD(R)=T
830 FOR RN=R1 TO N
840 X=B(RN,R)/B(R,R)
850 FOR I=R1 TO N
860 B(RN,I)=B(RN,I)-X*B(R,I)
870 NEXT I
880 FD(RN)=FD(RN)-X*FD(R)
890 NEXT RN
900 NEXT R
910 REM        ***COMPUTE ANSWER AND BACK
    SUBSTITUTE***
920 IF B(N,N)=0 THEN 1020
930 S(N)=FD(N)/B(N,N)
940 FOR R=N-1 TO 1 STEP -1
950 Y=0
```

```
960 FOR RN=R+1 TO N
970 Y=Y+B(R,RN)*S(RN)
980 NEXT RN
990 S(R)=(FD(R)-Y)/B(R,R)
1000 NEXT R
1010 RETURN                    'IF NO DIFFICULTIES
1020 ER=1
1030 PRINT "ERROR"
1040 RETURN
1050 REM      *****END GAUSS ELIMINATION MODULE*****
1060 REM      *****OUTPUT MODULE*****
1070 CLS
1080 IF N>5 THEN 1190
1090 IF X$="I" THEN PRINT "THE LOOP EQUATIONS ARE:"
     ELSE PRINT "THE NODE EQUATIONS ARE:"
1095 IF P$<>"H" THEN 1110
1100 IF X$="I" THEN LPRINT "THE LOOP EQUATIONS
     ARE:" ELSE LPRINT
     "THE NODE EQUATIONS ARE:"
1110 FOR R=1 TO N
1120 FOR C=1 TO N
1130 IF A(R,C)>=0 THEN PRINT
     "+";A(R,C);"(";X$;C;")"; ELSE PRINT
     A(R,C);"(";X$;C;")" ;
1135 IF P$<>"H" THEN 1150
1140 IF A(R,C)>=0 THEN
     LPRINT"+";A(R,C);"(";X$;C;")"; ELSE LPRINT
     A(R,C);"(";X$;C;")";
1150 NEXT C
1160 PRINT "=";F(R)
1165 IF P$<>"H" THEN 1180
1170 LPRINT "=";F(R)
1180 NEXT R
1190 PRINT
1195 IF P$<>"H" THEN 1210
1200 LPRINT
1210 IF ER=1 THEN GOTO 230
1220 PRINT "SOLUTION:"
1225 IF P$<>"H" THEN 1240
1230 LPRINT "SOLUTION:"
1240 PRINT
1245 IF P$<>"H" THEN 1260
1250 LPRINT
1260 FOR R=1 TO N
1270 PRINT X$;"(";R;")";S(R);
1275 IF P$<>"H" THEN 1290
1280 LPRINT X$;"(";R;")=";S(R);
1290 IF X$="I" THEN PRINT "amps" ELSE PRINT "volts"
1295 IF P$<>"H" THEN 1310
1300 IF X$="I" THEN LPRINT "amps" ELSE LPRINT
     "volts"
1310 NEXT R
1320 RETURN
1330 REM      *****END OUTPUT MODULE*****
```

EXAMPLE 7.17

Use the BASIC program for Gauss' Elimination Method to solve the loop equations obtained for the circuit in Examples 7.1 and 7.5.

Solution

Following are the input data and the output of the program:

```
THE LOOP EQUATIONS ARE:
+ 3 (I 1 )-2 (I 2 )= 10
-2 (I 1 )+ 5 (I 2 )=-5
SOLUTION:
I( 1 )= 3.636364 amps
I( 2 )= .4545456 amps
```

EXAMPLE 7.18

Use the BASIC program for Gauss' Elimination Method to solve the loop equations obtained from the circuit in Examples 7.2 and 7.8.

Solution

Following is the program printout:

```
THE LOOP EQUATIONS ARE:
+ 3 (I 1 )-2 (I 2 )+ 0 (I 3 )= 10
-2 (I 1 )+ 9 (I 2 )-4 (I 3 )=-5
+ 0 (I 1 )-4 (I 2 )+ 9 (I 3 )=-10
SOLUTION:
I( 1 )= 3.018868 amps
I( 2 )=-.4716981 amps
I( 3 )=-1.320755 amps
```

EXAMPLE 7.19

Use the BASIC program for Gauss' Elimination Method to solve the loop equations for the circuit in Examples 7.3 and 7.6.

Solution

The output of the program is:

```
THE LOOP EQUATIONS ARE:
+ 4 (I 1 )-1 (I 2 )=5
-1 (I 1 )+ 3 (I 2 )=-15
```

```
SOLUTION:
I( 1 )=0 amps
I( 2 )=-5 amps
```

EXAMPLE 7.20

Use the BASIC program for Gauss' Elimination Method to solve the loop equations obtained for the circuit in Examples 7.4 and 7.8.

Solution

The program printout is:

```
THE LOOP EQUATIONS ARE:
+ 14 (I 1 )-3 (I 2 )+ 0 (I 3 )-4 (I 4 )= 20
-3 (I 1 )+ 14 (I 2 )-5 (I 3 )-6 (I 4 )=-70
+ 0 (I 1 )-5 (I 2 )+ 12 (I 3 )-3 (I 4 )=-55
-4 (I 1 )-6 (I 2 )-3 (I 3 )+ 21 (I 4 )= 50
SOLUTION
I( 1 )=-.9772541 amps
I( 2 )=-9.040841 amps
I( 3 )=-8.76029 amps
I( 4 )=-1.639759 amps
```

Figure 7.11a shows the circuit with the original loop current directions (see Examples 7.4 and 7.8). The values of the loop currents just obtained are shown inserted in the circuit in Figure 7.11b. Notice that since all the values came out negative, all the loop currents are drawn

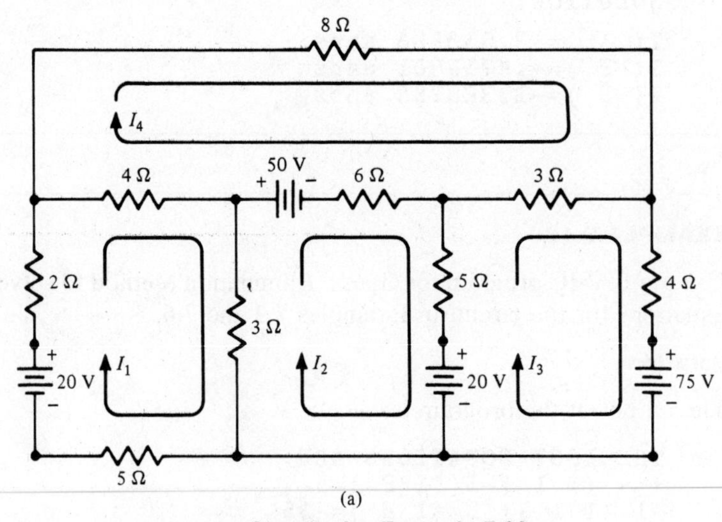

Figure 7.11 Circuits for Example 7.20.

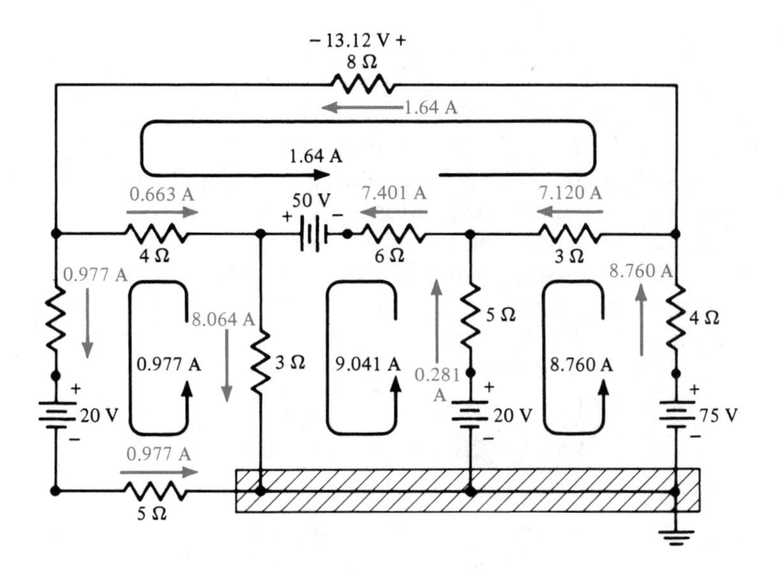

(b)

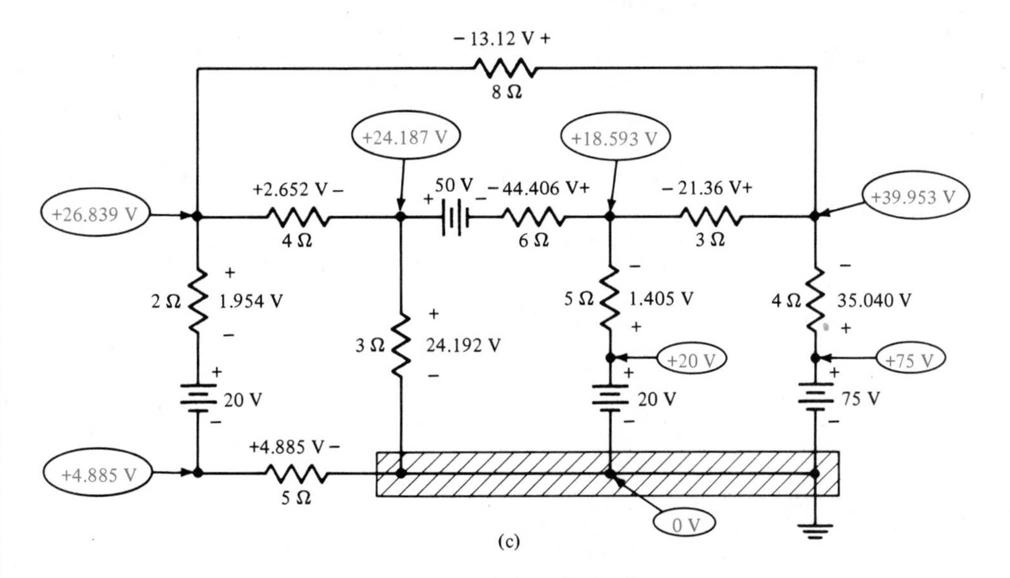

(c)

Figure 7.11 (continued)

opposite to the ones originally assumed. Figure 7.11b also shows the current through each resistive element. Each one of these currents is found by superimposing the loop currents that go through that element. The superposition, of course, applies to both magnitude and direction. The 6-Ω resistor, for example, has 9.041 A right to left and 1.64 A left to right. These superimpose to yield 7.401 A right to left.

Each one of these element currents being maintained through a resistor causes a voltage drop across that resistor. Each voltage drop is computed by Ohm's Law. Figure 7.11c shows each resistor labeled with its respective voltage drop. Figure 7.11c also shows the voltage at each node with respect to the reference ground node. Each one of these node voltages may be found by going to that node from ground and, in the process, adding voltage rises and subtracting voltage drops. The voltage at node V_1, for example, may be found by starting at ground and going up through the 3-Ω resistor and to the left through the 4-Ω resistor. Going up through the 3-Ω resistor gives a 24.192-V rise, whereas going left through the 4-Ω resistor gives another rise of 2.652 V. These two rises combine for a total of 24.839 V at node V_1.

A check around each loop shows that KVL holds; therefore the solution is correct.

	Voltage Rises	*Voltage Drops*
Loop 1	4.885	2.652
	20	24.192
	1.954	26.884
	26.839	
Loop 2	24.192	50
	44.406	20
	1.045	70
	70.003	
Loop 3	20	1.405
	21.36	75
	35.04	76.405
	76.40	
Loop 4	50	21.36
	2.652	44.406
	13.12	65.766
	65.772	

EXAMPLE 7.21

Use the BASIC program for Gauss' Elimination Method to solve the node equations obtained for the circuit in Example 7.10.

Solution

The output of the program is:

```
THE NODE EQUATIONS ARE:
+ 1.83 (V 1 )-.333 (V 2 )= 10
-.333 (V 1 )+ .78 (V 2 )= 5
```

```
SOLUTION:
V( 1 )= 7.92 volts
V( 2 )= 13.61 volts
```

EXAMPLE 7.22

Use the BASIC program for Gauss' Elimination Method to solve the node equations obtained for the circuit in Examples 7.11 and 7.13.

Solution

The results of the program are:

```
THE NODE EQUATIONS ARE:
+ .875 (V 1 )-.25 (V 2 )+ 0 (V 3 )-.125 (V 4
)-.5 (V 5 )= 10
-.25 (V 1 )+ .75 (V 2 )-.167 (V 3 )+ 0 (V 4 )+
0 (V 5 )= 8.33
+ 0 (V 1 )-.167 (V 2 )+ .7 (V 3 )-.333 (V 4 )+
0 (V 5 )=-4.333
-.125 (V 1 )+ 0 (V 2 )-.333 (V 3 )+ .708 (V 4
)+ 0 (V 5 )= 18.75
-.5 (V 1 )+ 0 (V 2 ) + 0 (V 3 )+ 0 (V 4 )+ .7
(V 5 )=-10
SOLUTION:
V( 1 )= 26.84595 volts
V( 2 )= 24.19616 volts
V( 3 )= 18.59657 volts
V( 4 )= 39.96949 volts
V( 5 )= 4.889968 volts
```

Figure 7.12a shows the circuit with the node voltages as labeled in Examples 7.11 and 7.13. The node voltages just obtained via the computer

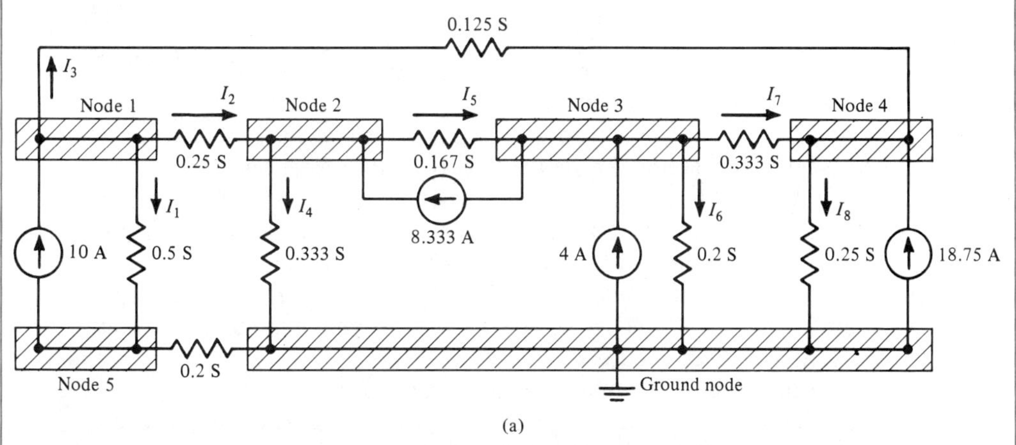

(a)

Figure 7.12 Circuits for Example 7.22.

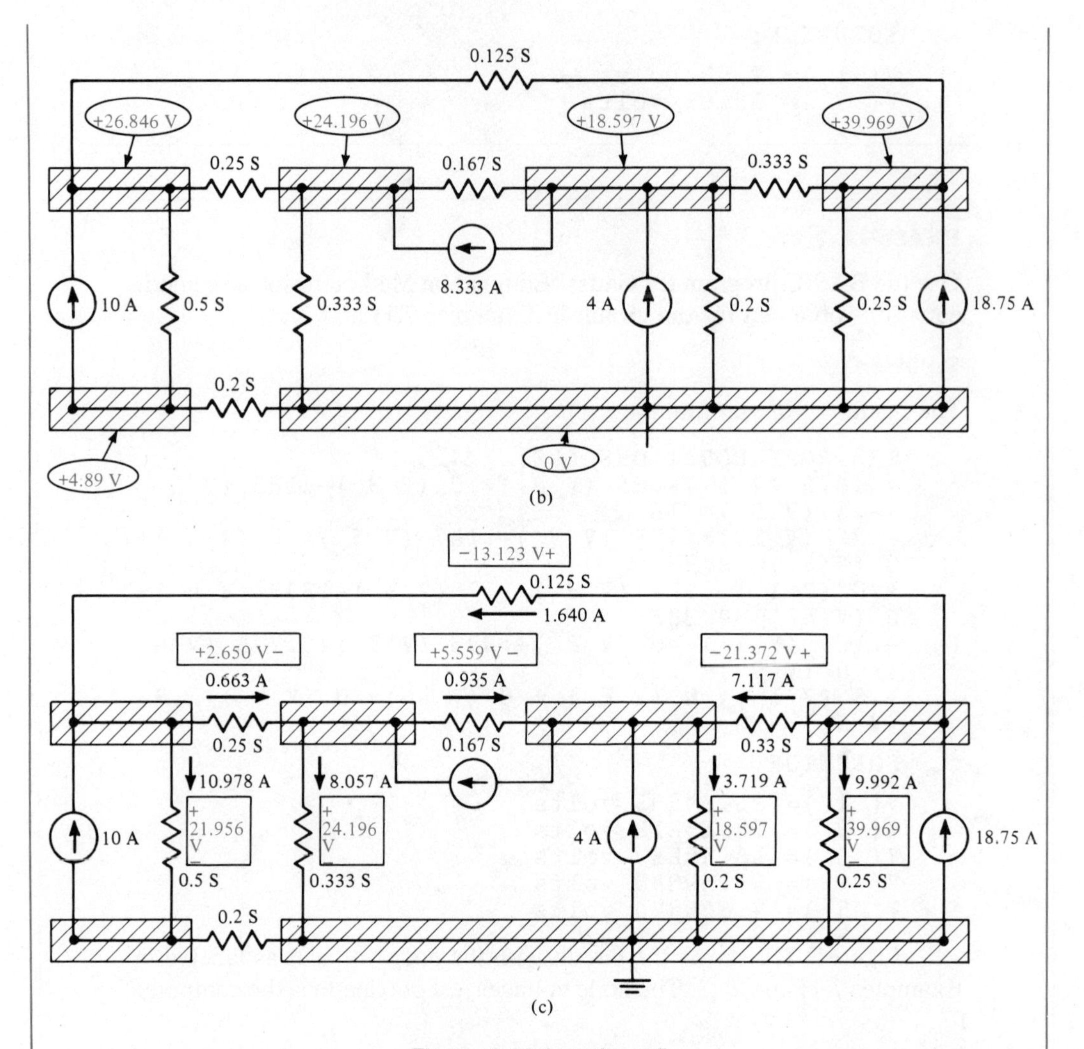

Figure 7.12 (continued)

solution are shown at the proper nodes in Figure 7.12b. Figure 7.12c shows the voltage drop across each resistor. Each one of the drops is obtained by observing the two node voltages between which the resistor is connected and subtracting the smaller one from the larger one. The polarity of each voltage drop is obtained by assigning the + to the larger node voltage and the − to the smaller node voltage. The voltage across the 0.167-S conductor is found by observing that it is connected between a 24.196-V node on its left and a 18.597-V node on its right. The voltage drop then is 24.196 − 18.597 = 5.599 V with the + on the left and the − on the right, as shown in Figure 7.12c.

Each individual conductor has a current through it. Each current may be found by Ohm's Law by multiplying the voltage drop across each element by the respective conductance. The 0.167-S conductance has a

5.599-V drop across it. Its current therefore is $5.599 \times 0.167 = 0.935$ A left to right. The direction of the current is through the element from the + to the − sign of the voltage drop. Figure 7.12c shows the current through each component with its respective direction.

KCL may be applied at each node to see if it is satisfied.

	Current Entering Node	*Current Leaving Node*
Node 1	10	10.978
	1.640	0.663
	11.640 A	11.641 A
Node 2	0.663	8.057
	8.333	0.935
	8.996 A	8.992 A
Node 3	0.935	8.333
	4	3.719
	7.117	12.052 A
	12.052 A	
Node 4	18.75 A	7.117
		9.992
		1.640
		18.749 A
Node 5	10.978 A	10
		0.978
		10.978 A
Ground	0.978 A	18.75
	8.057	4
	3.719	22.75 A
	9.992	
	22.746 A	

It is obvious that KCL holds. It is also interesting to compare the node voltages obtained in this example (shown in Figure 7.12b) with the node voltages obtained in Example 7.20 (shown in Figure 7.11c). Note that they agree very well.

SUMMARY

In this chapter classical as well as streamlined methods for obtaining loop and node equations were presented. Solution by Cramer's Rule and determinants as well as by Gauss' Elimination Method were illustrated. Finally, a computer program for solving multiple-loop or multiple-node circuits was offered.

SUMMARY OF IMPORTANT EQUATIONS

Generalized loop equations:

$$\begin{aligned}
+Z_{11}I_1 - Z_{12}I_2 - Z_{13}I_3 - \cdots - Z_{1N}I_N &= E_1 \\
-Z_{21}I_1 + Z_{22}I_2 - Z_{23}I_3 - \cdots - Z_{2N}I_N &= E_2 \\
-Z_{31}I_1 - Z_{32}I_2 + Z_{33}I_3 - \cdots - Z_{3N}I_N &= E_3 \\
&\vdots \\
-Z_{N1}I_1 - Z_{N2}I_2 - Z_{N3}I_3 - \cdots + Z_{NN}I_N &= E_N
\end{aligned} \tag{7.1}$$

The above set of simultaneous equations applies to any resistive network with N loops.

Generalized node equations:

$$\begin{aligned}
Y_{11}V_1 - Y_{12}V_2 - \cdots - Y_{1N}V_N &= I_1 \\
-Y_{21}V_1 + Y_{22}V_2 - \cdots - Y_{2N}V_N &= I_2 \\
&\vdots \\
-Y_{N1}V_1 - Y_{N2}V_2 - \cdots + Y_{NN}V_N &= I_N
\end{aligned} \tag{7.2}$$

The above set of simultaneous equations applies to any resistive network with one reference node and N additional nodes.

REVIEW QUESTIONS

7.1 Define a loop, or mesh.

7.2 Define a node.

7.3 Explain what loop equations are.

7.4 Explain how to obtain loop equations.

7.5 Explain what node equations are.

7.6 Explain how to obtain node equations.

7.7 Explain what a determinant is.

7.8 Name two methods of solving simultaneous equations.

PROBLEMS

7.1 Evaluate the following determinants:

a. $\begin{vmatrix} 1 & 2 \\ 3 & 4 \end{vmatrix}$ **b.** $\begin{vmatrix} 5 & -6 \\ 2 & -3 \end{vmatrix}$

c. $\begin{vmatrix} 0 & 3 \\ 4 & 5 \end{vmatrix}$ **d.** $\begin{vmatrix} 6 & -2 \\ -3 & 5 \end{vmatrix}$

7.2 Evaluate the following determinants:

a. $\begin{vmatrix} 1 & 2 & 3 \\ 4 & 5 & 6 \\ 7 & 8 & 9 \end{vmatrix}$ **b.** $\begin{vmatrix} -2 & -1 & 0 \\ 3 & 4 & -1 \\ 1 & 5 & 7 \end{vmatrix}$

(ans. −72, 45)

7.3 Using determinants and Cramer's Rule solve for x and y:

a. $3x + 4y = 10$
$x - 2y = 5$

b. $2x - 5y = -6$
$3x + 2y = 12$

7.4 Using determinants and Cramer's Rule solve for x, y, and z.

$$2x - 5y + 3z = 10$$
$$x - y + 5z = -5$$
$$3x + y - z = 4$$

(ans. 1.553, −2.461, −1.803)

7.5 For the circuit shown in Figure 7.13:

a. Use KVL to obtain the loop equations.
b. Use the Standard Notation Method to obtain the loop equations.
c. Use Cramer's Rule to solve the loop equations.
d. Compute the magnitude and direction of the current through each resistor.

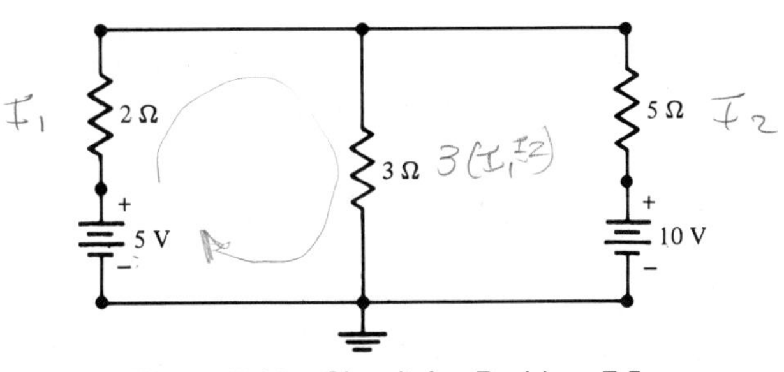

Figure 7.13 Circuit for Problem 7.5.

7.6 Repeat Problem 7.5 for the circuit shown in Figure 7.14. (ans. 1.125 A, −0.625 A)

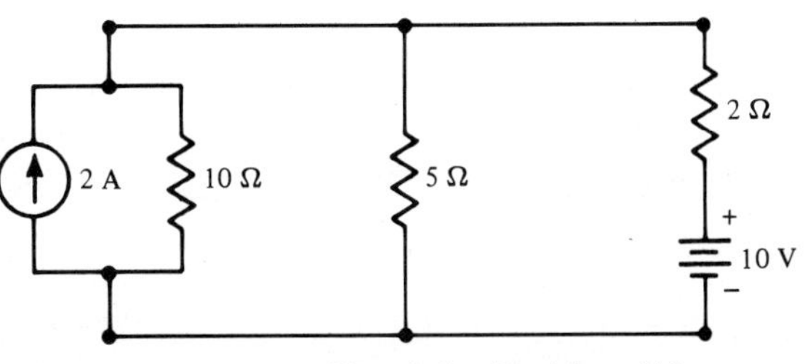

Figure 7.14 Circuit for Problem 7.6.

7.7 For the circuit shown in Figure 7.15:

a. Use the Standard Notation Method to obtain the loop equations.

b. Use Cramer's Rule to solve the equations.

c. Compute the magnitude and direction of the current through the resistor labeled R.

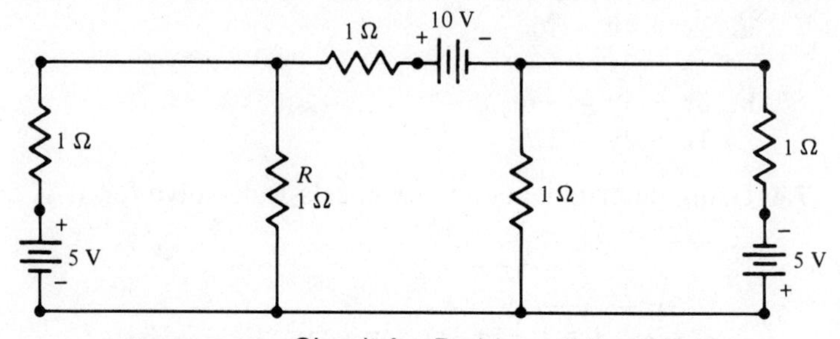

Figure 7.15 Circuit for Problems 7.7 and 7.10.

7.8 Repeat Problem 7.7 for the circuit shown in Figure 7.16. (ans. −0.355 A, 0.213 A, 0.816 A)

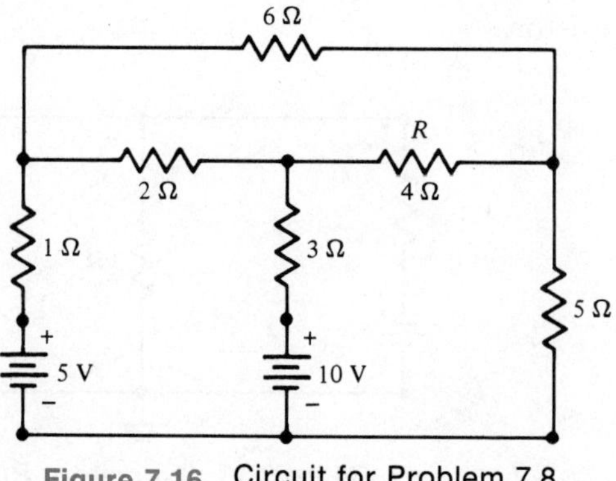

Figure 7.16 Circuit for Problem 7.8.

7.9 Repeat Problem 7.7 for the circuit shown in Figure 7.17.

7.10 For the circuit shown in Figure 7.15:

a. Use KCL to obtain the node equations.

b. Use the Standard Notation Method to obtain the loop equations. (ans. $3V_1 - V_2 = 15$, $-V_1 + 3V_2 = -5$)

c. Use Cramer's Rule to solve the equations. (ans. 5 V, 0 V)

d. Compute the current through the resistor labeled R. (ans. 5 A ↓)

7.11 Repeat Problem 7.10 for the circuit shown in Figure 7.17.

7.12 Use the Gauss Elimination Method to solve Problem 7.3. (ans. 4, 0.5)

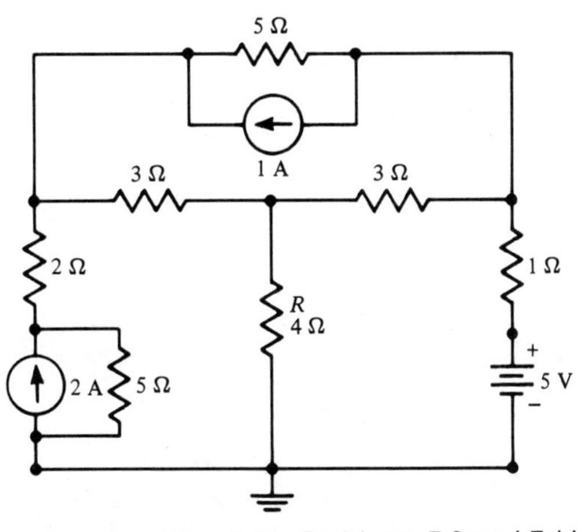

Figure 7.17 Circuit for Problems 7.9 and 7.11.

7.13 Use the Gauss Elimination Method to solve Problem 7.4.

7.14 Use the Gauss Elimination Method to solve the equations obtained in Problem 7.5. (ans. 0.323 A, −1.129 A)

PROBLEMS FOR COMPUTER SOLUTION

7.15 Use the BASIC program for Gauss' Elimination Method to solve the equations obtained in Problem 7.5.

7.16 Use the BASIC program for Gauss' Elimination Method to solve the equations obtained in Problem 7.3.

7.17 Use the BASIC program for Gauss' Elimination Method to solve the equations obtained in Problem 7.4.

7.18 Repeat Problem 7.15 for the equations obtained in Problem 7.6.

7.19 Repeat Problem 7.15 for the equations obtained in Problem 7.7.

7.20 Repeat Problem 7.15 for the equations obtained in Problem 7.8.

7.21 Repeat Problem 7.15 for the equations obtained in Problem 7.9.

CHAPTER

8

Equivalent Circuits in Circuit Analysis

GLOSSARY

equivalent circuit—a circuit with identical terminal characteristics as another circuit

Millman's Theorem—a theorem applicable to multisource circuits containing exclusively practical voltage sources connected in parallel, which allows the substitution by a single simple equivalent voltage source

Norton's equivalent circuit—an equivalent circuit consisting of an ideal current source and a resistor in parallel

Superposition Theorem—a theorem stating that the current in any element or voltage across any element of a linear multisource circuit (independent source) is equivalent to the algebraic sum of the voltage or current due to each source acting alone

Thevenin's equivalent circuit—an equivalent circuit represented by an ideal voltage source and a series resistance

INTRODUCTION

Mesh and nodal analysis techniques were studied in Chapter 7. These methods were necessary in order to analyze electric circuits that cannot be simplified into series and parallel sections. It was shown that nodal analysis was preferable over mesh analysis in some cases but the opposite was true in other cases. The choice of the method to use is based upon which method is simpler and will yield results in a shorter time.

In this chapter we examine additional techniques that simplify the circuit analysis process. We also show that circuits can be simplified by their *equivalent circuit* representation. Both the *Thevenin* and *Norton* equivalent circuits are very useful in analyzing practical circuits. The Superposition Theorem provides a very powerful means for solving multisource circuits.

8.1 THEVENIN'S THEOREM AND EQUIVALENT CIRCUIT

*An **equivalent circuit** is a circuit with terminal characteristics identical to another circuit.*

We dealt earlier with equivalent circuits when we compared series and parallel resistor combinations to simplify the analysis of circuits. In this section we reduce electric circuits to a single voltage source with its internal resistance. This reduced equivalent circuit is referred to as **Thevenin's equivalent circuit,** after the French scientist M. L. Thevenin.

Thevenin's Theorem states that:

> Any two-terminal linear dc network can be replaced by an equivalent circuit consisting of a voltage source and a series resistor.

The voltage of the source in the equivalent circuit is the potential drop across the open terminals, and the resistance value is the equivalent resistance, also measured between the open terminals. A two-terminal dc circuit and its Thevenin's equivalent circuit are shown in Figures 8.1a and 8.1b, respectively. The open terminals are designated as A and B. The double arrow denotes equivalence of the two circuits. The Thevenin equivalent voltage source is shown as E_{th} and the Thevenin resistance as R_{th}. It is obvious that the equivalent circuit is a much simpler circuit; hence, the simplified circuit can be used for further analysis.

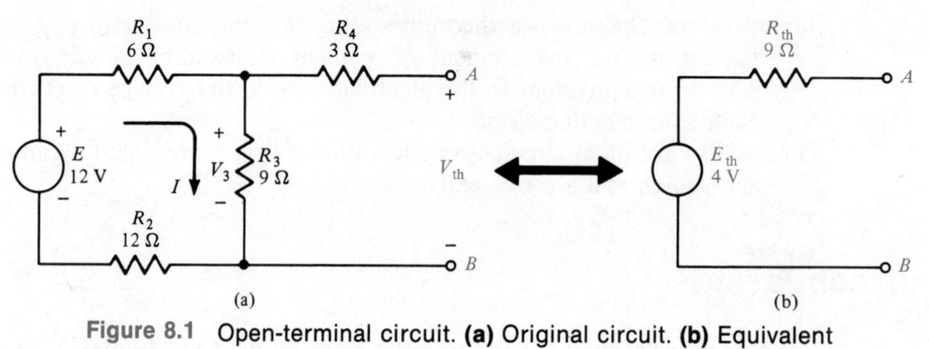

Figure 8.1 Open-terminal circuit. **(a)** Original circuit. **(b)** Equivalent circuit.

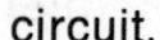

Example 8.1 illustrates the technique used in calculating the components of the equivalent circuit.

EXAMPLE 8.1

Draw Thevenin's equivalent circuit and calculate the component values of the circuit in Figure 8.1a.

Solution

The technique for calculating the equivalent circuit component values is as follows:

1. Determine the voltage drop across terminals A and B. This voltage drop is V_{th}.

   By Voltage Divider Rule, the voltage across the 9-Ω resistor is:

$$V_3 = \frac{(12)(9)}{6 + 9 + 12} = 4 \text{ V}$$

   Since there is no current through the 3-Ω resistor, the voltage between terminals A and B is also 4 V.

$$V_{th} = V_{AB} = V_3 = 4 \text{ V}$$

2. To calculate the resistance between terminals A and B, disconnect all voltage and current sources and replace them with their internal resistances. Then, calculate the resistance between terminals A and B. Keep in mind that an ideal voltage source has an internal resistance of 0 Ω, and an ideal current source has an infinite internal resistance. Therefore, ideal voltage sources are replaced by a short circuit and ideal current sources are replaced with an open circuit. In this circuit the 12-V source is replaced

with a short circuit and the resistance between A and B is:

$$R_{AB} = R_{th} = 3 + \frac{(6 + 12)9}{6 + 12 + 9}$$
$$= 9\ \Omega$$

The equivalent circuit is shown in Figure 8.1b.

EXAMPLE 8.2

Determine the current I_L in the load resistor R_L in the circuit of Figure 8.2a by applying Thevenin's Theorem.

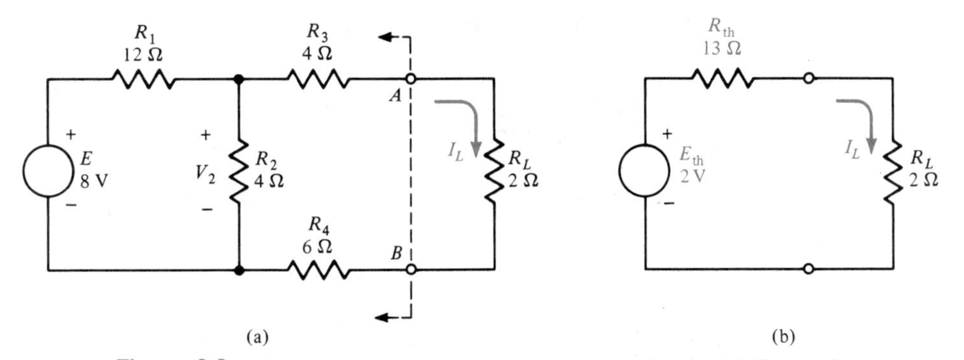

Figure 8.2 Circuit for Example 8.2. **(a)** Actual circuit. **(b)** Thevenin equivalent circuit with load.

Solution

The solution for the load current can be obtained by combining resistors in series and parallel to find the total resistance used than by using the Current Divider Rule. However, the circuit will be solved by reducing the circuit to a Thevenin's equivalent circuit.

First, the load is removed to make it a two-terminal circuit. The open terminal voltage, V_{AB}, is the voltage dropped across R_2:

$$V_{th} = V_{AB} = V_2 = \frac{(8)4}{4 + 12}$$
$$= 2\ \text{V}$$

The Thevenin resistance is calculated after replacing the voltage source with a short.

$$R_{th} = \frac{(12)(4)}{12 + 4} + 4 + 6 = 13\ \Omega$$

Thevenin's equivalent circuit is shown in Figure 8.2b with the load resistor R_L connected to it.

The current I_L is:

$$I_L = \frac{2}{13 + 2} = 0.133 \text{ A}$$

Example 8.2 shows the application of Thevenin's Theorem in the solution of a simple single-source circuit. The next example illustrates the application of Thevenin's Theorem to a multisource circuit.

EXAMPLE 8.3

Calculate the power to the load resistor R_L in the circuit of Figure 8.3a.

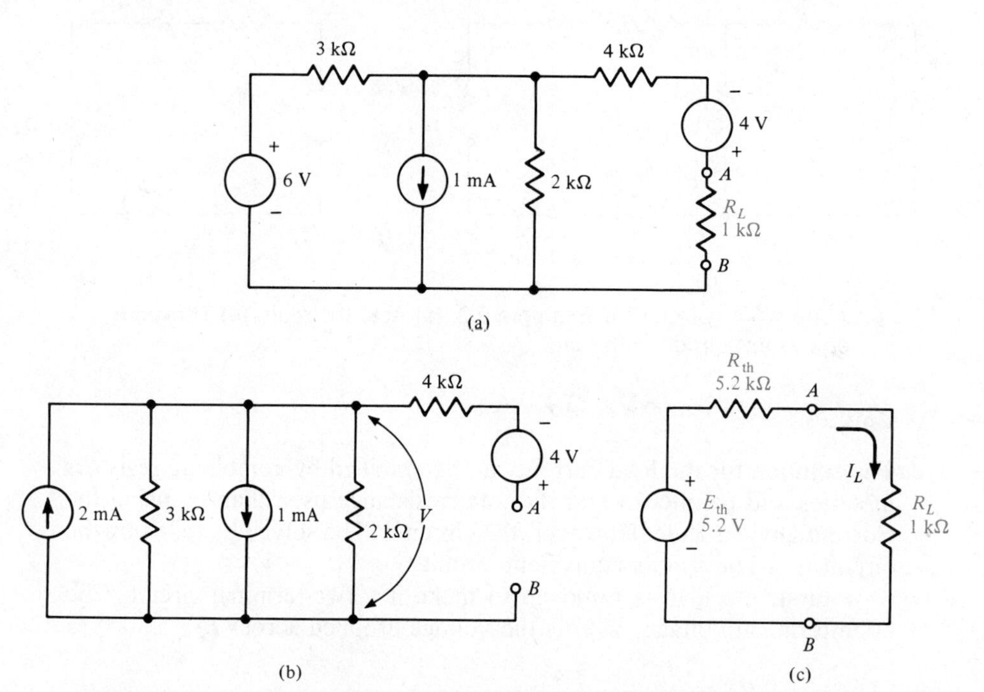

Figure 8.3 Circuit for Example 8.3. **(a)** Actual circuit. **(b)** Two-terminal circuit with source conversion. **(c)** Thevenin equivalent circuit with load.

Solution

Thevenin's Theorem will be used to solve the circuit. The load resistor R_L is disconnected, and the voltage across the open terminals A and B is calculated next.

Converting the 6-V source to a current source makes it easier to arrive at a solution. The resulting circuit is shown in Figure 8.3b. The voltage across the 2-kΩ resistor is calculated by writing a nodal equation.

$$\left(\frac{1}{3\ \text{k}\Omega} + \frac{1}{2\ \text{k}\Omega}\right) V = 2\ \text{mA} - 1\ \text{mA}$$

$$V = 1.2\ \text{V}$$

The voltage V_{AB} is simply

$$V_{th} = V_{AB} = 1.2\ \text{V} + 4\ \text{V} = 5.2\ \text{V}$$

Notice that there is no drop across the 4-kΩ resistor because there is no current through it.

The Thevenin's resistance is calculated after replacing the voltage sources with short circuits and open-circuiting the current source.

$$R_{AB} = R_{th} = \frac{(3\ \text{k}\Omega)(2\ \text{k}\Omega)}{3\ \text{k}\Omega + 2\ \text{k}\Omega} + 4\ \text{k}\Omega$$
$$= 5.2\ \text{k}\Omega$$

Thevenin's equivalent circuit is shown in Figure 8.3c with the load resistor reconnected.

The current through the load resistor from Thevenin's equivalent circuit is

$$I_L = \frac{5.2}{5.2\ \text{k}\Omega + 1\ \text{k}\Omega} = 0.839\ \text{mA}$$

The power to the load resistor is given by

$$P_L = I_L^2 R_L$$
$$= (0.839 \times 10^{-3})^2\,(1 \times 10^3) = 0.703\ \text{mW}$$

Example 8.4 illustrates the use of Thevenin's Theorem in a maximum power transfer application. Recall from Chapter 6 that the load gets maximum power if the load resistance is equal to the internal resistance of the source.

EXAMPLE 8.4

Calculate the value of the load resistor required to transfer maximum power from the circuit in Figure 8.4 to the load.

Solution

The entire circuit can be viewed as a practical voltage source with a load, R_L, connected to its terminals. Thevenin's equivalent resistance of the circuit represents the internal resistance of the practical source.

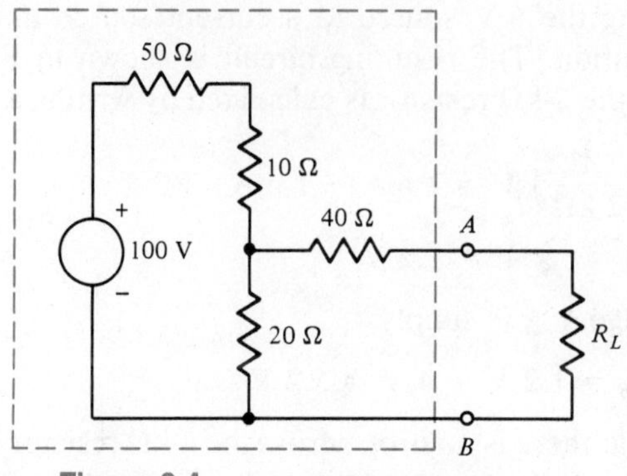

Figure 8.4 Circuit for Example 8.4.

To find R_{th}, disconnect R_L and replace the ideal voltage source with a short.

$$R_{th} = \frac{(50 + 10)(20)}{50 + 10 + 20} + 40$$
$$= 55\ \Omega$$

For maximum power transfer, the load resistance, R_L, should be the same as R_{th}, 55 Ω.

8.2 NORTON'S THEOREM AND EQUIVALENT CIRCUIT

Norton's Theorem is the parallel of Thevenin's Theorem and it states that:

> Any two-terminal linear dc network can be replaced by an equivalent circuit consisting of an ideal current source and a resistor in parallel.

This circuit is called **Norton's equivalent circuit.**

Recall that every practical voltage source has an equivalent practical current source and vice versa. Hence, Norton's circuit has an equivalent in Thevenin's circuit as shown in Figure 8.5. Recall from Chapter 6 that when a practical voltage

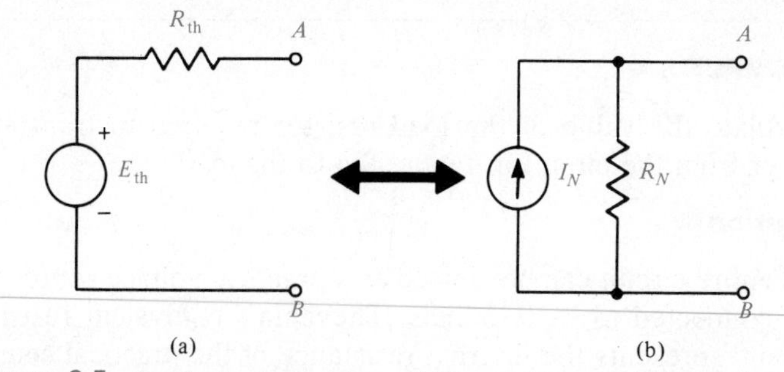

Figure 8.5 Equivalent circuits. **(a)** Thevenin's equivalent circuit. **(b)** Norton's equivalent circuit.

source is converted to a current source, the internal resistance of the two sources is of the same value; therefore,

$$R_{th} = R_N \tag{8.1}$$

where R_N is Norton's resistance. The value of the total current source is the ratio of the Thevenin voltage to the series resistance of the ideal voltage source:

$$I_N = \frac{V_{th}}{R_{th}} \tag{8.2}$$

where I_N is referred to as Norton's current.

EXAMPLE 8.5

In Example 8.1 the Thevenin equivalent circuit consists of a 4-V source and a 9-Ω resistor in series, as shown in Figure 8.6a. Draw Norton's equivalent circuit and determine the component values.

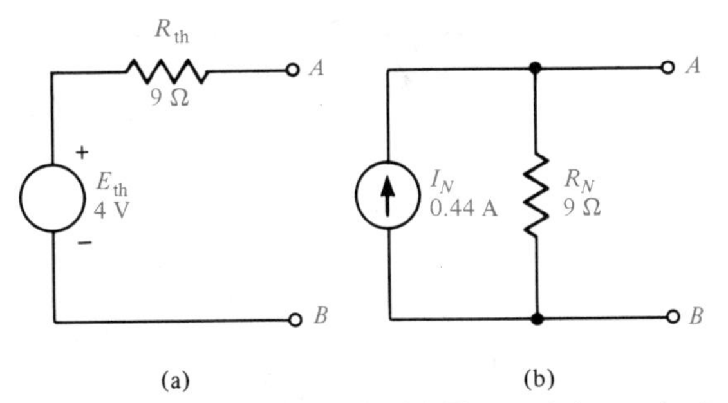

Figure 8.6 Circuits for Example 8.5. **(a)** Thevenin's equivalent circuit. **(b)** Norton's equivalent circuit.

Solution

By Eq. (8.1),

$$R_N = R_{th} = 9\ \Omega$$

and by Eq. (8.2),

$$I_N = \frac{4}{9} = 0.44\ \text{A}$$

Norton's equivalent circuit is shown in Figure 8.6b.

To determine Norton's equivalent circuit, the Norton current, I_N, is calculated by finding the *short-circuit* current in the original circuit, which is the current from terminal A to terminal B after the two terminals are shorted. Norton's resistance R_N is found the same way as finding Thevenin's resistance, namely, by replacing voltage and current sources by their equivalent resistances and calculating the resistance at the open terminals, A and B.

EXAMPLE 8.6

Determine Norton's equivalent circuit for the circuit in Figure 8.7a.

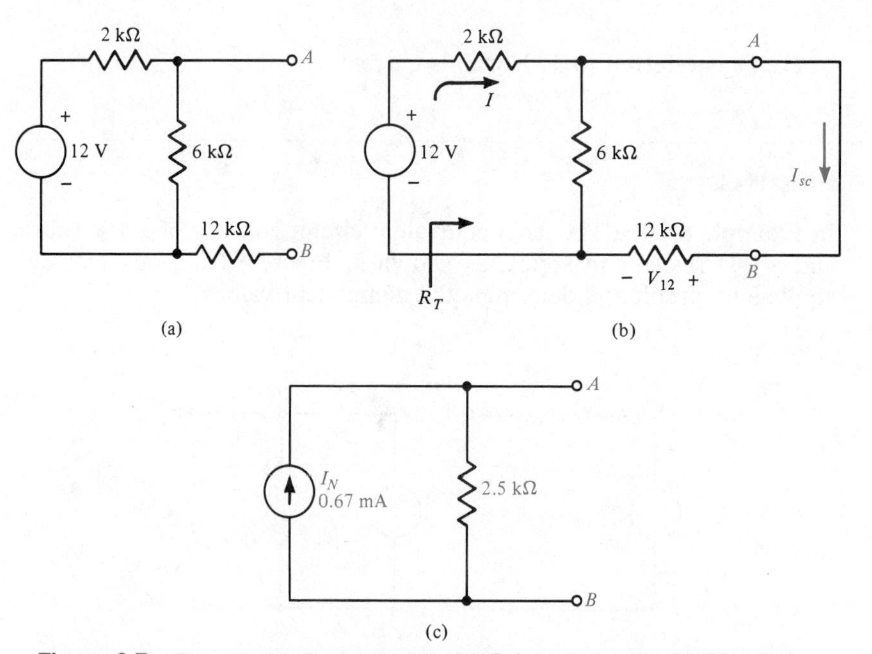

Figure 8.7 Circuits for Example 8.6. **(a)** Original circuit. **(b)** Circuit for calculating I_N. **(c)** Norton's equivalent circuit.

Solution

To calculate Norton's resistance, the 12-V source is replaced with a short and the resistance between A and B is:

$$R_{AB} = R_N = \frac{(2\ \text{k}\Omega)(6\ \text{k}\Omega)}{2\ \text{k}\Omega + 6\ \text{k}\Omega} + 12\ \text{k}\Omega = 13.5\ \text{k}\Omega$$

Terminals A and B are shorted in Figure 8.7b to calculate the short-circuit current, I_{SC}, which is also I_N, as shown in Figure 8.7c. The current, I_{SC}, is the current through the 12-kΩ resistor. The total resistance is:

$$R_T = \frac{(12\text{ k}\Omega)(6\text{ k}\Omega)}{12\text{ k}\Omega + 6\text{ k}\Omega} + 2\text{ k}\Omega$$
$$= 6\text{ k}\Omega$$

and the total current I is:

$$I = \frac{12}{6\text{ k}\Omega} = 2\text{ mA}$$

By the Current Divider Rule,

$$I_{12} = I_{SC} = 2\text{ mA}\,\frac{6\text{ k}\Omega}{6\text{ k}\Omega + 12\text{ k}\Omega}$$
$$= 0.67\text{ mA}$$

and

$$I_N = I_{SC} = 0.67\text{ mA}$$

Norton's equivalent circuit is shown in Figure 8.7c.

EXAMPLE 8.7

Calculate the current in the load resistor using Norton's equivalent circuit for the circuit in Figure 8.8a.

Solution

First, the load resistor is disconnected and a short is connected between terminals A and B. Then, the current source with its parallel resistance is converted to a voltage source. The resulting circuit is given in Figure 8.8b. The short-circuit current is the current in the series circuit. Kirchhoff's voltage equation is:

$$-10 + 5 + 20I_{SC} + 10I_{SC} = 0$$

Then

$$I_{SC} = \frac{10 - 5}{20 + 10} = 0.167\text{ A}$$

Next, the current source is replaced by an open circuit, and the voltage source is replaced by a short. The resistance measured at the open terminals A and B is:

$$R_N = 10 + 20 = 30\ \Omega$$

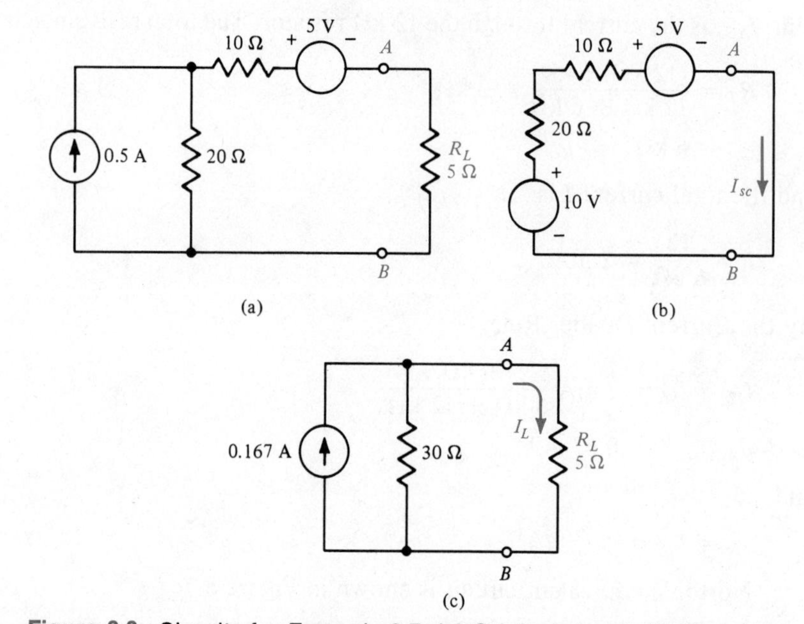

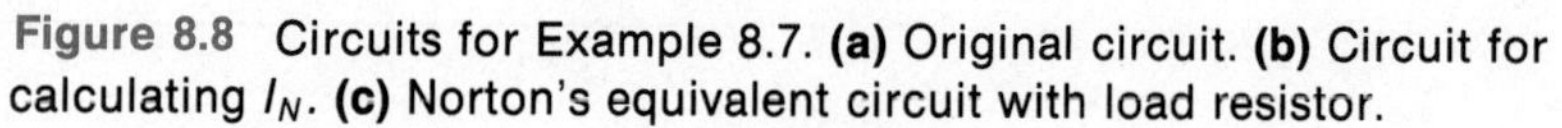

Figure 8.8 Circuits for Example 8.7. **(a)** Original circuit. **(b)** Circuit for calculating I_N. **(c)** Norton's equivalent circuit with load resistor.

Norton's equivalent circuit with the load reconnected is given in Figure 8.8c. The load current is calculated by the Current Divider Rule:

$$I_L = 0.167 \frac{30}{30 + 5} = 0.143 \text{ A}$$

The next two examples illustrate the method for calculating Norton's and Thevenin's equivalent circuits if the original circuit contains dependent sources.

EXAMPLE 8.8

Determine Norton's equivalent circuit for the circuit in Figure 8.9a.

Solution

Short-circuiting terminals *A* and *B* produces a two-mesh circuit with the following equations, assuming the loop currents as shown:

$$V_X = 10I_1$$

$$0.2\,(10)\,I_1 + 10I_1 + 20I_1 - 20I_2 = -20$$

$$-20I_1 + 20I_2 + 5I_2 = 20$$

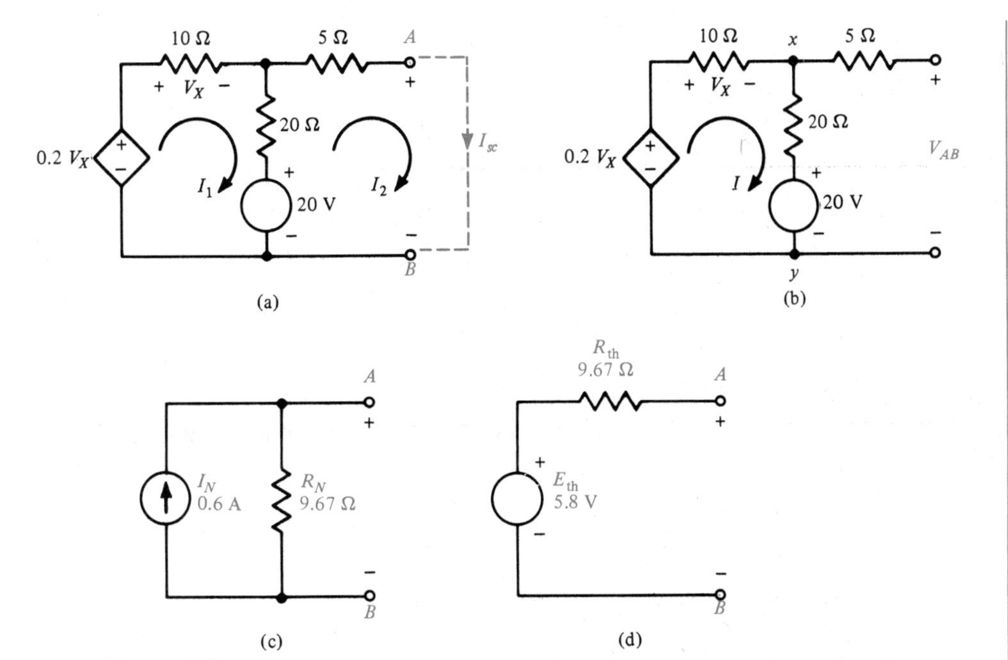

Figure 8.9 Circuit for Example 8.8. **(a)** Original circuit. **(b)** Open-circuit voltage calculation. **(c)** Norton's equivalent circuit. **(d)** Thevenin's equivalent circuit.

which reduces to

$$32I_1 - 20I_2 = -20$$
$$-20I_1 + 25I_2 = 20$$

Solving for I_2, which is also I_{SC}, by any of the methods in Chapter 7 gives

$$I_{SC} = I_2 = 0.6 \text{ A}$$

To calculate Norton's resistance, we find the open-circuit voltage between terminals A and B. Norton's resistance is then calculated from the ratio of V_{AB} to I_{SC}. The open-circuit voltage V_{AB} is the voltage between nodes x and y, as shown in Figure 8.9b. The KVL equation around the closed loop is:

$$0.2V_X - 20 = 10I + 20I$$

where $V_X = 10I$. Hence

$$(0.2)(10I) - 20 = 10I + 20I$$
$$-20 = (10 + 20 - 2)I$$
$$I = \frac{-20}{28} = -0.71 \text{ A}$$

The open-circuit voltage is:

$$V_{AB} = V_{XY} = 20 + (-0.71 \times 20) = 5.8 \text{ V}$$

$$R_N = \frac{V_{AB}}{I_{SC}} = \frac{5.8}{0.6} = 9.67 \ \Omega$$

The Norton equivalent circuit is shown in Figure 8.9c.

Thevenin's equivalent of the circuit in Figure 8.9a can be easily obtained from its Norton equivalent in Figure 8.9c.

$$R_N = R_{\text{th}} = 9.67 \ \Omega$$

Thevenin's voltage is calculated by applying techniques of source transformation:

$$V_{\text{th}} = I_N R_N$$

$$= 0.6 \times 9.67 = 5.8 \text{ V}$$

Thevenin's equivalent circuit is shown in Figure 8.9d.

8.3 SUPERPOSITION THEOREM

The **Superposition Theorem** can be used as a method of analysis in multisource circuits. The application of this theorem allows us to examine the effect of each individual source on the circuit. The theorem is stated as follows:

> The current in any element or voltage across any element of a linear multisource circuit (independent sources) is equivalent to the algebraic sum of the voltage or current due to each source acting alone.

The procedure for applying the Superposition Theorem to a circuit is as follows:

1. Replace all sources but one with their internal resistances. (Ideal voltage sources are replaced with a short circuit, and ideal current sources are replaced with an open circuit.)
2. Calculate the current or voltage of interest.
3. Repeat Steps 1 and 2 for all sources, leaving one active source connected to the circuit at a time.
4. Algebraically add the current or voltage values obtained due to each source to obtain the actual current or voltage of that element.

Superposition, or the process of *superposing,* applies to currents and voltages in the circuit. It is not valid for calculating power. Power is a square relationship; hence, it is not a linear relationship of current or voltage.

EXAMPLE 8.9

Determine the value of current I_X in the circuit in Figure 8.10a using the Superposition Theorem.

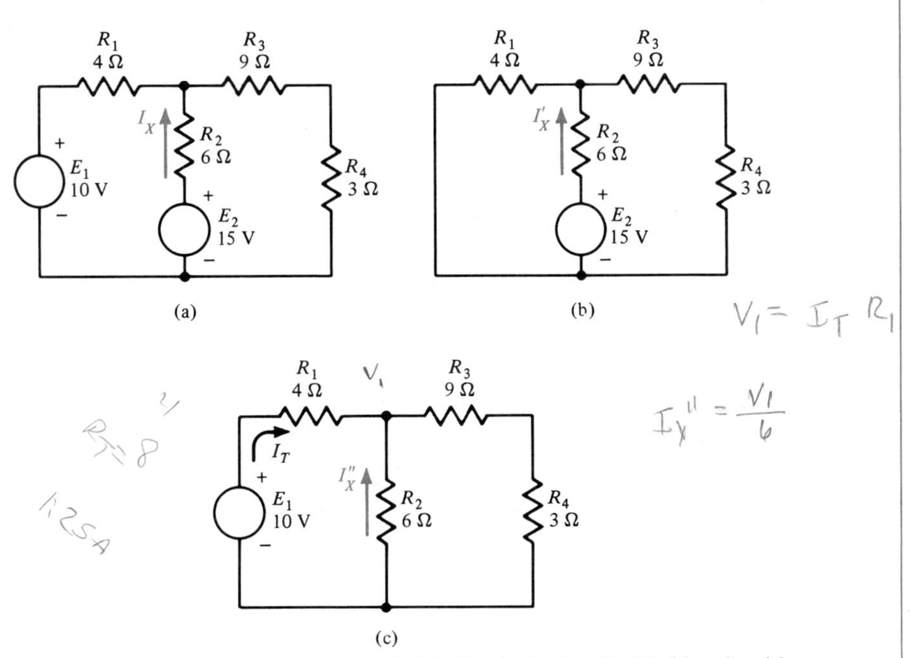

Figure 8.10 Circuit for Example 8.9. **(a)** Original circuit. **(b)** Circuit with 10-V source removed. **(c)** Circuit with 15-V source removed.

Solution

Following the procedure just outlined, the 10-V source is replaced by a short, and the only active source left is the 15-V source, as shown in Figure 8.10b. I_X denotes the unknown current due to the 15-V source only.

The total resistance connected to the 15-V source is:

$$R'_T = (R_3 + R_4)\|R_1 + R_2$$

$$= \frac{(9 + 3)4}{9 + 3 + 4} + 6 = 9\ \Omega$$

Then

$$I'_X = \frac{15}{9} = 1.67\ \text{A}$$

The effect of the 10-V source is calculated by replacing the 15-V source with a short and calculating I''_X, as shown in Figure 8.10c. The total resistance of the new circuit as "seen" by the source is:

$$R''_T = (R_3 + R_4)\|R_2 + R_1$$
$$= \frac{(9 + 3)(6)}{9 + 3 + 6} + 4 = 8\ \Omega$$

The total current is $I''_T = 10/8 = 1.25$ A, and the unknown current is found by the Current Divider Rule:

$$I''_X = -1.25\,\frac{(9 + 3)}{9 + 3 + 6} = -0.833\ \text{A}$$

Finally,

$$I_X = I'_X + I''_X$$
$$= 1.67 + (-0.833) = 0.833\ \text{A}$$

Example 8.10 illustrates the application of the Superposition Theorem to mixed sources.

EXAMPLE 8.10

Calculate the voltage drop, V_X, in the circuit of Figure 8.11a using the Superposition Theorem. Also determine the power dissipated by the 500-Ω resistor.

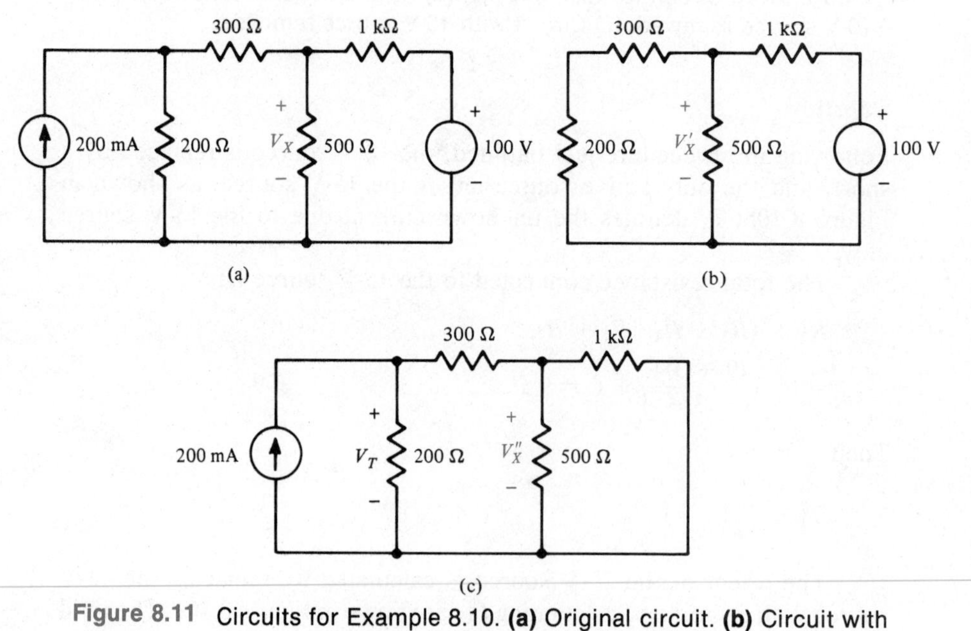

Figure 8.11 Circuits for Example 8.10. **(a)** Original circuit. **(b)** Circuit with 100-V source only. **(c)** Circuit with 200-mA source only.

Solution

Replacing the current source with an open circuit results in the circuit of Figure 8.11b. The voltage, V'_X, is the voltage across the parallel combination of:

$$R_P = (200 + 300) \| 500$$

$$= \frac{(200 + 300)(500)}{200 + 300 + 500} = 250\ \Omega$$

and V'_X is calculated by the Voltage Divider Rule:

$$V'_X = 100 \frac{250}{250 + 1000}$$

$$= 20\ \text{V}$$

Eliminating the 100-V source from the original circuit and replacing it with a short yields the circuit in Figure 8.11c. The voltage V''_X due to the current source only is determined next.

The total resistance across the current source is:

$$R''_T = (200) \| (300 + 500 \| 1\ \text{k}\Omega) = 152\ \Omega$$

The voltage across the source is:

$$V_T = 200 \times 10^{-3} \times 152 = 30.4\ \text{V}$$

and the voltage V''_X is determined by the Voltage Divider Rule:

$$V''_X = 30.4 \frac{1\ \text{k}\Omega \| 500}{(1\ \text{k}\Omega \| 500) + 300}$$

$$= 30.4 \frac{333.3}{333.3 + 300}$$

$$= 16\ \text{V}$$

The actual voltage V_X is:

$$V_X = V'_X + V''_X$$

$$= 20 + 16 = 36\ \text{V}$$

The power to the 500-Ω resistor is given by

$$P_X = \frac{V_X^2}{R}$$

$$= \frac{(36)^2}{500} = 2.59\ \text{W}$$

As mentioned earlier, power is not a linear relationship of voltage; hence, it cannot be calculated using the Superposition Theorem. The method for calculating the power dissipated by an element is to find the *actual* current or voltage of the element and then apply the corresponding squared relationship of current and voltage to calculate the power, as is shown in Example 8.10. Mathematically,

$$P_X = \frac{(V'_X + V''_X)^2}{R} \neq \frac{(V'_X)^2}{R} + \frac{(V''_X)^2}{R} \tag{8.3}$$

EXAMPLE 8.11

For the multisource circuit in Figure 8.12a, calculate the current I_X by the Superposition Theorem.

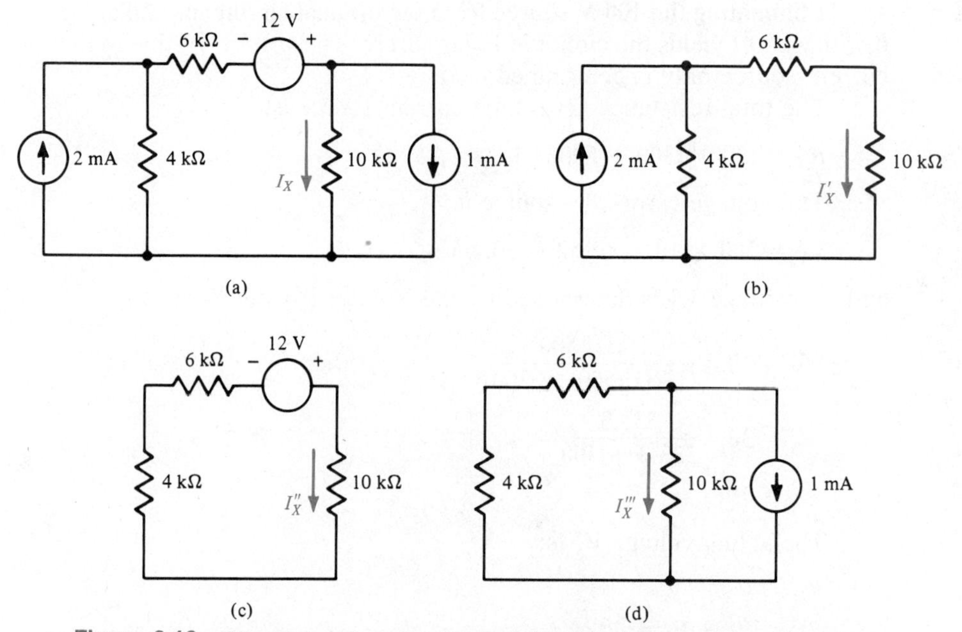

Figure 8.12 Circuit for Example 8.11. Using the Superposition Theorem to analyze the circuit. **(a)** Original circuit. **(b)** The effect of the 2-mA current source on the circuit. **(c)** The effect of the 12-V voltage source on the circuit. **(d)** The effect of the 1-mA current source on the circuit.

Solution

In this circuit two sources are eliminated at a time to calculate the effect that each source has on the circuit.

1. The 12-V and the 1-mA sources are eliminated; the resulting circuit is shown in Figure 8.12b. The current I'_X is now calculated

by the Current Divider Rule:

$$I'_X = 2 \times 10^{-3} \left(\frac{4 \times 10^3}{4 \times 10^3 + 16 \times 10^3}\right)$$
$$= 0.4 \times 10^{-3} = 0.4 \text{ mA}$$

2. Next, the 2-mA and the 1-mA sources are eliminated, resulting in the circuit of Figure 8.12c. The current I''_X is:

$$I''_X = \frac{12}{4 \times 10^3 + 6 \times 10^3 + 10 \times 10^3}$$
$$= 0.6 \times 10^{-3} = 0.6 \text{ mA}$$

3. Eliminating the 12-V and the 2-mA sources yields the circuit in Figure 8.12d, and the current I'''_X is found by applying the Current Divider Rule:

$$I'''_X = -1 \text{ mA} \frac{6 \times 10^3 + 4 \times 10^3}{6 \times 10^3 + 4 \times 10^3 + 10 \times 10^3}$$
$$= -0.5 \text{ mA}$$

Notice the negative value due to the opposite direction of the unknown current, I'''_X, with respect to the source. Finally

$$I_X = I'_X + I''_X + I'''_X$$
$$= 0.4 \text{ mA} + 0.6 \text{ mA} + (-0.5 \text{ mA}) = 0.5 \text{ mA}$$

8.4 MILLMAN'S THEOREM

Jacob Millman proposed a method for analyzing a certain type of multisource circuit by converting it to an equivalent circuit containing an ideal voltage source with its internal resistance in series. The circuit that lends itself to this equivalent circuit conversion must be constructed of practical voltage sources connected between two major nodes, as shown in Figure 8.13.

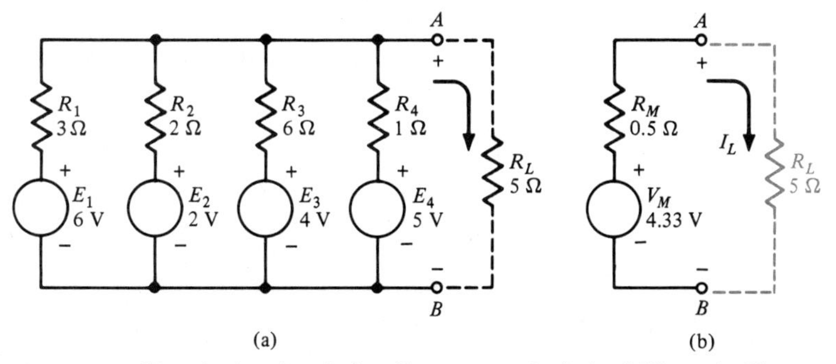

Figure 8.13 Circuit that lends itself to an analysis by Millman's Theorem. **(a)** Original circuit. **(b)** Equivalent circuit.

Millman's Theorem is stated as follows:

A multisource circuit containing exclusively practical voltage sources connected in parallel can be replaced by a simple equivalent voltage source.

The value of the equivalent ideal voltage source is the ratio of the sum of the short-circuit currents of the individual sources to the sum of the conductances in each original branch. Mathematically, it is expressed as follows:

$$V_M = \frac{\dfrac{E_1}{R_1} + \dfrac{E_2}{R_2} + \dfrac{E_3}{R_3} + \cdots + \dfrac{E_N}{R_N}}{\dfrac{1}{R_1} + \dfrac{1}{R_2} + \dfrac{1}{R_3} + \cdots + \dfrac{1}{R_N}} \tag{8.4}$$

where N is the number of parallel branches.

Millman's equivalent resistance, R_M, is the parallel combination of the resistances as measured between terminals A and B, with the ideal voltage sources replaced by a short circuit:

$$R_M = \frac{1}{\dfrac{1}{R_1} + \dfrac{1}{R_2} + \dfrac{1}{R_3} + \cdots + \dfrac{1}{R_N}} \tag{8.5}$$

EXAMPLE 8.12

In the circuit of Figure 8.13a, the values of the components are as shown.

a. Determine Millman's equivalent circuit.

b. If a 5-Ω load resistor is connected between terminals A and B, calculate the power dissipated by it.

Solution

a. By Eq. (8.4),

$$V_M = \frac{\dfrac{6}{3} + \dfrac{2}{2} + \dfrac{4}{6} + \dfrac{5}{1}}{\dfrac{1}{3} + \dfrac{1}{2} + \dfrac{1}{6} + \dfrac{1}{1}} = 4.33 \text{ V}$$

Millman's resistance is given by Eq. (8.5)

$$R_M = \frac{1}{\frac{1}{3} + \frac{1}{2} + \frac{1}{6} + \frac{1}{1}} = 0.5\ \Omega$$

The equivalent circuit is shown in Figure 8.13b.

b. Connecting the 5-Ω load resistor between terminals A and B, the current through the load is calculated as follows:

$$I_L = \frac{4.33}{0.5 + 5} = 0.787\ \text{A}$$

In many electronic circuits it is necessary to calculate the voltage at one point with respect to a reference point when a number of sources are connected to the same point. An example of it is an application of the operational amplifier (op-amp) used with multiple inputs. The op-amp is a very commonly used integrated circuit device. The schematic diagram of an op-amp circuit is shown in Figure 8.14a. The circuit is capable of supplying an output voltage, V_o, proportional to the sum of the input voltages.

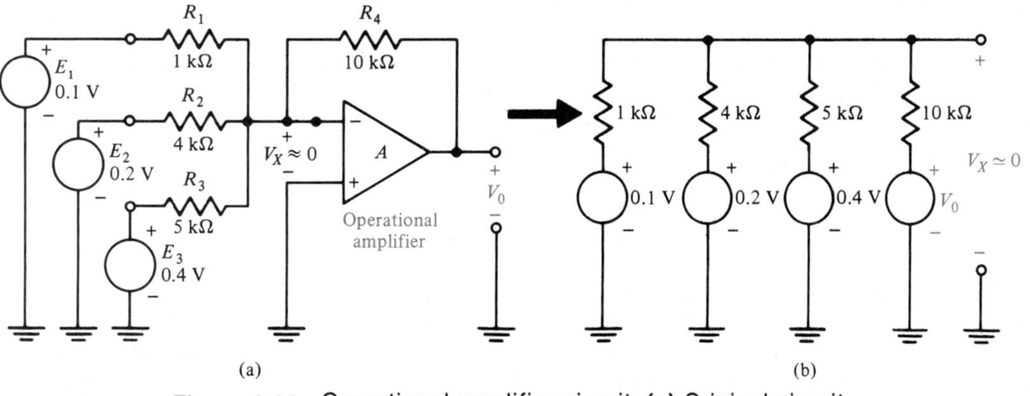

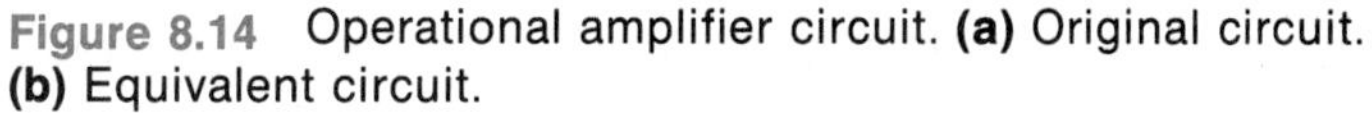

Figure 8.14 Operational amplifier circuit. **(a)** Original circuit. **(b)** Equivalent circuit.

EXAMPLE 8.13

Calculate the output voltage, V_o, in the circuit of Figure 8.14a.

Solution

The circuit diagram shows three input sources, E_1, E_2, and E_3, connected to the input of the amplifier, and the output voltage, V_o, is also *fed back* to the input side of the amplifier. The resistance of the op-amp between point X, referred to as the *summing point,* to ground is very high, and the voltage V_X is very small (approximately zero). Hence, the circuit for calculating V_X can be visualized as four practical voltage sources in parallel, as shown in Figure 8.14b. Assume the voltage V_X to be approximately zero; since it is the voltage measured across a very high resistance (the input of the amplifier), Eq. (8.4) gives

$$\frac{0.1/1 \times 10^3 + 0.2/4 \times 10^3 + 0.4/5 \times 10^3 + V_o/10 \times 10^3}{1/1 \times 10^3 + 1/4 \times 10^3 + 1/5 \times 10^3 + 1/10 \times 10^3} = V_M = V_X \cong 0$$

$$4.6 + \frac{2V_o}{31} \simeq 0$$

$$V_o = -2.3 \text{ V}$$

8.5 DELTA/WYE (Δ/Y) AND WYE/DELTA (Y/Δ) TRANSFORMATIONS

So far in this chapter we have presented various theorems used in analyzing electric circuits. Various methods of analysis were derived from these theorems, which simplified calculations. In this section we once again study a technique that is very useful in analyzing certain circuit configurations because it makes it possible to convert a circuit that otherwise would be analyzed by mesh or nodal methods to a simple series-parallel–type circuit.

The two component connections shown in Figure 8.15 are Y and Δ. The Y connection is sometimes also referred to as a T connection, since the two upper arms can be moved to resemble the letter *T,* and the inverted Δ (∇) is referred to as a π (pi) connection because it can be made to resemble π.

Any Y-connected network can be transformed to an equivalent Δ network and vice versa. In order for the two configurations to be equivalent, which means to exhibit the same characteristics when connected to a source between any of its terminals, the Y or the Δ configuration must have the same resistance when measured between corresponding terminals. Hence, the resistance between terminals A and B of the Y must be the same as between A and B of the Δ, and so forth.

From Table 8.1 and Figure 8.15c,

$$\boxed{R_A + R_C = \frac{R_1R_2 + R_1R_3}{R_1 + R_2 + R_3}} \qquad (8.6)$$

Similarly,

$$\boxed{R_A + R_B = \frac{R_1R_2 + R_2R_3}{R_1 + R_2 + R_3}} \qquad (8.7)$$

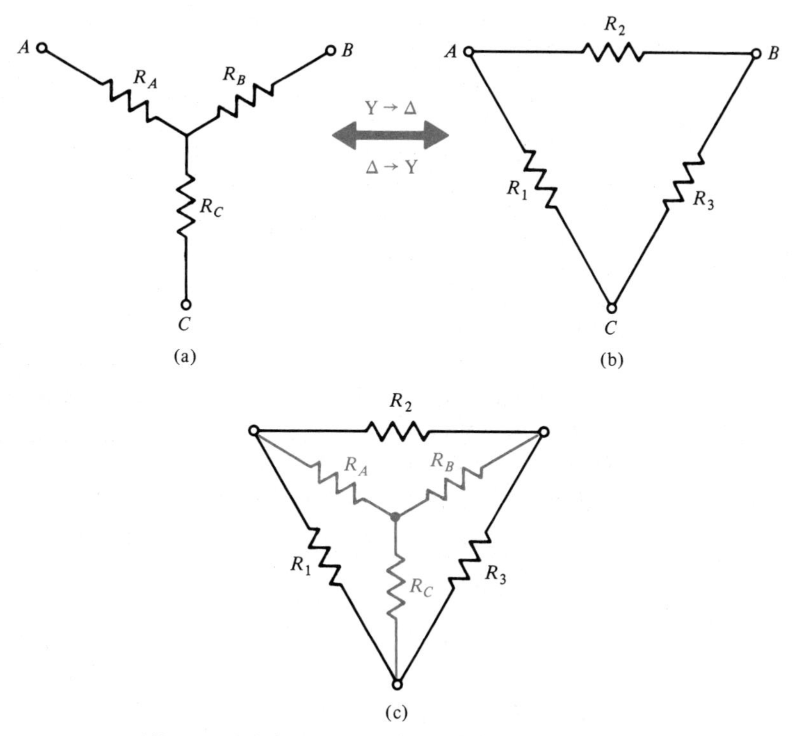

Figure 8.15 Wye and delta connections. **(a)** Wye. **(b)** Delta. **(c)** Delta and wye combined representation.

and

$$R_B + R_C = \frac{R_1R_3 + R_2R_3}{R_1 + R_2 + R_3} \tag{8.8}$$

Adding Eqs. (8.6) and (8.7) and subtracting Eq. (8.8) gives

$$(R_A + R_C) + (R_A + R_B) - (R_B + R_C)$$
$$= \frac{(R_1R_2 + R_1R_3) + (R_1R_2 + R_2R_3) - (R_1R_3 + R_2)}{R_1 + R_2 + R_3}$$

which results in

$$2R_A = \frac{2R_1R_2}{R_1 + R_2 + R_3}$$

or

$$R_A = \frac{R_1R_2}{R_1 + R_2 + R_3} \tag{8.9}$$

Similar algebraic manipulations result in

$$R_B = \frac{R_2 R_3}{R_1 + R_2 + R_3} \tag{8.10}$$

and

$$R_C = \frac{R_1 R_3}{R_1 + R_2 + R_3} \tag{8.11}$$

Equations (8.9), (8.10), and (8.11) are used to transform a Δ-configured connection to a Y connection.

To convert a Y connection to a Δ, additional algebraic manipulations with Eqs. (8.9), (8.10), and (8.11) are required, resulting in

$$R_1 = \frac{R_A R_B + R_B R_C + R_A R_C}{R_B} \tag{8.12}$$

$$R_2 = \frac{R_A R_B + R_B R_C + R_A R_C}{R_C} \tag{8.13}$$

$$R_3 = \frac{R_A R_B + R_B R_C + R_A R_C}{R_A} \tag{8.14}$$

From Eqs. (8.9) through (8.14) and Figure 10.15c, it becomes apparent that:

1. To convert a Δ connection to a Y connection, multiply the two Δ resistors on each side of the Y resistor that is being considered, and divide by the sum of the Δ resistances.
2. To convert a Y connection to a Δ, add the products of Y resistors taken two at a time and divide by the *opposite* Y resistor.

Y/Δ correspondences *are* listed in Table 8.1 below.

TABLE 8.1
Y/Δ Corresponding resistances

Terminals	Y	Δ
A–C	$R_A + R_C$	$R_1 \| (R_2 + R_3)$
A–B	$R_A + R_B$	$R_2 \| (R_1 + R_3)$
B–C	$R_B + R_C$	$R_3 \| (R_1 + R_2)$

The next three examples illustrate the transformation methods.

EXAMPLE 8.14

Transform the Δ connection shown in Figure 8.16a to a Y connection.

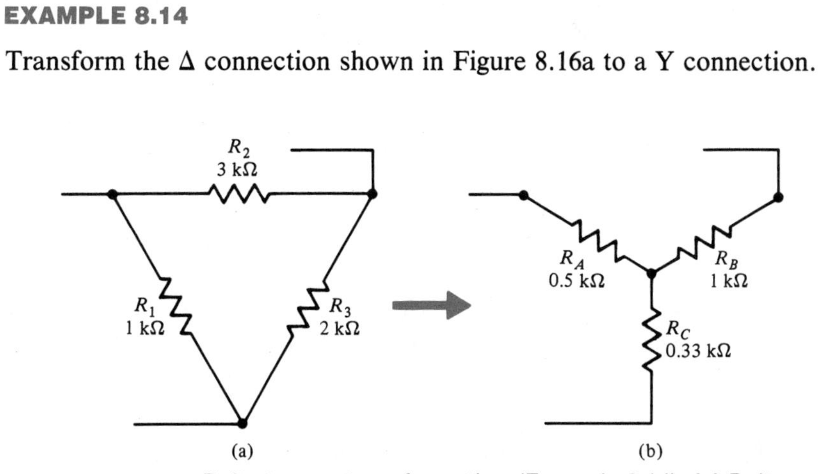

Figure 8.16 Delta to wye transformation (Example 8.14). **(a)** Delta connection. **(b)** Wye connection.

Solution

The Y component values are found by applying Eqs. (8.9), (8.10), and (8.11):

$$R_A = \frac{(1 \times 10^3)(3 \times 10^3)}{1 \times 10^3 + 3 \times 10^3 + 2 \times 10^3} = 0.50\ \text{k}\Omega$$

$$R_B = \frac{(3 \times 10^3)(2 \times 10^3)}{1 \times 10^3 + 3 \times 10^3 + 2 \times 10^3} = 1.0\ \text{k}\Omega$$

$$R_C = \frac{(1 \times 10^3)(2 \times 10^3)}{1 \times 10^3 + 3 \times 10^3 + 2 \times 10^3} = 0.33\ \text{k}\Omega$$

The corresponding Y configuration is shown in Figure 8.16b.

EXAMPLE 8.15

The circuit in Figure 8.17a is referred to as a bridge, also called the *Wheatstone bridge*. Calculate the total current supplied by the battery.

Solution

This problem cannot be solved by simply combining resistors in series or in parallel; however, it can be readily solved by mesh analysis. In this example two Δ's can be easily identified (shown as Δ_1 and Δ_2 on the

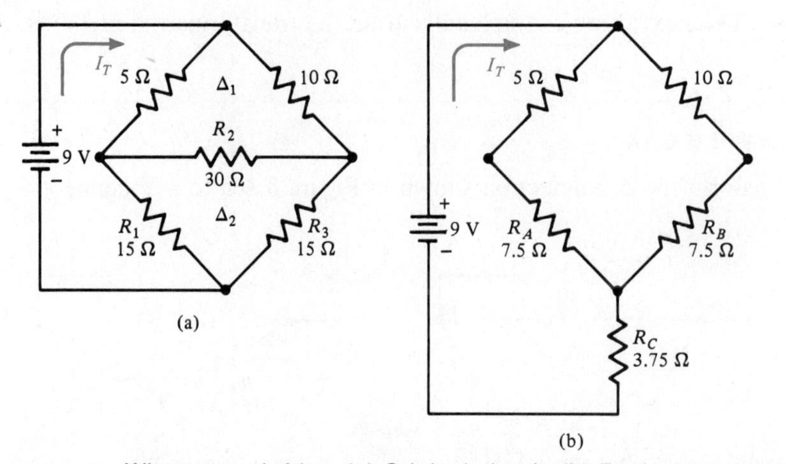

Figure 8.17 Wheatstone bridge. **(a)** Original circuit. **(b)** Equivalent circuit.

diagram), and we convert Δ_2 to a Y and solve the equivalent circuit as a simple series-parallel resistance problem. The equivalent circuit is shown in Figure 8.17b.

The values for R_A, R_B, and R_C are calculated again from Eqs. (8.9), (8.10), and (8.11):

$$R_A = \frac{(15)(30)}{15 + 30 + 15} = 7.5\ \Omega$$

$$R_B = \frac{(30)(15)}{15 + 30 + 15} = 7.5\ \Omega$$

$$R_C = \frac{(15)(15)}{15 + 30 + 15} = 3.75\ \Omega$$

The total resistance of the equivalent circuit is:

$$R_T = (5 + 7.5)\|(10 + 7.5) + 3.75$$

$$= \frac{(12.5)(17.5)}{12.5 + 17.5} + 3.75 = 11.04\ \Omega$$

Hence

$$I_T = \frac{9}{11.04} = 0.815\ \text{A}$$

EXAMPLE 8.16

Repeat Example 8.15, transforming a Y to a Δ.

Solution

Careful inspection of the circuit in Figure 8.17a reveals that there are two possible Y connections in this circuit. The circuit is redrawn in Figure 8.18a. Converting Y_1 to a Δ results in an equivalent circuit drawn in Figure 8.18b.

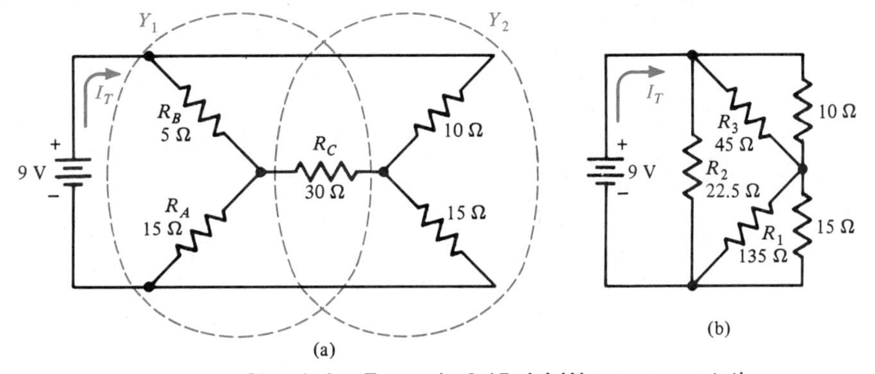

Figure 8.18 Circuit for Example 8.15. **(a)** Wye representation. **(b)** Equivalent circuit.

The resistance values for the Δ connections are calculated from Eqs. (8.12), (8.13), and (8.14):

$$R_1 = \frac{(15)(5) + (5)(30) + (15)(30)}{5} = \frac{675}{5} = 135\ \Omega$$

$$R_2 = \frac{675}{30} = 22.5\ \Omega$$

$$R_3 = \frac{675}{15} = 45\ \Omega$$

The total resistance is calculated next from the equivalent circuit.

$$\begin{aligned} R_T &= (45\,\|\,10 + 135\,\|\,15)\,\|\,22.5 \\ &= (8.18 + 13.5)\,\|\,22.5 = 11.04\ \Omega \end{aligned}$$

Therefore, the total current is:

$$I_T = \frac{9}{11.04} = 0.815\text{ A}$$

Note that this is the same result obtained in Example 8.15, as was expected.

The listings of two computer programs used in converting a Δ to a Y and a Y to a Δ are given next. The Δ/Y conversion was executed with Example 8.14 values, and the Y/Δ program was executed with values in Example 8.16.

```
0   REM  THIS PROGRAM TRANSFORMS A DELTA TO A WYE
0   REM  DELTA COMPONENTS- R1,R2,R3
0   REM  WYE COMPONENTS-RA,RB,RC
0   INPUT "R1=";R1
0   INPUT "R2=";R2
0   INPUT "R3=";R3
0 R = R1 + R2 + R3
0 RA = R1 * R2 / R
0 RB = R2 * R3 / R
00 RC = R1 * R3 / R
10  PRINT "RA=";RA;" OHMS"
20  PRINT "RB=";RB;" OHMS"
30  PRINT "RC=";RC;" OHMS"
40  END
```

```
RUN                                  (Example 8.14)
1=1E3
2=3E3
3=2E3
A=500 OHMS
B=1000 OHMS
C=333.333334 OHMS
```

```
0   REM  THIS PROGRAM TRANSFORMS A WYE TO A DELTA
0   REM  DELTA COMPONENTS-R1,R2,R3
0   REM  WYE COMPONENTS-RA,RB,RC
0   INPUT "RA= ";RA
0   INPUT "RA= ";RB
0   INPUT "RC= ";RC
0 R = RA * RB + RB * RC + RA * RC
0 R1 = R / RB
0 R2 = R / RC
00 R3 = R / RA
10  PRINT "R1= ";R1;" OHMS"
20  PRINT "R2= ";R2;" OHMS"
30  PRINT "R3= ";R3;" OHMS"
40  END
```

```
RUN                                  (Example 8.16)
A= 15
B= 5
C= 30
1= 135 OHMS
2= 22.5 OHMS
3= 45 OHMS
```

The Δ/Y and Y/Δ transformation method is very useful in the study of ac power circuits.

SUMMARY

Circuit theorems were introduced to aid in the understanding of the behavior of various circuits and to develop techniques for circuit analysis that simplify the solution for various electric quantities.

Thevenin's and *Norton's* theorems prove that a circuit can be drastically simplified to an equivalent practical source. This source contains an ideal voltage source with its internal resistance, called the Thevenin voltage, V_{th}, and resistance, R_{th}, or Norton's equivalent circuit, which contains an ideal current source with a parallel internal resistance, referred to as Norton's equivalent current, I_N, and resistance, R_N. Those two theorems are very useful in power circuits in matching loads for maximum power transfer and in simplifying the analysis of electronic circuits.

The *Superposition* Theorem provides a valuable tool in the analysis of multi-source linear circuits by providing a method of examining the effect that each source individually has on the circuit.

Millman's Theorem makes it possible to evaluate the voltage between common points when multiple practical sources are connected between them. It is an important tool in the analysis of electronic circuits.

Finally, the equations for transforming Δ/Y- and Y/Δ-connected configurations were derived. It was shown that many circuits can be converted to single series-parallel configurations by using the transformation method, where otherwise mesh or other more complicated analysis techniques would be used.

SUMMARY OF IMPORTANT EQUATIONS

$$R_{th} = R_N \tag{8.1}$$

$$I_N = \frac{V_{th}}{R_{th}} \tag{8.2}$$

$$V_M = \frac{\dfrac{E_1}{R_1} + \dfrac{E_2}{R_2} + \dfrac{E_3}{R_3} + \cdots + \dfrac{E_N}{R_N}}{\dfrac{1}{R_1} + \dfrac{1}{R_2} + \dfrac{1}{R_3} + \cdots + \dfrac{1}{R_N}} \tag{8.4}$$

$$R_M = \frac{1}{\dfrac{1}{R_1} + \dfrac{1}{R_2} + \dfrac{1}{R_3} + \cdots + \dfrac{1}{R_N}} \tag{8.5}$$

$$R_A = \frac{R_1 R_2}{R_1 + R_2 + R_3} \tag{8.9}$$

$$R_B = \frac{R_2 R_3}{R_1 + R_2 + R_3} \tag{8.10}$$

$$R_C = \frac{R_1 R_3}{R_1 + R_2 + R_3} \tag{8.11}$$

$$R_1 = \frac{R_A R_B + R_B R_C + R_A R_C}{R_B} \tag{8.12}$$

$$R_2 = \frac{R_A R_B + R_B R_C + R_A R_C}{R_C} \tag{8.13}$$

$$R_3 = \frac{R_A R_B + R_B R_C + R_A R_C}{R_A} \tag{8.14}$$

REVIEW QUESTIONS

8.1 Under what conditions are two circuits equivalent?

8.2 State Thevenin's Theorem in your own words.

8.3 What is the open-circuit voltage in Thevenin's equivalent circuit?

8.4 Describe the similarities of Thevenin's and Norton's equivalent circuits.

8.5 What is meant by "eliminating" the source in a circuit? Explain for voltage and current sources.

8.6 What is the relation between the short-circuit current and the open-terminal voltage between points *A* and *B* in a circuit?

8.7 State the Superposition Theorem.

8.8 Why is it not possible to apply the Superposition Theorem to calculate power in a circuit element?

8.9 What are the similarities between Millman's resistance and Thevenin's resistance?

8.10 Under what circumstances is Millman's Theorem used?

8.11 What is the purpose of transforming a Y configuration to a Δ and vice versa?

8.12 When transforming a Δ with equal resistance values, will the resultant Y elements also be equal?

8.13 Repeat Question 8.12 for a Y/Δ conversion.

PROBLEMS

8.1 Solve for the current, I_L, through the load in Figure 8.19 using Thevenin's Theorem.

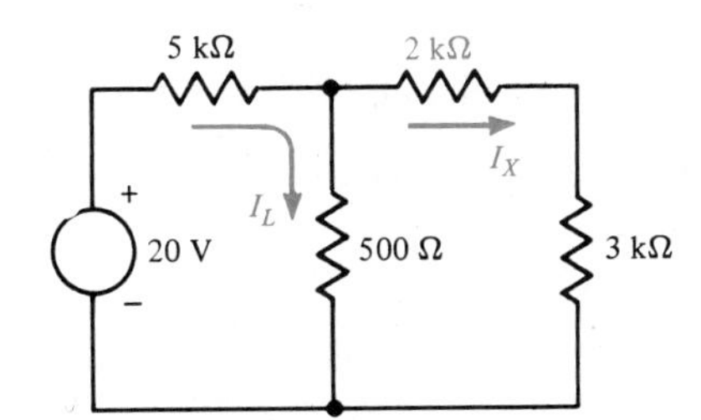

Figure 8.19 Circuit for Problems 8.1, 8.2, and 8.10.

8.2 Determine the current through the 2-kΩ resistor in Figure 8.19 using Thevenin's Theorem. (ans. 0.33 mA)

8.3 Calculate the voltage drop across the load resistor, R_L, in the circuit of Figure 8.20 using Thevenin's equivalent circuit.

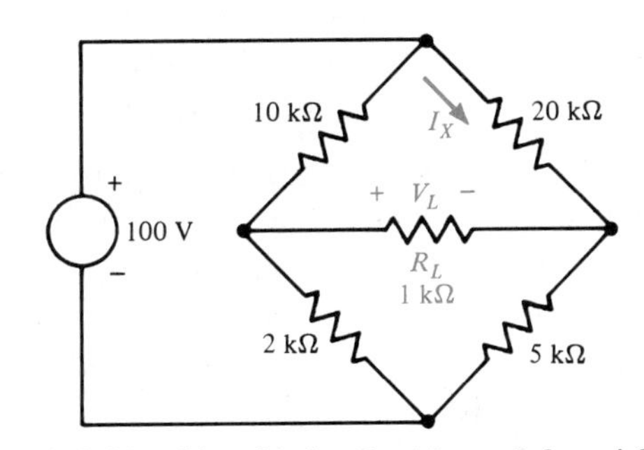

Figure 8.20 Circuits for Problems 8.3 and 8.4.

8.4 Calculate the current through the 20-kΩ resistor in the circuit of Figure 8.20 using Thevenin's Theorem. (ans. 3.87 mA)

8.5 Determine Thevenin's equivalent circuit between terminals *A* and *B* of the circuit in Figure 8.21.

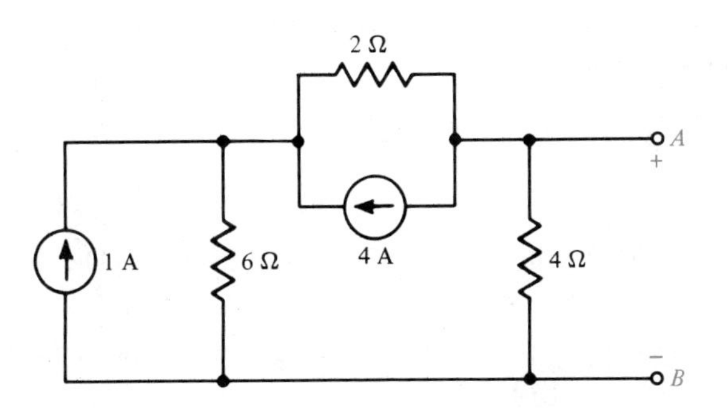

Figure 8.21 Circuit for Problems 8.5 and 8.6.

8.6 A load resistor of 4 Ω is connected between terminals A and B of the circuit in Figure 8.21. Calculate the power dissipated in the load. (ans. 4 W)

8.7 Determine Thevenin's equivalent circuit between terminals A and B for the circuit in Figure 8.22.

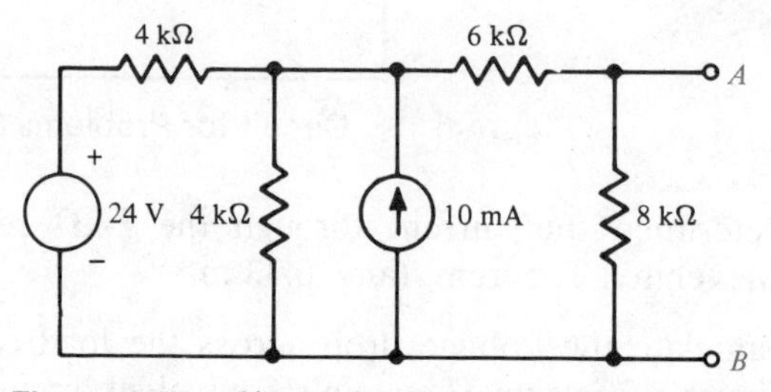

Figure 8.22 Circuit for Problems 8.7 and 8.8.

8.8 A load resistor of 1 kΩ is connected between terminals A and B of the circuit in Figure 8.22. Calculate the voltage drop across the load. (ans. 3.49 mA)

8.9 In the circuit of Figure 8.23, calculate the load current, I_L, using Thevenin's Theorem.

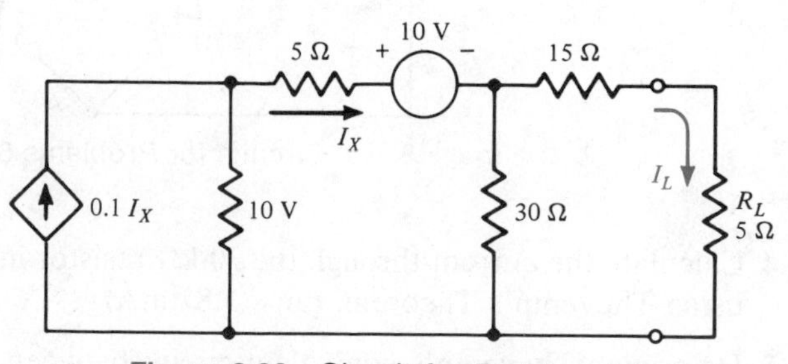

Figure 8.23 Circuit for Problem 8.9.

8.10 Repeat Problem 8.1 using Norton's Theorem. (ans. 3.33 mA)

8.11 Repeat Problem 8.3 using Norton's Theorem.

8.12 Repeat Problem 8.9 using Norton's Theorem. (ans. 0.23 A)

8.13 Determine Norton's equivalent circuit for the circuit in Figure 8.22.

8.14 Calculate the load current through the 5-Ω resistor in the circuit of Figure 8.24. (ans. −3.11 A)

8.15 Find the value of R_L in the circuit of Figure 8.25 for maximum power to be delivered to it.

8.16 Find the maximum power delivered to the load resistor R_L in Figure 8.25. (ans. 0.564 W)

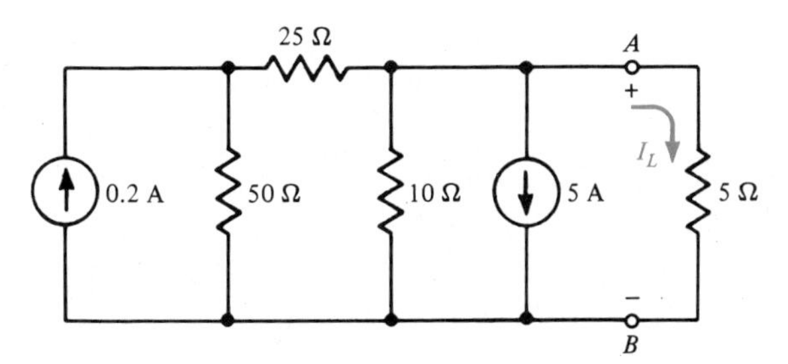

Figure 8.24 Circuit for Problems 8.13 and 8.14.

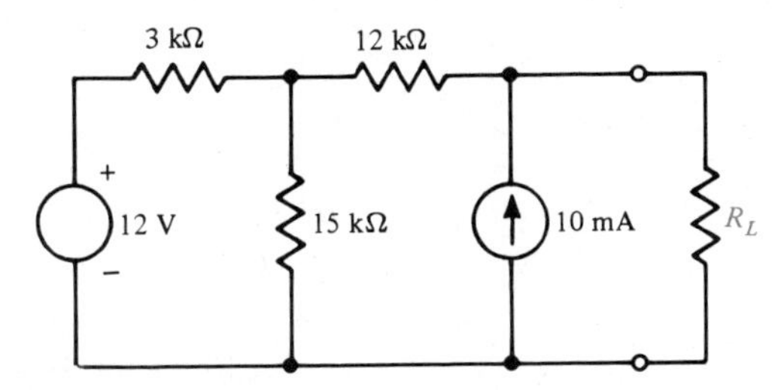

Figure 8.25 Circuit for Problems 8.15 and 8.16.

8.17 Solve for the current I_L in the circuit of Figure 8.24 by the Superposition Theorem.

8.18 Find the voltage V_X in the circuit of Figure 8.26 by the Superposition Theorem. (ans. 4.8 V)

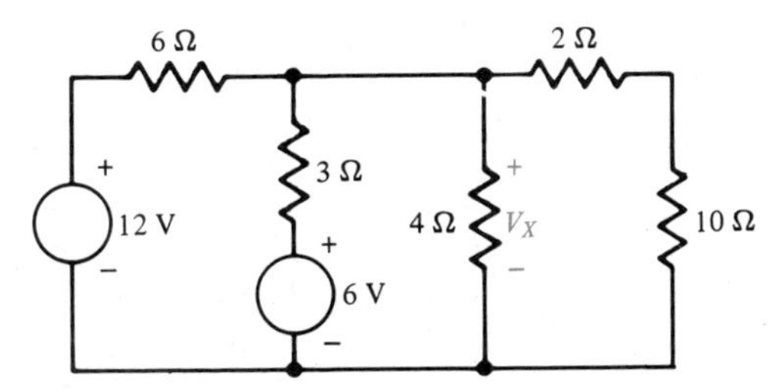

Figure 8.26 Circuit for Problems 8.18 and 8.19.

8.19 Repeat Problem 8.18 using Thevenin's equivalent circuit.

8.20 Determine the current I in Figure 8.27 using the Superposition Theorem. (ans. 9.87 mA)

8.21 Repeat Problem 8.20 using Norton's Theorem.

8.22 Calculate the current I_L in the circuit of Figure 8.28 using Millman's Theorem. (ans. 0.55 mA)

8.23 Repeat Problem 8.22 using the Superposition Theorem.

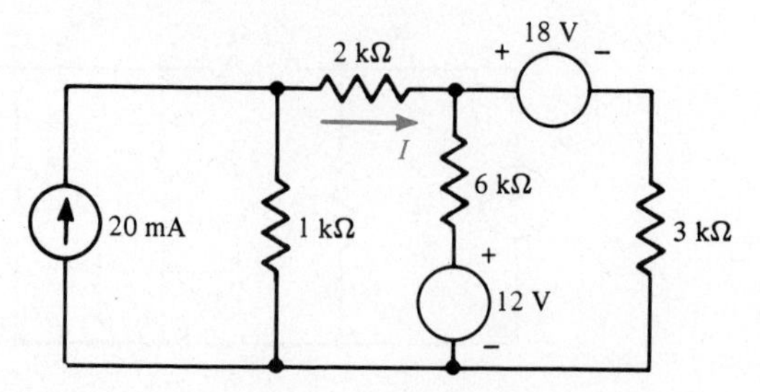

Figure 8.27 Circuit for Problems 8.20 and 8.21.

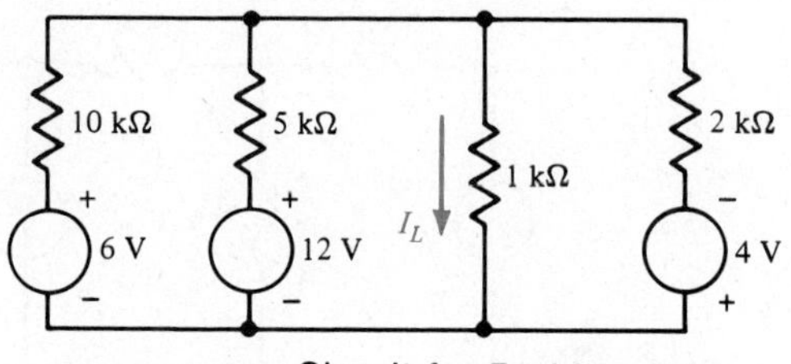

Figure 8.28 Circuit for Problem 8.22.

8.24 Determine voltage V_o in the circuit of Figure 8.29 using Millman's Theorem. (ans. −0.752 V)

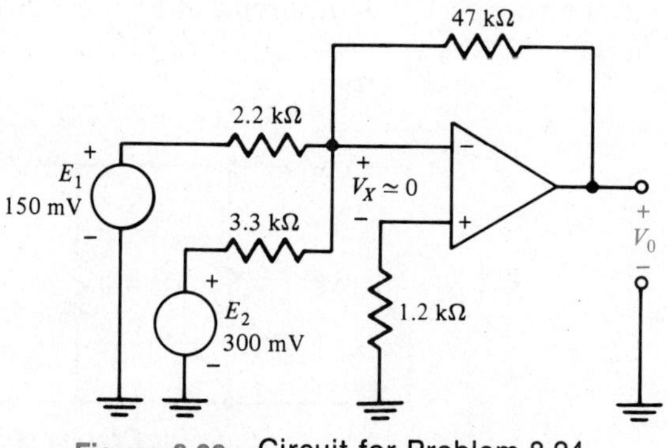

Figure 8.29 Circuit for Problem 8.24.

8.25 Solve for the current in resistor R_2 of the circuit in Figure 8.30 using Y/Δ transformation.

8.26 Solve for the current, I_X, in the circuit of Figure 8.30 using Δ/Y transformation. (ans. 1.59 mA)

8.27 Use Thevenin's Theorem to solve for the voltage, V_X, across the 1-kΩ resistor in Figure 8.30.

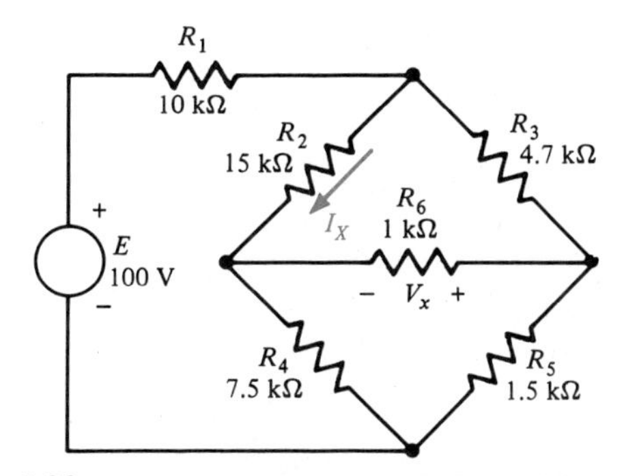

Figure 8.30 Circuit for Problems 8.25, 8.26, and 8.27.

8.28 Use a Y/Δ transformation to solve for the power dissipated by the load resistor, R_L, in the circuit of Figure 8.31. (ans. 1.86 W)

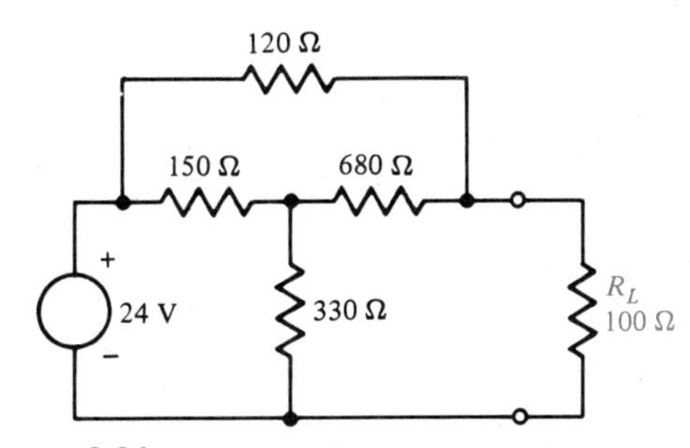

Figure 8.31 Circuit for Problems 8.28 and 8.29.

8.29 What should the value of R_L be in the circuit of Figure 8.31 for maximum power to be dissipated in the load resistor, R_L?

8.30 Use three different analysis methods to solve for the current I_X. Refer to the circuit of Figure 8.32. (ans. 1.92 A)

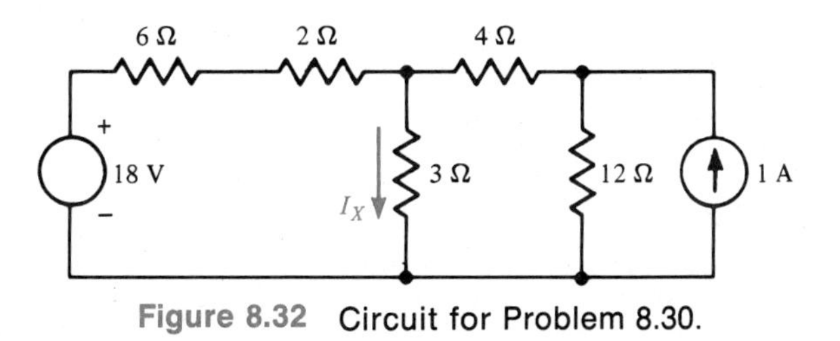

Figure 8.32 Circuit for Problem 8.30.

SECTION III

ALTERNATING CURRENT FUNDAMENTALS

CHAPTER
9

The Sinusoid and Other Time-Varying Functions

GLOSSARY

ac—acronym for alternating current
ac current—same as alternating current
ac voltage—same as alternating voltage
angular velocity (ω)—the rate at which an angle changes measure in radians per second (rad/s); same as radian frequency
average value—the mean value over a period of time
complex numbers—numbers having a real and an imaginary part
complex plane—a plane described by a real and an imaginary axis
cycle—one repetition of a periodic waveform
cycles per second (cps)—an older unit for frequency
degree—angular measurement (a complete circle contains 360°)
effective value—value indicating the energy content of a waveform; same as RMS
frequency—the number of repetitions per second
gradian—angular measurement (a complete circle contains 400 grad)
hertz (Hz)—unit of frequency
in phase—not having any time or angular difference
instantaneous value—the value of a varying waveform at a particular instant in time
lag—when a waveform starts after the reference
lead—when a waveform starts before the reference
magnetic field—the region of "altered space" that will interact with the poles of a magnet
peak value—the maximum value a varying waveform achieves
peak-to-peak value—the value between the negative and the positive peaks
period (T)—the time in seconds needed for one repetition of a repeating waveform
phase angle—the angular difference between a waveform and its reference
phasor—a line rotating at a constant speed; a number that represents a quantity that can be described by a phasor
phasor notation—representation of electrical components by their describing phasors
polar form—the form of a complex number having a magnitude and an angle
radian (rad)—angular measurement (a complete circle contains 2π rad)
radian frequency (ω)—same as angular velocity ($\omega = 2\pi f$)
rectangular form—the form of a complex number having a real and an imaginary part ($a + jb$)
reference waveform—a waveform to which others are compared
RMS value—root mean square value; same as effective value
waveshape—the form of a varying voltage or current

INTRODUCTION

Most electrical energy today is generated and transferred from place to place in the form of sinusoidally varying voltages and currents. It is imperative that the technologist be very familiar with this waveform and all the terminology used to describe it.

This chapter treats the analysis of sinusoidal functions and defines and discusses the related concepts of time and frequency, phase relationships, peak values, peak-to-peak values, and RMS values. Since some signal waveforms are not sinusoidal, other waveforms such as square and triangular are also examined for completeness. Phasors and complex numbers, which are the mathematical tools used to handle ac circuits, are also presented.

9.1 AC AND DC CONCEPTS

A current that repetitively and regularly reverses direction after the end of a fixed period of time is called an **alternating current.** If such an alternating current is maintained through a resistor, it causes a voltage drop that reverses its polarity with the same repetition and regularity as the alternating current. Such a voltage that changes its polarity repetitively and regularly is called an **alternating voltage.** It would seem reasonable to call an alternating current **ac** and an alternating voltage *av,* but through the strange wisdom of the evolution of electrical terminology, an alternating current is referred to as an *ac current,* which means an alternating current current. An alternating voltage is referred to as an *ac voltage,* which means an alternating current voltage. This terminology may initially cause confusion, but after a while a technologist begins to accept it. Even though this kind of terminology is redundant, it is nevertheless in popular use today.

The way in which an alternating current reverses direction or an alternating voltage reverses polarity determines the name given to the alternating current or voltage. If the repetitive reversals of the current direction or the repetitive reversals of the voltage polarity are viewed over a time interval that allows the observation of several reversals or alternations, the changing magnitude of the current or voltage depicts and characterizes the alternations. The shape that is produced is referred to as the **waveshape** of the alternating current or voltage. From the many waveshapes that are possible for alternating currents and voltages, Figure 9.1 shows three very common ones: namely, a square wave, a triangular wave, and a sinusoidal wave.

Currents or voltages whose values remain constant all the time are referred to as direct currents and direct voltages. The abbreviation for direct current is dc. The same redundant terminology that exists for ac also exists for dc. A direct current is referred to as a dc current, and a direct voltage is referred to as a dc voltage. Figure 9.2a shows a 15-V positive dc voltage. Figure 9.2b shows a −20-mA dc current.

Currents or voltages that neither alternate in direction nor remain constant are referred to as *pulsating* direct currents or voltages. Figure 9.3 shows examples of this type of waveform. Pulsating waveforms may be thought of as superpositions of alternating and direct currents or voltages.

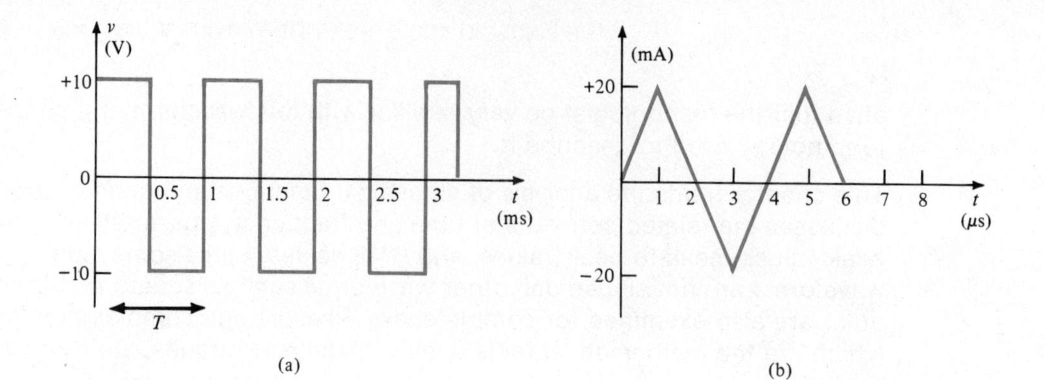

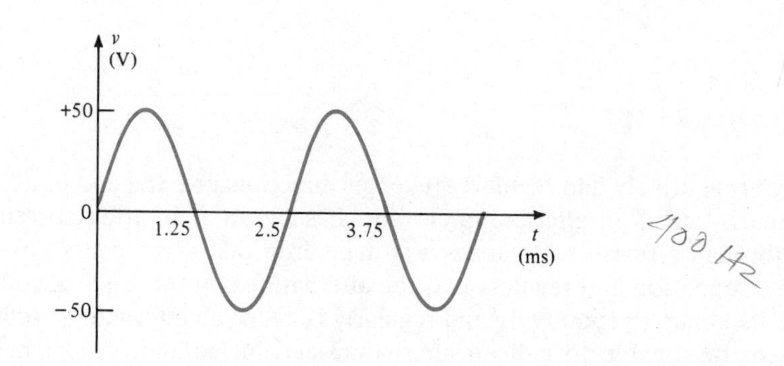

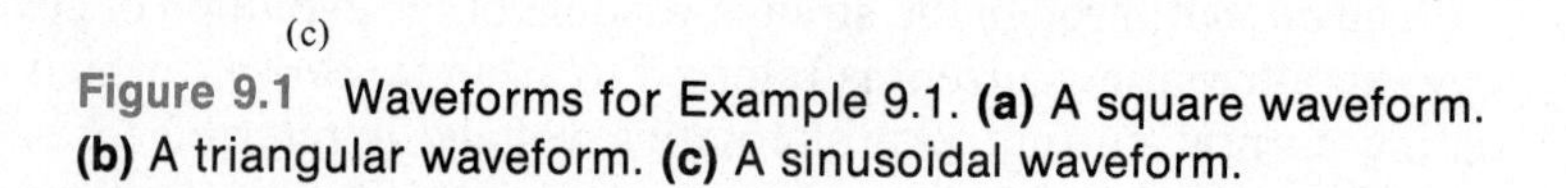
Figure 9.1 Waveforms for Example 9.1. **(a)** A square waveform. **(b)** A triangular waveform. **(c)** A sinusoidal waveform.

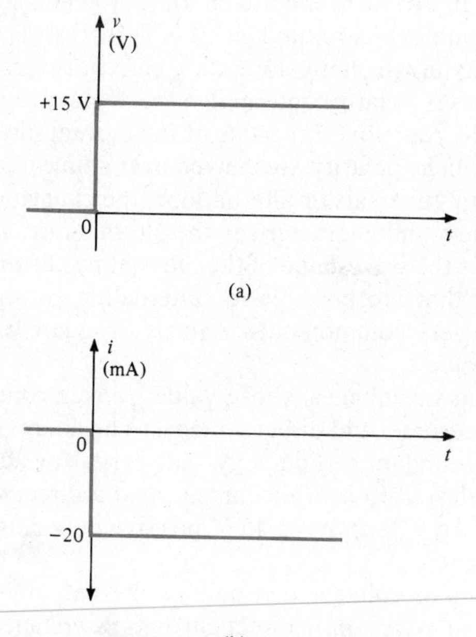

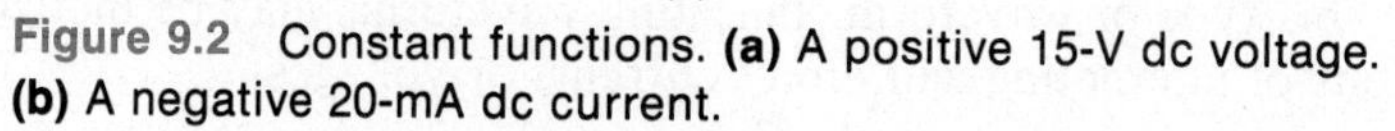
Figure 9.2 Constant functions. **(a)** A positive 15-V dc voltage. **(b)** A negative 20-mA dc current.

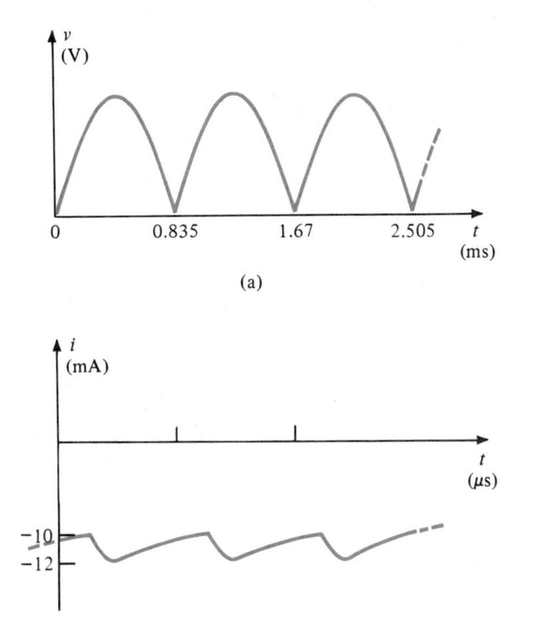

Figure 9.3 Pulsating waveforms. **(a)** Positive pulsating voltage waveform. **(b)** Negatively pulsating current waveform with a negative dc level.

9.2 PERIOD AND FREQUENCY

The length of time that it takes for a repetitive waveform to complete one repetition is defined to be the ***period*** *of the waveform.*

The abbreviation for period is $\boldsymbol{T}$. Figure 9.1a shows an alternating voltage square wave that is at +10 V for the first 0.5 ms and then reverses its polarity to −10 V for the next 0.5 ms. These two alternations form one repetition of the waveform. The period therefore is 1 ms (0.5 ms + 0.5 ms). Figure 9.1b shows a triangular alternating current waveform. The current value rises linearly from 0 to +20 mA in 1 μs. The current value then falls linearly to zero in one more microsecond. The current then reverses direction and falls linearly in value to −20 mA in another microsecond. The current then goes linearly back to zero in another microsecond. The process then starts again, and it repeats over and over. The period of this triangular alternating current waveshape is 4 μs because it takes this long to complete two alternations (1 repetition). Figure 9.1c shows an alternating voltage varying repetitively in a sinusoidal shape. It takes 0.625 ms for the voltage to reach +50 V from zero and another 0.625 ms to fall back down to zero. The voltage then changes polarity, and it takes another 0.625 ms to drop to −50 V and an additional 0.625 ms to climb back to zero and complete the repetition. The period of the sinusoid then is 2.5 ms (0.625 ms + 0.625 ms + 0.625 ms + 0.625 ms). The sinusoidal waveshape occurs quite commonly in physical systems and is a very important waveshape to analyze. Later it is shown how an alternating current and an alternating voltage with sinusoidally varying magnitudes may be produced.

Another name for the *repetition* of a waveshape is **cycle.** Thus the period of 1 repetition or 1 cycle of the square voltage waveform shown in Figure 9.1a is 1 ms.

The number of repetitions or cycles that a periodic waveform goes through in 1 s is defined as the ***frequency*** *of the waveform.*

Frequency is represented by **f** and is measured in **hertz** (Hz) (it used to be measured in **cycles per second,** or cps). A repetitive waveform has a frequency of 1 Hz if it repeats once in 1 s. If 2 cycles occur in 1 s (2 repetitions in 1 s) then the frequency is 2 Hz. If 1000 cycles occur in 1 s, then the frequency is 1000 Hz, or 1 kHz.

It should be apparent that if 2 cycles occur in 1 s, then each cycle lasts 0.5 s. Therefore, the period of a 2-Hz waveform is 0.5 s. It also should be obvious that if there are 1000 cycles in 1 s, then each repetition lasts $\frac{1}{1000}$ of a second, or 0.001 s, or 1 ms. Therefore, the period of a 1000-Hz waveform is 1 ms. Repeating this reasoning it becomes apparent that the period is the reciprocal of the frequency. The relationship between the period and frequency of a repetitive waveform is:

$$\boxed{f = \frac{1}{T}} \qquad \text{(9.1a)}$$

$$\boxed{T = \frac{1}{f}} \qquad \text{(9.1b)}$$

where: f => frequency in hertz (Hz)
T => period in seconds (s)

EXAMPLE 9.1

For the waveforms shown in Figure 9.1, determine:

a. The number of cycles shown

b. The period

c. The frequency of each waveform

Solution

a. The square wave repeats 3 times, the triangular wave repeats $1\frac{1}{2}$ times, and the sinusoid repeats 2 times.

b. The period for the square wave is 1 ms ($T_1 = 1$ ms); the triangular wave has a period of 4 μs ($T_2 = 4\ \mu$s). The sinusoid has a period of 2.5 ms ($T_3 = 2.5$ ms).

c. The frequency for each waveform is:

$$f_1 = \frac{1}{T_1} = \frac{1}{1\text{ ms}} = 1000\text{ Hz} = 1\text{ kHz}$$

$$f_2 = \frac{1}{T_2} = \frac{1}{4\ \mu\text{s}} = 250{,}000\text{ Hz} = 250\text{ kHz}$$

$$f_3 = \frac{1}{T_3} = \frac{1}{2.5\text{ ms}} = 400\text{ Hz}$$

EXAMPLE 9.2

Compute the period of a 60-Hz sinusoidal alternating voltage.

Solution

$$T = \frac{1}{f} = \frac{1}{60} = 0.0167\text{ s} = 16.7\text{ ms}$$

9.3 DEGREES, RADIANS, AND GRADIANS

The space between two straight lines meeting at a point is referred to as an angle. An angle may be measured in **degrees** *(°),* **radians** *(rad), or* **gradians** *(grad).*

Figure 9.4a shows two lines that intersect to divide a plane into four equal parts. The point at which the lines intersect is called the *origin*. The portion of the horizontal line extending to the right of the origin represents the positive x axis. The portion of the horizontal line extending to the left of the origin represents the negative x axis. The portion of the vertical line extending up from the origin represents the positive y axis, and the portion of the vertical axis extending down from the origin represents the negative y axis. The plane is said to be divided into four equal quadrants, as shown in Figure 9.4b.

The four resulting angles are said to be *right angles* and the two intersecting lines are said to be *orthogonal* (at right angles) to one another.

Each *right angle* is defined to contain 90°. Figure 9.4c shows a circle superimposed on the orthogonal set of axes. Note, from observing the diagram, that a circle contains 360°.

Another system of angular measurement defines a *right angle* to contain *100 grad*. From observing the diagram in Figure 9.4d, it should be apparent that a circle contains 400 grad.

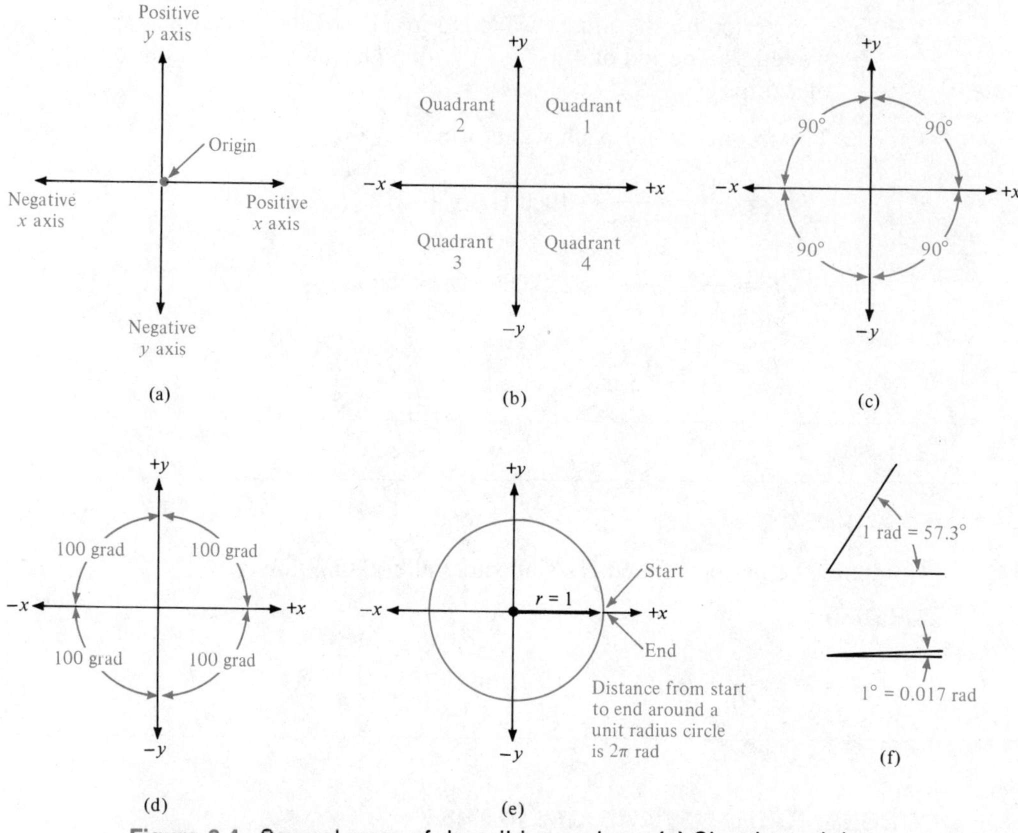

Figure 9.4 Several ways of describing a plane. **(a)** Showing origin and positive and negative x and y axes. **(b)** Showing four quadrants. **(c)** Showing 90° per quadrant. **(d)** Showing 100 grad per quadrant. **(e)** Showing 2π rad as the circumference of a circle with unity radius. **(f)** Showing the comparison between 1 rad and 1°.

Figure 9.4e shows a circle with a unit radius. The circumference of this circle (circumference $= 2\pi r$) is 2π ($\pi = 3.14\ldots$) times the unit radius. The angular distance around the circle is defined to be 2π rad.

Comparing Figures 9.4c, 9.4d, and 9.4e it should be clear that 360°, 400 grad, and 2π rad are all equivalent to one another.

$$360° = 400 \text{ grad} = 2\pi \text{ rad}$$

Gradians are not in popular use in electrical work; therefore, only degrees and radians are discussed from now on.

Since 360° are equivalent to 2π rad,

$$2\pi \text{ rad} = 360°$$

$$1 \text{ rad} = \frac{360°}{2\pi} = \frac{360°}{6.28} = 57.3°$$

$$1° = \frac{2\pi}{360°} = \frac{6.28}{360°} = 0.017 \text{ rad}$$

Therefore,

$$1 \text{ rad} = 57.3^\circ$$

$$1^\circ = 0.017 \text{ rad}$$

Figure 9.4f shows an angle of 1 rad (57.3°) and an angle of 1° (0.017 rad).

EXAMPLE 9.3

Convert to radians:

a. $a_1 = 30^\circ$

b. $a_2 = 45^\circ$

c. $a_3 = 250^\circ$

Solution

a. $a_1 = 30^\circ \text{ (0.017 rad/degree)} = 0.523 \text{ rad}$

b. $a_2 = 45^\circ \text{ (0.017 rad/degree)} = 0.785 \text{ rad}$

c. $a_3 = 250^\circ \text{ (0.017 rad/degree)} = 4.361 \text{ rad}$

EXAMPLE 9.4

Convert to degrees:

a. $a_1 = 2 \text{ rad}$

b. $a_2 = 3.3 \text{ rad}$

c. $a_3 = 4 \text{ rad}$

Solution

a. $a_1 = 2 \text{ rad } (57.3^\circ/\text{rad}) = 114.6^\circ$

b. $a_2 = 3.3 \text{ rad } (57.3^\circ/\text{rad}) = 189.1^\circ$

c. $a_3 = 4 \text{ rad } (57.3^\circ/\text{rad}) = 229.2^\circ$

9.4 GENERATION OF AN ALTERNATING VOLTAGE

Any magnet has a north and a south pole. By experimenting with magnets, the basic law of magnetism may be arrived at: *Like poles repel each other, whereas unlike poles attract each other.* These magnetic forces of attraction or repulsion are invisible. It is, therefore, helpful to represent these forces as imaginary lines. This region of influence around a magnet is called a **magnetic field.** The lines of force are assumed to go from the north pole through the surrounding region, into the south pole, through the magnet, and then back out of the north pole. If the magnet is bent so that the north and south poles come close together, the magnetic field is intensified within the gap between the two poles. Figure 9.5a shows a bar magnet with the imaginary lines of force. Figure 9.5b shows a magnet bent in the form of a horseshoe with the magnetic field concentrated in the gap between the poles. A magnetic field is the area around a magnet where magnetic forces can be detected.

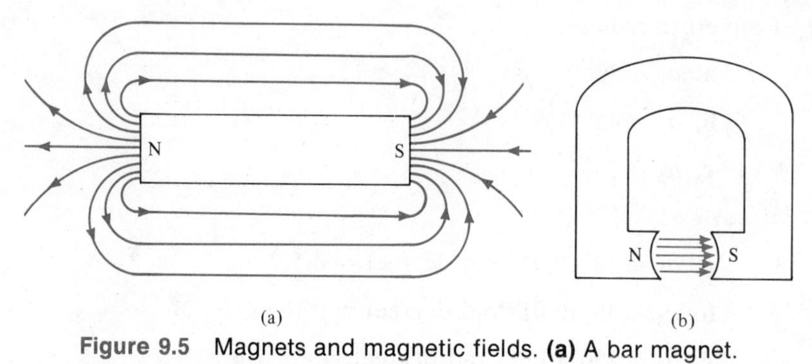

Figure 9.5 Magnets and magnetic fields. **(a)** A bar magnet. **(b)** A horseshoe magnet.

When a conductor moves within a magnetic field, a voltage is induced (generated) across the conductor. The magnitude of the voltage depends on the angle at which the conductor crosses the lines of force. Crossing at 0°, there is no voltage induced; crossing at 90°, a maximum voltage is induced. Crossing at intermediate angles, intermediate voltages are induced. The polarity of the induced voltage depends on the direction in which the conductor crosses the lines of force. Crossing one way induces a positive voltage, whereas crossing in the opposite direction induces a negative voltage.

An alternating sinusoidal voltage generator consists of a conductor rotating in a magnetic field. Figure 9.6a shows the north and south poles of a horseshoe magnet with a conductor placed in the magnetic field. Picture the field parallel to the paper with the conductor protruding out of the paper in a perpendicular fashion. Figure 9.6b shows the magnitude and polarity of the induced voltage as the conductor is rotated 360° through the magnetic field in the direction shown.

Note that at 0° the conductor is moving parallel to the magnetic field, resulting in no induced voltage. Between 0° and 180° the conductor is moving up and then down. At 90° the conductor is moving directly across and perpendicular to the magnetic field, causing the maximum induced voltage. At 180°, the motion of the conductor is again parallel to the magnetic field, inducing no voltage. Between 180° and 360°, the conductor is moving down and then up through the magnetic field. This induces a voltage opposite in polarity to the one between 0° and 180° induced when it was moving up and then down. At 270° the induced voltage is maximum because the conductor is moving perpendicular to the magnetic field. At

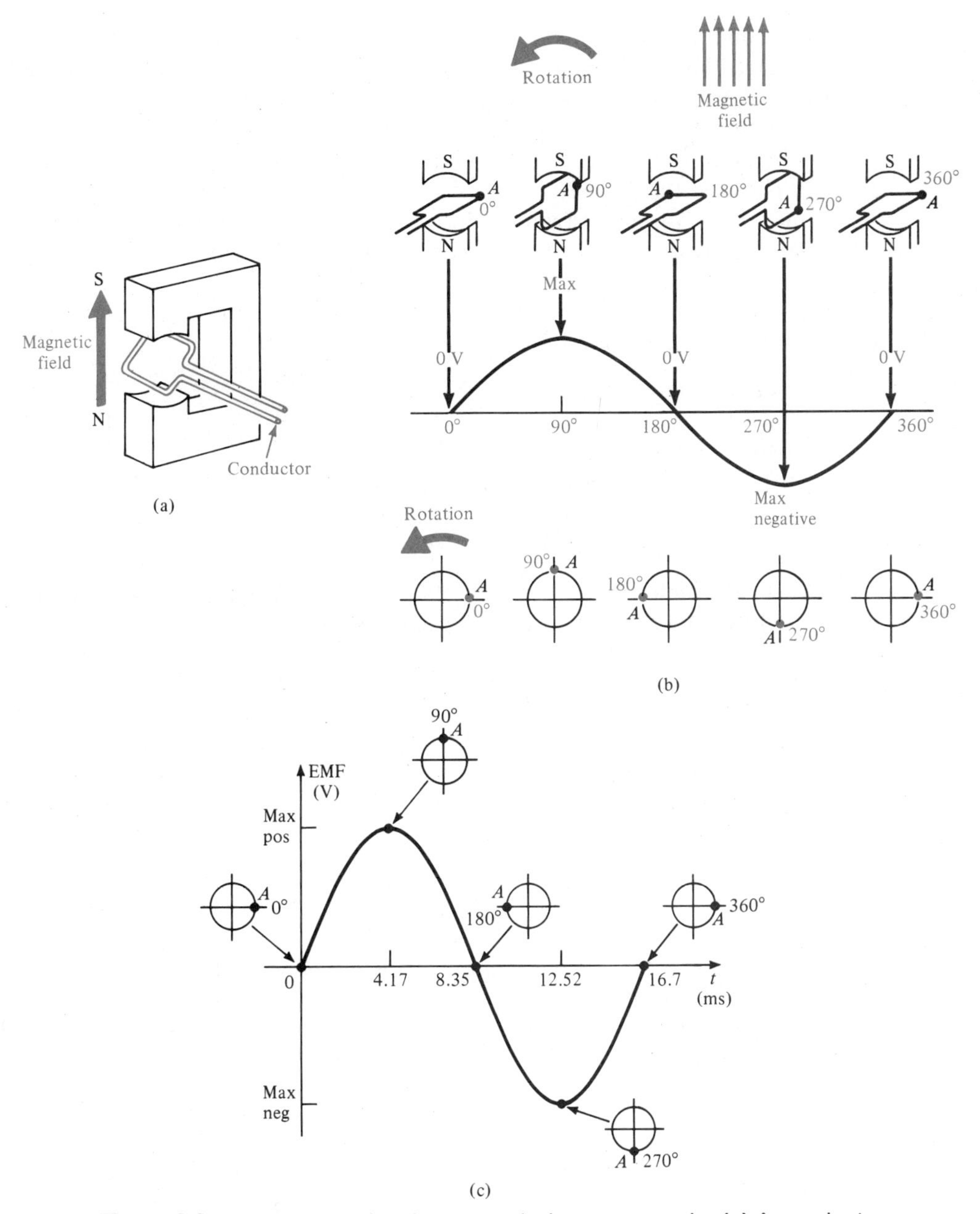

Figure 9.6 Generation of a sine wave during an ac cycle. **(a)** A conductor in a magnetic field. **(b)** The conductor rotates and interacts with the magnetic field generating the voltage waveform shown. **(c)** Curve showing the correlation between time and induced voltage during voltage generation.

360° the conductor is again moving parallel to the magnetic field, therefore inducing no voltage. The waveform in Figure 9.6b shows the instantaneous magnitude of the voltage as the conductor rotates once. It is clear that this waveform represents 1 repetition, or 1 cycle, and lasts 360°. It should be noted that it takes time for the conductor to be rotated through the 360°. The time required for 1 rotation (1 cycle) is called the **period,** which is abbreviated as T. Figure 9.6c shows the same sinusoidally varying waveform drawn as the magnitude of the induced voltage varies with respect to the time that it takes the conductor to reach a certain point in its rotation.

By comparing Figures 9.6b and 9.6c, the relationship between the angle rotated and the time required for the rotation may be observed to be:

θ = angle rotated
= 2π rad   for one rotation
t = time required
= T seconds   for one rotation

9.4.1 Angular Velocity (Radian Frequency)

The velocity at which the conductor is rotating is called the **angular velocity** and is represented by the Greek letter ω (lowercase omega).

$$\boxed{\omega = \frac{\theta}{t}} \tag{9.2}$$

$$\boxed{\omega = \frac{2\pi}{T}} \tag{9.3}$$

where: ω => angular velocity in radians per second (rad/s)
π => constant (3.14)
T => period in seconds (s)

Since $f = 1/T$, then

$$\boxed{\omega = 2\pi f} \tag{9.4}$$

where: ω => radian frequency (rad/s)
π => constant (3.14)
f => frequency in hertz (Hz)

In this last form of the relationship, ω is referred to as the **radian frequency.**

9.4.2 Instantaneous Values, Peak Values, and Peak-to-Peak Values

The magnitude of the voltage at any one time or at any angle is called the **instantaneous value.** The instantaneous value of a voltage is usually represented in an equation by v, whereas the instantaneous value of a current is represented by i.

*The **peak** value is the **maximum** value the waveform achieves.*

Note that in an alternating waveform, there is always a positive maximum and (an equal but opposite in polarity) negative maximum (also called a minimum). There is, therefore, a positive peak and a negative peak.

*The **peak-to-peak value** is the value from the positive maximum, or peak, to the negative maximum, or peak.*

The peak-to-peak value is twice the peak value.

$$V_{PP} = 2V_P \tag{9.5a}$$

$$I_{PP} = 2I_P \tag{9.5b}$$

EXAMPLE 9.5

The diagram in Figure 9.7a shows a sinusoidal voltage variation. The horizontal axis is labeled in degrees of rotation, the equivalent radian angles, and the time (in milliseconds) it takes the conductor to rotate to that angle.

Obtain the instantaneous values of the voltage at:

a. $a_1 = 40°$

b. $a_2 = 2$ rad

c. $t = 66.67$ ms

Solution

a. At 40° the instantaneous value is $v = 6.4$ V.

b. At 2 rad, $v = 8.6$ V.

c. At 66.67 ms, $v = -8.6$ V.

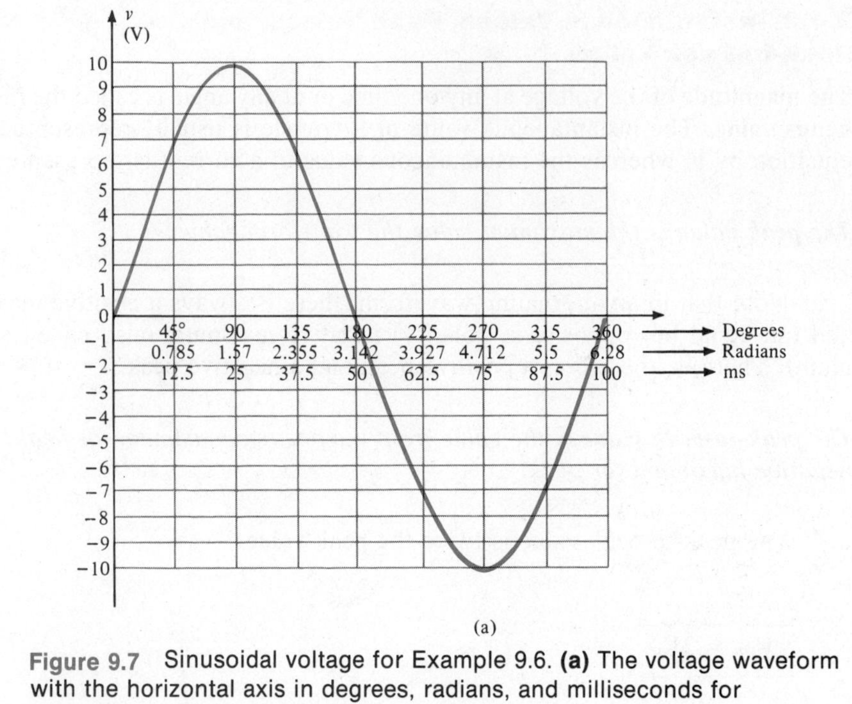

Figure 9.7 Sinusoidal voltage for Example 9.6. **(a)** The voltage waveform with the horizontal axis in degrees, radians, and milliseconds for comparison. **(b)** Four cycles of the waveform.

EXAMPLE 9.6

For the waveform shown in Figure 9.7a, obtain the period, the frequency, the radian frequency, the peak value, and the peak-to-peak value.

Solution

$$T = 100 \text{ ms}$$

$$f = \frac{1}{T} = \frac{1}{100 \text{ ms}} = \frac{1}{0.1} = 10 \text{ Hz}$$

$$\omega = 2\pi f = (6.28)(10) = 62.8 \text{ rad/s}$$

$$V_P = 10 \text{ V}$$

$$V_{PP} = 20 \text{ V}$$

EXAMPLE 9.7

Draw four cycles of the sinusoid shown in Figure 9.7a.

Solution

The four cycles are shown in Figure 9.7b.

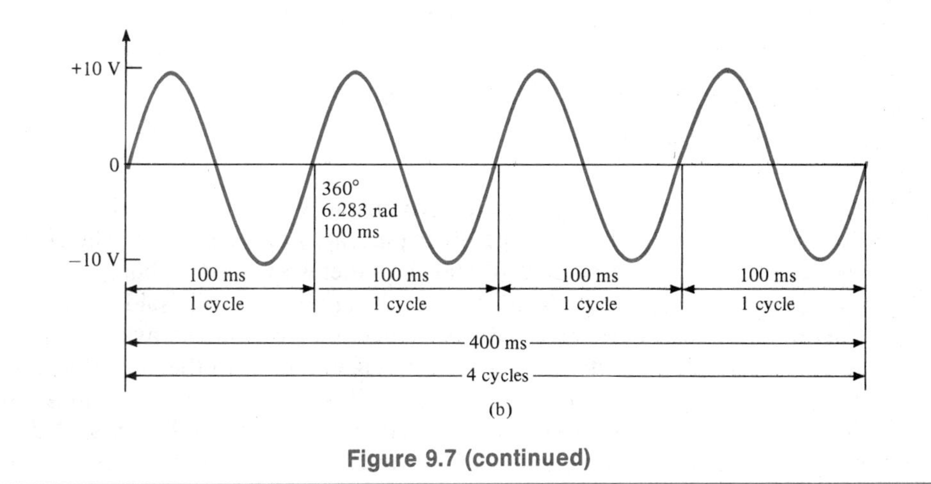

Figure 9.7 (continued)

EXAMPLE 9.8

Find the period, the frequency, the radian frequency, the peak current, and the peak-to-peak current of the triangular current waveform shown in Figure 9.8.

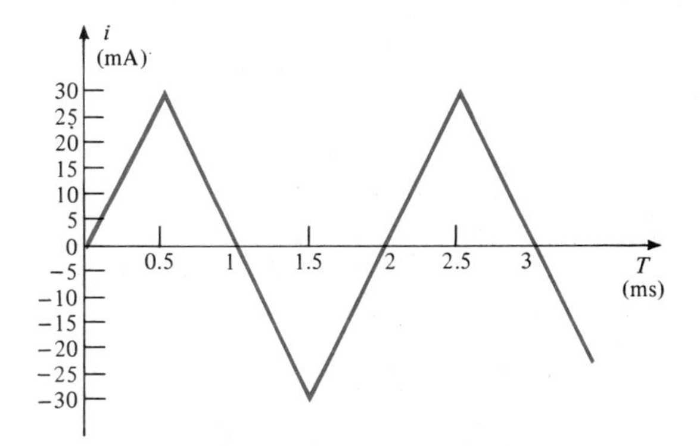

Figure 9.8 Triangular waveform for Example 9.8.

Solution

$$T = 2 \text{ ms}$$

$$f = \frac{1}{T} = \frac{1}{2 \text{ ms}} = \frac{1}{0.002} = 500 \text{ Hz}$$

$$\omega = 2\pi f = (6.28)(500) = 3140 \text{ Hz} = 3.14 \text{ kHz}$$

$$I_P = 30 \text{ mA}$$

$$I_{PP} = 60 \text{ mA}$$

9.5 PHASE RELATIONSHIPS

When comparing alternating current or voltage waveforms, one of the waveforms is chosen as the reference and the other ones are compared to it. The **reference waveform** is defined to start at 0° and the other waveform is compared to this reference. If the other waveform starts at the *same time,* it is said to be **in phase** with reference. If it starts *before* the reference, then the other waveform is said to **lead** the reference. If it starts *after* the reference, then the other waveform is said to **lag** the reference. The number of degrees of lead or lag is called the **phase angle.** The phase angle is usually denoted by the Greek letter ϕ (phi). Figure 9.9a shows a reference sinusoid labeled *A* and another waveform labeled *B*. Since *B* starts at the

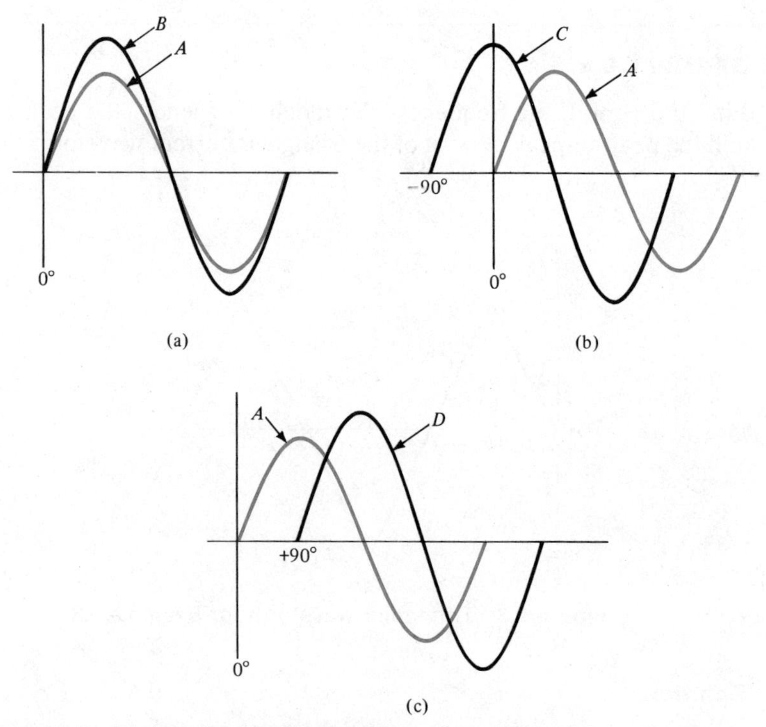

Figure 9.9 Phase relationships. **(a)** Waveforms *B* and *A* are in phase. **(b)** Waveform *C* leads waveform *A* by 90°. **(c)** Waveform *D* lags waveform *A* by 90°.

same time as reference A, waveform B is in phase with reference A. Figure 9.9b shows a reference waveform labeled A and another waveform labeled C. Since C starts 90° before the reference, waveform C leads the reference by 90°. Figure 9.9c shows a reference waveform labeled A and another waveform labeled D. Since D starts 90° after the reference, waveform D lags the reference by 90°.

9.6 THE SINUSOIDAL EQUATION

The equation of a sinusoidal voltage waveform is as follows:

$$\boxed{v = V_P \sin(\omega t + \phi)} \tag{9.6}$$

where: v => instantaneous value of the voltage at time t (V)
V_P => peak value of the sinusoid (V)
ω => radian frequency ($2\pi f$) (rad/s)
t => time that the voltage is required (s)
ϕ => phase angle (degrees)

The phase angle ϕ is positive if the waveform is leading the reference; it is negative if the waveform is lagging the reference.

The equation of a sinusoidal current waveform is as follows:

$$\boxed{i = I_P \sin(\omega t + \phi)} \tag{9.7}$$

where: i => instantaneous value of the current at time t (A)
I_P => peak value of the current (A)
ω, t, ϕ have the same meaning as before

Figure 9.10a shows a sinusoidal current waveform that starts at 0° and peaks at 10 A. Since the phase angle is 0°, the equation is:

$$i = 10 \sin(\omega t) \quad \text{A}$$

Figure 9.10b shows a sinusoidal voltage waveform that starts at −60° and peaks at 50 V. Since the voltage waveform starts *before* the 0° reference, it *leads* the reference and the phase angle is positive ($\phi = +60°$). The equation is:

$$v = 50 \sin(\omega t + 60°) \quad \text{V}$$

Figure 9.10c shows a sinusoidal current waveform that starts at +40° and peaks at 30 A. Since the waveform starts *after* the 0° reference, it *lags* the reference and the phase angle is negative ($\phi = -40°$). The equation is:

$$i = 30 \sin(\omega t - 40°) \quad \text{A}$$

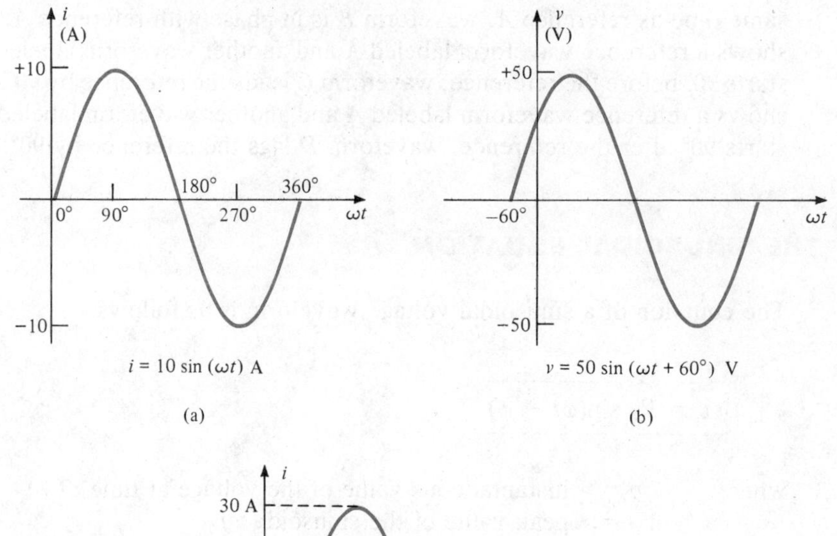

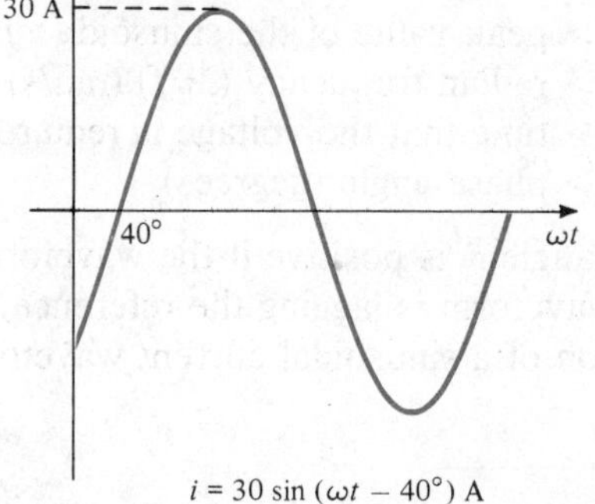

(c)

Figure 9.10 Sinusoidal waveforms. **(a)** Starting at 0°. **(b)** Starting at −60°. **(c)** Starting at +40°.

9.7 AVERAGE VALUES

The **average** (or mean) **value** of a waveform is the average value of the waveform over a period. Geometrically, this can be interpreted to be representative of the area under the waveform for one period, divided by the length of the period.

$$\text{Average value} = \frac{\text{area under curve}}{\text{period}} \tag{9.8}$$

For voltages and currents, the average value also corresponds to the value of the dc content of the waveform. *A pure alternating voltage or current waveform has by definition a zero dc content and therefore a zero average.*

EXAMPLE 9.9

Find the average, or dc, values of the waveforms shown in Figures 9.11a to 9.11c.

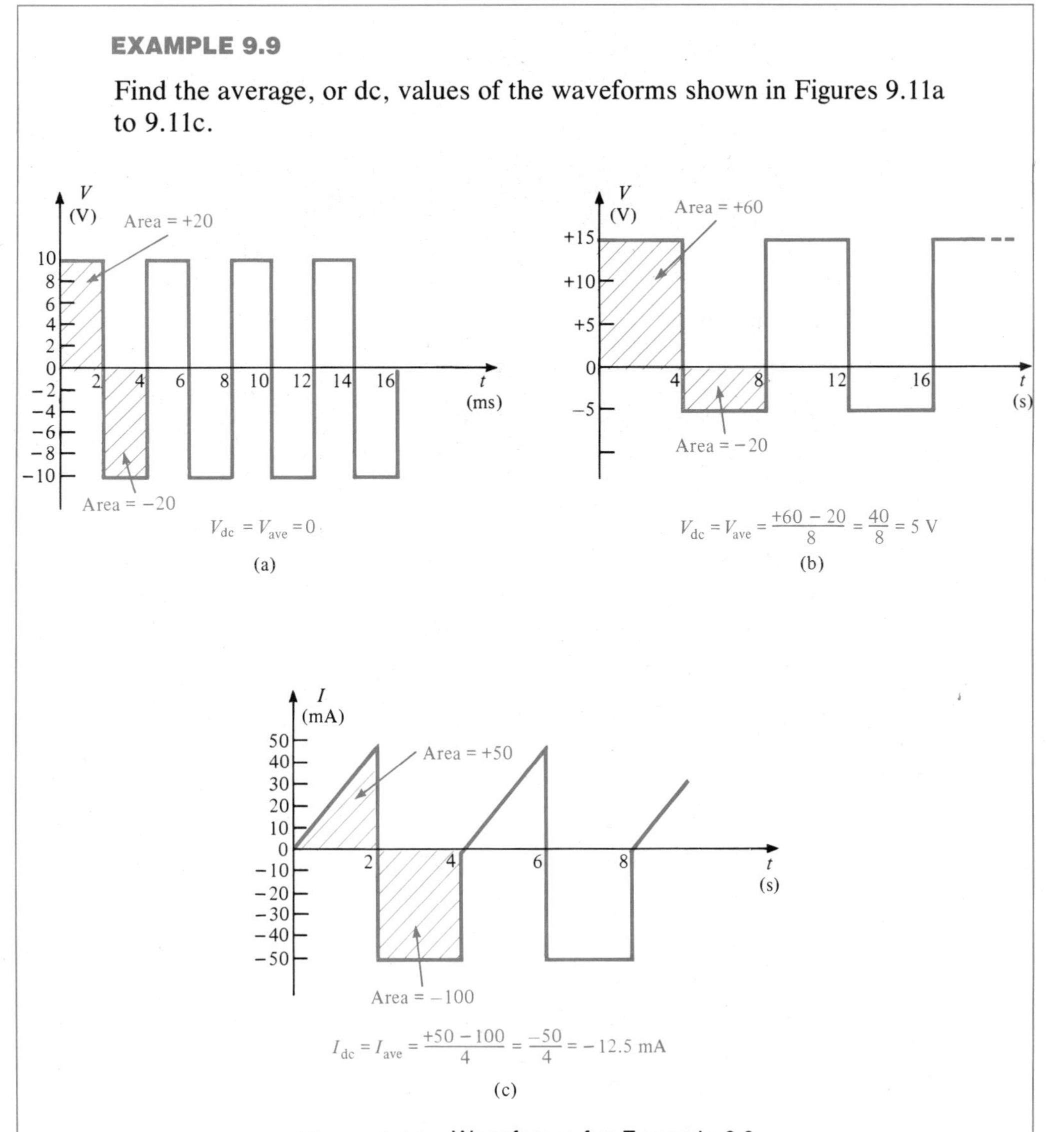

Figure 9.11 Waveforms for Example 9.9.

Solution

a. The area above the axis for one period is:

$$(10\ \text{V})(2\ \text{ms}) = +20\ \text{V} \cdot \text{ms}$$

The area below the axis (considered negative area) is:

$$(-10\ \text{V})(2\ \text{ms}) = -20\ \text{V} \cdot \text{ms}$$

Adding the area above the axis and the area below the axis results in a net cancellation. *The average, or dc, voltage is therefore zero.*

b. The area above the axis is:

$$(15 \text{ V})(4 \text{ s}) = +60 \text{ V} \cdot \text{s}$$

The area below the axis is:

$$(-5 \text{ V})(4 \text{ s}) = -20 \text{ V} \cdot \text{s}$$

The total area is:

$$+60 \text{ V} \cdot \text{s} - 20 \text{ V} \cdot \text{s} = 40 \text{ V} \cdot \text{s}$$

One period is 8 s. The average value is:

$$V_{dc} = V_{ave} = \frac{40 \text{ V} \cdot \text{s}}{8 \text{ s}} = 5 \text{ V}$$

c. The area above the axis is obtained by multiplying the base by the height and dividing by 2:

$$\frac{(50 \text{ mA})(2 \text{ s})}{2} = +50 \text{ mA} \cdot \text{s}$$

The area below the axis is:

$$(-50 \text{ mA})(2 \text{ s}) = -100 \text{ mA} \cdot \text{s}$$

The total area is:

$$50 \text{ mA} \cdot \text{s} - 100 \text{ mA} \cdot \text{s} = -50 \text{ mA} \cdot \text{s}$$

One period is 4 s. The average value is:

$$I_{dc} = I_{ave} = \frac{-50 \text{ mA} \cdot \text{s}}{4 \text{ s}} = -12.5 \text{ mA}$$

9.8 EFFECTIVE OR RMS VALUES

Even though symmetrically alternating voltages and currents have no average, or dc values, they do nevertheless transmit electrical energy. This is obvious from the fact that an alternating current, when maintained through a resistor, causes the resistor to become hot. There is, therefore, an energy-related value of alternating voltage or current quantities. The following experiment clarifies this concept.

Figure 9.12a shows a source of dc current connected to a switch and an immersion heater. The heater is immersed in a container with water and a thermometer is placed in the water. The thermometer shows that the water is at room temperature. Figure 9.12b shows the same setup except the current source is now alternating instead of direct. Both switches are closed and the water is heated for 2 min. Both thermometers are checked, and the rise in temperature is observed to be the same, suggesting that the amount of energy delivered by the dc system is the same as the amount delivered by the ac system. Both did *effectively* the same amount of work. Checking the magnitude of the current sources, it is found that the sinusoidal alternating current source has a peak value of 1.414 A, whereas the dc current source has a value of 1 A. Since they did the same amount of work, it is said that the **effective value** of the sinusoid is also 1 A (even though its peak value is 1.414 A). Since $1.414 = \sqrt{2}$, the effective value of a sinusoid is equal to the peak value divided by $\sqrt{2}$. Since 1/1.414 is equal to 0.707, the effective value of a sinusoid may also be stated as 0.707 times its peak value.

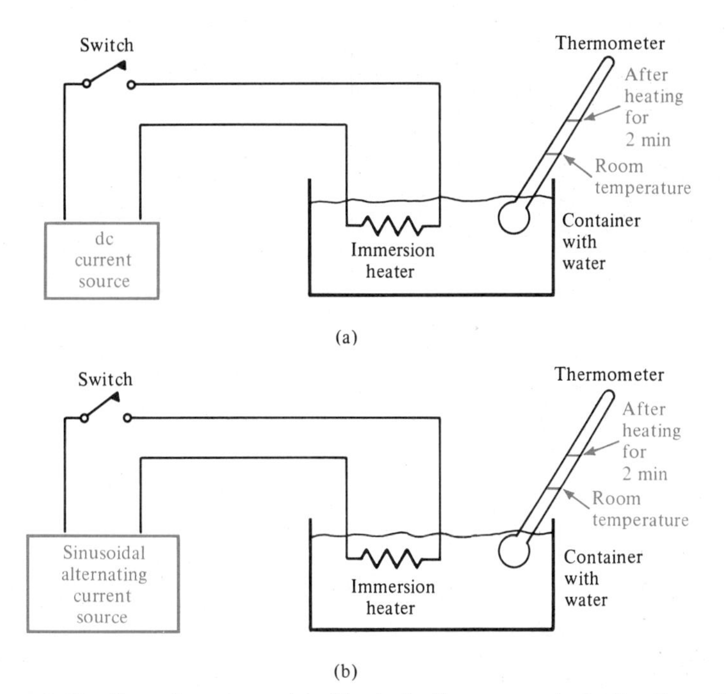

Figure 9.12 Experiment used to illustrate the concept of effective value. **(a)** With a dc energy source. **(b)** With an ac energy source.

For a sinusoid:

$$\boxed{\text{Effective value} = 0.707 \cdot (\text{peak value})} \quad \textbf{(9.9a)}$$

$$\boxed{\text{Peak value} = 1.414 \cdot (\text{effective value})} \quad \textbf{(9.9b)}$$

Mathematically, the effective value of any alternating waveform may be computed by performing the following three steps:

1. *Square* the magnitude of the waveform.
2. Find the average (*mean*).
3. Take the *square root*.

These three steps suggest the name root mean square, or RMS. Another name for the *effective value* of an alternating waveform is the **RMS value.**

The computation of RMS values involves areas under the squares of waveforms. Figure 9.13 shows certain waveforms and the areas under both the waveforms and their squares. These values may be derived by integral calculus.

Ac voltages and currents are always stated in RMS unless otherwise specified.

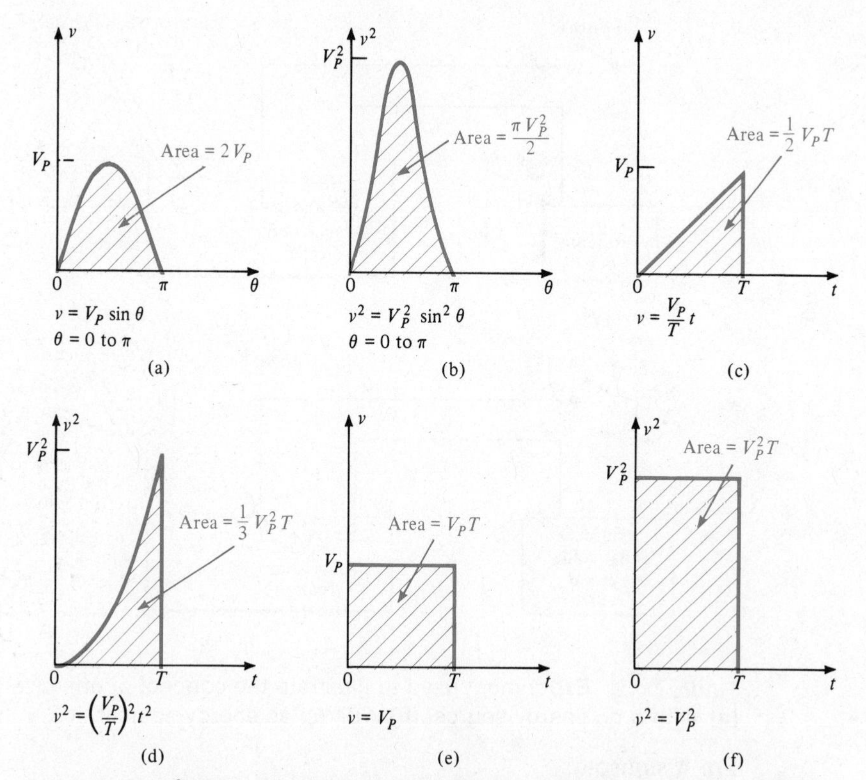

Figure 9.13 Areas under several waveshapes. **(a)** Half a sinusoid. **(b)** Half a squared sinusoid. **(c)** A triangular waveform. **(d)** A triangular waveform squared. **(e)** Half a square wave. **(f)** Half a square wave squared.

EXAMPLE 9.10

Compute the RMS, or effective, value of the sinusoid shown in Figure 9.14a.

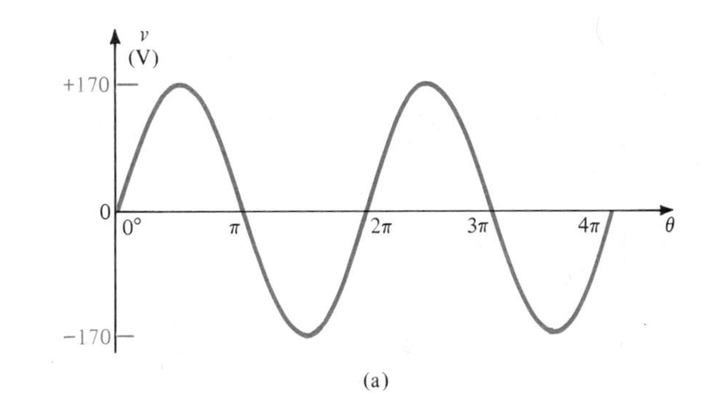

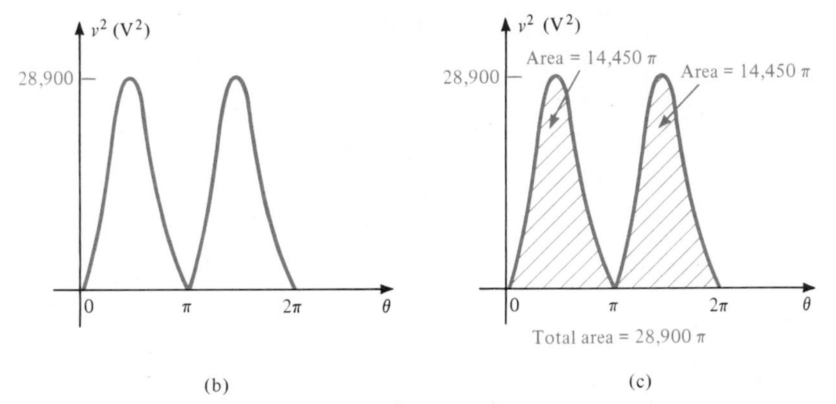

Figure 9.14 Diagrams for Example 9.10.

Solution

First, the waveform has to be squared. Figure 9.14b shows the waveform squared. Since the peak value is 170 V, the peak value squared is 28,900 V^2. The area under half a sinusoid squared is $\pi V_P^2/2$. The area under each half-curve is therefore $14{,}450\pi$. The total area is $28{,}900\pi$ (as shown in Figure 9.14c). The average, or mean, is:

$$\text{Mean} = \frac{28{,}900\pi}{2\pi} = 14{,}500$$

The RMS value is then the square root of this mean:

$$V_{\text{RMS}} = \sqrt{14{,}500} = 120 \text{ V}$$

$$V_{\text{eff}} = V_{\text{RMS}} = 120 \text{ V}$$

The same result could have been obtained by simply multiplying the peak value by 0.707:

$$V_{\text{RMS}} = 0.707V_P = (0.707)(170) = 120 \text{ V}$$

EXAMPLE 9.11

Find the effective, or RMS, value of the sawtooth waveform shown in Figure 9.15a.

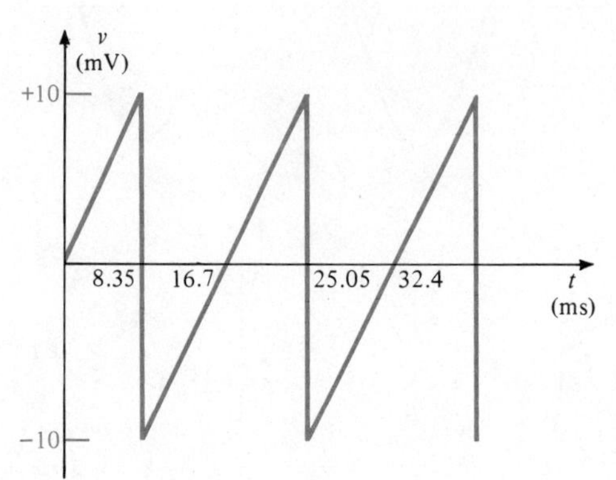

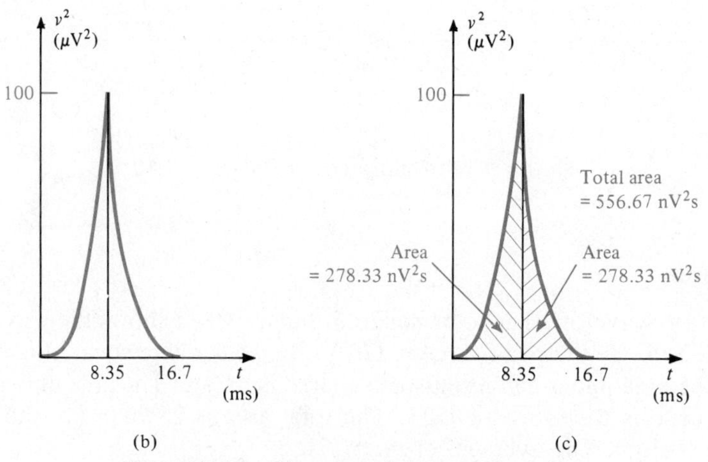

Figure 9.15 Diagrams for Example 9.11.

Solution

First the waveform must be squared. The squared waveform is shown in Figure 9.15b. Since the peak of the sawtooth is 10 mV, the peak of the squared sawtooth is 100 μV^2. The area under half the squared waveform is (Figure 9.13d) $V_P^2 T/3$, or (100 μV^2)(8.35 ms)/3 = 278.33 nV^{2}s. Since there are two equal areas, the total area is 556.67 nV^{2}s. The mean is this value divided by the period:

$$\text{Mean} = \frac{556.67 \text{ nV}^2\text{s}}{16.7 \text{ ms}} = 33.33\ \mu\text{V}^2$$

The RMS value is the square root of this mean:

$$V_{\text{RMS}} = \sqrt{33.33\ \mu\text{V}^2} = 5.77\ \text{mV}$$

If the symbols themselves are manipulated instead of numerical values being substituted, the following result is achieved:

$$\text{Mean} = \frac{2V_P^2T}{3(2T)} = \frac{V_P^2}{3}$$

$$V_{\text{RMS}} = \frac{V_P}{\sqrt{3}}$$

This states that the effective value of a sawtooth is equal to its peak value divided by the square root of 3 ($\sqrt{3} = 1.732$).

Using this relationship, the preceding problem may be solved simply as

$$V_{\text{RMS}} = \frac{10\ \text{mV}}{1.732} = 5.77\ \text{mV}$$

EXAMPLE 9.12

Find the effective value of the square current waveform shown in Figure 9.16a.

Solution

First the waveform must be squared. Figure 9.16b shows the waveform squared. Note the peak value of the waveform squared is $(50\ \text{mA})^2 = 2500\ \mu\text{A}^2$. Since each half of the waveform is a rectangle, the area is (Figure 9.13f) $5000\ \mu\text{A}^2$ s. The total area is $10{,}000\ \mu\text{A}^2$ s. The mean is this value divided by the period:

$$\text{Mean} = \frac{10{,}000\ \mu\text{A}^2\text{s}}{4\ \text{s}} = 2500\ \mu\text{A}^2$$

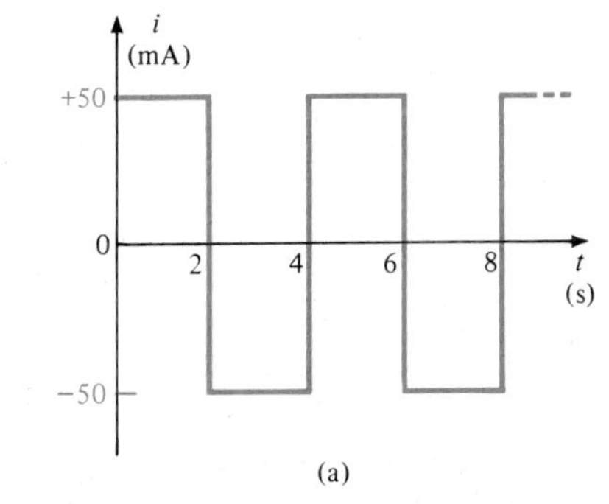

Figure 9.16 Diagrams for Example 9.12.

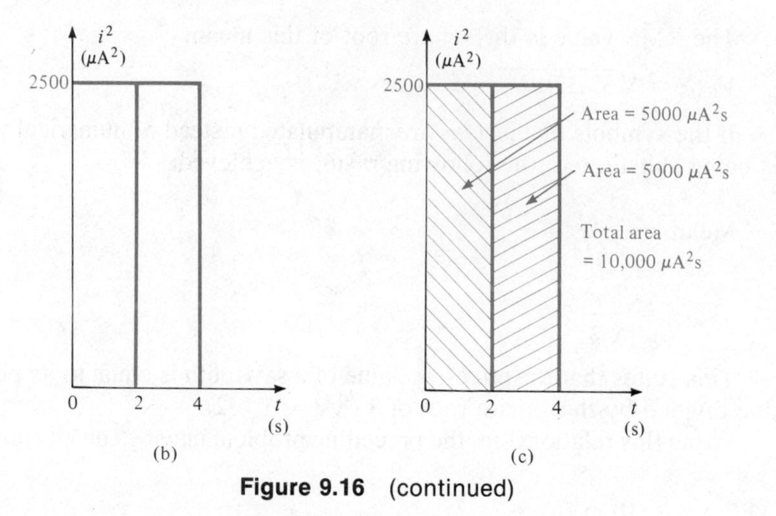

Figure 9.16 (continued)

The RMS value is the square root of this mean:

$$I_{\text{RMS}} = \sqrt{2500\ \mu\text{A}^2} = 50\ \text{mA}$$

Note that the effective value of a square wave whose positive half-cycle is equal to the negative half-cycle, is equal to the peak value of the square waveform.

9.9 PHASOR NOTATION

It would be very tedious and time-consuming to draw the sinusoid for each alternating voltage and current showing magnitudes and phase relationships or even to draw them as time functions. Technicians and engineers use **phasor notation** to handle sinusoidal voltage and current waveforms of the same frequency.

Figure 9.17a shows a line (V_P) rotating counterclockwise about the origin O with constant velocity (ω). The length of its projection on the horizontal axis varies with time as the cosine ($V_P \cos\omega t$). Here, t is the amount of time elapsed since the line V_P was in a horizontal position. *Such a rotating line is called a **phasor.***

To represent a current such as $I_P \cos(\omega t + \phi)$, another phasor advanced by an angle ϕ is drawn. This phasor has an angle $\omega t + \phi$ and is shown in Figure 9.17b.

Figure 9.17c shows the voltage phasor and the current phasor drawn on the same set of axes for easy comparison. Both lines are rotating at the same constant velocity and, since they turn indefinitely, the current I_P always leads the voltage V_P by the phase angle ϕ. The two lines separated by the phase angle ϕ contain all the pertinent information. The rest of the diagram is superfluous and can be left out without losing any information. Figure 9.17d shows the two phasors separated by the phase angle. Since the orientation of the diagram is also unimportant, as

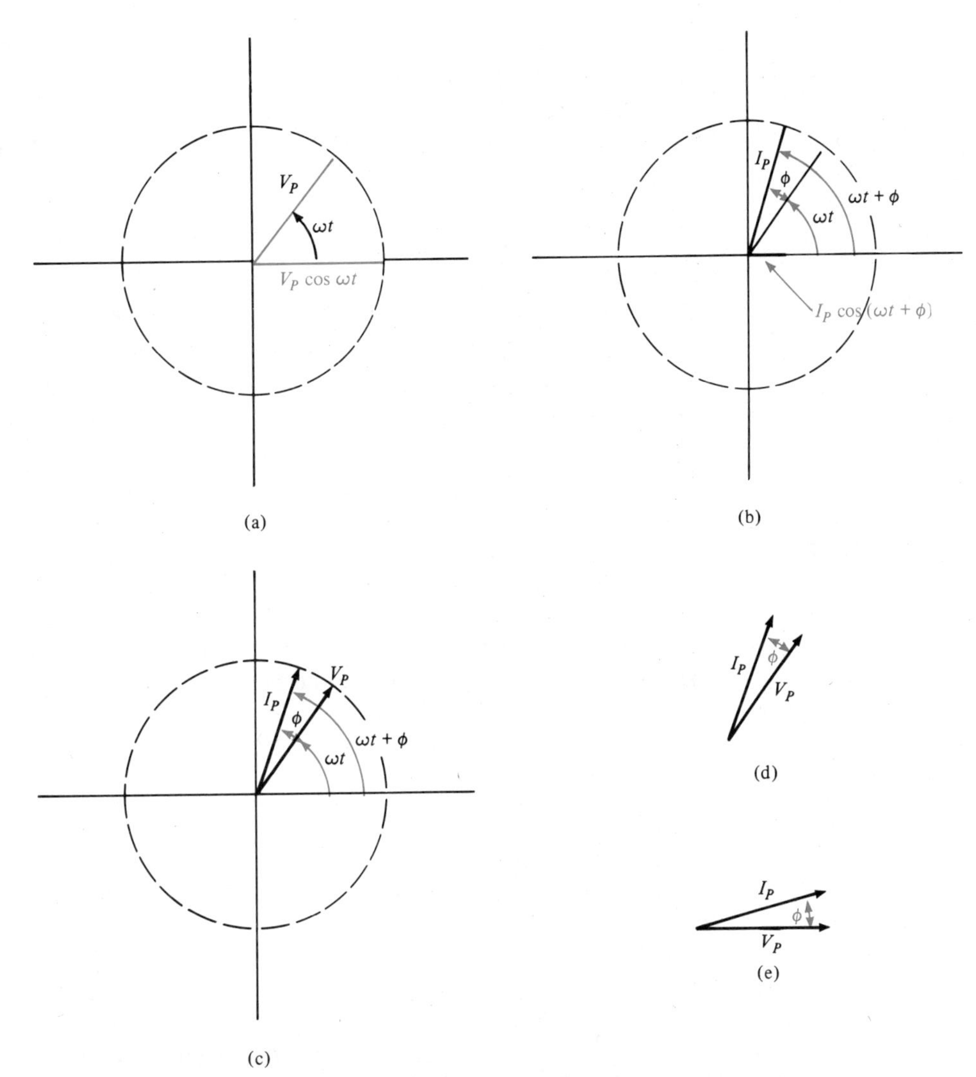

Figure 9.17 Illustration of phasor concepts. **(a)** A rotating line. **(b)** Another rotating line with a phase angle. **(c)** Voltage and current in phasor diagram. **(d)** Rotating line shown without axes. **(e)** Rotating line at the instant $t = 0$.

long as the two lines are pictured rotating synchronously about their common origin, any picture of them is merely a "snapshot" at a particular instant of time. Usually the snapshot is taken when the reference phasor is on top of the positive horizontal axis. Figure 9.17e shows the two phasors with the voltage phasor V_P as the reference.

Here the phasors are represented by their peak value (V_P, I_P). However, since ac voltages and currents are usually measured in RMS, or effective, values, it is customary to use RMS, or effective, values when representing alternating voltages or currents as phasors.

The phasor representation of a sinusoidal waveform consists of its RMS value and its phase angle written in the following notation for a voltage and a current:

$$\mathbf{V} = V_{\text{RMS}}\underline{/\phi^\circ}$$

$$\mathbf{I} = I_{\text{RMS}}\underline{/\phi^\circ}$$

In textbooks phasor quantities are printed as bold letters. When written by hand, a bar is placed above the symbol to indicate a phasor representation.

The following are examples of phasors:

$$\mathbf{V} = 10\underline{/30^\circ}\text{ V}$$

$$\mathbf{I} = 5\underline{/-45^\circ}\text{ A}$$

The first one is read as: "The voltage V is 10 V at 30°." The second is read as: "The current I is 5 A at −45°."

Phasors correspond completely to vectors and follow all the rules of vector algebra. They may be transformed to rectangular form and they may be handled with all the rules of vector algebra. Since vectors follow the rules of mathematics described by **complex numbers,** in order to handle phasors, one must be familiar with complex numbers.

9.10 COMPLEX NUMBERS AND PHASORS

A plane may be divided into four quadrants, as shown in Figure 9.18a. Each quadrant is seen to contain 90°, or $\pi/2$ rad. In this kind of angular measurement system, positive angles are defined as counterclockwise rotations and negative angles are clockwise rotations, as shown in Figure 9.18b. Figure 9.18c shows a plane divided into four quadrants with the angles in a negative direction.

Figure 9.18d shows another way to divide the plane. Here the plane is divided by means of a positive and a negative x axis and a positive and a negative y axis.

Figure 9.18e shows still another way of dividing the plane. Here the horizontal axis is called the *real axis,* and the vertical axis is called the *imaginary axis*. In this system a number on the real axis is represented by the letter a, whereas a number on the imaginary axis is represented by the letter b. In order to distinguish numbers on the real axis from numbers on the imaginary axis, numbers on the imaginary axis are "flagged," or accompanied by the letter j. The plane formed by the real and imaginary axes is called a **complex plane.**

All these methods are simply ways of locating a point on the plane. Figure 9.19a shows the **rectangular form** of locating a point on a plane. It directs us to go 3 units in the positive real direction and then 4 units in the positive imaginary direction. If this number represented a voltage, then the notation is:

$$\mathbf{V} = 3 + j4\text{ V}$$

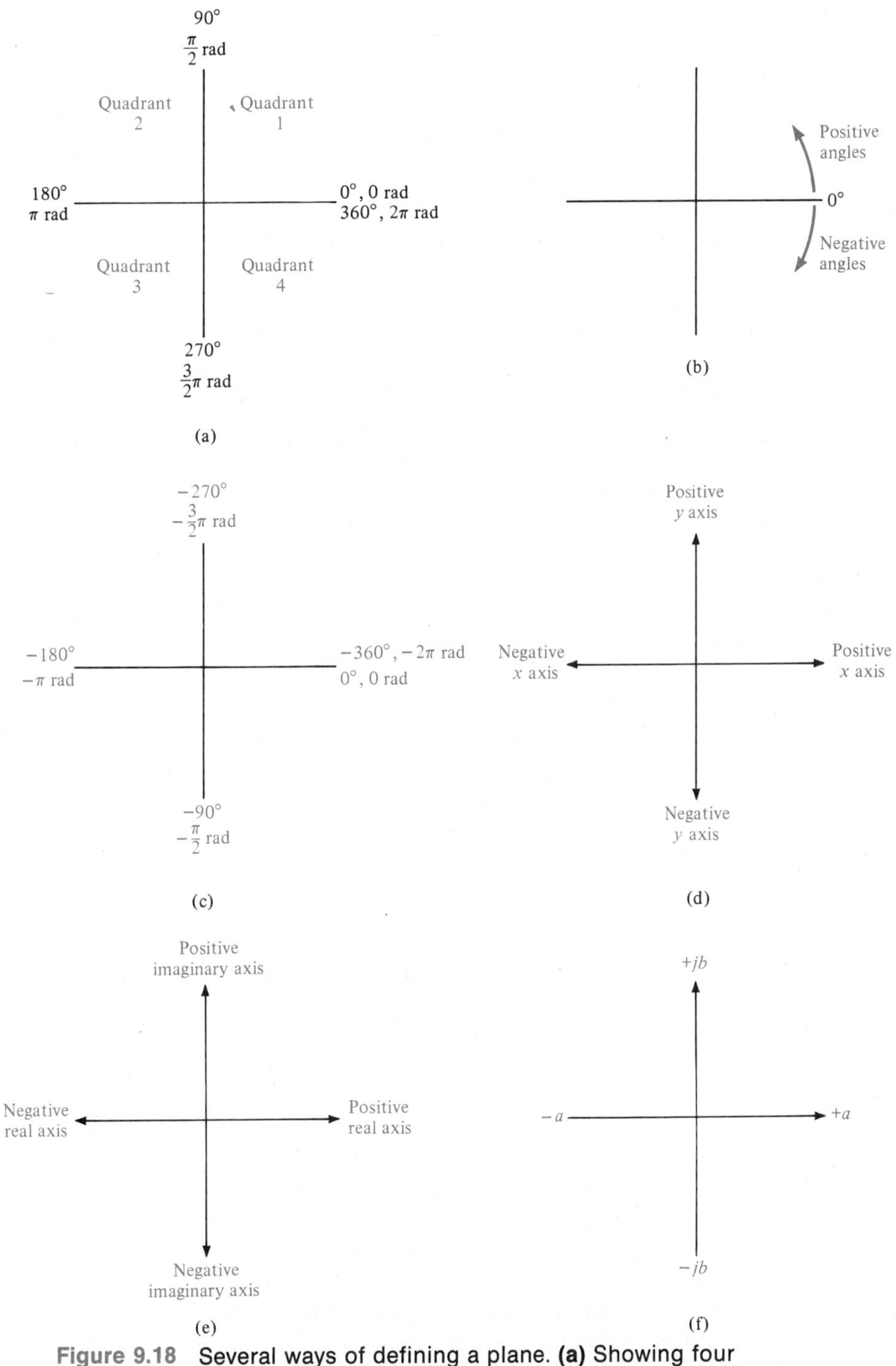

Figure 9.18 Several ways of defining a plane. **(a)** Showing four quadrants, degrees, and radians. **(b)** Standard method of defining positive and negative angles. **(c)** Correlating degrees and radians in a plane. **(d)** Defining a plane with *x* and *y* axes. **(e)** Defining a plane with real and imaginary axes. **(f)** Showing the complex plane with rectangular notation.

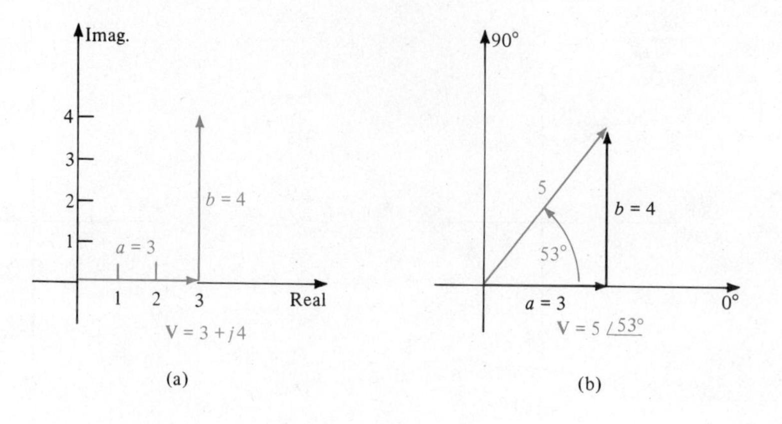

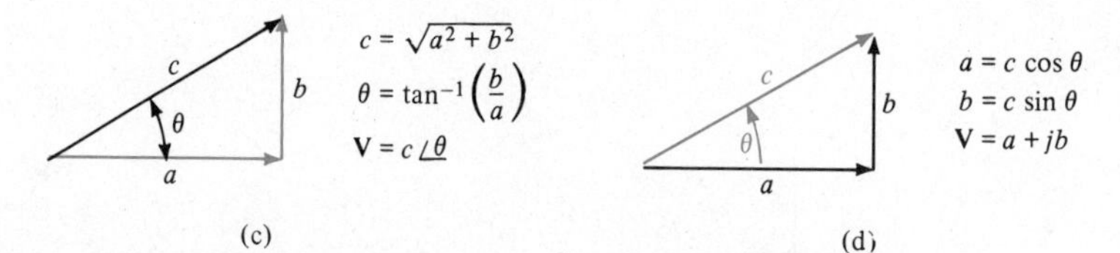

Figure 9.19 Conversion between complex number forms. **(a)** Representation of a complex number in rectangular form. **(b)** Representation of a complex number in polar form. **(c)** Conversion from rectangular to polar form. **(d)** Conversion from polar to rectangular form.

The real part of the complex number is 3 and the imaginary part is 4. Note that the letter *j* only indicates the imaginary part of the number but it is not included in it.

The same point on the plane could be reached (and therefore the same voltage could be described) by first turning 53° from zero and then proceeding out 5 units. This is shown in Figure 9.19b and is referred to as the **polar form** of the number. If this number represents a voltage, then the notation is:

$$\mathbf{V} = 5\underline{/53^\circ}\ \text{V}$$

Here, 5 is called the magnitude (and represents the RMS, or effective, value of the voltage) and 53° is called the phase (and represents the phase angle of the sinusoid as compared to a reference).

Figure 9.19c shows how the Pythagorean Theorem may be applied to convert from rectangular to polar. Figure 9.19d shows how simple trigonometry may be used to convert from polar to rectangular.

Presently, there are many scientific calculators available that perform polar-to-rectangular and rectangular-to-polar conversions, and some actually perform complex number operations. Up to now, books such as this one had to spend several pages explaining complex number arithmetic. Since the advent of these inexpensive calculators, the authors feel the student is better served by purchasing one of these calculators and following the instructions supplied with it.

All the examples in this book will assume the use of a scientific calculator capable of performing polar-to-rectangular and rectangular-to-polar conversions.

In electrical work the most common phasor operations that are essential are addition, subtraction, multiplication, and division. Addition and subtraction are more easily handled in rectangular form, whereas multiplication and division are less cumbersome in polar form.

Addition is handled by adding the real parts and adding the imaginary parts as follows:

$$\begin{array}{rl} & \mathbf{V}_1 => a_1 \qquad + jb_1 \\ + & \underline{\mathbf{V}_2 => a_2 \qquad + jb_2} \\ & \mathbf{V}_T => (a_1 + a_2) + j(b_1 + b_2) \end{array}$$

Subtraction is handled by subtracting the real parts and subtracting the imaginary parts as follows:

$$\begin{array}{rl} & \mathbf{V}_1 => a_1 \qquad + jb_1 \\ - & \underline{\mathbf{V}_2 => a_2 \qquad + jb_2} \\ & \mathbf{V}_T => (a_1 - a_2) + j(b_1 - b_2) \end{array}$$

Multiplication is handled by multiplying the magnitudes of the two numbers and adding the angle as follows:

$$\begin{array}{rl} & \mathbf{V}_1 => V_1\angle\phi_1 \\ \times & \underline{\mathbf{V}_2 => V_2\angle\phi_2} \\ & \mathbf{V}_T => V_1V_2\angle\phi_1 + \phi_2 \end{array}$$

Division is handled by dividing the magnitudes and subtracting the angle of the denominator (bottom) from the angle of the numerator (top) as follows:

$$\frac{\mathbf{V}_1\angle\phi_1}{\mathbf{V}_2\angle\phi_2} = \frac{V_1}{V_2}\angle\phi_1 - \phi_2$$

Recall that phasors are usually expressed in RMS and that the magnitude is represented by capital letters.

The following examples illustrate these rules.

EXAMPLE 9.13

a. Convert $V_1 = 35 \sin(377t + 30°)$ V and $V_2 = 50 \sin(377t + 20°)$ V to phasor notation.

b. Add the two voltages.

c. Convert the resulting voltage to sinusoidal form.

Solution

a. By observation, $V_{1P} = 35$ V and $V_{2P} = 50$ V.

$$V_{1\text{RMS}} = 0.707(35) = 24.75 \text{ V}$$

$$V_{2\text{RMS}} = 0.707(50) = 35.35 \text{ V}$$

$$\mathbf{V_1} = 24.75\angle 30° \text{ V} \quad \text{and} \quad \mathbf{V_2} = 35.35\angle 20° \text{ V}$$

b. Phasors are easily added when they are expressed in rectangular form.

$$\mathbf{V_1} = 24.75\angle 30° = 21.43 + j12.38 \text{ V}$$

$$\mathbf{V_2} = 35.35\angle 20° = 33.22 + j12.09 \text{ V}$$

$$\mathbf{V_T} = 54.65 + j24.47 \text{ V}$$

c. To express the resulting voltage in sinusoidal form, the polar form is obtained first:

$$\mathbf{V_T} = 54.65 + j24.47 = 59.88\angle 24.12° \text{ V}$$

Since $V_{T_{\text{RMS}}} = 59.88$ V, then $V_{T_P} = 1.414(59.88) = 84.67$ V. The sinusoidal form is:

$$v_T = 84.67 \sin(377t + 24.12°) \text{ V}$$

EXAMPLE 9.14

Multiply $53.03\angle 50°$ and $17.68\angle -100°$.

Solution

To multiply complex numbers, multiply the magnitudes and add the angles:

$$(53.03)(17.68) = 937.57$$

$$\phi_v + \phi_i = 50° - 100° = -50°$$

The result is $937.57\angle -50°$.

SUMMARY

In this chapter the differences between dc and ac voltages and currents were examined. The generation and characteristics of sinusoidal waveforms were presented. The definition of degrees, radians, and gradians was given. Waveform characteristics, such as period (T) and frequency (f), as well as their units, seconds (s) and hertz (Hz), were introduced.

The average (dc) and the effective (RMS) values of time-varying waveforms were defined. The RMS value of sinusoidal waveforms was shown to be 0.707 times the peak value. Finally, phasors and complex numbers were introduced as mathematical tools necessary to handle ac circuit analysis.

SUMMARY OF IMPORTANT EQUATIONS

$$f = \frac{1}{T} \quad \text{Hz} \tag{9.1a}$$

$$T = \frac{1}{f} \quad \text{s} \tag{9.1b}$$

$$\omega = \frac{\phi}{t} \quad \text{rad/s} \tag{9.2}$$

$$\omega = \frac{2\pi}{T} \quad \text{rad/s} \tag{9.3}$$

$$\omega = 2\pi f \quad \text{rad/s} \tag{9.4}$$

$$V_{PP} = 2V_P \quad \text{V} \tag{9.5a}$$

$$I_{PP} = 2I_P \quad \text{A} \tag{9.5b}$$

$$v = V_P \sin(\omega t + \phi) \quad \text{V} \tag{9.6}$$

$$i = I_P \sin(\omega t + \phi) \quad \text{A} \tag{9.7}$$

$$\text{Average value} = \frac{\text{area under curve}}{\text{period}} \tag{9.8}$$

For a sinusoid:

$$\text{Effective value} = 0.707 \cdot (\text{peak value}) \tag{9.9a}$$

$$\text{Peak value} = 1.414 \cdot (\text{effective value}) \tag{9.9b}$$

REVIEW QUESTIONS

9.1 Define alternating current.

9.2 Define alternating voltage.

9.3 Define a pulsating direct current.

9.4 Define the period of a waveform.

9.5 Define the frequency of a waveform.

9.6 Define the unit hertz.

9.7 Compare degrees, radians, and gradians.

9.8 State the basic law of magnetism.

9.9 Explain the term magnetic field.

9.10 Define angular velocity and radian frequency.

9.11 Explain and compare the terms instantaneous value, peak value, and peak-to-peak value.

9.12 Define phase angle and explain the difference between a leading and a lagging waveform.

9.13 State the standard form of a sinusoidal voltage equation.

9.14 Explain what is meant by the average value of a waveform.

9.15 Explain what is meant by the effective, or RMS, value of a waveform.

9.16 Define a phasor.

9.17 Explain what is meant by phasor notation.

9.18 Explain what is meant by a complex number and compare the rectangular and the polar forms.

9.19 Explain the special significance of the letter *j* in complex numbers.

PROBLEMS

9.1 Compute the frequency for each period.
a. 16.7 ms
b. 1 ms
c. 2.5 ms
d. 50 μs

9.2 Compute the period for each frequency.
a. 50 Hz
b. 100 Hz
c. 2 MHz
d. 880 kHz
(ans. 20 ms, 10 ms, 500 ms, 1.14 ms)

9.3 A sinusoidal voltage has a 30-V peak value.
a. Compute the peak-to-peak value.
b. Compute the RMS value.

9.4 For the sinusoid shown in Figure 9.20, find:
a. V_P
b. V_{PP}
c. V_{RMS}
(ans. 15 V, 30 V, 10.61 V)

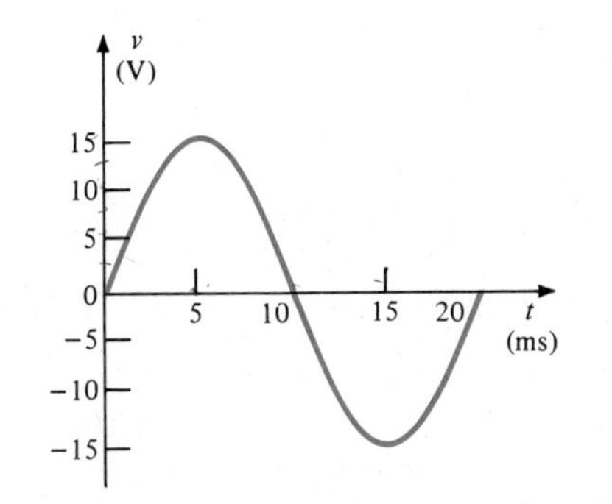

Figure 9.20 Waveform for Problem 9.4.

9.5 Convert to radians:
a. 30°
b. 45°
c. 90°

9.6 Convert to gradians:
a. 30°
b. 45°
c. 90°
(ans. 33.3 grad, 50 grad, 100 grad)

9.7 Convert to degrees:
a. 0.785 rad
b. 2.09 rad
c. 300 grad

9.8 For the waveform shown in Figure 9.21 find:
a. The instantaneous value at 30°
b. V_P
c. V_{PP}

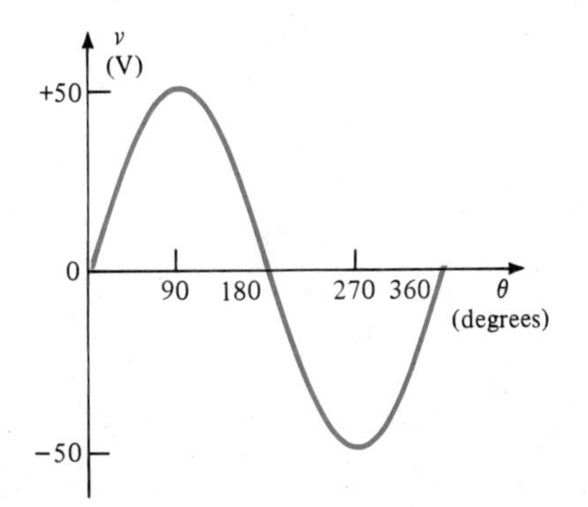

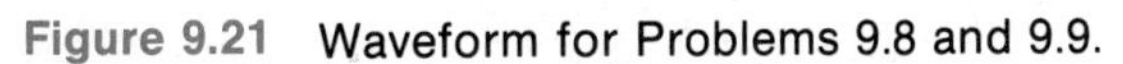
Figure 9.21 Waveform for Problems 9.8 and 9.9.

d. V_{RMS}
e. T
f. f
g. ω
(ans. 25 V, 50 V, 100 V, 35.35 V, 100 ms, 10 Hz, 62.8 rad/s)

9.9 Write the equation for the sinusoid in Problem 9.8.

9.10 **a.** In Figure 9.22a, does v lead or lag i?
b. In Figure 9.22b, state whether the waveform leads or lags a reference starting at 0° and by how much.
(ans. v leads i, v leads by 20°)

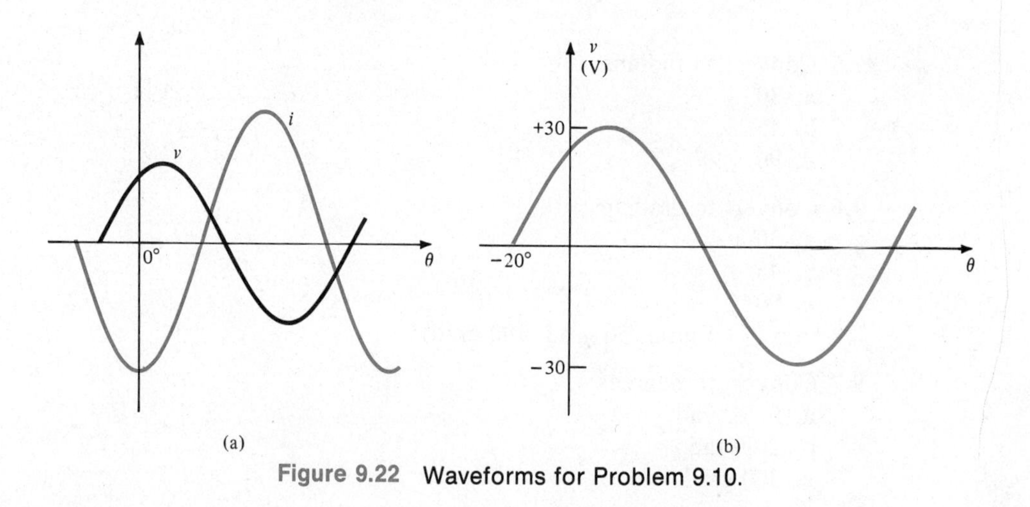

Figure 9.22 Waveforms for Problem 9.10.

9.11 Sketch the current $i = 20 \sin(377t + 30°)$ mA.

9.12 Compute the average value of the waveform shown in Figure 9.23.
(ans. 3.75 V)

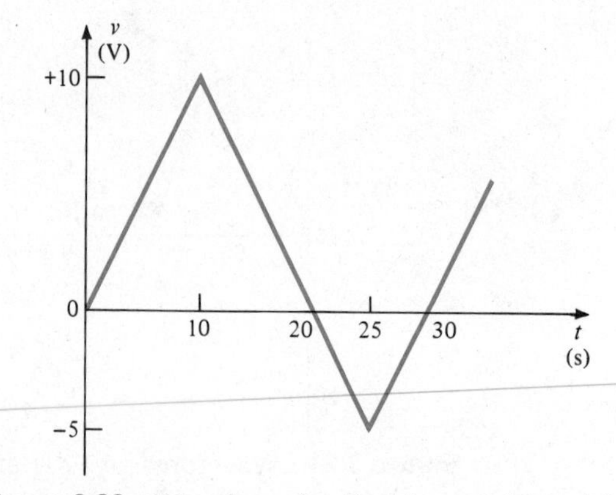

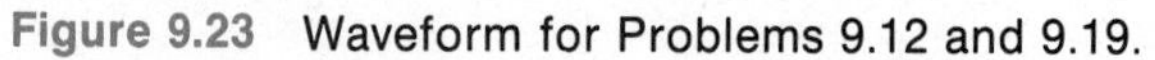

Figure 9.23 Waveform for Problems 9.12 and 9.19.

9.13 Write the phasor form of:
a. $v = 180 \sin(377t + 0^\circ)$ V
b. $i = 42.42 \sin(314t - 30^\circ)$ mA

9.14 Write the equation of the sinusoids represented by:
a. $50\angle 45^\circ$ V
b. $100\angle -20^\circ$ mA
[ans. $70.7 \sin(\omega t + 45^\circ)$ V, $141.4 \sin(\omega t - 20^\circ)$ mA]

9.15 Compute the RMS value of the square wave shown in Figure 9.1a.

9.16 Write the phasor form of:
a. $v = 60 \sin(6280t - 50^\circ)$ V
b. $i = 105 \sin(3140t + 20^\circ)$ mA
(ans. $42.42\angle -50^\circ$ V, $74.24\angle +20^\circ$ mA)

9.17 Write the equation of the sinusoids at 200 Hz represented by:
a. $50\angle -30^\circ$ mV
b. $75\angle 60^\circ$ mA

Problems 9.18 through 9.23 require solutions using calculus methods. See Appendix F.

9.18 Compute the RMS value of the triangular waveform shown in Figure 9.1b. (ans. 11.55 mA)

9.19 Compute the RMS value of the waveform shown in Figure 9.23.

9.20 Compute the average value of the waveform shown in Figure 9.24. (ans. 5.73 V)

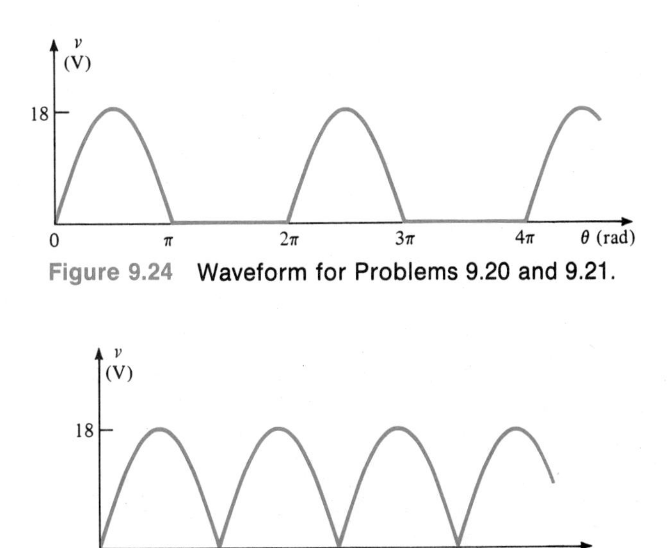

Figure 9.24 Waveform for Problems 9.20 and 9.21.

Figure 9.25 Waveform for Problems 9.22 and 9.23.

9.21 Compute the RMS value of the half-wave rectified sinusoid shown in Figure 9.24.

9.22 Compute the average value of the waveshape shown in Figure 9.25. (ans. 11.46 V)

9.23 Compute the RMS value of the full-wave rectified sinusoid shown in Figure 9.25.

CHAPTER
10

Capacitors and Inductors

GLOSSARY

admittance (Y)—the ability of ac circuits to allow ac current and the phase between the voltage and current; measured in siemens (S)

B_C—capacitive susceptance; measured in siemens (S)

B_L—inductive susceptance; measured in siemens (S)

capacitance—property of matter that describes the ability to store separated particles of opposite charge; opposition to changes in voltage; measured in farads (F)

capacitor—a circuit component built to show the property of capacitance
decaying exponential—the time representation of a capacitor's discharge or the collapse of current in an inductor
dielectric—the insulating material between the plates of a capacitor
dielectric constant—the ratio of capacitance of a capacitor with the given dielectric to the capacitance of a capacitor having air as the dielectric
electric field—energetic region of space surrounding a charged body
electrolytic capacitor—large-value capacitors that are usually polarized
farad (F)—unit of measurement for capacitance
henry (H)—unit of measurement for inductance
impedance (Z)—opposition of ac circuits to ac current and the angle between the voltage and current
inductance (*L*)—opposition to changes in current; measured in henrys (H)
inductor—a circuit component built to show the property of inductance
inertia—the opposition of an object to changes in velocity
reactance (*X*)—opposition of circuits to ac current
rising exponential—the time representation of the charging of a capacitor or the current buildup in an inductor
self-inductance—same as inductance
steady-state response—response of a system after all the initial disturbances and variations have subsided
susceptance (*B*)—ability of inductors or capacitors to allow ac current; measured in siemens (S); $B = 1/X$
time constant—RC for capacitors; L/R for inductors; the time needed for a capacitor to charge to 63.2% of its final value; the time needed for an inductor's current to build up to 63.2% of its final value
transient response—the behavior of a system from the time it is disturbed to the time it achieves its steady-state response
trimmer capacitor—a variable capacitor used to make small capacitance adjustments in electronic equipment
tuning capacitor—a capacitor used in radio receivers to tune in a particular radio station
tuning slug—an iron screw that changes the value of an inductor when screwed into or out of an inductor's core
variable capacitor—a capacitor whose value may be changed
working voltage (WVDC)—a capacitor rating that describes the maximum voltage that can safely be applied across a capacitor

INTRODUCTION

Capacitors and inductors are electrical energy-storing devices that are essential in both dc and ac applications. Timing circuits, high-voltage spark production, fluorescent light operation, and ac motor control are some of the many areas where capacitors and inductors are very important for proper operation.

This chapter discusses the concepts of capacitance and inductance as well as practical capacitors and practical inductors. The behavior of capacitors and inductors under both dc and ac conditions is examined. Reactance, impedance, susceptance, and admittance are studied. Transient analysis of *RC*, *RL*, and *RLC* circuits is presented.

10.1 CAPACITANCE AND CAPACITORS

***Capacitance** is the property of matter that describes the ability to store separated particles of opposite charge.*

Any two conductors separated by an insulator exhibit the property of capacitance. Capacitance may also be described as the property of a circuit that opposes changes in voltage. A device used to store these separated charges is called a **capacitor.**

Figure 10.1a shows a capacitor in its simplest form. It consists of two metal conductive plates separated by an insulating material (usually called a **dielectric**). When connected to a battery, the voltage forces electrons onto one plate, making it negative, while it pulls them off the other plate, making it positive. Once the charges have been separated on the two plates, they cannot flow through the dielectric, and therefore a voltage and an *electric field* results between the two plates. The capacitor is then said to be *charged.* An electric field is the energetic region of space surrounding a charged body. Between the oppositely charged plates of a capacitor, the electric field is relatively uniform. A charged object placed in an electric field experiences a force (see Coulomb's Law in Chapter 2). Figure 10.1b shows the circuit symbol for a capacitor.

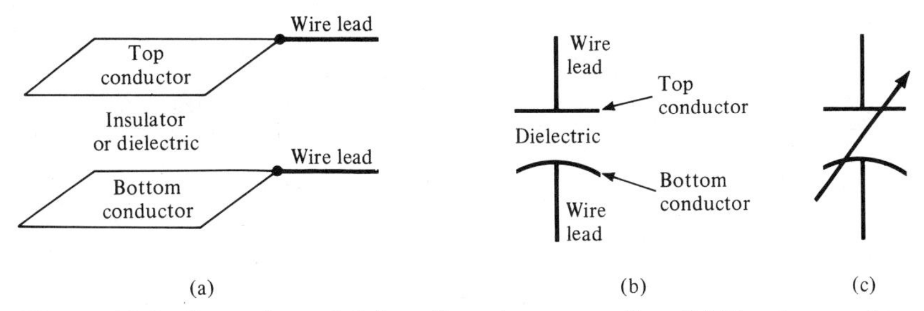

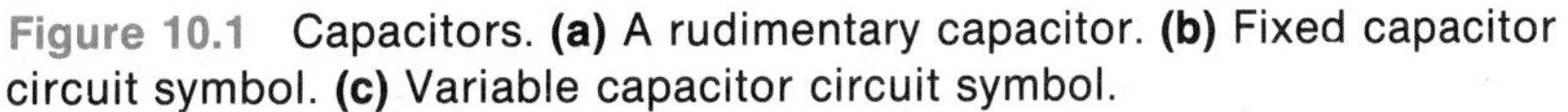
Figure 10.1 Capacitors. **(a)** A rudimentary capacitor. **(b)** Fixed capacitor circuit symbol. **(c)** Variable capacitor circuit symbol.

The symbol for either capacitance or a capacitor is C. Capacitance is measured in a unit called the **farad** (F).

It was discovered that for the same capacitor, the ratio of the amount of charge deposited to the resulting voltage across the capacitor plates was always a constant. This ratio was defined to be the capacitance value:

$$C = \frac{Q}{V} \tag{10.1a}$$

where: C => capacitance in farads (F)
Q => charge deposited in coulombs (C)
V => resulting voltage (V)

A capacitor is said to have a capacitance of 1 F if it shows 1 V of potential difference after 1 C of electrons is moved from one plate to the other.

The farad is a unit that is far too large for most electrical and electronic applications. The **microfarad** (μF), which represents one millionth of a farad, and the **picofarad** (pF), which represents a millionth of a millionth of a farad, are much more convenient and, therefore, more popular capacitance units.

The storage of the separated charges on the plates of a capacitor is in effect the storage of energy. If the electrons that were separated from the protons on the top plate were allowed to go back through an external resistor connected between the two plates, the resistor would get hot, showing that indeed there was energy stored in the charged capacitor.

The equation showing the dependence of the capacitor-charging current on the capacitor value and the derivative of the capacitor voltage with respect to time may be stated as:

$$i_C = C\frac{dv_C}{dt} \tag{10.1b}$$

For a mathematical derivation of this equation using calculus, see Appendix F.

The amount of energy stored in a charged capacitor is:

$$W = 0.5CV^2 \tag{10.2}$$

where: W => energy in joules (J)
C => capacitance in farads (F)
V => voltage across capacitor in volts (V)

EXAMPLE 10.1

Compute the amount of charge stored in a 10-μF capacitor that measures 12 V across its terminals. Also compute the energy stored in the capacitor.

Solution

$$Q = CV = (10\ \mu F)(12\ V) = 120\ \mu C$$

$$W = 0.5CV^2 = 0.5(10\ \mu F)(12\ V)^2 = 720\ \mu J$$

Capacitors are rated in terms of their capacitance value and the maximum voltage that the dielectric can withstand without breaking down. The **working voltage** (usually abbreviated as WV or VW) is the recommended maximum voltage at which the capacitor should be operated. Manufacturers usually specify this rating as WVDC. To assure safe operation, a capacitor should always have a WVDC rating that is at least 50% higher than the highest dc, or peak if ac is used, voltage to which it may be subjected in the circuit.

EXAMPLE 10.2

A capacitor is to be placed in a circuit where the biggest expected dc voltage is 15 V and the biggest expected ac voltage is 15 V RMS. Compute the WVDC rating needed for this capacitor.

Solution

$$V_{RMS} = 15\ V$$

$$V_P = 1.414(15) = 21.21\ V$$

$$\text{Maximum capacitor voltage} = 15\ V\ dc + 21.21\ V\ pk = 36.21\ V$$

$$\text{Safety margin voltage} = (0.5)(36.21) = 18.1\ V$$

$$\text{WVDC rating} = 36.21 + 18.1 = 54.31\ V$$

A capacitor should be chosen with a WVDC rating above 54 V.

Capacitors may be fixed or variable. The circuit symbol for a fixed capacitor is shown in Figure 10.1b; Figure 10.1c shows the symbol for a **variable capacitor.**

Since a capacitor is a device that is expressly constructed and inserted into circuits to introduce a desired capacitance, a look into the relationship between the construction, the materials, and the capacitance value is very instructive. Typical dielectrics are air, wax-impregnated paper, plastic, mica, ceramic, glass, and some oils.

The capacitance of a capacitor depends on (1) the overlapping area of the plates, (2) the type of dielectric used, and (3) the thickness of the dielectric. The bigger the plate area is, the bigger the capacitance value. Each dielectric is classified according to its **dielectric constant.** The higher the dielectric constant is, usually the higher the resulting capacitance value. The thinner the dielectric is, the closer are the plates; this results in higher capacitance values. As just mentioned, all that is necessary to make a capacitor is two metallic plates that are separated by an insulating material and encased in a cardboard or plastic case for protection. To increase the plate area and therefore the capacitance, several plates may be sandwiched between several sheets of dielectric. Alternate plates are then connected together, thus effectively enlarging the plate area and increasing the capacitance. These mica- and ceramic-type capacitors generally have quite small capacitance values.

Larger capacitance values are achieved by using **electrolytic capacitors.** This type of capacitor consists of two conductive electrodes, with the anode (positive terminal) having a metal oxide film formed on it. The film acts as a dielectric. An electrolyte (a substance in which a passage of electric current is accompanied by a chemical reaction), in the form of an acid or a salt, usually forms and maintains this layer of metal oxide. The electrolyte usually serves as the anode (positive terminal). It is imperative that electrolytic (sometimes called *polarized*) capacitors be connected with the positive terminal to the more positive side of the circuit connection and the negative terminal to the more negative side. If an electrolytic capacitor is improperly connected, a chemical reaction occurs, with gases as by-products. Since the capacitor is encapsulated in an airtight container, these gases build internal pressure, and if the pressure gets high enough, the capacitor will explode. To avoid dangerous explosions, these capacitors usually have "weak points" called *blowholes* built into them. Figure 10.2a shows the internal structure of a typical electrolytic capacitor. Figure 10.2b shows the cross-sectional view of the capacitor. Figure 10.2c shows the schematic symbol for an electrolytic capacitor.

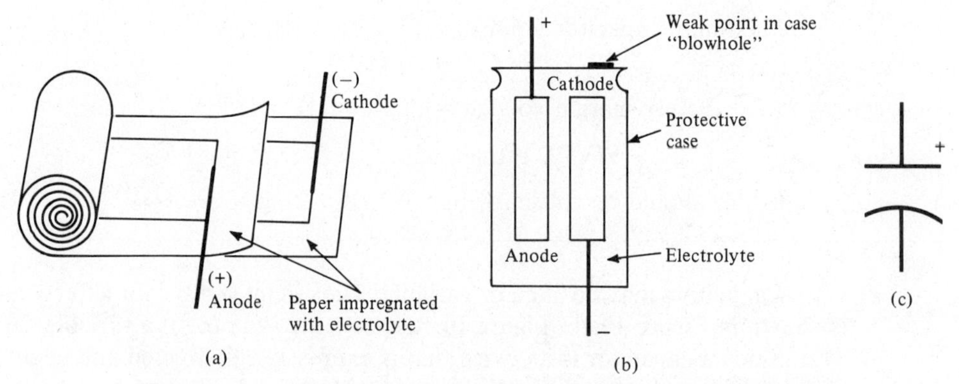

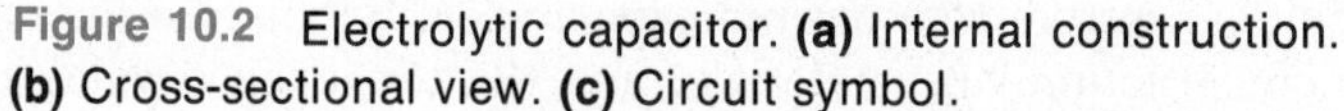
Figure 10.2 Electrolytic capacitor. **(a)** Internal construction. **(b)** Cross-sectional view. **(c)** Circuit symbol.

The capacitance of a capacitor may be varied by changing the effective area in common between the two plates or by changing the distance between the two plates. Figure 10.3a shows a **trimmer capacitor,** in which two spring-type plates are separated by a dielectric and held together by an adjustable screw. Turning the screw allows the distance between the two plates—and, therefore, the capacitance—to be varied. Figure 10.3b shows a ceramic trimmer capacitor. A semicircular silver plate is deposited on the bottom of a ceramic disk to form the fixed plate. The movable semicircular plate is a piece of metal that can be rotated by turning a screw and is placed on top of the ceramic disk, which serves as the dielectric. By turning the screw and therefore rotating the top plate, the area between the two plates—and, therefore, the capacitance—may be changed. Figure 10.3c shows what is commonly referred to as a **tuning capacitor** because it is used to tune in radio stations in AM receivers. Two sets of metal plates are arranged so that the rotor, or movable plates, moves between the stator, or nonmovable plates. Air is the dielectric. As the position of the rotor is changed, the area between the plates is changed and therefore the capacitance is changed.

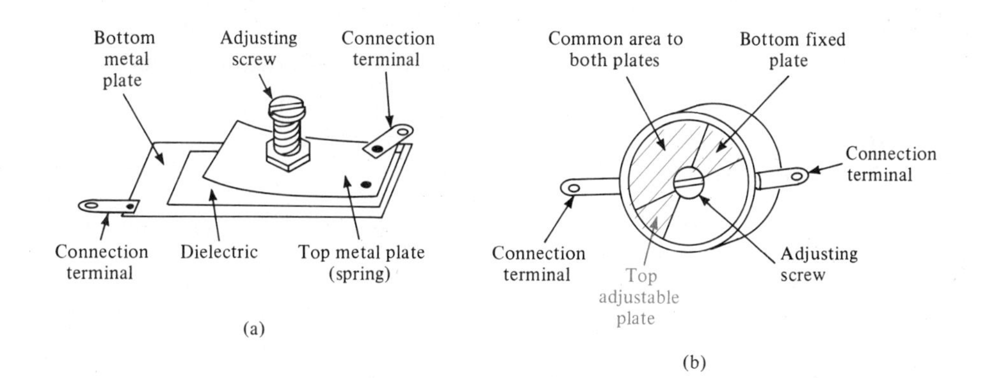

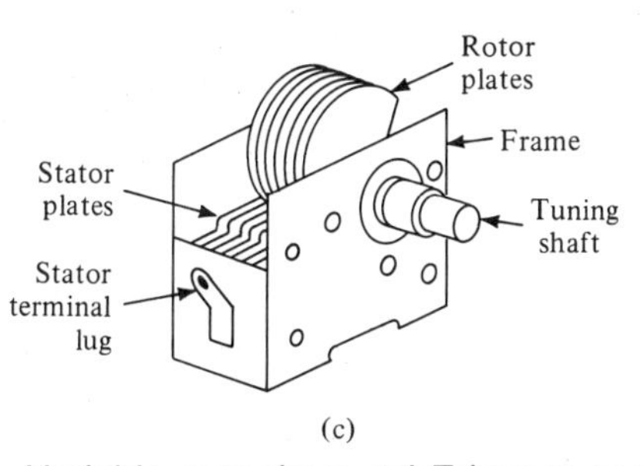

Figure 10.3 Variable capacitors. **(a)** Trimmer capacitor. **(b)** Ceramic trimmer capacitor. **(c)** Tuning capacitor.

Capacitors may be combined in series or in parallel. When placed in parallel, the plate area is effectively increased, therefore increasing the capacitance value. Capacitors in parallel thus add.

$$C_T = C_1 + C_2 + \cdots \quad \text{in parallel} \tag{10.3}$$

The voltage across capacitors in parallel is the same.

Capacitors in series combine as follows:

$$\frac{1}{C_T} = \frac{1}{C_1} + \frac{1}{C_2} + \cdots \tag{10.4}$$

The charge supplied to each capacitor in series is the same as the charge to all the series capacitors. The voltage across each series capacitor can then be observed to be:

$$V_1 = \frac{Q}{C_1}, \qquad V_2 = \frac{Q}{C_2}, \ldots$$

where Q is the charge supplied to the equivalent total capacitance.

EXAMPLE 10.3

a. Compute the equivalent capacitance of the configuration shown in Figure 10.4a.

b. Compute the equivalent capacitance of the configuration shown in Figure 10.4b.

c. Compute the equivalent capacitance of the configuration shown in Figure 10.4c.

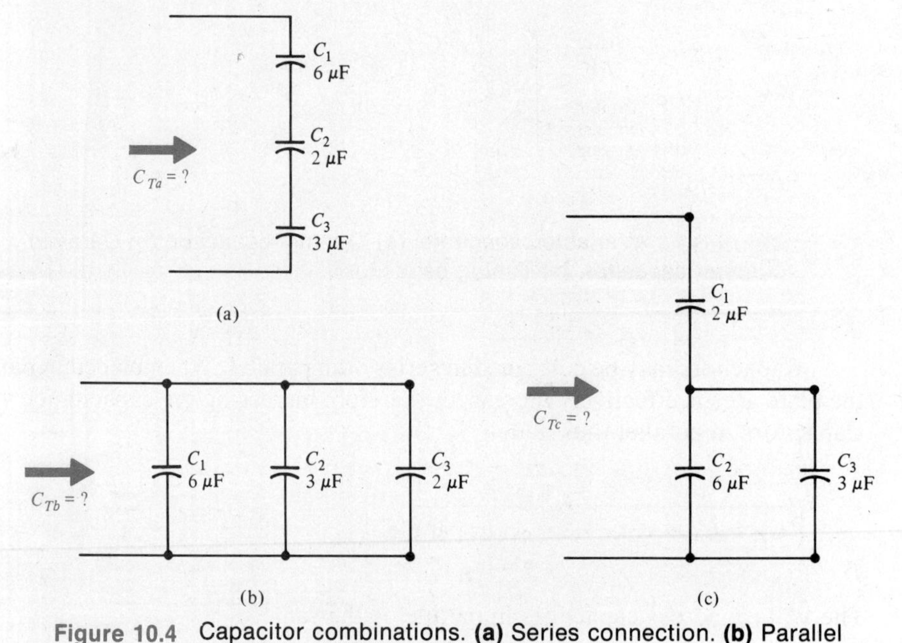

Figure 10.4 Capacitor combinations. **(a)** Series connection. **(b)** Parallel connection. **(c)** Series-parallel connection.

Solution

a. Figure 10.4a shows three capacitors in series.

$$\frac{1}{C_{Ta}} = \frac{1}{6\ \mu\text{F}} + \frac{1}{2\ \mu\text{F}} + \frac{1}{3\ \mu\text{F}}$$

$$= 0.167\ \text{MF}^{-1} + 0.5\ \text{MF}^{-1} + 0.333\ \text{MF}^{-1} = 1\ \text{MF}^{-1}$$

$$C_{Ta} = \frac{1}{1\ \text{MF}^{-1}} = 1\ \mu\text{F}$$

b. Figure 10.4b shows three capacitors in parallel.

$$C_{Tb} = 6\ \mu\text{F} + 3\ \mu\text{F} + 2\ \mu\text{F} = 11\ \mu\text{F}$$

c. Figure 10.4c shows capacitor C_2 in parallel with capacitor C_3. This combination is then in series with capacitor C_1.

$$C_X = C_2 + C_3 = 6\ \mu\text{F} + 3\ \mu\text{F} = 9\ \mu\text{F}$$

$$\frac{1}{C_{Tc}} = \frac{1}{C_X} + \frac{1}{C_1}$$

$$= \frac{1}{9\ \mu\text{F}} + \frac{1}{2\ \mu\text{F}}$$

$$= 0.111\ \text{MF}^{-1} + 0.5\ \text{MF}^{-1} = 0.611\ \text{MF}^{-1}$$

$$C_{Tc} = \frac{1}{0.611\ \text{MF}^{-1}} = 1.64\ \mu\text{F}$$

10.2 INDUCTANCE AND INDUCTORS

Inductance *is the characteristic property of an electrical component or circuit that opposes starting, stopping, or changing current.*

Inductance describes the opposition to changes in current. The symbol for inductance is L and the unit for inductance is the **henry** (H). Since the henry is a large unit, the **millihenry** (mH) and the **microhenry** (μH) are more popular units. Figure 10.5a shows the circuit symbol for a fixed-value **inductor.** The symbol for a variable inductor is shown in Figure 10.5b.

Since an inductor is made of a length of wire, besides showing an inductive effect, it also behaves like a resistor due to the resistance of the wire. If a voltage is applied across a resistor, the current immediately becomes E/R. If a voltage is applied across a coil, the current does not immediately become E/R but reaches this value only after a certain time delay. The property of an inductor or coil that causes this time delay is called *inductance*.

Recall from Chapter 9 that if a coil is cut by a magnetic field, a voltage is induced across the coil. If the magnetic field is produced by a current through a conductor, the induced voltage tends to set up a current in such a direction as to

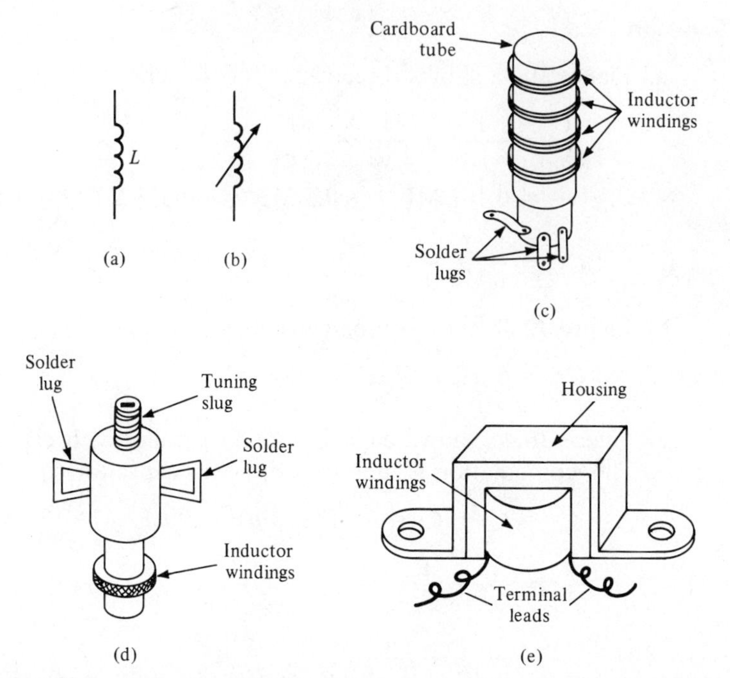

Figure 10.5 Inductors. **(a)** Fixed inductor circuit symbol. **(b)** Variable inductor circuit symbol. **(c)** Air core inductor. **(d)** Powered iron core inductor. **(e)** Sheet-iron core inductor.

oppose any changes in the existing magnetic field. For this reason it is now called a countervoltage (CV) or counter EMF (CEMF). Consider the inductor shown in Figure 10.6a. Refer to Figure 10.6b and notice that a current I is forced into the first loop. As the current is being established in the first loop, it generates an expanding magnetic field, which cuts loop 2 and loop 3 of the inductor. This action induces a CV across loops 2 and 3, which tends to set up a current (I_{CV}) in opposition to the current that is being forced through loop 1. Since the current I is being forced into the inductor, it overcomes this opposing current. Figure 10.6c shows the current I being forced and being established through loop 2, where it generates an expanding magnetic field, which cuts loop 3 of the inductor. This action induces a CV across loop 3, which tends to set up a current (I_{CV}) in opposition to the current that is being forced through loops 1 and 2. Notice that the CV induced is becoming smaller as the current I establishes itself through more loops. Figure 10.6d shows the current I completely established through all the loops of the inductor. Since the current is now constant, the induced CV collapses to zero because the magnetic field is not expanding anymore through the loops of the inductor. It should be observed that the current through the inductor was changed from zero to a constant value I, but it took time to overcome the effects of the CV that was induced by the inductor itself. This is the reason this self-induced opposition to current increase is called **self-inductance** (*inductance* for short). A decrease in current causes a similar but opposite effect in the inductor.

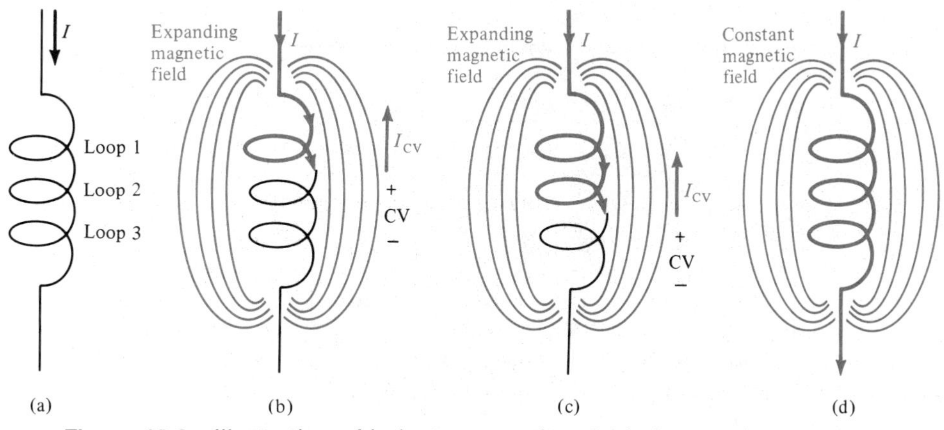

Figure 10.6 Illustration of inductor operation. **(a)** Inductor with three loops. **(b)** Current is established through the first loop. **(c)** Current is established through the first two loops. **(d)** Current is established through all three loops.

Thus no matter whether a 10-mA or a 10-A current is maintained through an inductor, there will be no induced voltage unless the current is either increasing or decreasing.

The following physical analogy should clarify the concept of inductance. Anyone who ever had to push a stalled automobile knows that it takes more work to start the auto moving than to keep it moving. The auto (or any mass) possesses the property of **inertia,** which is the characteristic of matter that opposes changes in velocity. It should be obvious that the auto may be brought to a certain velocity from standstill, but it takes time. Keeping the auto rolling is then not too difficult. However, to stop the car takes time and, again, more energy than to keep it moving. Thus inductance in electrical terms may be compared to inertia in physical terms. Figure 10.7 illustrates this physical analogy.

As mentioned, the unit of inductance is the henry, H. From the discussion so far, it should be clear that this unit involves a CV induced by a change in current that took a certain period of time to change from one value to another. In equation form,

$$\boxed{L = \frac{V}{\dfrac{\Delta I}{\Delta t}}} \qquad \textbf{(10.5a)}$$

where: L => inductance in henrys (H)
V => induced countervoltage in volts (V)
ΔI => change in current in amps (A)
Δt => time needed for the current change in seconds (s)

An inductor has a 1-H inductance if 1 V of countervoltage is induced by a change in current of 1 A in 1 s.

Figure 10.7 Illustration of inertia. **(a)** A car standing still opposes any attempt to start it moving. **(b)** Once the car is moving, it is easier to keep it moving. **(c)** A moving car opposes any attempt to slow it down and stop it.

The relationship may be stated in terms of derivatives. For a mathematical explanation using calculus, see Appendix F.

$$v_L = L\frac{\mathrm{d}(i_L)}{\mathrm{d}t} \tag{10.5b}$$

The main factors controlling the inductance of a coil are the number of turns of the conductor, the ratio of the cross-sectional area of the coil to its length, and the permeability of the material used to wind the coils on (the core).

$$L = \frac{N^2 \mu A}{l} \tag{10.6}$$

where: L => inductance in henrys (H)
N => number of turns
μ => permeability of the core in henrys per meter (H/m)
A => area of core (m^2)
l => average length of the core (m)

The following observations may be made:

1. The inductance increases as the square of the number of turns. Doubling the number of turns quadruples the inductance.
2. The inductance is directly proportional to the cross-sectional area of the coil. Doubling the cross-sectional area doubles the inductance.
3. The inductance is inversely proportional to the length of the coil. Doubling the length halves the inductance.
4. The type of material inserted in the middle of the coil serves as the core and influences the inductance considerably. *Permeability* (μ) is the measure of how good a given material is as a medium for allowing the existence of and for concentrating a magnetic field. Iron, for example, has a much higher permeability than air.

It is this property that is used to make variable inductors. If an inductor is made so that an iron screw may be inserted into or removed from the core area, then the core is either more iron than air or more air than iron. In the first case the inductance of the coil is increased; in the latter case it is decreased. The iron screw is called the **tuning slug,** and the coil is referred to as a variable inductor.

Inductors used for high-frequency circuits usually have air cores, whereas low-frequency inductors have iron cores. Figures 10.5c, 10.5d, and 10.5e show an air core inductor, a powdered iron core inductor, and a sheet-iron core inductor, respectively.

Inductors may be connected in series or in parallel. Inductors combine similarly to resistors. Inductors in series add:

$$L_T = L_1 + L_2 + \cdots \qquad \text{in series} \tag{10.7}$$

Inductors in parallel combine like resistors in parallel:

$$\frac{1}{L_T} = \frac{1}{L_1} + \frac{1}{L_2} + \cdots \qquad \text{in parallel} \tag{10.8}$$

EXAMPLE 10.4

a. Compute the equivalent inductance for the configuration shown in Figure 10.8a.
b. Compute the equivalent inductance for the configuration shown in Figure 10.8b.
c. Compute the equivalent inductance for the configuration shown in Figure 10.8c.

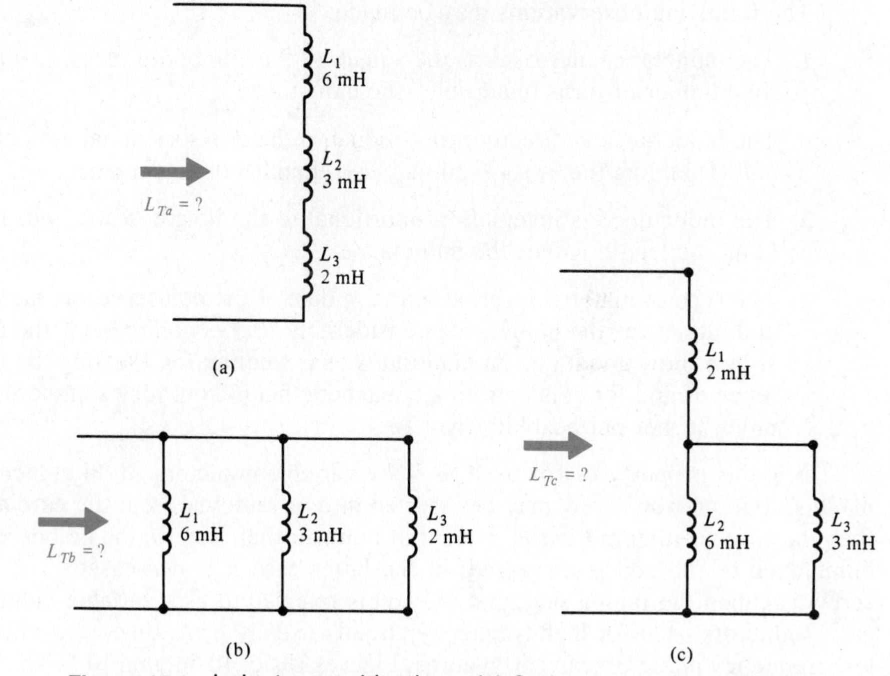

Figure 10.8 Inductor combinations. **(a)** Series connection. **(b)** Parallel connection. **(c)** Series-parallel connection.

Solution

a. Figure 10.8a shows three inductors in series. Since inductors in series add:

$$L_{Ta} = 6\text{ mH} + 3\text{ mH} + 2\text{ mH} = 11\text{ mH}$$

b. Figure 10.8b shows three inductors in parallel:

$$\frac{1}{L_{Tb}} = \frac{1}{6\text{ mH}} + \frac{1}{3\text{ mH}} + \frac{1}{2\text{ mH}}$$

$$= 0.167\text{ kH}^{-1} + 0.333\text{ kH}^{-1} + 0.5\text{ kH}^{-1} = 1\text{ kH}^{-1}$$

$$L_{Tb} = \frac{1}{1\text{ kH}^{-1}} = 1\text{ mH}$$

c. Figure 10.8c shows inductor L_2 in parallel with inductor L_3. This combination is then in series with inductor L_1.

$$\frac{1}{L_X} = \frac{1}{6\text{ mH}} + \frac{1}{3\text{ mH}} = 0.5\text{ kH}^{-1}$$

$$L_X = \frac{1}{0.5\text{ kH}^{-1}} = 2\text{ mH}$$

$$L_{Tc} = L_X + L_1 = 2\text{ mH} + 2\text{ mH} = 4\text{ mH}$$

It should be obvious by now that when a current is applied to an inductor, a magnetic field swells around the coil. When the coil is disconnected, a spark appears at the point where the circuit was broken and the magnetic field collapses. These events seem to indicate that there was energy stored in the magnetic field. The energy stored in a magnetic field of an inductor is:

$$\boxed{W = 0.5LI^2} \tag{10.9}$$

where: W => energy stored in joules (J)
L => inductance value in henrys (H)
I => existing steady inductor current in amps (A)

EXAMPLE 10.5

Compute the energy stored in the magnetic field of a 10-mH coil that has a steady 100-mA current through it.

Solution

$$W = 0.5LI^2 = 0.5(10 \text{ mH})(100 \text{ mA})^2 = 50\ \mu\text{J}$$

10.3 CAPACITIVE AND INDUCTIVE TIME CONSTANTS

As described earlier, it takes time for the voltage across a capacitor to change and it takes time for the current through an inductor to change.

The time (in seconds) it takes to charge an initially uncharged capacitor to 63.2% of its final voltage is called the charge time constant.

The time (in seconds) it takes for a charged capacitor to discharge to 36.8% (therefore losing 63.2%) of its initial voltage is called the discharge time constant.

Similarly, the time (in seconds) it takes to build up an inductor current from zero to 63.2% of its final value is called the buildup time constant.

The time (in seconds) it takes to decrease an established inductor current to 36.8% (therefore losing 63.2%) of its initial value is called the collapse time constant.

A **time constant** is usually represented by the Greek letter τ (tau).

The expression **steady state** refers to a condition in which circuit values remain essentially constant. This occurs after all initial fluctuations have settled down. These initial fluctuations are referred to as the **transient response** of the circuit.

For a capacitor C being charged or discharged through a resistor R, the time constant is simply the product of R and C.

$$\boxed{\tau = RC \qquad \text{(for a capacitor)}} \tag{10.10}$$

where: τ => time constant (s)
R => resistance (Ω)
C => capacitance (F)

For an inductor L having a current either being built up or collapsed through a resistor R (this could be the resistance of the coil itself), the time constant is the ratio of L to R.

$$\boxed{\tau = \frac{L}{R} \qquad \text{(for an inductor)}} \tag{10.11}$$

where: τ => time constant (s)
L => inductance (H)
R => resistance (Ω)

Complete charge or discharge of a capacitor or complete buildup or collapse of an inductor's current takes approximately five time constants.

For a mathematical explanation using a calculus derivation of Eqs. (10.12), (10.13), (10.14), and (10.15), see Appendix F.

Figure 10.9a shows the graph of the relationship that represents the charging of a capacitor. Expressed mathematically, this is:

$$\boxed{v_C = V_F(1 - e^{-t/\tau})} \tag{10.12}$$

where: v_C => capacitor voltage at any particular time (V)
V_F => final capacitor voltage (V)
e => 2.7183 (base of natural logarithms)
τ => RC time constant (s)
t => particular time at which v_C is desired (s)

Figure 10.9b shows the graph of the relationship that represents the discharging of a capacitor. Expressed mathematically this is:

$$\boxed{v_C = V_I e^{-t/\tau}} \tag{10.13}$$

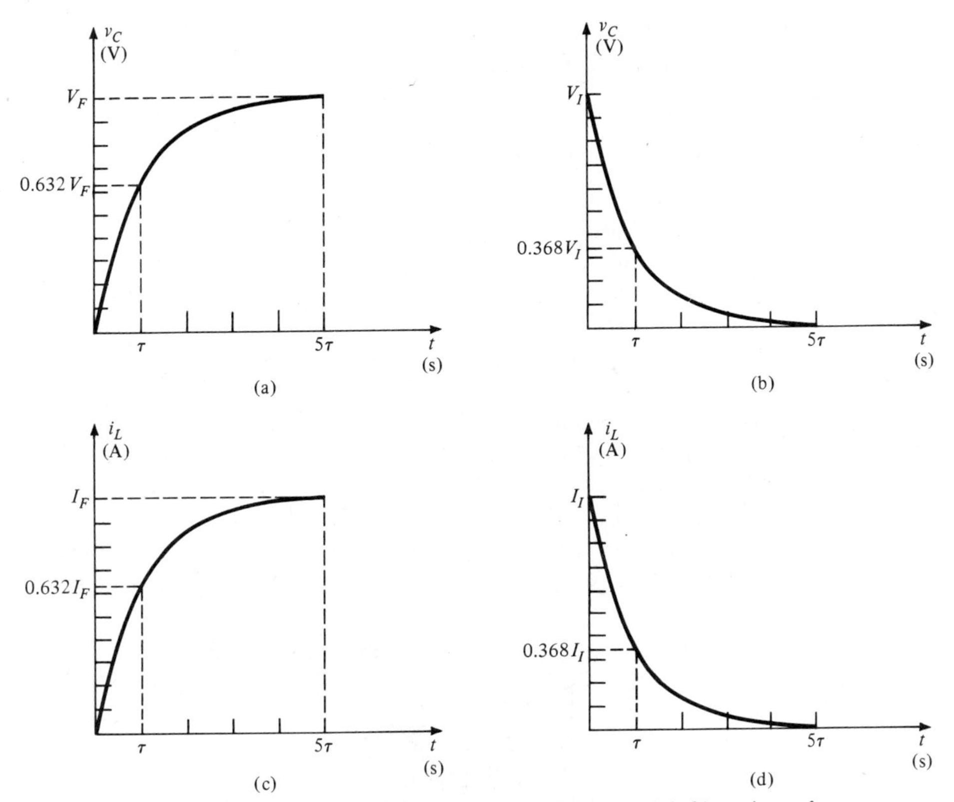

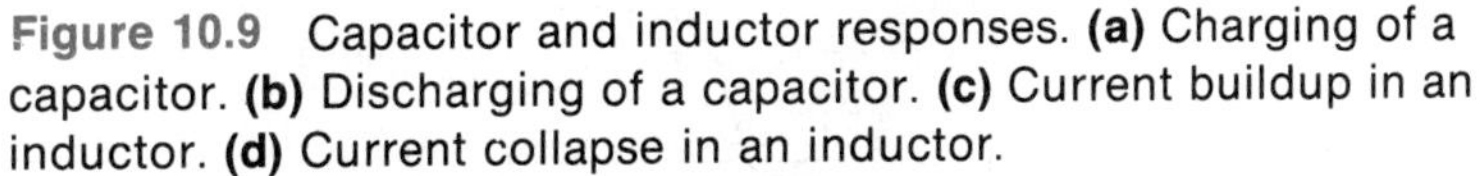

Figure 10.9 Capacitor and inductor responses. **(a)** Charging of a capacitor. **(b)** Discharging of a capacitor. **(c)** Current buildup in an inductor. **(d)** Current collapse in an inductor.

where: v_C => capacitor voltage at any particular time (V)
V_I => initial voltage (V)
e => 2.7183 (base of natural logarithms)
τ => RC time constant (s)
t => particular time at which v_C is desired (s)

Figure 10.9c shows the graph of the relationship that represents the build up of the current through an inductor. Expressed mathematically this is:

$$i_L = I_F\,(1 - e^{-t/\tau}) \qquad \textbf{(10.14)}$$

where: i_L => coil current at any particular time (A)
I_F => final coil current (A)
e => 2.7183 (base of natural logarithms)
τ => L/R time constant (s)
t => particular time at which i_L is desired (s)

Figure 10.9d shows the graph of the relationship that represents current collapse through a coil. Expressed mathematically this is:

$$i_L = I_I e^{-t/\tau} \tag{10.15}$$

where: i_L => coil current at any particular time (A)
I_I => initial coil current (A)
e => 2.7183 (base of natural logarithms)
τ => L/R time constant (s)
t => particular time at which i_L is desired (s)

These types of equations are referred to as exponential equations and occur quite frequently in electrical and other physical systems. The curve representing the charging of a capacitor and the curve representing the current buildup in an inductor are referred to as a **rising exponential.** The curve representing capacitor discharge and inductor current collapse is referred to as a **decaying exponential.** A fairly accurate graph of the "universal" rising exponential is shown in Figure

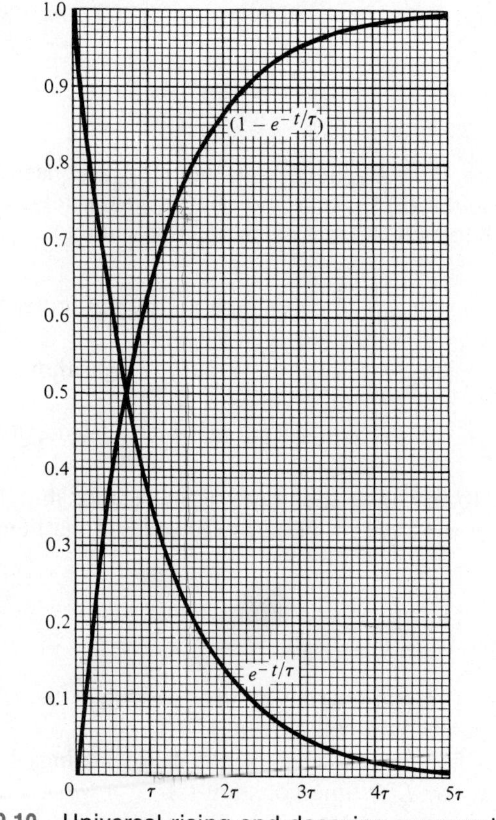

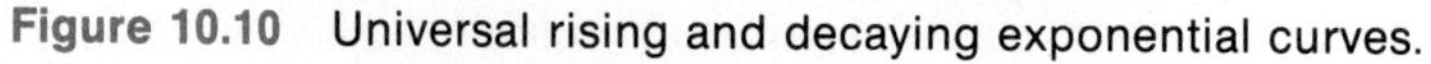

Figure 10.10 Universal rising and decaying exponential curves.

10.10. Figure 10.10 also shows the falling or decaying exponential. These graphs may be used instead of solving the above equations by calculator.

An uncharged capacitor has no voltage across it and starts charging at a rate limited by the resistance in the charging path. Since it has no voltage, an *uncharged capacitor* may be looked at as equivalent to a *short circuit*. As the capacitor charges, it is in its transient condition, and its voltage is increasing. When the capacitor is charged completely, it is in its steady-state condition, and the charging current is zero. Zero current suggests that a *charged capacitor* may be looked at as equivalent to an *open circuit*. The final capacitor voltage and the charging path resistance may be obtained by treating the capacitor as a load, removing it, and finding the Thevenin voltage and the equivalent Thevenin resistance at the open load terminals. The Thevenin voltage is the final capacitor voltage and the Thevenin resistance is the charging path resistance to be used in the time constant calculations.

EXAMPLE 10.6

a. In Figure 10.11a, switch 1 is closed and switch 2 is open. Compute the charging time constant, the initial surge-charging current, and the final capacitor voltage.

b. After 50 s switch 1 is opened and switch 2 is left open. After an additional 50 s, switch 2 is closed and the capacitor is allowed to discharge through the 0.5-MΩ resistor. State the voltage across the capacitor while both switches were open, and compute the discharging time constant and the initial surge of the discharging current.

c. Sketch the capacitor current and voltage for 200 s after switch 1 is initially closed.

Solution

a. The charging path resistance is 1 MΩ and the capacitor value is 10 μF; therefore, by Eq. (10.12) the charging time constant is:

$$\tau_C = (1\ \text{M}\Omega)(10\ \mu\text{F}) = 10\ \text{s}$$

Since the capacitor is uncharged when switch 1 is initially closed, it looks like a short circuit. The initial surge of the charging current, therefore, is limited only by the charging path resistor:

$$I_{\text{surge}} = \frac{10\ \text{V}}{1\ \text{M}\Omega} = 10\ \mu\text{A} \qquad \text{charge}$$

When the capacitor is finally fully charged, it looks like an open circuit. The charging current has stopped bringing the resistor voltage to zero. Therefore, the capacitor voltage must be the 10 V supplied by the battery.

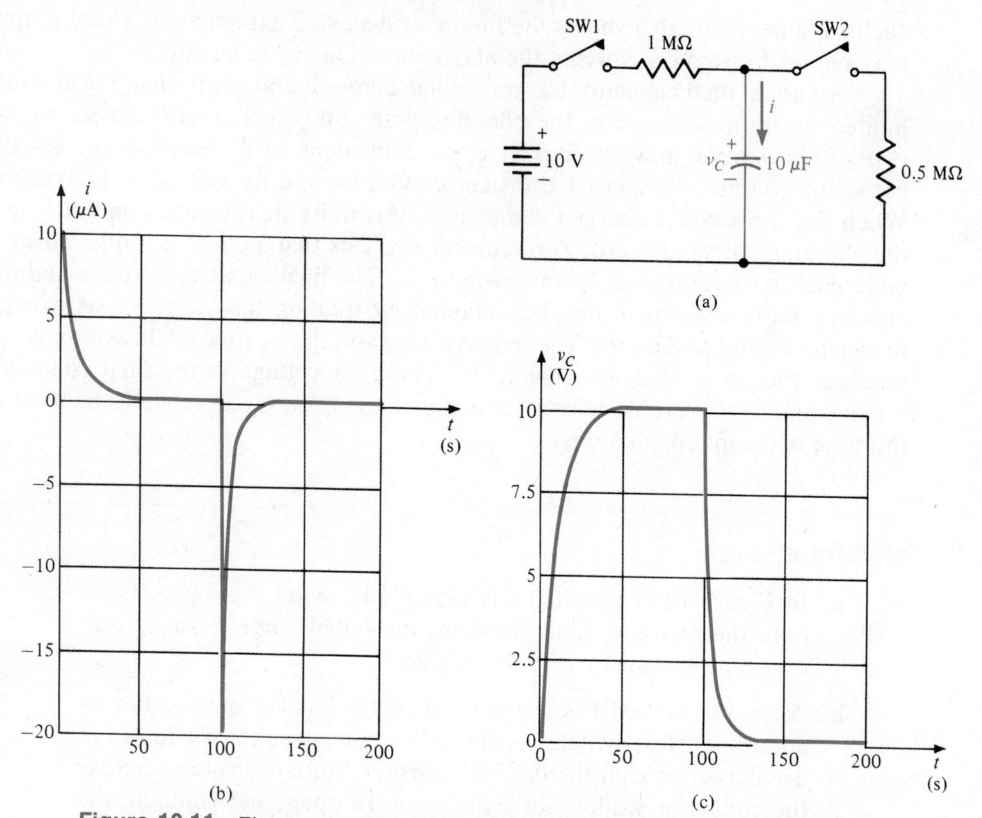

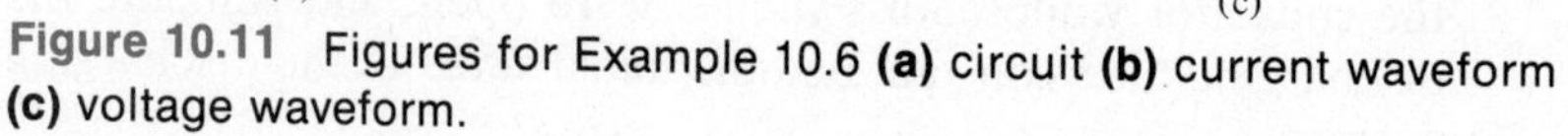

Figure 10.11 Figures for Example 10.6 **(a)** circuit **(b)** current waveform **(c)** voltage waveform.

b. Since it takes five time constants for the capacitor to charge fully, the capacitor charges to 10 V in 50 s. At this time switch 1 is opened. This leaves the capacitor charged with no discharge path. Therefore, the capacitor maintains its 10-V potential difference. Fifty additional seconds later, switch 2 is closed. This now provides a discharge path for the capacitor. The discharge path resistance is 0.5 MΩ, yielding a discharge time constant:

$$\tau_d = (0.5\ \text{M}\Omega)(10\ \mu\text{F}) = 5\ \text{s}$$

The initial surge of the discharging current is obtained by Ohm's Law as:

$$I_{\text{surge}} = \frac{10\ \text{V}}{0.5\ \text{M}\Omega} = 20\ \mu\text{A} \qquad \text{discharge}$$

The discharging current is considered to be negative, since it is in the opposite direction from the charging current, which is labeled positive.

c. Figure 10.11b shows a sketch of the capacitor current for the first 200 s. Note that the charging current initially surges to 10 μA and then decays to zero in five charging time constants. While the capacitor is not connected to anything (between 50 and 100 s), the current is zero. When the capacitor starts discharging (at 100 s) the discharge current surges to -20 μA and then it decays to zero in five discharging time constants.

Figure 10.11c shows how the capacitor voltage changes for the first 200 s. At $t = 0$, when switch 1 is initially closed, the capacitor is not charged and it therefore shows zero voltage. As the capacitor charges, the current is bringing electrons from the top plate to the bottom plate, therefore separating the electrons from the protons, leaving the positively charged protons on the top plate and depositing the negatively charged electrons on the bottom plate. As the current causes more charged particles to separate, the voltage increases, reaching the maximum in five charging time constants (50 s). While the capacitor is not connected to anything, the current is zero and the capacitor cannot charge or discharge. The voltage between 50 and 100 s therefore remains at 10 V. At 100 s, switch 2 is closed and the discharging current surges to -20 μA, indicating that the electrons are beginning to go back and neutralize the positive charge on the top plate. As the separated charges are reunited, the voltage decreases, reaching zero in 25 s (five discharging time constants). From 125 s on, the capacitor is discharged, and therefore both its voltage and current remain zero.

EXAMPLE 10.7

Sketch the charging voltage and current curves for the 10-μF capacitor shown in the circuit in Figure 10.12a. The switch is closed at $t = 0$. Prior to $t = 0$, the capacitor was uncharged.

Solution

To find the initial charging surge current, the uncharged capacitor may be replaced by a short, as shown in Figure 10.12b. Since the uncharged capacitor is shorting out its parallel 1-MΩ resistor, the initial surge current is:

$$I_{\text{surge}} = \frac{10\ \text{V}}{1\ \text{M}\Omega} = 10\ \mu\text{A}$$

To find the charging path resistance and the final capacitor voltage, Thevenin's Theorem may be applied. Figure 10.12c shows the capacitor removed, the open-circuit terminals labeled x and y, and the voltage source replaced by a short.

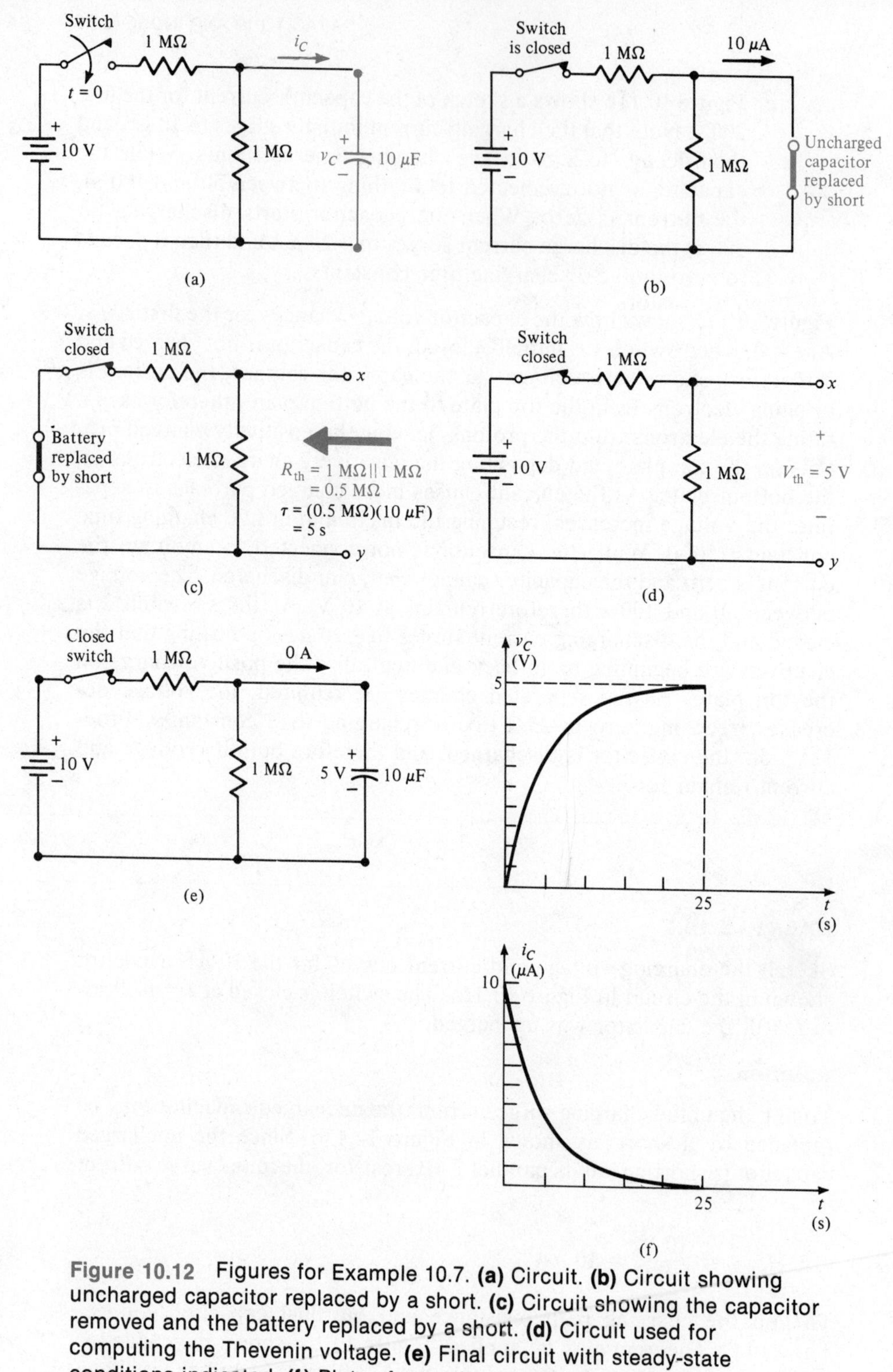

Figure 10.12 Figures for Example 10.7. **(a)** Circuit. **(b)** Circuit showing uncharged capacitor replaced by a short. **(c)** Circuit showing the capacitor removed and the battery replaced by a short. **(d)** Circuit used for computing the Thevenin voltage. **(e)** Final circuit with steady-state conditions indicated. **(f)** Plots of capacitor current and voltage.

This allows the computation of Thevenin's resistance as:

$$R_{th} = 1\ M\Omega \| 1\ M\Omega = 0.5\ M\Omega$$

Figure 10.12d shows the circuit with the capacitor removed. The Thevenin voltage may be computed by simple voltage division:

$$V_{th} = 0.5(10\ V) = 5\ V$$

Since both resistors are equal, each one gets half the battery voltage, or 5 V.

The charging time constant is:

$$\tau_C = (0.5\ M\Omega)(10\ \mu F) = 5\ s$$

The capacitor will charge completely in five time constants, or 25 s. The final capacitor voltage is 5 V. Figure 10.12e shows the circuit with the final voltage labeled across the capacitor. Figure 10.12f shows the graphs of the capacitor voltage and current.

An inductor with *zero current* through it may be represented by an equivalent *open circuit;* an inductor with an *established current* through it may be represented by the resistance of the inductor's wire. For a pure inductor, this resistance is zero (*short circuit*). The CV initially surges to a maximum value and then decreases to zero. Here Thevenin's Theorem may be applied to compute this initial maximum voltage surge and the building-up time constant. The Thevenin resistance is the R to be used in the L/R relationship and the Thevenin voltage is the initial surge voltage. The inductor's steady-state current may be obtained by replacing the inductor by a short circuit (Norton's Theorem) and computing the current through it.

EXAMPLE 10.8

For the circuit shown in Figure 10.13a, sketch the following:

a. Total voltage v_T

b. Inductor current i_L

c. Resistor voltage v_R

d. Inductor voltage v_L

The special double switch is made up of two double-pole, double-throw switches and is used here so that SW2 closes the *instant* SW1 opens.

Solution

When both SW1 and SW2 are open, all four required quantities are zero. Figure 10.13b shows that this condition exists for 1 s. It should be clear

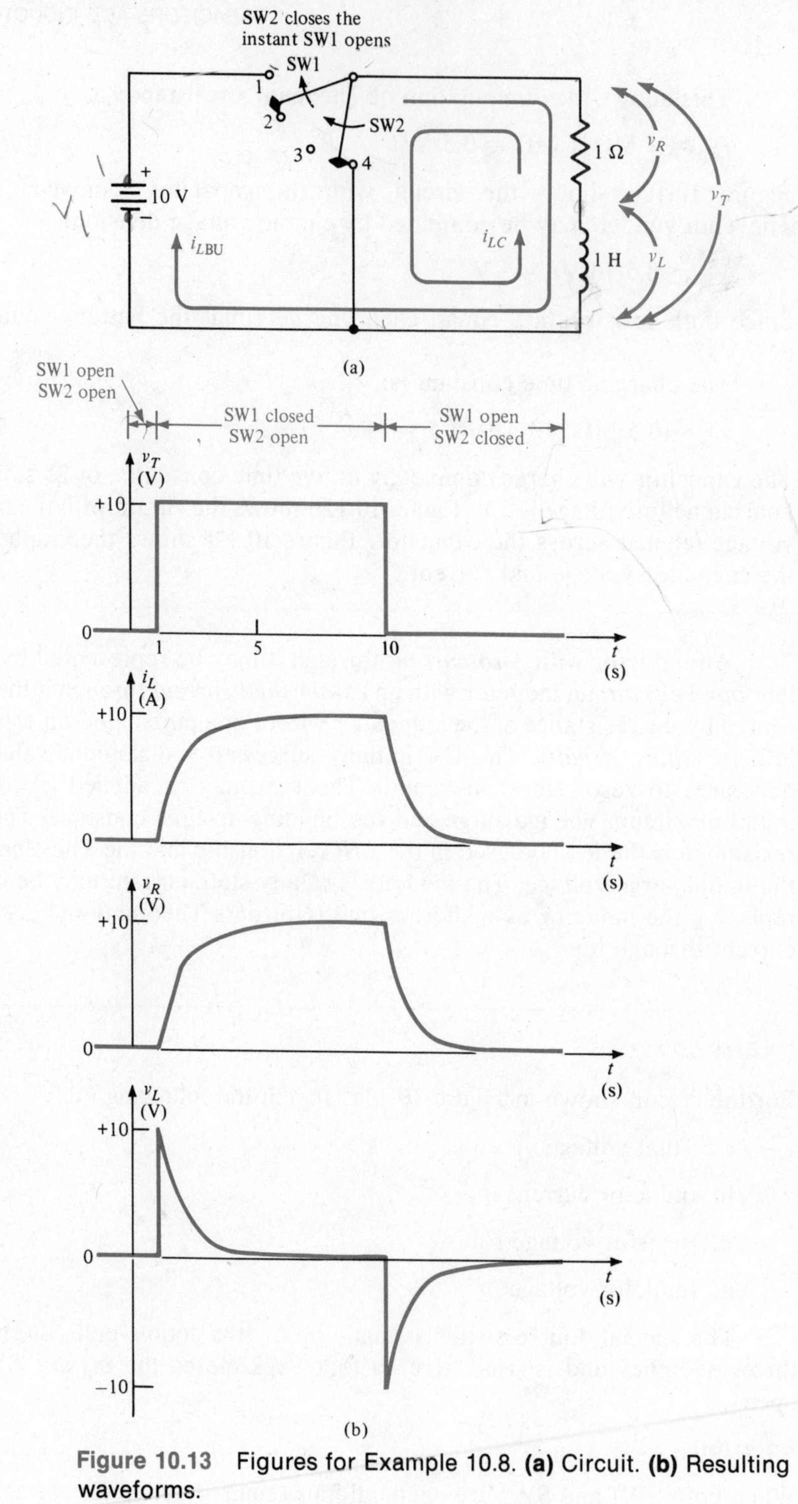

Figure 10.13 Figures for Example 10.8. **(a)** Circuit. **(b)** Resulting waveforms.

from the circuit that both the buildup time constant (τ_{BU}) and the collapse time constant (τ_C) involve the same 1-Ω resistor and they both are 1 s, as shown in Figure 10.13a.

At 1 s, SW1 is closed, and instantaneously the following occur:

1. The total voltage v_T jumps to +10 V.
2. The coil current is zero but starts increasing at a very fast rate.
3. The resistor voltage follows the coil current, by Ohm's Law.
4. The inductor back voltage surges to +10 V and starts decreasing.

At 2 s (one time constant after SW1 is closed), the quantities are at the following values:

1. v_T is still +10 V because the battery is still connected across the series *RL* combination.
2. i_L has increased to 6.32 A (63.2% of steady state).
3. v_R has increased to 6.32 V (also 63.2% of steady state).
4. v_L has decreased to 3.68 V (36.8% of the initial surge).

At 6 s (five time constants after SW1 is closed), the quantities have reached steady state and are at the following values:

1. v_T is at +10 V because the battery is still connected.
2. i_L has increased to the 10-A steady-state value (since the back coil voltage has collapsed, the coil looks like a short, and the current is limited only by the resistor).
3. v_R has increased to +10 V (following the inductor current and satisfying Ohm's Law and Kirchhoff's Voltage Law).
4. v_L has collapsed to zero, indicating that the magnetic field has stopped expanding and has reached a constant steady-state condition.

All quantities remain in this steady-state condition until the switch is operated at 10 s. At this time, SW2 closes at the *instant* that SW1 opens. The quantities under observation show the following:

1. v_T drops abruptly to zero because the battery is disconnected.
2. i_L starts collapsing at a very fast rate because the battery has been removed.
3. v_R starts falling, following the current and Ohm's Law.
4. v_L surges to −10 V because the magnetic field starts collapsing and, therefore, cutting the inductor coils.

At 11 s (one time constant after SW2 is closed and the current collapse has begun), the quantities under observation are at the following levels:

1. v_T is zero because the battery is disconnected.
2. i_L has collapsed to 3.68 A (36.8% of its initial value).
3. v_R has decreased to 3.68 V, following the current (also 36.8% of its initial value).
4. v_L has collapsed in magnitude to 36.8% of its surge value (it is still shown negative because it is opposite to the polarity shown in the diagram) and is shown as −3.68 V on the sketch.

At 15 s (five time constants after SW2 is closed), all quantities once again reach a steady-state condition. Since there is no energy source connected and the magnetic field of the coil has totally collapsed, all the quantities are at zero. Figure 10.13b shows the sketch of four quantities under observation.

EXAMPLE 10.9

For the circuit shown in Figure 10.14, compute the initial and the final inductor voltage and current when the switch is closed.

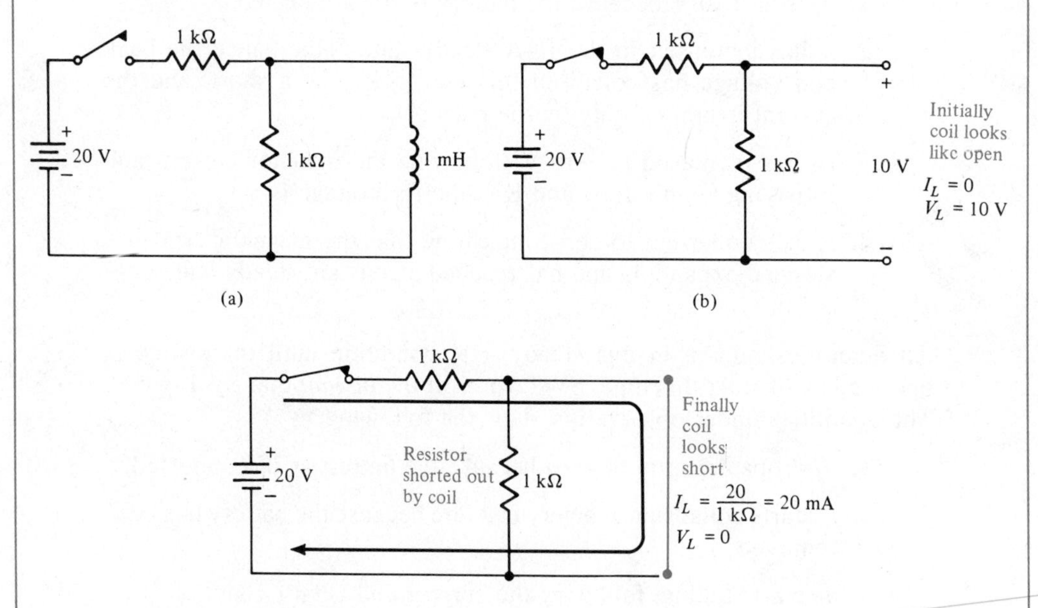

Figure 10.14 Figures for Example 10.9. **(a)** Circuit. **(b)** Circuit showing coil acting like an open circuit. **(c)** Circuit showing coil acting like a short circuit in steady state.

Solution

As soon as the switch is closed, the coil's inductance opposes any change in current through the coil and builds up a back voltage trying to keep the coil current at zero. The coil therefore acts like an open circuit. Figure 10.14b shows the coil under this condition. Here it is obvious that the 20-V source divides equally between the two series resistors. The initial coil back-voltage surge therefore is the voltage across its parallel resistor (10 V). Eventually, the current through the coil builds up and its associated magnetic field stops expanding, causing the back voltage to decrease to zero. Under this steady-state condition, the coil looks like a short circuit. Figure 10.14c shows the resulting circuit. Note that the resistor parallel to the coil is shorted out by the coil that now acts like a short circuit. The steady-state coil current is then only limited by the series 1-kΩ resistor. The resulting current is 20 V/1 kΩ = 20 mA. The steady-state inductor voltage is zero.

Inductance was previously defined as:

$$L = \frac{V}{\dfrac{\Delta I}{\Delta t}}$$

This relationship may be rearranged to show the voltage induced across a coil by a collapsing magnetic field:

$$\boxed{V = L\frac{\Delta I}{\Delta t}} \tag{10.16}$$

where:
V => induced voltage (V)
L => inductance of coil (H)
ΔI => change in current (A)
Δt => change in time (s)

This voltage, which is many times bigger than the applied voltage and is produced by the collapsing magnetic field when the current is abruptly cut off, is sometimes called the *inductive kick*.

EXAMPLE 10.10

The switch in the circuit shown in Figure 10.15a is closed and the current through the coil is allowed to build up to its 12-A steady-state value. Compute the inductive kick (value of the induced voltage) at the moment the switch is opened if the current decreased 5 mA in 1 μs. Describe the circuit action.

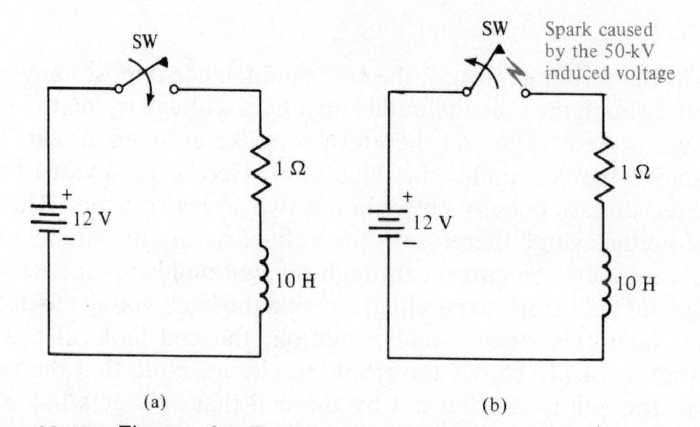

Figure 10.15 Figures for Example 10.10. **(a)** Circuit. **(b)** Circuit showing switch contacts sparking.

Solution

The switch is closed, the current builds up to 12 A (12 V/1 Ω) and the magnetic field is at a steady maximum value. At the instant the switch is opened, the current tends to collapse to zero, causing the magnetic field to collapse. The collapsing magnetic field cuts the coil windings, and a voltage is induced across the coil. This value is calculated by Eq. (10.16):

$$V = (10)\,\frac{5\ \text{mA}}{1\ \mu\text{s}} = (10)(5\ \text{kV}) = 50\ \text{kV}$$

Since the switch is in the process of being opened, this induced voltage appears across the switch, forcing a spark to jump between the opening switch contacts. The energy that was stored in the magnetic field dissipates in the form of heat and light energy in the form of the momentary spark between the switch contacts. Figure 10.15b shows the switch contacts sparking. The energy dissipated in this spark can seriously damage the switch contacts or break down the insulation between the coil windings. In inductive circuits, therefore, either special contact switches are used or other devices (such as arcing contacts or diodes) are used across the coil to dissipate this collapsing magnetic field energy.

The inductive kick is not always a disadvantage. The development of a very large inductive voltage kick across the points of a distributor in a car engine from a mere 12-V battery is the key feature of an ignition system.

A simple experiment showing that a magnetic field does store energy may be performed using the circuit shown in Figure 10.16. Here, a coil in parallel with a neon bulb is connected to a 12-V battery through a switch. A short time after the switch is closed, the inductor current is at maximum (limited only by the resis-

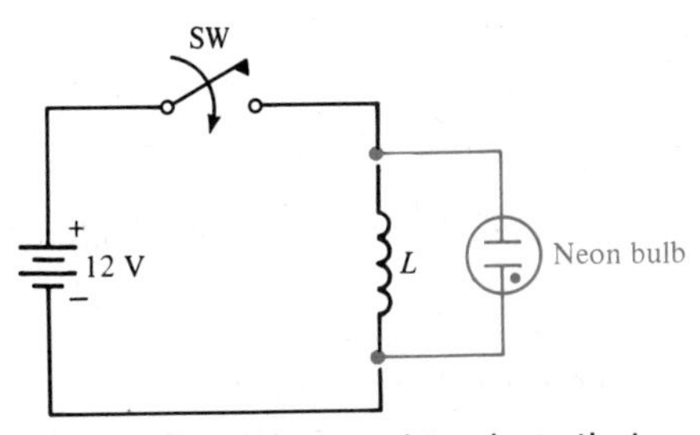

Figure 10.16 Experimental setup used to show that magnetic fields do store energy.

tance of the coil wire) and the magnetic field around the inductor is steady. The voltage across the coil and the neon bulb is 12 V, and the neon bulb is not lit because it needs at least 80 V for the neon gas to ionize and the bulb to light.

When the switch is open, the neon bulb turns on for an instant. This shows that the induced voltage across the coil when the switch was opened must have been at least 80 V.

10.4 REACTANCE, SUSCEPTANCE, IMPEDANCE, AND ADMITTANCE

***Reactance** is the opposition of inductive and capacitive components to alternating current.*

***Susceptance** is the ability of inductive and capacitive components to allow alternating current.*

***Impedance** is the opposition of electrical components to alternating current and the phase shift they cause between current and voltage.*

***Admittance** is the ability of electrical components to allow alternating current and the phase shift they cause between the voltage and the current.*

Capacitance is the property of electrical components that describes their *opposition to changes in voltage,* whereas *inductance* is the property of electrical components that describes their *opposition to changes in current*. The reaction of capacitors and inductors to abrupt, nonrepetitive changes was examined previously. Recall that for the voltage to change across a capacitor, charge must be first moved from one plate to the other through the surrounding circuit. This movement of charge is electric current, and it is clear that in a capacitor, the current occurs before the voltage. By observing Figures 10.11b and 10.11c, it should be evident that the *current* is *maximum* when the *voltage* is *zero;* when the *current* drops to *zero,* the *voltage* is at *maximum*. This suggests that if the applied voltage was changing sinusoidally and periodically, the current would be maximum when the voltage was zero and it would be zero when the voltage was maximum in both the positive and negative sense of the labeled voltage and current directions. Figure 10.17a shows a picture of this relationship.

The observation to be made here is that the capacitor current leads the capacitor voltage by 90°.

Similarly, recall that before a current is established through an inductor, an induced voltage is generated across the inductor. By observing Figure 10.13b, it should be evident that the *voltage* is *maximum* when the *current* is *zero,* and when

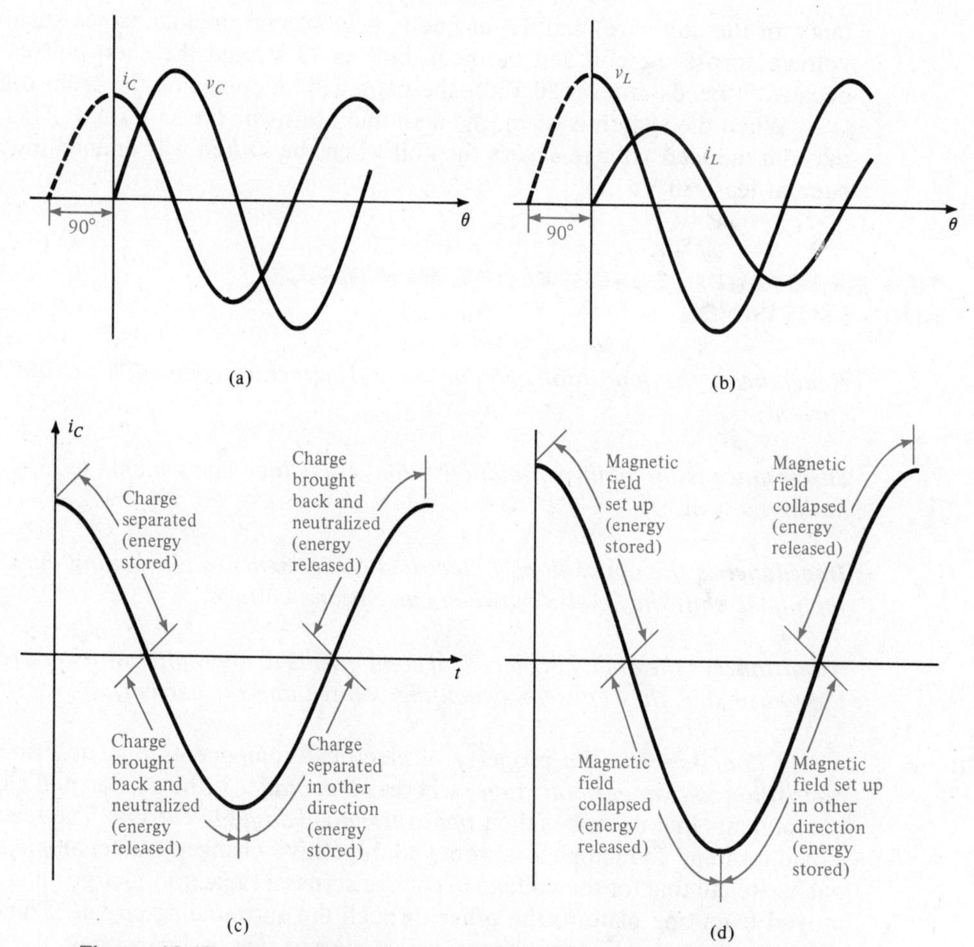

Figure 10.17 Phase differences. **(a)** Capacitive voltage and current sinusoids showing phase difference. **(b)** Inductive voltage and current showing phase difference. **(c)** Showing the quarter cycles when charges are separated and the quarter cycles when charges are brought back together in a capacitor. **(d)** Showing the quarter cycles when the magnetic field is set up and the quarter cycles when it collapses.

the *voltage* decreases to *zero* the *current* is at *maximum*. This suggests that if the applied current was changing sinusoidally and periodically, the voltage would be maximum when the current was zero and it would be zero when the current was maximum in both the positive and the negative sense of the labeled voltage and current directions. Figure 10.17b shows a picture of this relationship.

The observation to be made here is that the inductor voltage leads the inductor current by 90°.

It was already discussed how the time needed for the voltage across a capacitor to change depends on the value of the capacitor. The bigger the capacitor value, the more time it takes to charge and therefore to change its voltage. If a sinusoidal voltage is applied to a capacitive circuit, the capacitor would be charged and discharged, first one way, then the other, by a sinusoidal current that leads the voltage by 90°. Sinusoids, however, have periods (the time it takes for one repetition), and the period of a specific sinusoid may or may not be long enough to allow the capacitor to charge or discharge completely. Figure 10.17c shows that the first quarter cycle of current separates the charge (resulting in voltage), the second quarter cycle of current brings the charge back to the other plate, therefore neutralizing the charge (canceling the voltage), and the third and fourth quarter cycles repeat the same action but in the other direction. If the period is long enough to allow complete charge and discharge, the capacitor will charge to the total voltage. However, if the period is not long enough, the capacitor will not charge completely and the capacitor voltage will be less. The period may be so short, in fact, that the capacitor will have no chance at all to charge and show no voltage at all. This implies that as the period decreases, the capacitor looks more and more like a short circuit (zero voltage). The response of a capacitor to such repetitive charging and discharging is called *capacitive reactance* and is represented by X_C. Capacitive reactance is then proportional to the period and the capacitor value. Since period and frequency are inversely related, then capacitive reactance is inversely proportional to the frequency and capacitor value:

$$\boxed{X_C = \frac{1}{\omega C}} \qquad \textbf{(10.17a)}$$

or

$$\boxed{X_C = \frac{1}{2\pi f C} = \frac{0.159}{fC}} \qquad \textbf{(10.17b)}$$

where: X_C => capacitive reactance (Ω)
ω => radian frequency, $2\pi f$ (rad/s)
C => capacitor value (F)
f => frequency (Hz)

Capacitive reactance is the opposition that a capacitor has to alternating current, whereas capacitive susceptance is the ability of a capacitor to allow alternating current.

$$\boxed{B_C = \frac{1}{X_C}} \qquad (10.18)$$

where: B_C => capacitive susceptance (S)
X_C => capacitive reactance (Ω)

Since there is a 90° phase shift between the voltage and the current, the terms *capacitive impedance* and *capacitive admittance* were coined. Capacitive impedance shows the amount of reactance and the phase shift between the voltage and current; capacitive admittance shows the amount of susceptance and the phase between the voltage and current. The symbol for capacitive impedance is $\mathbf{Z_C}$. The symbol for capacitive admittance is $\mathbf{Y_C}$. Capacitive impedance has the form

$$\boxed{\mathbf{Z_C} = X_C\angle{-90^\circ}} \qquad (10.19a)$$

Since this is a phasor in polar form, it may be expressed in rectangular form as:

$$\boxed{\mathbf{Z_C} = -jX_C} \qquad (10.19b)$$

where: $\mathbf{Z_C}$ => capacitive impedance (Ω)
$-j$ => alternate representation of −90°
X_C => capacitive reactance (Ω)
−90° => phase angle between voltage and current

Capacitive admittance has the form:

$$\boxed{\mathbf{Y_C} = \frac{1}{\mathbf{Z_C}}} \qquad (10.20a)$$

$$\boxed{\mathbf{Y_C} = +jB_C} \qquad (10.20b)$$

where: $\mathbf{Y_C}$ => capacitive admittance in siemens (S)
$\mathbf{Z_C}$ => capacitive impedance (Ω)
$+j$ => alternate representation of $+90°$
B_C => capacitive susceptance in siemens (S)

It was discussed previously how it takes time for a current to be established through an inductor. If a sinusoidal current is applied through an inductor, the inductor reacts with a back voltage that is also sinusoidal but leads the current by 90°. If the applied sinusoidal current has a long enough period, the magnetic field changes sinusoidally and, therefore, back voltage has time to increase and subside in each quarter cycle (as shown in Figure 10.17d). If the period is not long enough, the induced back voltage does not have time to subside completely and is, therefore, constantly present at a certain value. If the period is extremely short, the induced back voltage does not have time to subside at all. This suggests that as the period gets smaller, the back voltage increases. A larger voltage indicates a larger opposition to current. The ultimate opposition to current is an open circuit. Since frequency is the inverse of the period, as the period decreases, the frequency increases. As the frequency increases, therefore, an inductor shows more and more back voltage and thus more opposition to alternating current.

Inductive reactance is the opposition of an inductor to alternating current.

Inductive reactance is directly proportional to the frequency and the inductor value. Inductive reactance is represented by X_L. Inductive reactance is expressed as:

$$X_L = \omega L \tag{10.21}$$

where: X_L => inductive reactance (Ω)
ω => radian frequency, $2\pi f$ (rad/s)
L => inductance (H)

Inductive susceptance is the ability of an inductor to maintain alternating current.

Inductive susceptance is represented by B_L. Inductive susceptance is the reciprocal of reactance and is expressed as:

$$B_L = \frac{1}{X_L} \tag{10.22}$$

where: B_L => inductive susceptance (S)
X_L => inductive reactance (Ω)

It was already shown how the back voltage leads the inductor current by 90°. The quantity that incorporates this information is called *inductive impedance.* Inductive impedance is represented by $\mathbf{Z_L}$ and may be expressed as:

$$\mathbf{Z_L} = X_L\angle 90^\circ \qquad (10.23a)$$

Since this is a phasor in polar form, it can be converted to rectangular form as:

$$\mathbf{Z_L} = +jX_L \qquad (10.23b)$$

where: $\mathbf{Z_L}$ => inductive impedance (Ω)
j => alternate representation of 90°
90° => phase angle between voltage and current

Inductive admittance is the reciprocal of inductive impedance and may be represented as:

$$\mathbf{Y_L} = \frac{1}{\mathbf{Z_L}} \qquad (10.24a)$$

$$\mathbf{Y_L} = B_L\angle -90^\circ \qquad (10.24b)$$

$$\mathbf{Y_L} = -jB_L \qquad (10.24c)$$

where: $\mathbf{Y_L}$ => inductive admittance (S)
$\mathbf{Z_L}$ => inductive impedance (Ω)
$-j$ => alternate representation of 90°
−90° => phase angle between voltage and current

EXAMPLE 10.11

Compute the impedance and admittance of a 1-μF capacitor at the following frequencies:

a. $f = 60$ Hz

b. $f = 400$ Hz

c. $f = 1$ kHz

d. $f = 20$ kHz

Solution

a. $\omega = 2\pi f = (2)(3.14)(60 \text{ Hz}) = 376.8 \text{ rad/s}$

By Eq. (10.17a):

$$X_C = \frac{1}{(376.8 \text{ rad/s})(1\ \mu\text{F})} = 2.654 \text{ k}\Omega$$

By Eq. (10.19a):

$$\mathbf{Z_C} = 2.654 \underline{/-90^\circ} \text{ k}\Omega \qquad \text{polar form}$$

By Eq. (10.19b):

$$\mathbf{Z_C} = -j2.654 \text{ k}\Omega \qquad \text{rectangular form}$$

By Eq. (10.20a):

$$\mathbf{Y_C} = \frac{1}{2.654 \text{ k}\Omega \underline{/-90^\circ}} = 0.377 \underline{/90^\circ} \text{ mS} \qquad \text{polar form}$$

By Eq. (10.20b):

$$\mathbf{Y_C} = j0.377 \text{ mS} \qquad \text{rectangular form}$$

b. $\omega = (2)(3.14)(400 \text{ Hz}) = 2512 \text{ rad/s} = 2.512 \text{ krad/s}$

$$X_C = \frac{1}{(2.512 \text{ krad/s})(1\ \mu\text{F})} = 398\ \Omega$$

$$B_C = \frac{1}{X_C} = \frac{1}{0.398 \text{ k}\Omega} = 2.513 \text{ mS}$$

$$\begin{aligned}\mathbf{Z_C} &= 398 \underline{/-90^\circ}\ \Omega \qquad \text{polar form}\\ &= -j398\ \Omega \qquad \text{rectangular form}\end{aligned}$$

$$\begin{aligned}\mathbf{Y_C} &= 2.513 \underline{/90^\circ} \text{ mS} \qquad \text{polar form}\\ &= +j2.513 \text{ mS} \qquad \text{rectangular form}\end{aligned}$$

c. $\omega = (2)(3.14)(1 \text{ kHz}) = 6.28 \text{ krad/s}$

$$X_C = \frac{1}{(6.28 \text{ krad/s})(1\ \mu\text{F})} = 159\ \Omega$$

$$B_C = \frac{1}{0.159 \text{ k}\Omega} = 6.289 \text{ mS}$$

$$\begin{aligned}\mathbf{Z_C} &= 159 \underline{/-90^\circ}\ \Omega \qquad \text{polar form}\\ &= -j159\ \Omega \qquad \text{rectangular form}\end{aligned}$$

$$\begin{aligned}\mathbf{Y_C} &= 6.289 \underline{/90^\circ} \text{ mS} \qquad \text{polar form}\\ &= +j6.289 \text{ mS} \qquad \text{rectangular form}\end{aligned}$$

d. $\omega = (2)(3.14)(20 \text{ kHz}) = 125.6 \text{ krad/s}$

$$X_C = \frac{1}{(125.6 \text{ krad/s})(1\ \mu\text{F})} = 7.96\ \Omega$$

$$B_C = \frac{1}{7.96\ \Omega} = 0.127 \text{ S}$$

$$\mathbf{Z_C} = 7.96\angle{-90^\circ}\ \Omega \quad \text{polar form}$$
$$= -j7.96\ \Omega \quad \text{rectangular form}$$

$$\mathbf{Y_C} = 0.127\angle{90^\circ} \text{ S} \quad \text{polar form}$$
$$= +j0.127 \text{ S} \quad \text{rectangular form}$$

This example makes it clear that as the frequency increases, the reactance and impedance of a capacitor decrease while the susceptance and admittance increase. As the frequency increases, the capacitor looks more and more like a short circuit.

EXAMPLE 10.12

Compute the impedance and admittance of a 1-mH inductor at the following frequencies:

a. $f = 60$ Hz

b. $f = 400$ Hz

c. $f = 1$ kHz

d. $f = 20$ kHz

Solution

a. $\omega = 2\pi f = (2)(3.14)(60 \text{ Hz}) = 376.8 \text{ rad/s}$

By Eq. (10.21):

$$X_L = (376.8 \text{ rad/s})(1 \text{ mH}) = 0.3768\ \Omega$$

By Eq. (10.22):

$$B_L = \frac{1}{0.3768\ \Omega} = 2.654 \text{ S}$$

By Eq. (10.23a):

$$\mathbf{Z_L} = 0.3768\angle{90^\circ}\ \Omega \quad \text{polar form}$$

By Eq. (10.23b):

$$\mathbf{Z_L} = j0.3768\ \Omega \quad \text{rectangular form}$$

By Eq. (10.24b):

$$\mathbf{Y_L} = B_L\angle{-90^\circ} = 2.654\angle{-90^\circ} \text{ S} \quad \text{polar form}$$

By Eq. (10.24c):

$$\mathbf{Y_L} = -jB_L = -j2.654 \text{ S} \qquad \text{rectangular form}$$

b. $\omega = (2)(3.14)(400 \text{ Hz}) = 2512 \text{ rad/s} = 2.512 \text{ krad/s}$

$$X_L = (2.512 \text{ krad/s})(1 \text{ mH}) = 2.512\ \Omega$$

$$\mathbf{Z_L} = 2.512\angle 90^\circ\ \Omega \qquad \text{polar form}$$
$$= +j2.512\ \Omega \qquad \text{rectangular form}$$

$$\mathbf{Y_L} = \frac{1}{\mathbf{Z_L}} = \frac{1}{2.512\angle 90^\circ\ \Omega} = 0.398\angle -90^\circ \text{ S} \qquad \text{polar form}$$
$$= -j0.398 \text{ S} \qquad \text{rectangular form}$$

c. $\omega = (2)(3.14)(1 \text{ kHz}) = 6.28 \text{ krad/s}$

$$X_L = (6.28 \text{ krad/s})(1 \text{ mH}) = 6.28\ \Omega$$

$$B_L = \frac{1}{X_L} = \frac{1}{6.28\ \Omega} = 0.159 \text{ S}$$

$$\mathbf{Z_L} = 6.28\angle 90^\circ\ \Omega \qquad \text{polar form}$$
$$= +j6.28\ \Omega \qquad \text{rectangular form}$$

$$\mathbf{Y_L} = 0.159\angle -90^\circ \text{ S} \qquad \text{polar form}$$
$$= -j0.159 \text{ S} \qquad \text{rectangular form}$$

d. $\omega = (2)(3.14)(20 \text{ kHz}) = 125.6 \text{ krad/s}$

$$X_L = (125.6 \text{ krad/s})(1 \text{ mH}) = 125.6\ \Omega$$

$$B_L = \frac{1}{0.1256 \text{ k}\Omega} = 7.962 \text{ mS}$$

$$\mathbf{Z_L} = 125.6\angle 90^\circ\ \Omega \qquad \text{polar form}$$
$$= +j125.6\ \Omega \qquad \text{rectangular form}$$

$$\mathbf{Y_L} = 7.962\angle -90^\circ \text{ mS} \qquad \text{polar form}$$
$$= -j7.962 \text{ mS} \qquad \text{rectangular form}$$

This example makes it clear that as the frequency increases, the inductive reactance and impedance also increase while the inductive susceptance and admittance decrease. As the frequency increases, the inductor looks more and more like an open circuit.

SUMMARY

In this chapter capacitance (C) was defined as the ratio of the charge stored (Q) to the voltage produced (V). The farad (F) and its alternate expressions microfarad (μF) and picofarad (pF) were defined as the units of capacitance. Practical capacitors and their specifications were examined. Fixed, variable, trimmer, and electrolytic capacitors were described.

Inductance (L) was defined as the characteristic property of electrical components that oppose starting, stopping, or simply changing current. The henry (H) was defined as the unit of inductance. The millihenry (mH) and the microhenry (μH) were defined as more practical and more popular inductance units.

Rules for combining capacitors and inductors in series and parallel as well as relations describing the energy stored in electric and magnetic fields were presented. Charging and discharging of capacitors as well as the buildup and collapse of current in inductors were examined.

Behavior of capacitors and inductors in ac circuits was presented. Reactance, susceptance, impedance, and admittance were defined. Impedance and admittance were identified as phasors and, therefore, were noted to follow all the rules of phasor algebra and complex numbers.

SUMMARY OF IMPORTANT EQUATIONS

$$C = \frac{Q}{V} \quad \text{F} \quad (10.1a)$$

$$i_C = C\frac{dv_C}{dt} \quad \text{A} \quad (10.1b)$$

$$W = 0.5CV^2 \quad \text{J} \quad (10.2)$$

In parallel:

$$C_T = C_1 + C_2 + \cdots \quad \text{F} \quad (10.3)$$

In series:

$$\frac{1}{C_T} = \frac{1}{C_1} + \frac{1}{C_2} + \cdots \quad \text{F}^{-1} \quad (10.4)$$

$$L = \frac{V}{\dfrac{\Delta I}{\Delta t}} \quad \text{H} \quad (10.5a)$$

$$v_L = L\frac{d(i_L)}{dt} \quad \text{V} \quad (10.5b)$$

$$L = \frac{N^2\mu A}{l} \quad \text{H} \quad (10.6)$$

In series:

$$L_T = L_1 + L_2 + \cdots \quad \text{H} \quad (10.7)$$

In parallel:

$$\frac{1}{L_T} = \frac{1}{L_1} + \frac{1}{L_2} + \cdots \quad \text{H}^{-1} \quad (10.8)$$

$$W = 0.5LI^2 \quad \text{J} \quad (10.9)$$

$$\tau = RC \quad \text{s} \quad (10.10)$$

$$\tau = \frac{L}{R} \quad \text{s} \quad (10.11)$$

Capacitor charge: $$v_C = V_F\,(1 - e^{-t/\tau}) \tag{10.12}$$

Capacitor discharge: $$v_C = V_I e^{-t/\tau} \tag{10.13}$$

$$i_L = I_F\,(1 - e^{-t/\tau}) \tag{10.14}$$

$$i_L = I_I e^{-t/\tau} \tag{10.15}$$

$$V = L\,\frac{\Delta I}{\Delta t} \quad \text{V} \tag{10.16}$$

$$X_C = \frac{1}{\omega C} \quad \Omega \tag{10.17a}$$

$$X_C = \frac{1}{2\pi fC} = \frac{0.159}{fC} \quad \Omega \tag{10.17b}$$

$$B_C = \frac{1}{X_C} \quad \Omega \tag{10.18}$$

$$\mathbf{Z_C} = X_C\underline{/-90^\circ} \quad \Omega \tag{10.19a}$$

$$\mathbf{Z_C} = -jX_C \quad \Omega \tag{10.19b}$$

$$\mathbf{Y_C} = \frac{1}{\mathbf{Z_C}} \quad \text{S} \tag{10.20a}$$

$$\mathbf{Y_C} = +jB_C \quad \text{S} \tag{10.20b}$$

$$X_L = \omega L \quad \Omega \tag{10.21}$$

$$B_L = \frac{1}{X_L} \quad \text{S} \tag{10.22}$$

$$\mathbf{Z_L} = X_L\underline{/90^\circ} \quad \Omega \tag{10.23a}$$

$$\mathbf{Z_L} = +jX_L \quad \Omega \tag{10.23b}$$

$$\mathbf{Y_L} = \frac{1}{\mathbf{Z_L}} \quad \text{S} \tag{10.24a}$$

$$\mathbf{Y_L} = B_L\underline{/-90^\circ} \quad \text{S} \tag{10.24b}$$

$$\mathbf{Y_L} = -jB_L \quad \text{S} \tag{10.24c}$$

REVIEW QUESTIONS

10.1 Define capacitance.

10.2 Define the units F, μF, and pF.

10.3 Explain what is meant by WVDC.

10.4 Explain the construction of an electrolytic capacitor.

10.5 Explain what is meant by a trimmer capacitor.

10.6 Define inductance.

10.7 Compare inductance and inertia.

10.8 Explain how a tuning slug varies the value of an inductor.

10.9 Define a time constant.

10.10 Explain the difference between a transient and a steady-state response.

10.11 Explain what is meant by a rising exponential and a decaying exponential.

10.12 Explain the difference between reactance and susceptance.

10.13 Define impedance and admittance.

PROBLEMS

10.1 Two capacitors, having values 5 μF and 10 μF, are connected in parallel across a 12-V supply. Compute:
a. Total capacitance
b. Voltage across each capacitor
c. Charge on each capacitor
d. Energy stored by each capacitor

10.2 Compute the total capacitance of the two capacitors from Problem 10.1 if the capacitors are connected in series. (ans. 3.33 μF)

10.3 Compute the minimum value of WVDC for the 10-μF capacitor in Problem 10.1.

10.4 Compute the minimum value of WVDC for the 5-μF capacitor in Problem 10.2. (ans. 12 V)

10.5 A 6-mH and a 3-mH coil are connected in parallel. Compute the value of a third parallel coil if the total desired inductance is 1.5 mH.

10.6 Find the inductance of a coil if in 1 s the current rose by 2 A and the back EMF was 20 V. (ans. 10 H)

10.7 For the circuit shown in Figure 10.18, the capacitor is initially not charged.

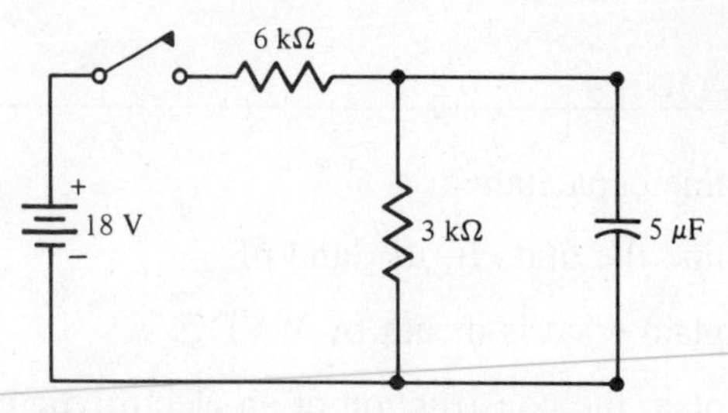

Figure 10.18 Circuit for Problem 10.7.

Compute:

a. Charging time constant

b. Time needed to reach maximum voltage after the switch is closed

c. Length of time it takes for the capacitor voltage to reach 3 V

10.8 The capacitor in Figure 10.19 is charged to 10 V, as shown. How long after the switch is opened would the capacitor voltage reach zero? (ans. 500 ms)

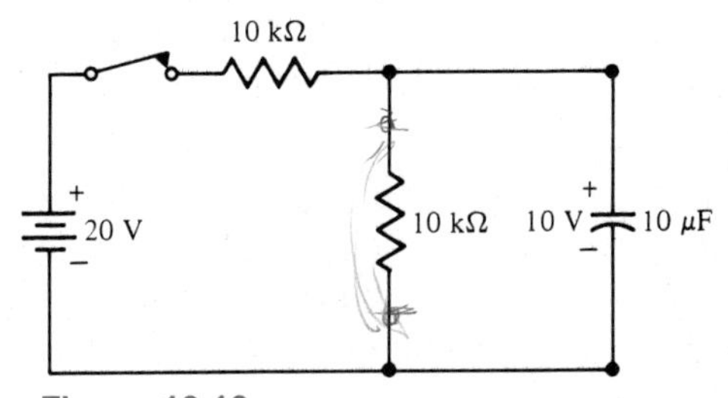

Figure 10.19 Circuit for Problem 10.8.

10.9 For the circuit shown in Figure 10.20, calculate the inductor current 150 μs after the switch is closed.

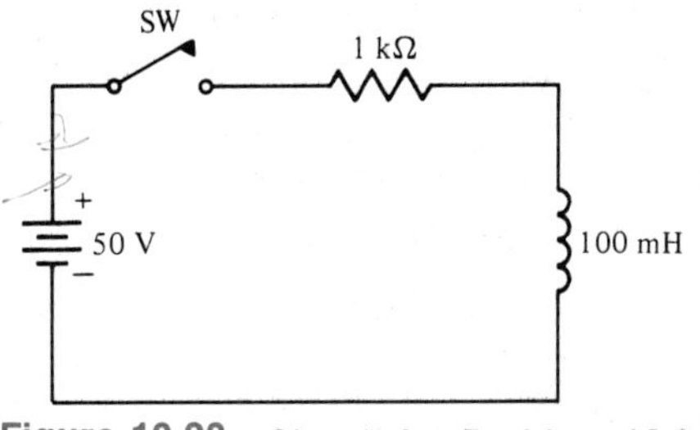

Figure 10.20 Circuit for Problem 10.9.

10.10 An inductor and a 2.2-kΩ resistor are connected in series across a 12-V supply. The inductor current is to reach 2.73 mA within 637 μs after the circuit is connected. Determine the required inductor value. (ans. 3.14 mH)

10.11 Compute the initial surge current for the capacitor in Figure 10.18.

10.12 Use your calculator to find the value of $e^{-t/RC}$ for the following values of R, C, and t:

a. $R = 5\ \Omega$, $C = 0.5\ \mu F$, and $t = 1$ s

b. $R = 1\ k\Omega$, $C = 200\ \mu F$, and $t = 4$ s

c. $R = 1\ M\Omega$, $C = 5\ \mu F$, and $t = 5$ s

(ans. 0, 2.06×10^{-9}, 0.368)

10.13 Use your calculator to find the value of $e^{-t(L/R)}$ for the following values of R, L, and t:
a. $R = 15\ \Omega$, $L = 10$ H, and $t = 1$ s
b. $R = 10\ \Omega$, $L = 10$ H, and $t = 2$ s
c. $R = 5\ \Omega$, $L = 10$ H, and $t = 3$ s

10.14 A 10-H coil has a 5-Ω resistor connected in series. The coil and resistor combination is connected across a 100-V dc source.
a. Compute the initial current at the instant the circuit is closed.
b. Compute the current at 2, 4, 6, 8, and 10 s after the circuit is closed.
c. Compute the voltage across the coil 2 s after the circuit is closed.
(ans. 0; 12.6 A, 17.2 A, 19 A, 19.6 A, 19.8 A; 37 V)

10.15 A series RC ($R = 1$ kΩ, $C = 1\ \mu$F) circuit is connected across a voltage source that feeds it a 10-V pulse for 1 s (after that the source may be considered a short circuit). Accurately plot the voltage across the resistor for the first 1 ms and then for the next 3 ms.

10.16 A 10-A current exits through a 10-H coil the instant before the switch is opened. If a 5-Ω resistor is placed across the coil as the switch is opened, find the current 4 s later. (ans. 1.3 A)

10.17 A 20-H coil and a 10-Ω resistor are placed in series across a 80-V dc line. Find the steady-state current and the time needed to reach it.

10.18 A 5-kΩ resistor and a 400-μF capacitor are in series across a 120-V dc source. One second after the circuit is connected, find:
a. The current
b. The resistor voltage
c. The capacitor voltage
(ans. 14.4 mA, 72 V, 48 V)

10.19 Determine the reactance of a 5-μF capacitor at:
a. 20 Hz
b. 50 Hz
c. 60 Hz
d. 20 kHz

10.20 Determine the susceptance of the 5-μF capacitor at each frequency shown in Problem 10.19. (ans. 629 μS, 1.58 mS, 1.88 mS, 629 mS)

10.21 Write the polar form of the impedance for the 5-μF capacitor for each frequency shown in Problem 10.19.

10.22 Write the rectangular form of the impedance for each frequency shown in Problem 10.19 for the 5-μF capacitor. (ans. $-j\,1.59$ kΩ, $-j\,634\ \Omega$, $-j\,531\ \Omega$, $-j1.59\ \Omega$)

10.23 Repeat Problem 10.21 for the admittance.

10.24 Repeat Problem 10.22 for the admittance. (ans. $j629\ \mu$S, $j1.58$ mS, $j1.88$ mS, $j629$ mS)

10.25 Compute the reactance of a 10-mH inductor at the frequency specified in Problem 10.19.

10.26 Determine the susceptance of the 10-mH coil at the frequencies specified in Problem 10.19. (ans. 796 mS, 318 mS, 265 mS, 796 μS)

10.27 Determine the polar form of the 10-mH inductor's impedance at the various frequencies stated in Problem 10.19.

10.28 Determine the rectangular form of the impedance obtained in Problem 10.27. (ans. $j1.256\ \Omega$, $j3.14\ \Omega$, $j3.768\ \Omega$, $j1.256\ \text{k}\Omega$)

10.29 Determine the polar form of the admittance of the 10-mH coil at the frequencies shown in Problem 10.19.

10.30 Determine the rectangular form of the admittance of the 10-mH coil for each frequency specified in Problem 10.19. (ans. $-j796$ mS, $-j318$ mS, $-j265$ mS, $-j796$ mS)

CHAPTER 11

Ac Circuits

GLOSSARY

parallel ac circuit—ac circuit in which all the impedances are connected between the same two nodes

series ac circuit—ac circuit in which all the impedances are connected in string fashion

series-parallel ac circuit—ac circuit containing both series and parallel combinations of impedances

INTRODUCTION

All the rules of circuit analysis developed in Chapters 5 and 7 for dc circuits may be applied to ac circuits by simply changing from the algebra of real numbers to that of complex numbers and observing the following substitutions:

Dc analysis	*Ac analysis*
R	$\mathbf{Z}$
E	$\mathbf{E}$
V	$\mathbf{V}$
I	$\mathbf{I}$

In ac circuit analysis all impedances and alternating voltages and currents must be treated as phasor quantities. This leads to many more calculations when analyzing an ac impedance network compared to a similar dc resistance network.

In this chapter, series, parallel, and series-parallel ac networks are treated. Mesh and nodal analyses for ac networks are also presented by conventional as well as computer methods.

11.1 AC CIRCUIT LAWS

For ac circuits Ohm's Law is restated in terms of phasors:

$$\boxed{\mathbf{I} = \frac{\mathbf{E}}{\mathbf{Z}}} \tag{11.1}$$

where: $\mathbf{I}$ => phasor current (A)
$\mathbf{E}$ => phasor voltage (V)
$\mathbf{Z}$ => impedance (Ω)

Kirchhoff's Voltage Law (KVL) for ac circuits may be stated as:

In any closed ac circuit, the phasor sum of the voltage drops equals the phasor sum of the voltage rises.

Kirchhoff's Current Law (KCL) for ac circuits may be stated as:

The phasor sum of the currents entering a node equals the phasor sum of the currents leaving the node.

It is evident that these laws require multiplication, division, addition, and subtraction of phasors. Since phasors are complex numbers, multiplication and division are easily carried out in polar form, whereas addition and subtraction are simpler in rectangular form. To this end, all impedances, admittances, voltages, and currents should always be stated in both polar and rectangular form, and the appropriate form should be used.

11.2 SERIES AC CIRCUITS

A **series ac circuit** is a circuit in which all the impedances are connected end to end in string fashion (Figure 11.1). In a series ac circuit the current is the same through all the components. The total equivalent impedance of a series ac circuit is:

$$\mathbf{Z_T} = \mathbf{Z_1} + \mathbf{Z_2} + \mathbf{Z_3} + \cdots + \mathbf{Z_N} \tag{11.2}$$

where $\mathbf{Z_N}$ is the value of the Nth impedance in the string.

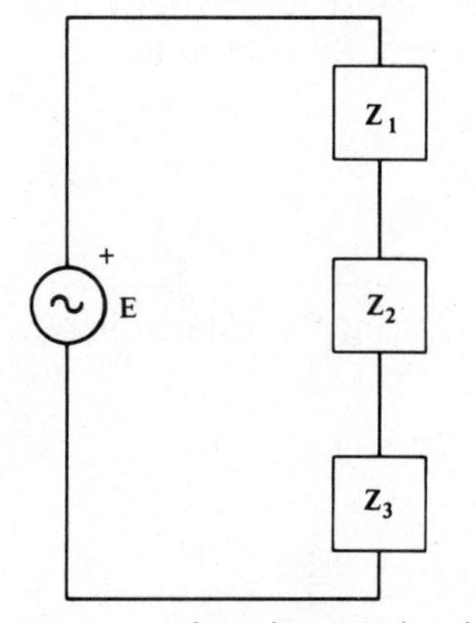

Figure 11.1 A series ac circuit.

EXAMPLE 11.1

For the circuit shown in Figure 11.2a determine:

a. The phasor form of the voltage source

b. The reactance and impedance of each component

c. The total impedance of the circuit

d. The circuit current

e. The voltage across the resistor

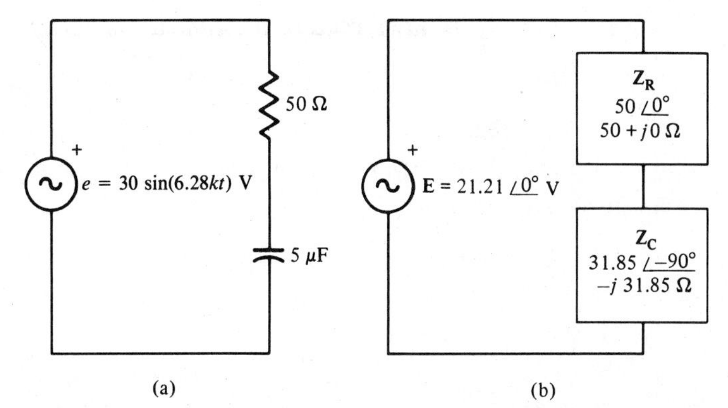

Figure 11.2 Diagrams for Example 11.1. **(a)** Circuit diagram. **(b)** Phasor representation.

f. The voltage across the capacitor

g. If the source voltage equals the resistor voltage plus the capacitor voltage

Solution

a. The equation for the source voltage is:

$$e = 30 \sin(6.28 \times 10^3 t) \quad \text{V}$$

Comparing it to the standard form,

$$e = V_P \sin(\omega t + \phi) \quad \text{V}$$

the following may be recognized:

$$V_P = 30 \text{ V}$$

$$\omega = 6.28 \text{ krad/s}$$

$$\phi = 0°$$

$$V_{\text{RMS}} = 0.707 V_P = (0.707)(30) = 21.21 \text{ V}$$

The phasor form of the source is:

$$\mathbf{E} = V_{\text{RMS}} \angle \phi = 21.21 \angle 0° \text{ V}$$

b. The impedance of the resistor is:

$$\mathbf{Z_R} = R + j0 = 50 + j0 \ \Omega \qquad \text{rectangular form}$$
$$= R \angle 0° = 50 \angle 0° \ \Omega \qquad \text{polar form}$$

The reactance of the capacitor is:

$$X_C = \frac{1}{\omega C} = \frac{1}{(6.28 \text{ krad/s})(5 \ \mu\text{F})} = 31.85 \ \Omega$$

The impedance of the capacitor is:

$$\mathbf{Z_C} = -jX_C = -j31.85\ \Omega \qquad \text{rectangular form}$$

$$= X_C\angle{-90^\circ} = 31.85\angle{-90^\circ}\ \Omega \qquad \text{polar form}$$

Figure 11.2b shows the circuit representation with all the components in phasor notation.

c. The total impedance of the circuit is simply the addition of the two series impedances:

$$\mathbf{Z_T} = \mathbf{Z_R} + \mathbf{Z_C}$$
$$= 50 - j31.85\ \Omega \qquad \text{rectangular form}$$
$$= 59.28\angle{-32.5^\circ}\ \Omega \qquad \text{polar form}$$

d. The circuit current is computed by applying Ohm's Law:

$$\mathbf{I} = \frac{\mathbf{E}}{\mathbf{Z_T}} = \frac{21.21\angle{0^\circ}}{59.28\angle{-32.5^\circ}} = 0.358\angle{32.5^\circ}\text{ A}$$
$$= 358\angle{32.5^\circ}\text{ mA} \qquad \text{polar form}$$
$$= 302 + j192\text{ mA} \qquad \text{rectangular form}$$

e. The voltage across the resistor may be computed by applying Ohm's Law:

$$\mathbf{V_R} = \mathbf{IZ_R}$$
$$= (0.358\angle{32.5^\circ})(50\angle{0^\circ})$$
$$= 17.9\angle{32.5^\circ}\text{ V} \qquad \text{polar form}$$
$$= 15 + j9.6\text{ V} \qquad \text{rectangular form}$$

The voltage across the capacitor may be computed by applying Ohm's Law again:

$$\mathbf{V_C} = \mathbf{IZ_C}$$
$$= (0.358\angle{32.5^\circ})(31.85\angle{-90^\circ})$$
$$= 11.40\angle{-57.5^\circ}\text{ V} \qquad \text{polar form}$$
$$= 6.12 - j9.61\text{ V} \qquad \text{rectangular form}$$

f. By KVL:

$$\begin{array}{lrl} & \mathbf{V_R} => 15 & + j9.6 \\ + & \mathbf{V_C} => 6.12 & - j9.6 \\ \hline & \mathbf{E} => 21.12 & - j0.01 \end{array}$$

which is close enough to the source voltage.

EXAMPLE 11.2

Repeat Example 11.1 for the series *RL* circuit shown in Figure 11.3a.

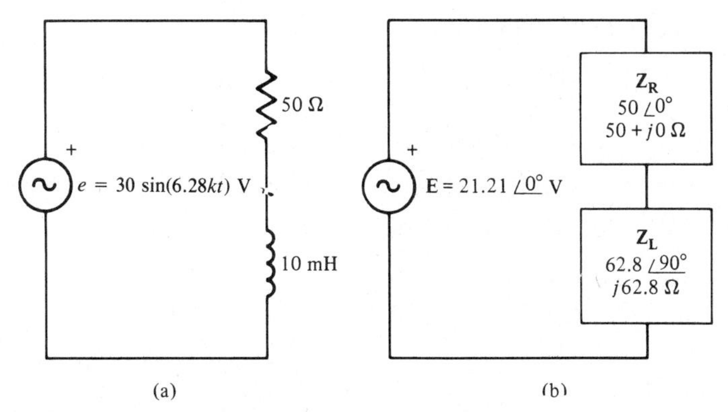

Figure 11.3 Diagrams for Example 11.2. **(a)** Circuit diagram. **(b)** Phasor representation.

Solution

a. From Example 11.1: $\mathbf{E} = 21.21\angle 0°$ V.

b. From Example 11.1: $\mathbf{Z_R} = 50\ \Omega$.

The reactance of the coil is:

$$X_L = \omega_L = (6.28\text{ krad/s})(10\text{ mH}) = 62.8\ \Omega$$

The impedance of the coil is:

$$\begin{aligned} \mathbf{Z_L} &= jX_L = j62.8\ \Omega && \text{rectangular form} \\ &= X_L\angle 90° = 62.8\angle 90°\ \Omega && \text{polar form} \end{aligned}$$

Figure 11.3b shows the circuit with all the components replaced by their equivalent phasor form.

c. The total impedance is the sum of the two series impedances:

$$\begin{aligned} \mathbf{Z_T} &= \mathbf{Z_R} + \mathbf{Z_L} \\ &= 50 + j62.8\ \Omega && \text{rectangular form} \\ &= 80.27\angle 51.5°\ \Omega && \text{polar form} \end{aligned}$$

d. The circuit current may be obtained by using Ohm's Law:

$$\mathbf{I} = \frac{\mathbf{E}}{\mathbf{Z_T}} = \frac{21.21\angle 0°}{80.27\angle 51.5°} = 0.264\angle -51.5°\text{ A}$$

$$\begin{aligned} &= 264\angle 51.5°\text{ mA} && \text{polar form} \\ &= 164.34 - j206.61\text{ mA} && \text{rectangular form} \end{aligned}$$

e. The resistor voltage is:

$$\begin{aligned} \mathbf{V_R} &= \mathbf{IZ_R} \\ &= (0.264\angle -51.5°)(50) \\ &= 13.2\angle -51.5°\text{ V} && \text{polar form} \\ &= 8.22 - j10.33\text{ V} && \text{rectangular form} \end{aligned}$$

The inductor voltage is:

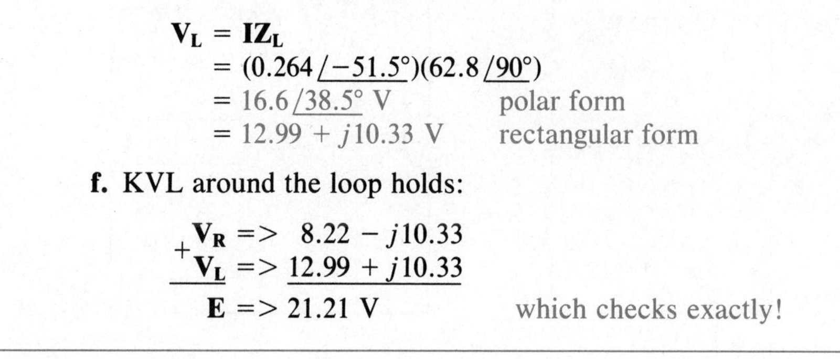

$$\begin{aligned}\mathbf{V_L} &= \mathbf{IZ_L}\\ &= (0.264\,\underline{/-51.5^\circ})(62.8\,\underline{/90^\circ})\\ &= 16.6\,\underline{/38.5^\circ}\text{ V} && \text{polar form}\\ &= 12.99 + j10.33\text{ V} && \text{rectangular form}\end{aligned}$$

f. KVL around the loop holds:

$$\begin{array}{rl} \mathbf{V_R} \Rightarrow & 8.22 - j10.33\\ +\ \mathbf{V_L} \Rightarrow & 12.99 + j10.33\\ \hline \mathbf{E} \Rightarrow & 21.21\text{ V} \end{array}$$

which checks exactly!

EXAMPLE 11.3

Repeat Example 11.1 for the series *RLC* circuit shown in Figure 11.4a.

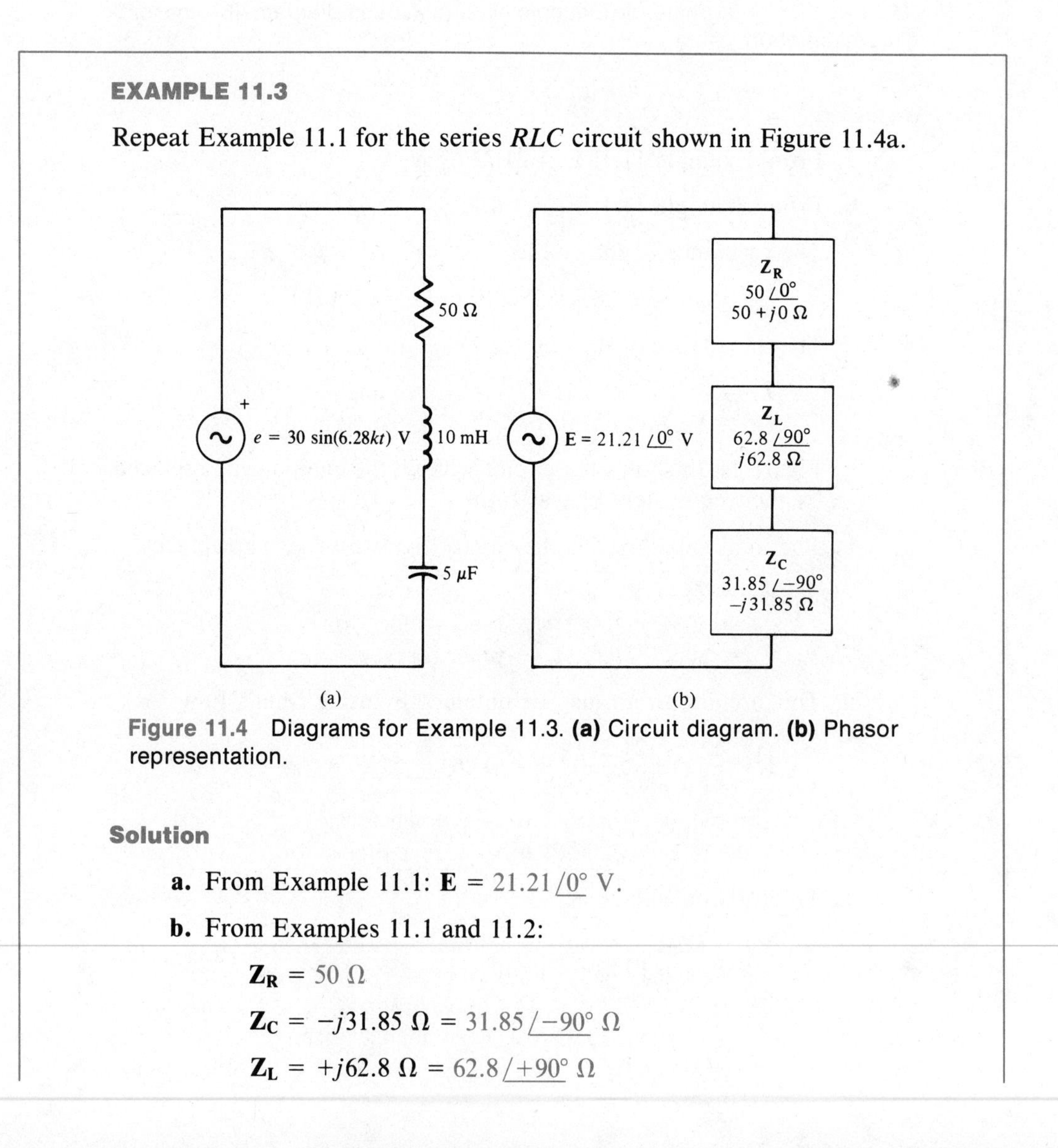

Figure 11.4 Diagrams for Example 11.3. **(a)** Circuit diagram. **(b)** Phasor representation.

Solution

a. From Example 11.1: $\mathbf{E} = 21.21\,\underline{/0^\circ}$ V.

b. From Examples 11.1 and 11.2:

$$\mathbf{Z_R} = 50\ \Omega$$

$$\mathbf{Z_C} = -j31.85\ \Omega = 31.85\,\underline{/-90^\circ}\ \Omega$$

$$\mathbf{Z_L} = +j62.8\ \Omega = 62.8\,\underline{/+90^\circ}\ \Omega$$

Figure 11.4b shows the circuit with the components represented in phasor notation.

c. The total impedance is:

$$\begin{aligned}\mathbf{Z_T} &= \mathbf{Z_R} + \mathbf{Z_L} + \mathbf{Z_C}\\ &= 50 + j62.8 - j31.85\\ &= 50 + j30.95\ \Omega \qquad \text{rectangular form}\\ &= 58.80\underline{/31.76^\circ}\ \Omega \qquad \text{polar form}\end{aligned}$$

d. The total current is:

$$\mathbf{I} = \frac{\mathbf{E}}{\mathbf{Z_T}} = \frac{21.21\underline{/0^\circ}}{58.80\underline{/31.76^\circ}} = 0.361\underline{/-31.76^\circ}\ \text{A}$$

$$\begin{aligned}&= 361\underline{/-31.76^\circ}\ \text{mA} \qquad \text{polar form}\\ &= 307 - j190\ \text{mA} \qquad \text{rectangular form}\end{aligned}$$

e. The voltage across the resistor is:

$$\begin{aligned}\mathbf{V_R} &= \mathbf{IZ_R}\\ &= (0.361\underline{/-31.76^\circ})(50)\\ &= 18.05\underline{/-31.76^\circ}\ \text{V} \qquad \text{polar form}\\ &= 15.35 - j9.5\ \text{V} \qquad \text{rectangular form}\end{aligned}$$

The voltage across the coil is:

$$\begin{aligned}\mathbf{V_L} &= \mathbf{IZ_L}\\ &= (0.361\underline{/-31.76^\circ})(62.8\underline{/90^\circ})\\ &= 22.67\underline{/58.24^\circ}\ \text{V} \qquad \text{polar form}\\ &= 11.93 + j19.28\ \text{V} \qquad \text{rectangular form}\end{aligned}$$

The voltage across the capacitor is:

$$\begin{aligned}\mathbf{V_C} &= \mathbf{IZ_C}\\ &= (0.361\underline{/-31.76^\circ})(31.85\underline{/-90^\circ})\\ &= 11.5\underline{/-121.76^\circ}\ \text{V} \qquad \text{polar form}\\ &= -6.05 - j9.78\ \text{V} \qquad \text{rectangular form}\end{aligned}$$

f. KVL holds:

$$\begin{aligned}&\mathbf{V_R} => 15.35 - j9.5\\ +\ &\mathbf{V_L} => 11.93 + j19.28\\ +\ &\mathbf{V_C} => \underline{-6.05 - j9.78}\\ &\mathbf{E} => 21.23 + j0\ \text{V} \qquad \text{which is close enough!}\end{aligned}$$

11.3 PARALLEL AC CIRCUITS

In a **parallel ac circuit,** all the impedances are connected between the same two nodes. Conversely, impedances are in parallel if they are connected between the same two nodes. The voltage across all impedances in parallel is the same. For an ac circuit containing all parallel impedances, it is easier to work with the corresponding admittances. Recall that admittance is the reciprocal of impedance:

$$\mathbf{Y} = \frac{1}{\mathbf{Z}} \tag{11.3}$$

where: $\mathbf{Y}$ => admittance (S)
$\mathbf{Z}$ => impedance (Ω)

Figure 11.5a shows an ac circuit with three impedances in parallel. Figure 11.5b shows the same circuit with the impedances replaced by their equivalent admittances.

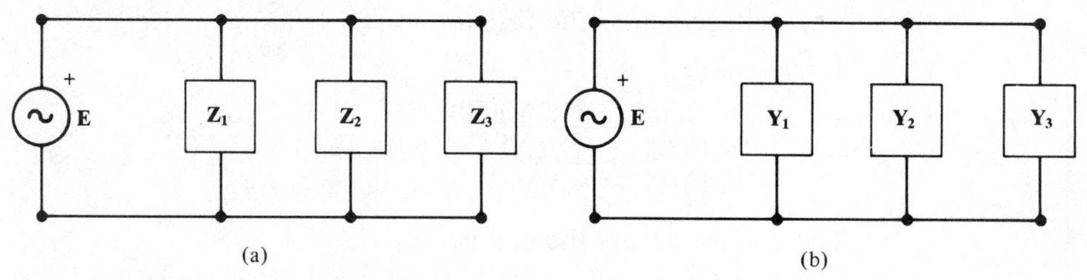

Figure 11.5 A parallel ac circuit. **(a)** Impedance representation. **(b)** Admittance representation.

Recall from dc circuits that conductances in parallel add. Similarly, admittances in ac circuits add when they are in parallel. Therefore, for a parallel ac circuit with *N* impedances in parallel:

$$\mathbf{Y_T} = \mathbf{Y_1} + \mathbf{Y_2} + \mathbf{Y_3} + \cdots + \mathbf{Y_N} \tag{11.4a}$$

$$\frac{1}{\mathbf{Z_T}} = \frac{1}{\mathbf{Z_1}} + \frac{1}{\mathbf{Z_2}} + \frac{1}{\mathbf{Z_3}} + \cdots + \frac{1}{\mathbf{Z_N}} \tag{11.4b}$$

A very popular notation for impedances in parallel is the symbol $\|$. For example, $\mathbf{Z_1} \| \mathbf{Z_2}$ is read as $\mathbf{Z_1}$ in parallel with $\mathbf{Z_2}$.

EXAMPLE 11.4

For the parallel ac circuit shown in Figure 11.6a determine:

a. The phasor representation for the source

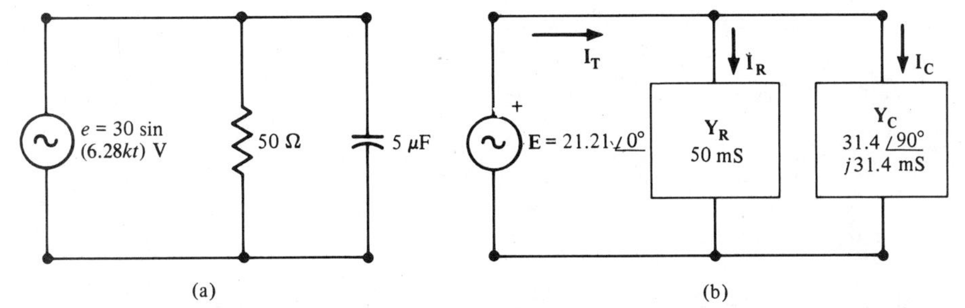

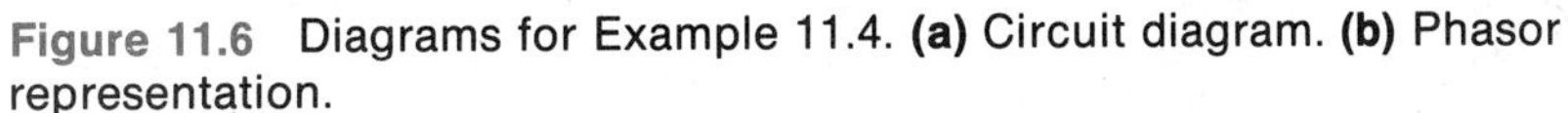

Figure 11.6 Diagrams for Example 11.4. **(a)** Circuit diagram. **(b)** Phasor representation.

b. The reactance, the susceptance, the impedance, and the admittance of each component

c. The total admittance and the total impedance of the circuit

d. The total circuit current

e. The current through each component

f. That KCL holds

Solution

a. From Example 11.1: $\mathbf{E} = 21.21\angle 0°$ V.

b. From Example 11.1: $X_C = 31.85\ \Omega$.

The susceptance of the capacitor is:

$$B_C = \frac{1}{X_C} = \frac{1}{31.85} = 31.4 \text{ mS}$$

The admittance of the capacitor is:

$$\begin{aligned} \mathbf{Y_C} &= jB_C \\ &= j31.4 \text{ mS} && \text{rectangular form} \\ &= 31.4\angle 90° \text{ mS} && \text{polar form} \end{aligned}$$

The admittance of the resistor is:

$$\mathbf{Y_R} = \frac{1}{R} = \frac{1}{50} = 20 \text{ mS}$$

c. The total admittance of the circuit is:

$$\begin{aligned} \mathbf{Y_T} &= \mathbf{Y_R} + \mathbf{Y_C} \\ &= 20 + j31.4 \text{ mS} && \text{rectangular form} \\ &= 37.23\angle 57.5° \text{ mS} && \text{polar form} \end{aligned}$$

The total impedance is:

$$\mathbf{Z_T} = \frac{1}{\mathbf{Y_T}} = \frac{1}{37.23\angle 57.5^\circ}$$

$= 26.86\angle -57.5^\circ\ \Omega$ polar form
$= 14.43 - j22.65\ \Omega$ rectangular form

d. The total circuit current is:

$$\mathbf{I_T} = \frac{\mathbf{E}}{\mathbf{Z_T}} = \frac{21.21}{26.86\angle -57.5^\circ}$$

$= 789.7\angle 57.5^\circ$ mA polar form
$= 424.31 + j666.02$ mA rectangular form

e. The current through the resistor is:

$\mathbf{I_R} = \mathbf{EY_R} = (21.21\angle 0^\circ)(20\text{ mA}\angle 0^\circ)$
$= 424.2\angle 0^\circ$ mA polar form
$= 424 + j0$ mA rectangular form

The current through the capacitor is:

$\mathbf{I_C} = \mathbf{EY_C} = (21.21\angle 0^\circ)(31.4\text{ mA}\angle 90^\circ)$
$= 666\angle 90^\circ$ mA polar form
$= 0 + j666$ mA rectangular form

f. KCL holds:

$\mathbf{I_R} \Rightarrow 424.2 + j0$
$+\ \mathbf{I_C} \Rightarrow 0 + j666$
$\mathbf{I_T} \Rightarrow 424.2 + j666$ mA which checks with part (d)

EXAMPLE 11.5

For the parallel *RL* combination shown in Figure 11.7a, determine:

a. The phasor representation for the source

b. The reactance, the susceptance, the impedance, and the admittance of each component

c. The total admittance and the total impedance of the circuit

d. The total circuit voltage

e. The current through each component

f. That KCL holds

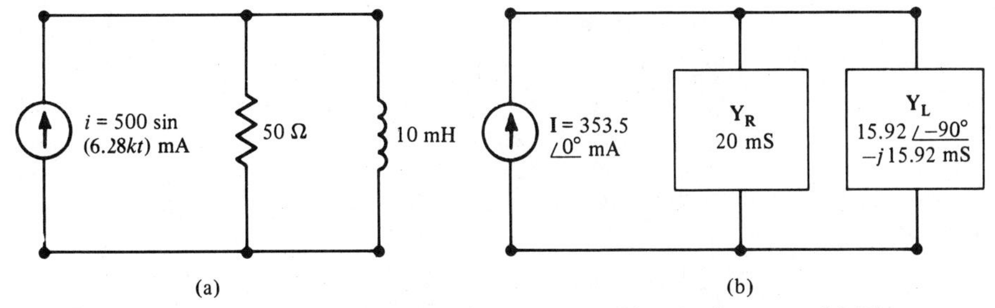

Figure 11.7 Diagrams for Example 11.5. **(a)** Circuit diagram. **(b)** Phasor representation.

Solution

a. The equation for the current source is:

$$i = 500 \sin(6.28\ kt)\ \text{mA}$$

Comparing it to the standard form,

$$i = I_P \sin(\omega t + \phi)\ \text{A}$$

the following may be recognized:

$$I_P = 500\ \text{mA}$$

$$\omega = 6.28\ \text{krad/s}$$

$$\phi = \underline{/0^\circ}$$

$$I_{\text{RMS}} = 0.707 I_P = (0.707)(500) = 353.5\ \text{mA}$$

The current source in phasor notation is:

$$\mathbf{I} = 353.5\underline{/0^\circ}\ \text{mA} \qquad \text{polar form}$$
$$= 353.5 + j0\ \text{mA} \qquad \text{rectangular form}$$

b. From Example 11.2: $X_L = 62.8\ \Omega$. The inductive susceptance is:

$$B_L = \frac{1}{X_L} = \frac{1}{62.8} = 15.92\ \text{mS}$$

The admittance of the inductor is:

$$\mathbf{Y_L} = -jX_L$$
$$= -j15.92\ \text{mS} \qquad \text{rectangular form}$$
$$= 15.92\underline{/-90^\circ}\ \text{mS} \qquad \text{polar form}$$

From Example 11.4: $\mathbf{Y_R} = 20$ mS. Figure 11.7b shows the circuit with the components replaced by their equivalent phasor notation.

c. The total circuit admittance is:

$$\mathbf{Y_T} = \mathbf{Y_R} + \mathbf{Y_L}$$
$$= 20 - j15.92 \text{ mS} \qquad \text{rectangular form}$$
$$= 25.56\angle{-38.53^\circ} \text{ mS} \qquad \text{polar form}$$

The total circuit impedance is:

$$\mathbf{Z_T} = \frac{1}{\mathbf{Y_T}} = \frac{1}{25.56 \times 10^{-3}\angle{-38.53^\circ}}$$
$$= 39.12\angle{38.53^\circ}\ \Omega \qquad \text{polar form}$$
$$= 30.60 + j24.37\ \Omega \qquad \text{rectangular form}$$

(ø sin) × 13.83

d. The resistor voltage and the inductor voltage are the same, since the components are in parallel:

$$\mathbf{V_R} = \mathbf{V_L} = \mathbf{IZ_T} = (0.3535\angle{0^\circ})(39.12\angle{38.53^\circ})$$
$$= 13.83\angle{38.53^\circ} \text{ V} \qquad \text{polar form}$$
$$= 10.82 + j8.62 \text{ V} \qquad \text{rectangular form}$$

e. The resistor current is:

$$\mathbf{I_R} = \mathbf{V_R Y_R} = (13.83\angle{38.53^\circ})(20 \times 10^{-3})$$
$$= 276.6\angle{38.53^\circ} \text{ mA} \qquad \text{polar form}$$
$$= 216.38 + j172.3 \text{ mA} \qquad \text{rectangular form}$$

The inductor current is:

$$\mathbf{I_L} = \mathbf{V_L Y_L} = (13.83\angle{38.53^\circ})(15.92 \times 10^{-3}\angle{-90^\circ})$$
$$= 220.2\angle{-51.65^\circ} \text{ mA} \qquad \text{polar form}$$
$$= 136.62 - j172.6 \text{ mA} \qquad \text{rectangular form}$$

f. KCL holds:

$$\mathbf{I_R} \Rightarrow 216.38 + j172.3$$
$$+\ \mathbf{I_L} \Rightarrow 136.62 - j172.6$$
$$\mathbf{I} \Rightarrow 353.00 - j0.3 \text{ mA} \qquad \text{KCL holds}$$

EXAMPLE 11.6

For the circuit shown in Figure 11.8a, determine:

a. The phasor representation for the source

b. The reactance, the susceptance, the impedance, and the admittance of each component

c. The total admittance and the total impedance of the circuit

d. The total circuit voltage

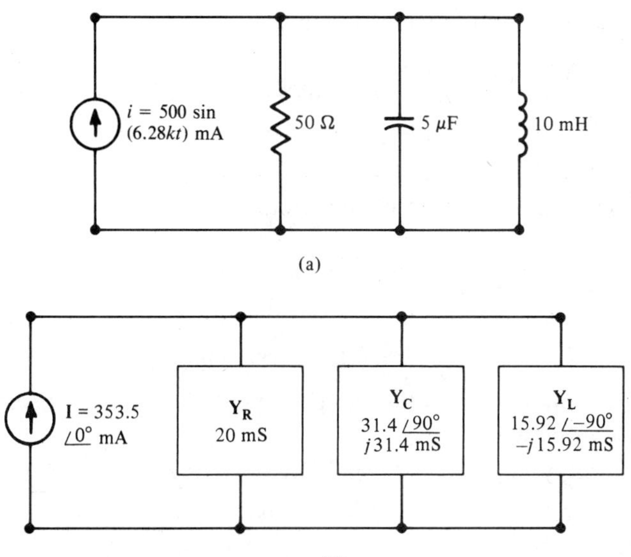

Figure 11.8 Diagrams for Example 11.6. **(a)** Circuit diagram. **(b)** Phasor representation.

e. The current through each component

f. That KCL holds

Solution

a. From Example 11.5: $\mathbf{I} = 353.5\,\underline{/0^\circ}$ mA.

b. From Example 11.4:

$$X_C = 31.85\ \Omega$$

$$B_C = 31.4\text{ mS}$$

$$\mathbf{Y_C} = j31.4\text{ mS} = 31.4\,\underline{/90^\circ}\text{ mS}$$

From Example 11.5:

$$X_L = 62.8\ \Omega$$

$$B_L = 15.92\text{ mS}$$

$$\mathbf{Y_L} = -j15.92\text{ mS} = 15.92\,\underline{/-90^\circ}\text{ mS}$$

$$\mathbf{Y_R} = 20\text{ mS}$$

Figure 11.8b shows the circuit with each component replaced by its phasor notation.

c. The total admittance of the circuit is:

$$\mathbf{Y_T} = \mathbf{Y_R} + \mathbf{Y_C} + \mathbf{Y_L}$$
$$= 20 + j31.4 - j15.92 \text{ mS}$$
$$= 20 + j15.48 \text{ mS} \quad \text{rectangular form}$$
$$= 25.29\angle 37.74° \text{ mS} \quad \text{polar form}$$

The total impedance of the circuit is:

$$\mathbf{Z_T} = \frac{1}{\mathbf{Y_T}} = \frac{1}{25.29 \times 10^{-3}\angle 37.74°}$$
$$= 39.54\angle -37.74° \ \Omega \quad \text{polar form}$$
$$= 31.27 - j24.20 \ \Omega \quad \text{rectangular form}$$

d. The voltage across the three elements is the same because they are in parallel:

$$\mathbf{V_R} = \mathbf{V_C} = \mathbf{V_L} = \mathbf{IZ_T} = (0.3535\angle 0°)(39.54\angle -37.74°)$$
$$= 13.98\angle -37.74° \text{ V} \quad \text{polar form}$$
$$= 11.05 - j8.56 \text{ V} \quad \text{rectangular form}$$

e. The current through the resistor is:

$$\mathbf{I_R} = \mathbf{V_R Y_R} = (13.98\angle -37.74°)(20 \times 10^{-3})$$
$$= 279.6\angle -37.74° \text{ mA} \quad \text{polar form}$$
$$= 221.11 - j171.12 \text{ mA} \quad \text{rectangular form}$$

The capacitor current is:

$$\mathbf{I_C} = \mathbf{V_C Y_C} = (13.98\angle -37.74°)(31.4 \times 10^{-3}\angle 90°)$$
$$= 439\angle 52.26° \text{ mA} \quad \text{polar form}$$
$$= 268.7 + j347.16 \text{ mA} \quad \text{rectangular form}$$

The current through the inductor is:

$$\mathbf{I_L} = \mathbf{V_L Y_L} = (13.98\angle -34.74°)(15.92 \times 10^{-3}\angle -90°)$$
$$= 222.6\angle -127.74° \text{ mA} \quad \text{polar form}$$
$$= -136.25 - j176.03 \text{ mA} \quad \text{rectangular form}$$

f. KCL holds:

$$\begin{array}{rl} \mathbf{I_R} \Rightarrow & 221.11 - j171.12 \\ + \ \mathbf{I_C} \Rightarrow & 268.70 + j347.16 \\ + \ \mathbf{I_L} \Rightarrow & -136.25 - j176.03 \\ \hline \mathbf{I} \Rightarrow & 353.56 + j0.01 \text{ mA} \quad \text{KCL holds} \end{array}$$

11.4 SERIES-PARALLEL AC CIRCUITS

Most practical ac circuits contain both series and parallel combinations of impedances. These are referred to as **series-parallel ac circuits.** Equipped with KVL, KCL, and Ohm's Law, we are able to solve these circuits.

EXAMPLE 11.7

For the circuit shown in Figure 11.9a, determine:

a. The phasor form of the source

b. The impedance and admittance of each component

c. The total circuit impedance

d. The total circuit current

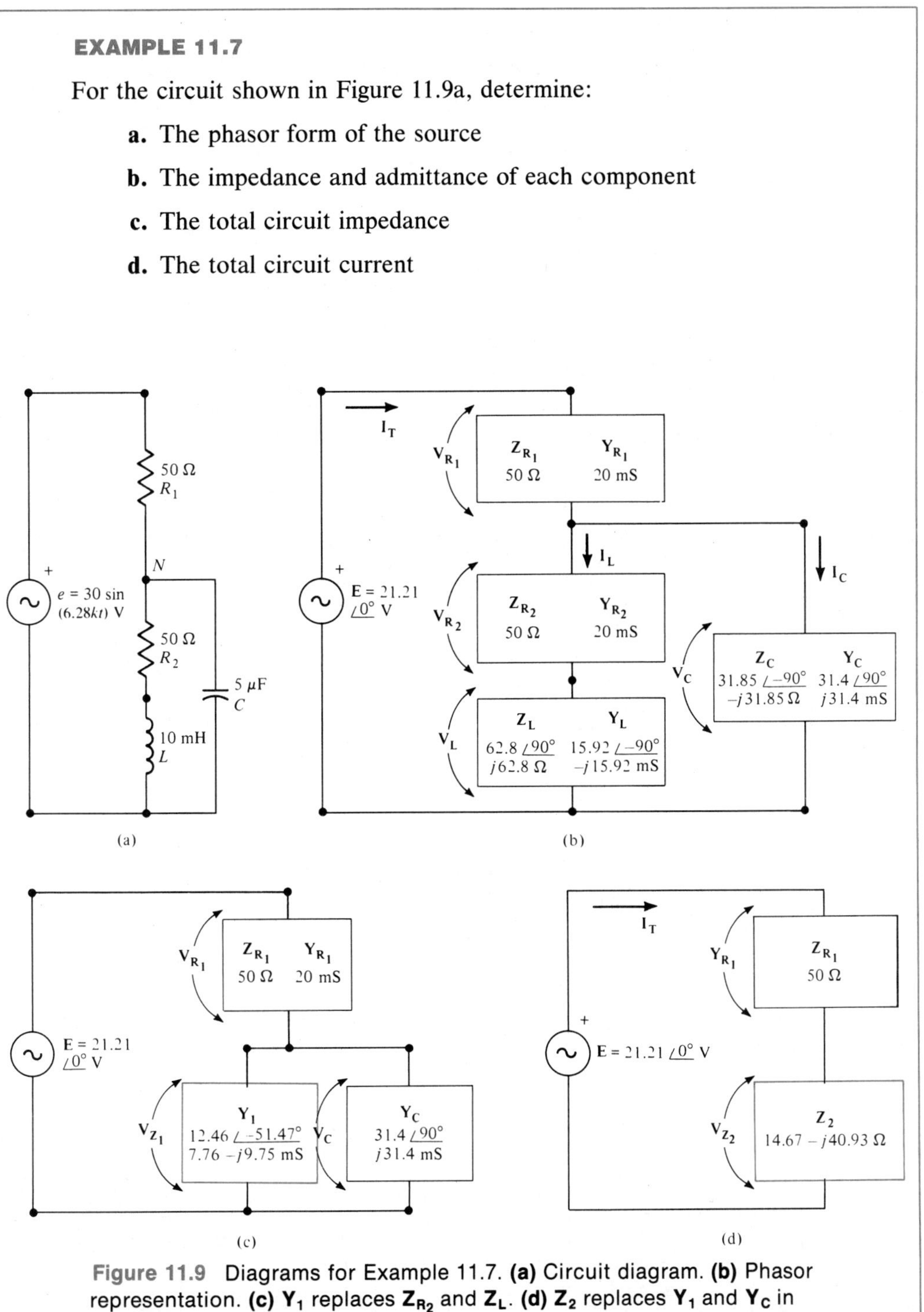

Figure 11.9 Diagrams for Example 11.7. **(a)** Circuit diagram. **(b)** Phasor representation. **(c)** $\mathbf{Y_1}$ replaces $\mathbf{Z_{R_2}}$ and $\mathbf{Z_L}$. **(d)** $\mathbf{Z_2}$ replaces $\mathbf{Y_1}$ and $\mathbf{Y_C}$ in parallel.

e. The voltage across R_1 and C

f. The capacitor current, the inductor current, and the current through R_2

g. That KCL holds at node N

Solution

a. From Example 11.1: $\mathbf{E} = 21.21$ V.

b. From Example 11.1:

$$\mathbf{Y_{R_1}} = \mathbf{Y_{R_2}} = 20 \text{ mS}$$

$$\mathbf{Z_C} = -j31.85\ \Omega = 31.85\underline{/-90^\circ}\ \Omega$$

From Example 11.4: $\mathbf{Y_C} = j31.4$ mS $= 31.4\underline{/90^\circ}$ mS.
From Example 11.2: $\mathbf{Z_L} = j62.8\ \Omega = 62.8\underline{/90^\circ}\ \Omega$.
From Example 11.5: $\mathbf{Z_L} = j62.8\ \Omega = 62.8\underline{/90^\circ}\ \Omega$.
Figure 11.9b shows the circuit with each component replaced by its phasor notation.

c. Inspection of Figure 11.9b indicates that the inductor and resistor R_2 are in series. This series combination is called $\mathbf{Z_1}$ and has the value:

$$\begin{aligned} \mathbf{Z_1} &= \mathbf{Z_{R_2}} + \mathbf{Z_L} \\ &= 50 + j62.8\ \Omega && \text{rectangular form} \\ &= 80.27\underline{/51.47^\circ}\ \Omega && \text{polar form} \end{aligned}$$

$$\begin{aligned} \mathbf{Y_1} &= \frac{1}{\mathbf{Z_1}} = \frac{1}{80.27\underline{/51.47^\circ}} \\ &= 12.46\underline{/-51.47}\text{ mS} && \text{polar form} \\ &= 7.76 - j9.75\text{ mS} && \text{rectangular form} \end{aligned}$$

Figure 11.9c shows the circuit with the equivalent series RL combination in place. Inspection of Figure 11.9c indicates that admittance $\mathbf{Y_1}$ is in parallel with admittance $\mathbf{Y_C}$:

$$\begin{aligned} \mathbf{Y_2} &= \mathbf{Y_1} + \mathbf{Y_C} \\ &= (7.76 - j9.75\text{ mS}) + j31.4\text{ mS} \\ &= 7.76 + j21.65\text{ mS} && \text{rectangular form} \\ &= 23\underline{/70.28^\circ}\text{ mS} && \text{polar form} \end{aligned}$$

$$\begin{aligned} \mathbf{Z_2} &= \frac{1}{\mathbf{Y_2}} = \frac{1}{23 \times 10^{-3}\underline{/70.28^\circ}} \\ &= 43.48\underline{/-70.28^\circ}\ \Omega && \text{polar form} \\ &= 14.67 - j40.93\ \Omega && \text{rectangular form} \end{aligned}$$

Figure 11.9d shows the circuit with impedance $\mathbf{Z_2}$ in place. Inspection of Figure 11.9d indicates that the total impedance is:

$$\begin{aligned}\mathbf{Z_T} &= \mathbf{Z_{R_1}} + \mathbf{Z_2}\\ &= 50 + (14.67 - j40.93)\\ &= 64.67 - j40.93\ \Omega \quad \text{rectangular form}\\ &= 76.53\angle{-32.33^\circ}\ \Omega \quad \text{polar form}\end{aligned}$$

d. The total current is:

$$\begin{aligned}\mathbf{I_T} &= \frac{\mathbf{E}}{\mathbf{Z_T}} = \frac{21.21\angle 0^\circ}{76.53\angle{-32.33^\circ}}\\ &= 277.15\angle 32.33^\circ\ \text{mA} \quad \text{polar form}\\ &= 234.19 + j148.23\ \text{mA} \quad \text{rectangular form}\end{aligned}$$

e. The voltage across $\mathbf{R_1}$ is:

$$\begin{aligned}\mathbf{V_{R_1}} &= \mathbf{I_T Z_{R_1}} = (0.277\angle 32.33^\circ)(50\angle 0^\circ)\\ &= 13.86\angle 32.33^\circ\ \text{V} \quad \text{polar form}\\ &= 11.71 + j7.41\ \text{V} \quad \text{rectangular form}\end{aligned}$$

The voltage across $\mathbf{Z_2}$ is:

$$\begin{aligned}\mathbf{V_{Z_2}} &= \mathbf{I_T Z_2} = (0.277\angle 32.33^\circ)(43.48\angle{-70.28^\circ})\\ &= 12.05\angle{-37.95^\circ}\ \text{V} \quad \text{polar form}\\ &= 9.5 - j7.41\ \text{V} \quad \text{rectangular form}\end{aligned}$$

A quick KVL check shows

$$\begin{aligned}&\mathbf{V_{R_1}} => 11.71 + j7.41\\ +\ &\mathbf{V_{Z_2}} => 9.5 - j7.41\\ \hline &\mathbf{E} => 21.21 + j0\ \text{V} \quad \text{which matches exactly}\end{aligned}$$

The voltage across the capacitor is the same as $\mathbf{V_{Z_2}}$:

$$\mathbf{V_C} = \mathbf{V_{Z_2}}$$

f. The capacitor current is:

$$\begin{aligned}\mathbf{I_C} &= \mathbf{V_C Y_C} = (12.05\angle{-37.95^\circ})(31.4 \times 10^{-3}\angle 90^\circ)\\ &= 378.37\angle 52.05^\circ\ \text{mA} \quad \text{polar form}\\ &= 232.69 + j298.36\ \text{mA} \quad \text{rectangular form}\end{aligned}$$

The inductor current and the current through resistor R_2 are the same since they are in series:

$$\begin{aligned}\mathbf{I_L} &= \mathbf{I_{R_2}} = \mathbf{V_{Z_2} Y_1} = (12.05\angle{-37.95^\circ})(12.46 \times 10^{-3}\angle{-51.47^\circ})\\ &= 150.14\angle{-89.42^\circ}\ \text{mA} \quad \text{polar form}\\ &= 1.52 - j150.13\ \text{mA} \quad \text{rectangular form}\end{aligned}$$

g. KCL holds at node N:

$$\begin{array}{rl} & \mathbf{I}_C \Rightarrow 232.69 + j298.36 \\ + & \mathbf{I}_L \Rightarrow \underline{\quad 1.52 - j150.13} \\ & \mathbf{I}_T \Rightarrow 234.31 + j148.23 \text{ mA} \end{array}$$

which checks with part (d)

11.5 LOOP ANALYSIS OF AC CIRCUITS

The procedure for loop analysis of ac circuits is exactly the same as for dc circuits except that voltages, currents, and impedances are now all complex numbers. The procedure for ac circuit analysis by loops is as follows.

RULE 1. Convert each source to its phasor representation and replace each circuit component by its impedance.

RULE 2. Replace each current source and its parallel impedance by the equivalent voltage source and its series impedance.

RULE 3. Label all phasor loop currents in a clockwise direction. Label each loop current as $\mathbf{I}_1$, $\mathbf{I}_2$, and so on.

RULE 4. By inspection write the following set of equations:

$$\mathbf{Z}_{11}\mathbf{I}_1 - \mathbf{Z}_{12}\mathbf{I}_2 - \mathbf{Z}_{13}\mathbf{I}_3 - \cdots - \mathbf{Z}_{1N}\mathbf{I}_N = \mathbf{E}_1 \qquad (11.5a)$$

$$-\mathbf{Z}_{21}\mathbf{I}_1 + \mathbf{Z}_{22}\mathbf{I}_2 - \mathbf{Z}_{23}\mathbf{I}_3 - \cdots - \mathbf{Z}_{2N}\mathbf{I}_N = \mathbf{E}_2 \qquad (11.5b)$$

$$\vdots$$

$$-\mathbf{Z}_{N1}\mathbf{I}_1 - \mathbf{Z}_{N2}\mathbf{I}_2 - \mathbf{Z}_{N3}\mathbf{I}_3 - \cdots + \mathbf{Z}_{NN}\mathbf{I}_N = \mathbf{E}_N \qquad (11.5c)$$

The above set of simultaneous equations applies to any impedance network with N loops. Notice that this format is similar to the loop equations discussed in Chapter 7 for dc analysis.

DESCRIPTION 1. $\mathbf{E}_1$ is the *total phasor voltage rise* due to voltage sources in the direction of $\mathbf{I}_1$ in loop 1. Following the direction of the loop current, if the voltage source is traversed minus to plus (− to +), then the voltage source contributes a voltage rise and the value is positive. Following the direction of the loop current, if the voltage source is traversed plus to minus (+ to −), then the voltage source contributes a voltage drop and the value is negative. $\mathbf{E}_1$ is the algebraic sum of these rises and drops (add the rises, subtract the drops). If there is a voltage source in a branch that is shared by two loops, then it is counted in both loops. The same description applies to voltages $\mathbf{E}_2$ through $\mathbf{E}_N$.

DESCRIPTION 2. $\mathbf{Z}_{11}$ is the sum of all the impedances in loop 1. $\mathbf{Z}_{22}$ is the sum of all the impedances in loop 2. $\mathbf{Z}_{NN}$ is the sum of all the impedances in loop N.

DESCRIPTION 3. $\mathbf{Z}_{12}$ is the common impedance between loop 1 and loop 2. $\mathbf{Z}_{LM}$ is the impedance common between loop L and loop M. If loops L and M do not touch, then $\mathbf{Z}_{LM} = 0$.

DESCRIPTION 4. The set of simultaneous equations is now ready for solution. Use an appropriate method to solve simultaneous equations.

Once the values of the loop currents are obtained, the current through each individual impedance is calculated by superimposing the contributing loop currents (phasor and direction). Notice that the only positive coefficients in the equations are the ones with both subscripts the same ($\mathbf{Z}_{11}$, for example).

Because of the complex numbers involved, it is much more laborious to analyze ac networks than comparable dc circuits. Computer methods are much more desirable than hand solution for the solution of simultaneous equations with complex coefficients. The first example is solved both by hand and by the computer program presented in this section. The second example would be so laborious by hand that only the computer solution is presented.

EXAMPLE 11.8

Obtain the loop equations for the ac circuit shown in Figure 11.10a.

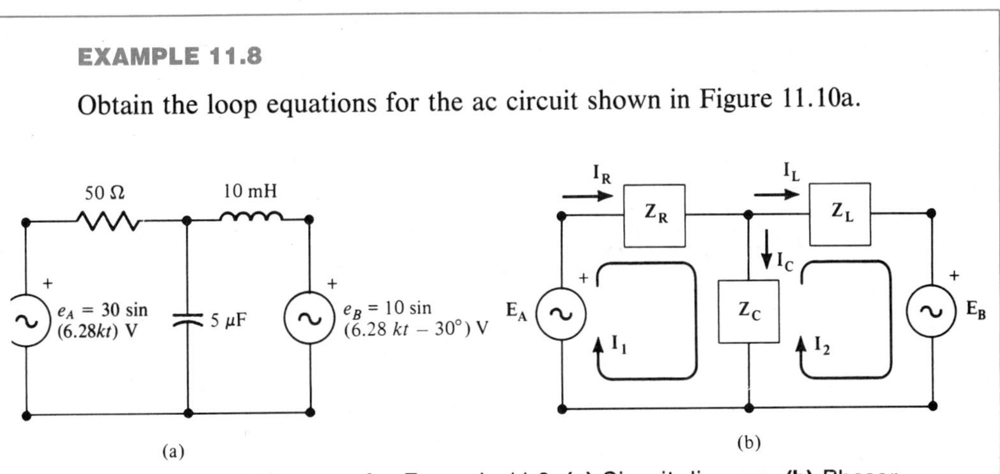

Figure 11.10 Diagrams for Example 11.8. **(a)** Circuit diagram. **(b)** Phasor representation showing loop currents, component currents, and component voltages.

Solution

Voltage source e_A converts to the phasor $\mathbf{E}_A$:

$$e_A = 30 \sin(6.28kt) \text{ V}$$

$\mathbf{E_A} = 21.21\underline{/0^\circ}$ V polar form
$= 21.21 + j0$ V rectangular form

Voltage source e_B converts to the phasor $\mathbf{E_B}$:

$$e_B = 10 \sin(6.28kt - 30^\circ) \text{ V}$$

$\mathbf{E_B} = 7.07\underline{/-30^\circ}$ V polar form
$= 6.12 - j3.54$ V rectangular form

From Example 11.3:

$\mathbf{Z_R} = 50\ \Omega$

$\mathbf{Z_C} = 31.85\underline{/-90^\circ}\ \Omega$ polar form
$= -j31.85\ \Omega$ rectangular form

$\mathbf{Z_L} = 62.8\underline{/90^\circ}\ \Omega$ polar form
$= j62.8\ \Omega$ rectangular form

Figure 11.10b shows the circuit with each voltage source replaced by its phasor form and each component replaced by its impedance.

Since there are no current sources, Rule 2 need not be applied.

Apply Rule 3. Figure 11.10b also shows each loop current labeled in a clockwise direction.

Apply Description 1:

$\mathbf{E_1} = 21.21\underline{/0^\circ}$ V polar form
$= 21.21 + j0$ V rectangular form

$\mathbf{E_2} = -7.07\underline{/-30^\circ}$ V polar form
$= -6.12 + j3.54$ V rectangular form

Apply Description 2:

$\mathbf{Z_{11}} = \mathbf{Z_R} + \mathbf{Z_C}$
$= 50 - j31.85\ \Omega$ rectangular form
$= 59.28\underline{/-32.5^\circ}\ \Omega$ polar form

$\mathbf{Z_{22}} = \mathbf{Z_C} + \mathbf{Z_L}$
$= -j31.85 + j62.8$
$= j30.95\ \Omega$ rectangular form
$= 30.95\underline{/90^\circ}\ \Omega$ polar form

Apply Description 3:

$\mathbf{Z_{12}} = -\mathbf{Z_C}$
$= j31.85\ \Omega$ rectangular form
$= 31.85\underline{/90^\circ}\ \Omega$ polar form

$\mathbf{Z_{21}} = -\mathbf{Z_C}$
$= j31.85\ \Omega$ rectangular form
$= 31.85\underline{/90^\circ}\ \Omega$ polar form

Apply Rule 4. *The loop equations in rectangular form are:*

$$(50 - j31.85)\mathbf{I}_1 + (j31.85)\mathbf{I}_2 = 21.21 + j0$$

$$(0 + j31.85)\mathbf{I}_1 + (j30.95)\mathbf{I}_2 = -6.12 + j3.54$$

The loop equations in polar form are:

$$(59.28\underline{/-32.5^\circ})\mathbf{I}_1 + (31.85\underline{/90^\circ})\mathbf{I}_2 = 21.21\underline{/0^\circ}$$

$$(31.85\underline{/90^\circ})\mathbf{I}_1 + (30.95\underline{/90^\circ})\mathbf{I}_2 = -7.07\underline{/-30^\circ}$$

EXAMPLE 11.9

Obtain the loop equations for the ac circuit shown in Figure 11.11a.

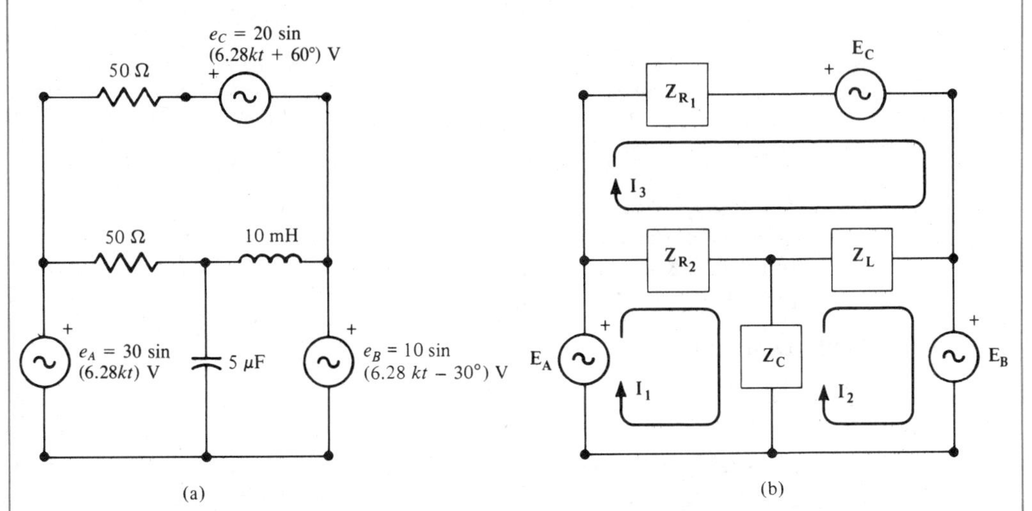

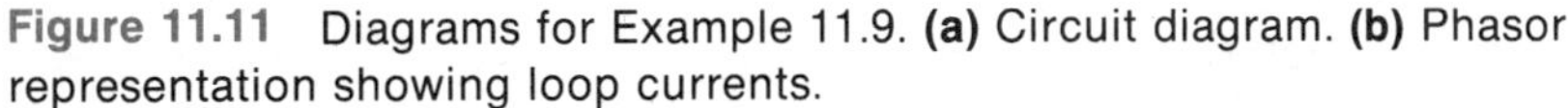

Figure 11.11 Diagrams for Example 11.9. **(a)** Circuit diagram. **(b)** Phasor representation showing loop currents.

Solution

The phasor form of e_C is $\mathbf{E}_C$:

$$e_C = 20 \sin(6.28kt + 60^\circ) \text{ V}$$

$$\mathbf{E}_C = 14.14\underline{/60^\circ} \text{ V} \qquad \text{polar form}$$
$$= 7.07 + j12.25 \text{ V} \qquad \text{rectangular form}$$

All the other sources and impedances are the same as in Example 11.8.

Figure 11.11b shows the circuit with each source replaced by its phasor form and each component replaced by its impedance.

Rule 2 need not be applied since there are no current sources.

Apply Rule 3. Each loop current was drawn clockwise as indicated in Figure 11.11b.

Apply Description 1:

$\mathbf{E_1} = 21.21 + j0$ V rectangular form
$= 21.21 \angle 0°$ V polar form

$\mathbf{E_2} = -6.12 + j3.54$ V rectangular form
$= -7.07 \angle -30°$ V polar form

$\mathbf{E_3} = -7.07 - j12.25$ V rectangular form
$= -14.14 \angle 60°$ V polar form

Apply Description 2:

$\mathbf{Z_{11}} = \mathbf{Z_{R_2}} + \mathbf{Z_C}$
$= 50 - j31.85\ \Omega$ rectangular form
$= 59.28 \angle -32.5°\ \Omega$ polar form

$\mathbf{Z_{22}} = \mathbf{Z_C} + \mathbf{Z_L}$
$= -j31.85 + j62.8$
$= j30.95\ \Omega$ rectangular form
$= 30.95 \angle 90°\ \Omega$ polar form

$\mathbf{Z_{33}} = \mathbf{Z_{R_1}} + \mathbf{Z_{R_2}} + \mathbf{Z_L}$
$= 50 + 50 + j62.8$
$= 100 + j62.8\ \Omega$ rectangular form
$= 118.08 \angle 32.13°\ \Omega$ polar form

Apply Description 3:

$\mathbf{Z_{12}} = -\mathbf{Z_C}$
$= j31.85\ \Omega$ rectangular form
$= 31.85 \angle 90°\ \Omega$ polar form

$\mathbf{Z_{13}} = -\mathbf{Z_{R_2}}$
$= -50\ \Omega$

$\mathbf{Z_{21}} = -\mathbf{Z_C}$
$= j31.5\ \Omega$ rectangular form
$= 31.5 \angle 90°\ \Omega$ polar form

$\mathbf{Z_{23}} = -\mathbf{Z_L}$
$= -j62.8\ \Omega$ rectangular form
$= 62.8 \angle -90°\ \Omega$ polar form

$\mathbf{Z_{31}} = -\mathbf{Z_{R_2}}$
$= -50\ \Omega$

$\mathbf{Z_{32}} = -\mathbf{Z_L}$
$= -j62.8\ \Omega$ rectangular form
$= 62.8 \angle -90°\ \Omega$ polar form

Apply Rule 4. *The loop equations in rectangular form are:*

$$(50 - j31.85)\mathbf{I}_1 + j31.85\mathbf{I}_2 - 50\mathbf{I}_3 = 21.21 + j0$$

$$(j31.85)\mathbf{I}_1 + j30.95\mathbf{I}_2 - j62.8\mathbf{I}_3 = -6.12 + j3.54$$

$$(-50)\mathbf{I}_1 - j62.80\mathbf{I}_2 + (100.08 + j62.8)\mathbf{I}_3 = -7.07 - j12.25$$

The loop equations in polar form are:

$$(59.28\underline{/-32.5^\circ})\mathbf{I}_1 + (31.85\underline{/90^\circ})\mathbf{I}_2 - (50)\mathbf{I}_3 = 21.21\underline{/0^\circ}$$

$$(31.85\underline{/90^\circ})\mathbf{I}_1 + (30.95\underline{/90^\circ})\mathbf{I}_2 + (62.8\underline{/-90^\circ})\mathbf{I}_3 = -7.07\underline{/-30^\circ}$$

$$(-50)\mathbf{I}_1 + (62.8\underline{/-90^\circ})\mathbf{I}_2 + (118.08\underline{/32.13^\circ})\mathbf{I}_3 = -14.14\underline{/60^\circ}$$

11.6 NODAL ANALYSIS OF AC CIRCUITS

The procedure for nodal analysis of ac circuits is the same as for dc circuits except that in ac circuits all computations involve complex numbers. The following systematic method is useful in writing node equations for ac circuits.

RULE 1. Convert each source to its phasor representation and replace each circuit component by its impedance.

RULE 2. Replace each voltage source and its series impedance with the equivalent current source and its parallel impedance. Replace all impedances by the equivalent admittances.

RULE 3. Choose a reference node and label it with the ground symbol. The reference node is usually the one with the most connections. This node will have 0 voltage, and all the other node voltages will be measured with respect to this node. Label all the other nodes as 1, 2, 3, and so on.

RULE 4. By inspection, write the following set of node equations:

$$\mathbf{Y}_{11}\mathbf{V}_1 - \mathbf{Y}_{12}\mathbf{V}_2 - \cdots - \mathbf{Y}_{1N}\mathbf{V}_N = \mathbf{I}_1 \quad (11.6a)$$

$$-\mathbf{Y}_{21}\mathbf{V}_1 + \mathbf{Y}_{22}\mathbf{V}_2 - \cdots - \mathbf{Y}_{2N}\mathbf{V}_N = \mathbf{I}_2 \quad (11.6b)$$

$$\vdots$$

$$-\mathbf{Y}_{N1}\mathbf{V}_1 - \mathbf{Y}_{N2}\mathbf{V}_2 - \cdots + \mathbf{Y}_{NN}\mathbf{V}_N = \mathbf{I}_N \quad (11.6c)$$

This set of simultaneous node equations applies to any ac network with one reference node and N additional nodes.

DESCRIPTION 1. $\mathbf{I}_1$ is the phasor sum of the current sources connected to node 1. Current sources supplying current to the node are positive. Current sources withdrawing current from the node are negative. $\mathbf{I}_2$ is the phasor sum of the current sources connected to node 2. $\mathbf{I}_N$ is the phasor sum of the current sources connected to node N.

DESCRIPTION 2. $\mathbf{Y}_{11}$ is the sum of all the admittances connected to node 1. $\mathbf{Y}_{22}$ is the sum of all the admittances connected to node 2. $\mathbf{Y}_{NN}$ is the sum of all the admittances connected to node N.

DESCRIPTION 3. $\mathbf{Y}_{12}$ is the admittance between node 1 and node 2. $\mathbf{Y}_{N2}$ is the admittance between node N and node 2. If there is no admittance element between two nodes (say A and B), then $\mathbf{Y}_{AB} = 0$.

DESCRIPTION 4. The set of node equations is now ready for solution. The results yield the phasor voltages at the respective nodes with respect to the reference node.

EXAMPLE 11.10

Obtain the node equations for the ac circuit shown in Figure 11.12a.

Solution

The phasor form of the voltage sources as well as the impedance of all the components were obtained in Example 11.9 and may be used here.

Figure 11.12b shows the circuit in phasor notation.

Apply Rule 2. The circuits in Figures 11.12c, 11.12d, and 11.12e show the conversion of each voltage source and its series impedance to the equivalent current source and its parallel impedance.

Apply Rule 3. The circuit in Figure 11.12f shows the current sources replacing the voltage sources. The figure also shows the reference node labeled with the ground symbol, and the other two nodes labeled as $\mathbf{V}_1$ and $\mathbf{V}_2$. Figure 11.12g shows the circuit with all the nodes and currents labeled and each impedance substituted by its equivalent admittance.

Apply Description 1:

$$\begin{aligned} \mathbf{I}_1 &= \mathbf{I}_A \\ &= 424.2 + j0 \text{ mA} && \text{rectangular form} \\ &= 424.2\angle 0^\circ \text{ mA} && \text{polar form} \end{aligned}$$

$$\begin{aligned} \mathbf{I}_2 &= \mathbf{I}_B + \mathbf{I}_C \\ &= (122.46 - j70.7) + (141.4 + j244.91) \\ &= 263.86 + j174.21 \text{ mA} && \text{rectangular form} \\ &= 316.18\angle 33.43^\circ \text{ mA} && \text{polar form} \end{aligned}$$

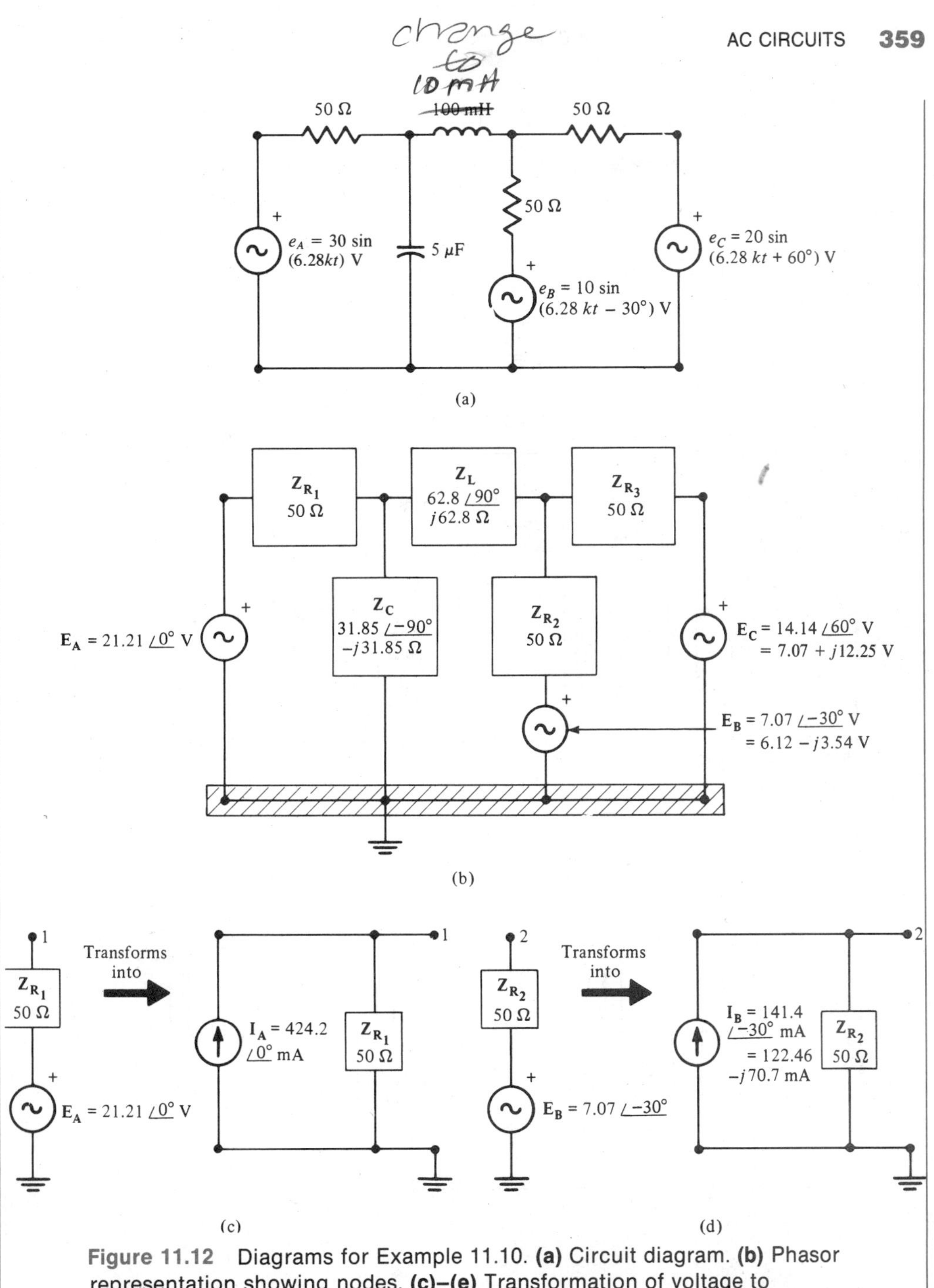

Figure 11.12 Diagrams for Example 11.10. **(a)** Circuit diagram. **(b)** Phasor representation showing nodes. **(c)–(e)** Transformation of voltage to current sources. **(f)** Impedance phasor representation. **(g)** Admittance phasor representation showing nodes and currents.

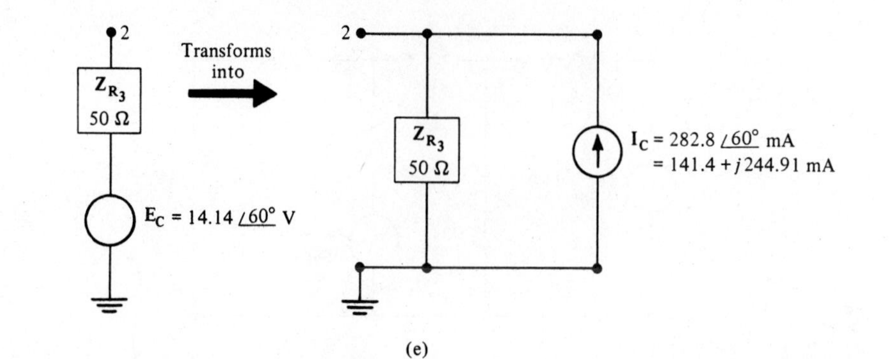

(e)

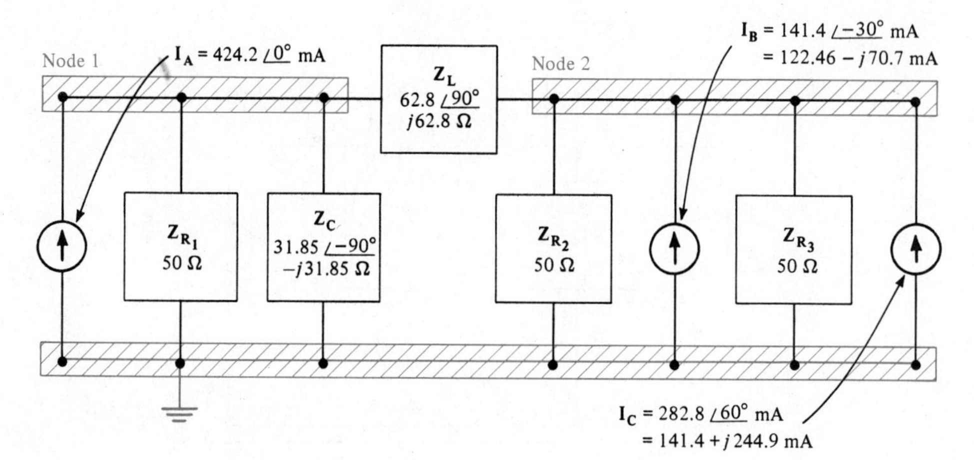

(f)

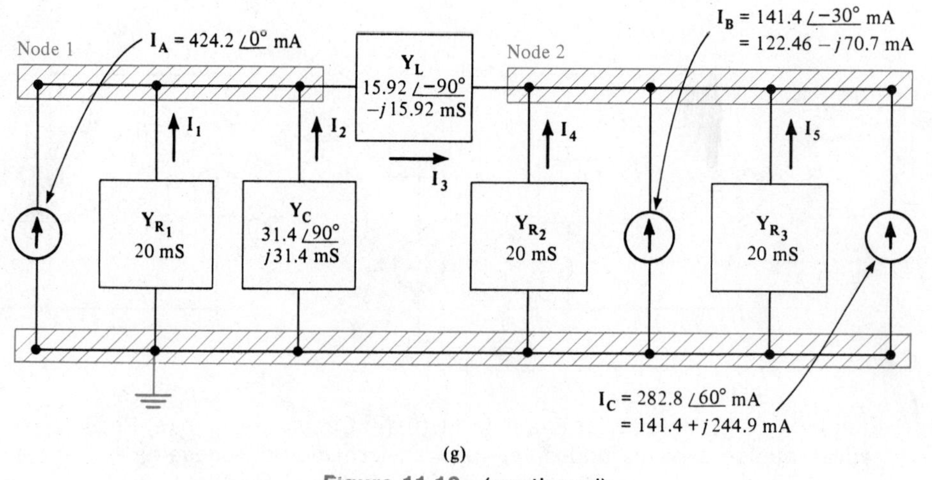

(g)

Figure 11.12 (continued)

Apply Description 2:

$$\begin{aligned}\mathbf{Y}_{11} &= \mathbf{Y}_{R1} + \mathbf{Y}_{C} + \mathbf{Y}_{L} \\ &= 20 + j31.4 - j15.92 \\ &= 20 + j15.48 \text{ mS} && \text{rectangular form} \\ &= 25.29\angle 37.74^\circ \text{ mS} && \text{polar form}\end{aligned}$$

$$\begin{aligned}\mathbf{Y}_{22} &= \mathbf{Y}_{L} + \mathbf{Y}_{R2} + \mathbf{Y}_{R3} \\ &= -j15.92 + 20 + 20 \\ &= 40 - j15.92 \text{ mS} && \text{rectangular form} \\ &= 43.05\angle -21.7^\circ \text{ mS} && \text{polar form}\end{aligned}$$

Apply Description 3:

$$\begin{aligned}\mathbf{Y}_{12} &= \mathbf{Y}_{21} = \mathbf{Y}_{L} \\ &= \mathbf{Y}_{21} = -j15.92 \text{ mS} && \text{rectangular form} \\ &= \mathbf{Y}_{21} = 15.92\angle -90^\circ \text{ mS} && \text{polar form}\end{aligned}$$

Apply Rule 4. *The rectangular form of the node equations is:*

$$(20 + j15.48)\mathbf{V}_1 - \quad (-j15.92)\mathbf{V}_2 = 424.20 + j0$$

$$-(-j15.92)\mathbf{V}_1 + (40 - j15.92)\mathbf{V}_2 = 263.86 + j174.21$$

The polar form of the node equations is:

$$(25.29\angle 37.74^\circ)\mathbf{V}_1 + \quad (15.92\angle 90^\circ)\mathbf{V}_2 = 424.2\angle 0^\circ$$

$$(15.92\angle 90^\circ)\mathbf{V}_1 + (43.05\angle -21.7^\circ)\mathbf{V}_2 = 316.18\angle 33.43^\circ$$

11.7 SOLUTION OF SIMULTANEOUS EQUATIONS WITH COMPLEX COEFFICIENTS BY CRAMER'S RULE AND BY COMPUTER

The use of Cramer's Rule presented in Chapter 7 may be used here with the substitution of phasors for real numbers.

EXAMPLE 11.11

Use Cramer's Rule to solve the equations obtained in Example 11.8. Solve for the voltage across and the current through each component.

Solution

The equations in polar form are:

$$(59.28\angle -32.5^\circ)\mathbf{I}_1 + (31.85\angle 90^\circ)\mathbf{I}_2 = 21.21\angle 0^\circ$$

$$(31.85\angle 90^\circ)\mathbf{I}_1 + (30.95\angle 90^\circ)\mathbf{I}_2 = -7.07\angle -30^\circ$$

$$\mathbf{D} = \begin{vmatrix} 59.28\,\underline{/-32.85^\circ} & 31.85\,\underline{/90^\circ} \\ 31.85\,\underline{/90^\circ} & 30.95\,\underline{/90^\circ} \end{vmatrix}$$

$= (59.28\,\underline{/-32.5^\circ})(30.95\,\underline{/90^\circ}) - (31.85\,\underline{/90^\circ})(31.85\,\underline{/90^\circ})$
$= 1835\,\underline{/57.5^\circ} - 1014.42\,\underline{/180^\circ}$
$= 985.79 + j1547.39 + 1014.42$
$= 2000.21 + j1547.39$ rectangular form
$= 2528.88\,\underline{/37.73^\circ}$ polar form

$$\mathbf{D_1} = \begin{vmatrix} 21.21\,\underline{/0^\circ} & 31.85\,\underline{/90^\circ} \\ -7.07\,\underline{/-30^\circ} & 30.95\,\underline{/90^\circ} \end{vmatrix}$$

$= (21.21\,\underline{/0^\circ})(30.95\,\underline{/90^\circ}) + (31.85\,\underline{/90^\circ})(7.07\,\underline{/-30^\circ})$
$= 656.45\,\underline{/90^\circ} + 225.18\,\underline{/60^\circ}$
$= j656.45 + (112.59 + j195.01)$
$= 112.59 + j851.46$ rectangular form
$= 858.87\,\underline{/82.47^\circ}$ polar form

$$\mathbf{D_2} = \begin{vmatrix} 59.28\,\underline{/-32.5^\circ} & 21.21\,\underline{/0^\circ} \\ 31.85\,\underline{/90^\circ} & -7.07\,\underline{/-30^\circ} \end{vmatrix}$$

$= (59.28\,\underline{/-32.5^\circ})(-7.07\,\underline{/-30^\circ}) - (21.21\,\underline{/0^\circ})(31.85\,\underline{/90^\circ})$
$= (-419.12\,\underline{/-62.5^\circ}) - (675.54\,\underline{/90^\circ})$
$= -193.53 - j303.78$ rectangular form
$= 360.19\,\underline{/-122.5^\circ}$ polar form

Apply Cramer's Rule:

$$\mathbf{I_1} = \frac{\mathbf{D_1}}{\mathbf{D}} = \frac{858.87\,\underline{/82.47^\circ}}{2528.88\,\underline{/37.73^\circ}}$$

$= 339.62\,\underline{/44.74^\circ}$ mA polar form
$= 241.23 + j239.06$ mA rectangular form

$$\mathbf{I_2} = \frac{\mathbf{D_2}}{\mathbf{D}} = \frac{360.19\,\underline{/-122.78^\circ}}{2528.88\,\underline{/37.73^\circ}}$$

$= 142.43\,\underline{/-160.51^\circ}$ mA polar form
$= -134.27 - j47.52$ mA rectangular form

The capacitor current may be computed as:

$\mathbf{I_C} = \mathbf{I_1} - \mathbf{I_2}$
$= (241.23 + j239.06) - (-134.27 - j47.52)$
$= 375.5 + j286.58$ mA rectangular form
$= 472.36\,\underline{/37.35^\circ}$ mA polar form

The voltage across each component may be found by Ohm's Law:

$\mathbf{V_R} = \mathbf{I_R Z_R} = (339.62\,\underline{/44.74^\circ}\text{ mA})(50\ \Omega)$
$= 16.98\,\underline{/44.74^\circ}$ V polar form
$= 12.06 + j11.95$ V rectangular form

$$\mathbf{V_C} = \mathbf{I_C Z_C} = (472.36\angle 37.35°\ \text{mA})(31.85\angle -90°\ \Omega)$$
$$= 15.04\angle -52.65°\ \text{V} \quad \text{polar form}$$
$$= 9.12 - j11.96\ \text{V} \quad \text{rectangular form}$$

$$\mathbf{V_L} = \mathbf{I_L Z_L} = (142.43\angle -160.51°)(62.8\angle 90°)$$
$$= 8.94\angle -70.51°\ \text{V} \quad \text{polar form}$$
$$= 2.98 - j8.43\ \text{V} \quad \text{rectangular form}$$

KVL around loop 1:

$$\begin{array}{rl} \mathbf{V_R} \Rightarrow & 12.06 + j11.95 \\ + \ \mathbf{V_C} \Rightarrow & 9.12 - j11.96 \\ \hline \mathbf{E_A} \Rightarrow & 21.18 - j0.01\ \text{V} \end{array}$$

which is close enough

KVL around loop 2:

$$\begin{array}{rl} \mathbf{V_L} \Rightarrow & 2.98 - j8.43 \\ + \ \mathbf{E_B} \Rightarrow & 6.12 - j3.54 \\ \hline \mathbf{V_C} \Rightarrow & 9.10 - j11.97\ \text{V} \end{array}$$

which checks

EXAMPLE 11.12

For the circuit shown in Figure 11.13a, find the node equations for nodes 1, 2, and 3.

Solution

Figure 11.13b shows the circuit in its phasor form with the components replaced by their admittance values.

Applying Description 1:

$$\mathbf{I_1} = 424.2 + j0 + 122.46 - j70.7$$
$$= 546.66 - j70.7\ \text{mA}$$

$$\mathbf{I_2} = -122.46 + j70.7 - 141.4 - j244.91$$
$$= -263.86 - j174.21\ \text{mA}$$

$$\mathbf{I_3} = 0$$

Applying Description 2:

$$\mathbf{Y_{11}} = 20 + 20 + 20$$
$$= 60\ \text{mS}$$

$$\mathbf{Y_{22}} = 20 + 20 + j31.4 - j15.92$$
$$= 40 - j15.48\ \text{mS}$$

$$\mathbf{Y_{33}} = 20 - j15.92\ \text{mS}$$

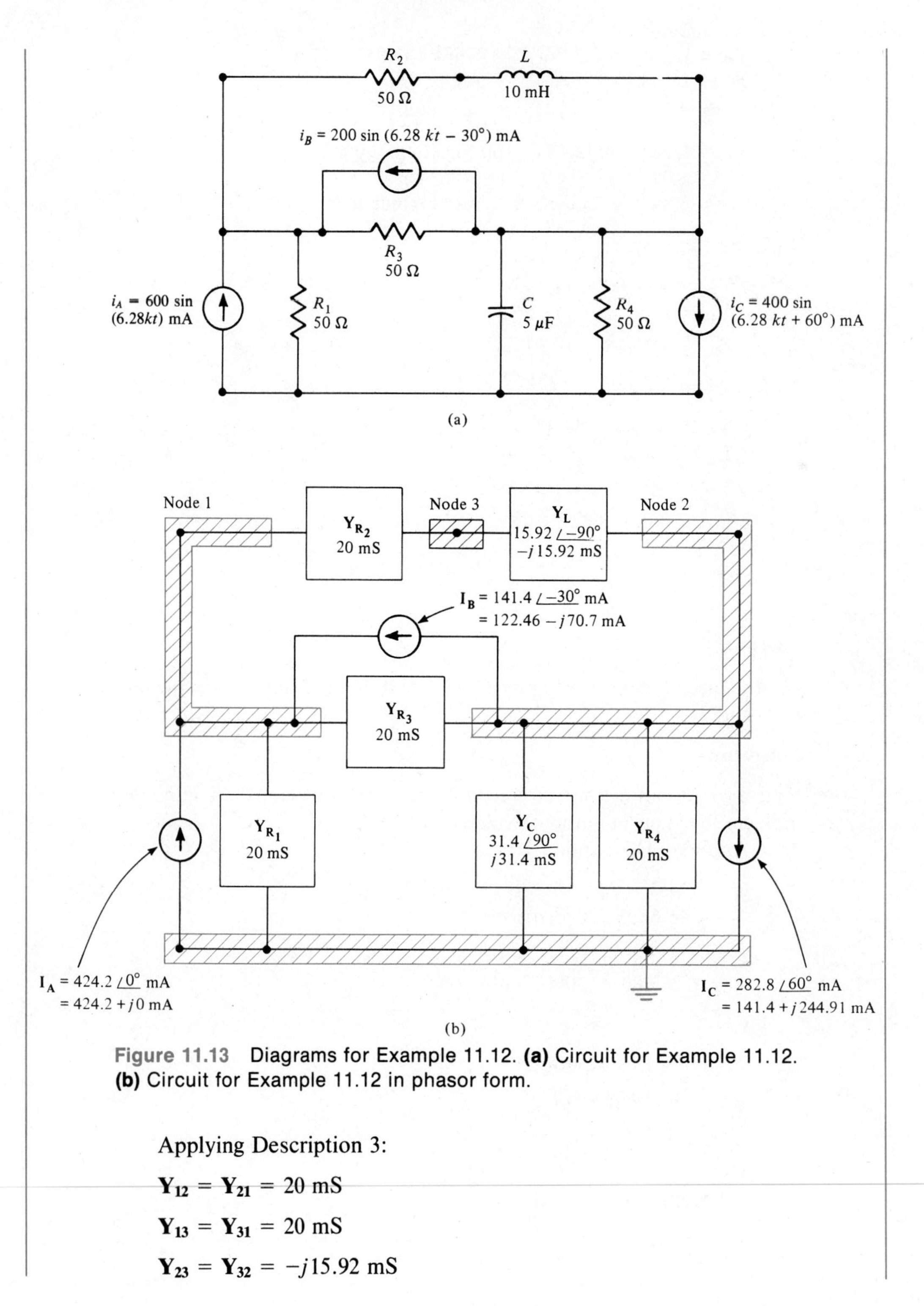

Figure 11.13 Diagrams for Example 11.12. **(a)** Circuit for Example 11.12. **(b)** Circuit for Example 11.12 in phasor form.

Applying Description 3:

$$\mathbf{Y}_{12} = \mathbf{Y}_{21} = 20 \text{ mS}$$

$$\mathbf{Y}_{13} = \mathbf{Y}_{31} = 20 \text{ mS}$$

$$\mathbf{Y}_{23} = \mathbf{Y}_{32} = -j15.92 \text{ mS}$$

Applying Rule 4, the node equations are:

$$(60 + j0)\mathbf{V}_1 + (-20 + j0)\mathbf{V}_2 + (-20 + j0)\mathbf{V}_3 = 546.66 - j70.7$$

$$(-20 + j0)\mathbf{V}_1 + (40 + j15.48)\mathbf{V}_2 + (0 + j15.92)\mathbf{V}_3 = -263.86 - j174.21$$

$$(-20 + j0)\mathbf{V}_1 + (0 + j15.92)\mathbf{V}_2 + (20 - j15.92)\mathbf{V}_3 = 0 + j0$$

The following BASIC program solves simultaneous equations with complex coefficients. All coefficients are entered in rectangular form and the answers are outputted in rectangular form. Following the program are some examples that use the program to solve the equations obtained in previous examples.

```
10 KEY OFF: CLS : COLOR 10
20     REM      *****GAUSS ELIMINATION*****
30     REM      by Ed Brumgnach and Jerry Sitbon
40     REM      ---------------------------------------
50     REM      *****SUMMARY OF SYMBOLS*****
60     REM      F(R)......ORIGINAL CONSTANTS COLUMN
70     REM      FD(R).....WORKING COEFFICIENT COLUMN
80     REM      A(R,C)....ORIGINAL COEFFICIENT ARRAY
90     REM      B(R,C)....WORKING COEFFICIENT ARRAY
100    REM      R.........EQUATION NUMBER OR ROW
       NUMBER
110    REM      C.........UNKNOWN NUMBER OR COLUMN
       NUMBER
120    REM      EC........EQUATION COUNTER
130    REM      MN........MAXIMUM NUMBER OF EQUATIONS
140    REM      N.........NUMBER OF EQUATIONS
150    REM      T.........TEMPORARY STORAGE
160    REM      S(R)......SOLUTION ARRAY
170    REM      BG........BIGGEST COEFFICIENT IN
       COLUMN
180    REM      ER........ERROR
190    REM      EQ$(R)....EQUATION ARRAY
200    REM      EQ$.......TEMPORARY EQUATION ARRAY
210    REM      *****END SUMMARY OF SYMBOLS*****
220    REM      *****SOLUTION OF SIMULTANEOUS
       EQUATIONS BY GAUSS ELIMINATION*****
230 GOSUB 280   'GO TO INPUT MODULE SUBROUTINE
240 GOSUB 1920    'GO TO CONVERSION MODULE
250 GOSUB 360     'GO TO SOLUTION MODULE
260 GOSUB 920     'GO TO OUTPUT MODULE
270 END
280    REM      *****INPUT MODULE*****
290 CLS
300 INPUT "FOR LOOP EQUATIONS TYPE: 'LE', FOR NODE
        EQUATIONS TYPE: 'NE'";ET$
310 IF ET$="le" THEN ET$="LE"
320 IF ET$="ne" THEN ET$="NE"
330 IF ET$="LE" THEN X$="I" : GOTO 1350 ' GO TO
      INPUT COEFFICIENTS MODULE
340 IF ET$="NE" THEN X$="V" : GOTO 1350 ' GO TO
      INPUT COEFFICIENTS MODULE
```

```
350 GOTO 300
360   REM      *****SOLUTION BY GAUSS ELIMINATION*****
370   REM         ***CREATE WORKING COEFFICIENT
      ARRAYS "B" AND "FD"***
380 PRINT "COMPUTING"
390 N=2*N
400 FOR R=1 TO N
410 FOR C=1 TO N
420 B(R,C)=A(R,C)
430 NEXT C
440 FD(R)=F(R)
450 NEXT R
460   REM         ***END COEFFICIENT CREATION***
470   REM         ***FIND BIGGEST COEFFICIENT IN THIS
                  COLUMN***
480 ER=0
490 FOR R=1 TO N-1
500 BG=ABS(B(R,R))
510 EC=R
520 R1=R+1
530 FOR RN=R1 TO N
540 IF ABS(B(RN,R))<BG THEN 570
550 BG=ABS(B(RN,R))
560 EC=RN
570 NEXT RN
580 IF BG=0 THEN 880
590 IF EC=R THEN 690
600   REM  ***BRING BIGGEST COEFFICIENT TO THE TOP***
610 FOR C=1 TO N
620 T=B(EC,C)
630 B(EC,C)=B(R,C)
640 B(R,C)=T
650 NEXT C
660 T=FD(EC)
670 FD(EC)=FD(R)
680 FD(R)=T
690 FOR RN=R1 TO N
700 X=B(RN,R)/B(R,R)
710 FOR I=R1 TO N
720 B(RN,I)=B(RN,I)-X*B(R,I)
730 NEXT I
740 FD(RN)=FD(RN)-X*FD(R)
750 NEXT RN
760 NEXT R
770  REM         ***COMPUTE ANSWER AND BACK SUBSTITUTE***
780 IF B(N,N)=0 THEN 880
790 S(N)=FD(N)/B(N,N)
800 FOR R=N-1 TO 1 STEP -1
810 Y=0
820 FOR RN=R+1 TO N
830 Y=Y+B(R,RN)*S(RN)
840 NEXT RN
850 S(R)=(FD(R)-Y)/B(R,R)
860 NEXT R
870 RETURN                 'IF NO DIFFICULTIES
```

```
880 ER=1
890 PRINT "ERROR"
900 RETURN
910 REM      *****END GAUSS ELIMINATION MODULE*****
920 REM      *****OUTPUT MODULE*****
930 CLS : INPUT"IF A HARD COPY IS DESIRED TYPE
       'H'";P$
940 IF P$="h" THEN P$="H"
950 N=N/2
960 IF N>5 THEN 1050
970 IF X$="I" THEN PRINT "THE LOOP EQUATIONS
    ARE: " ELSE PRINT "THE NODE EQUATIONS ARE:"
980 IF P$<>"H" THEN 1000
990 IF X$="I" THEN LPRINT "THE LOOP EQUATIONS
    ARE: " ELSE LPRINT "THE NODE EQUATIONS ARE:"
1000 CLS
1010 FOR R = 1 TO N
1020 PRINT "EQUATION";R;" ";EQ$(R)
1030 IF P$="H" THEN LPRINT "EQUATION";R;" ";EQ$(R)
1040 NEXT R
1050 PRINT
1060 IF P$<>"H" THEN 1080
1070 LPRINT
1080 IF ER=1 THEN GOTO 280
1090 PRINT "SOLUTION:"
1100 IF P$<>"H" THEN 1120
1110 LPRINT "SOLUTION:"
1120 PRINT
1130 IF P$<>"H" THEN 1150
1140 LPRINT
1150 FOR R = 1 TO N
1160 SR(R) = S(2*R-1)
1170 SI(R) = S(2*R)
1180 IF SI(R)<0 THEN SN$="-" ELSE SN$="+"
1190 SI$ = MID$(STR$(SI(R)),2) : SR$=STR$(SR(R))
1200 IF X$="I" THEN XU$="A" ELSE XU$="V"
1210 PRINT X$;"(";R;")= ";
1220 PRINT USING "####.##^^^^";SR(R);
1230 PRINT " "; SN$ ; " j";
1240 PRINT USING "####.##^^^^";ABS(SI(R));
1250 PRINT " "; UN$+XU$
1260 IF P$="H" THEN LPRINT X$;"(";R;")= ";
1270 IF P$="H" THEN LPRINT USING
     "####.##^^^^";SR(R);
1280 IF P$="H" THEN LPRINT " "; SN$ ; " j";
1290 IF P$="H" THEN LPRINT USING
     "####.##^^^^";ABS(SI(R));
1300 IF P$="H" THEN LPRINT " "; UN$+XU$
1310 NEXT R
1320 IF P$="H" THEN LPRINT CHR$(12);CHR$(7);
1330 RETURN
1340 REM      *****END OUTPUT MODULE*****
1350 CLS
1360 REM *****INPUT COMPLEX COEFFICIENTS
     MODULE*****
```

```
1370 INPUT "How many equations (2-5) ";N
1380 PRINT
1390 MN=5: IF N>MN THEN 300
1400 IF N<2 THEN 300
1410 PRINT "The coefficient of each equation will
         be entered one at a time"
1420 PRINT "                         in rectangular
     form."
1430 PRINT
1440 FOR R = 1 TO N
1450 PRINT "ENTER EQUATION ";CHR$(48+R) : REM ASCII
     OF "1" IS 49
1460 PRINT
1470 PRINT TAB(5);"Real Part";TAB(28);"Imaginary
     Part"
1480 PRINT STRING$(50, "-")
1490 FOR C = 1 TO N
1500 PRINT "AR(";CHR$(48+R);",";CHR$(48+C);")";
1510 INPUT;"= ",AR(R,C)
1520 PRINT TAB(25);
1530 PRINT "AI(";CHR$(48+R);",";CHR$(48+C);")";
1540 INPUT "= ",AI(R,C)
1550 NEXT C
1560 PRINT STRING$(50, "-")
1570 PRINT "FR(";CHR$(48+R);")";
1580 INPUT; "= ",FR(R)
1590 PRINT TAB(25);
1600 PRINT "FI(";CHR$(48+R);")";
1610 INPUT "= ",FI(R)
1620 REM Pretty Equation Output Module
1630 EQ$(R)="":EQ$=""
1640 FOR C = 1 TO N
1650 IF AR(R,C)<0 THEN SN$="-" ELSE SN$=""
1660 CH$=MID$(STR$(AR(R,C)),2)
1670 IF C=1 THEN EQ$ = "(" + SN$ + CH$ : GOTO 1690
1680 EQ$ = "+(" +SN$ + CH$
1690 IF AI(R,C)<0 THEN SN$="-" ELSE SN$="+"
1700 CH$=MID$(STR$(AI(R,C)),2)
1710 EQ$ = EQ$ + SN$ + "j" + CH$ + ")I" +
     CHR$(48+C)
1720 LONG = LEN(EQ$)
1730 EQ$ = STRING$(15-LONG," ") + EQ$
1740 EQ$(R) = EQ$(R) + EQ$ : EQ$=""
1750 NEXT C
1760 IF FR(R)<0 THEN SN$="-" ELSE SN$=""
1770 CH$=MID$(STR$(FR(R)),2)
1780 EQ$ = " =(" + SN$ + CH$
1790 IF FI(R)<0 THEN SN$="-" ELSE SN$="+"
1800 CH$=MID$(STR$(FI(R)),2)
1810 EQ$ = EQ$ + SN$ + "j" + CH$ + ")"
1820 EQ$(R) = EQ$(R) + EQ$
1830 BEEP : PRINT
1840 PRINT EQ$(R)
1850 PRINT
```

```
1860 INPUT "Is this the correct Equation? Press
            enter key for yes otherwise type NO! ",
            Q$
1870 IF Q$="NO" OR Q$="no" THEN CLS : GOTO 1450
1880 CLS
1890 NEXT R
1900 REM ******* END OF INPUT MODULE *******
1910 RETURN             ' GOES BACK TO 195
1920 REM  ******CONVERSION MODULE*******
1930 REM  *****complex coefficient to real
coefficient equations*****
1940 FOR R = 1 TO N
1950 FOR C = 1 TO N
1960 A(2*R-1,2*C-1) = AR(R,C)
1970 A(2*R,2*C-1) = AI(R,C)
1980 A(2*R-1,2*C) = -AI(R,C)
1990 A(2*R,2*C) = AR(R,C)
2000 NEXT C
2010 F(2*R-1) = FR(R)
2020 F(2*R) = FI(R)
2030 NEXT R
2040 RETURN
2050 REM *****END CONVERSION MODULE*****
2060 FOR R = 1 TO N
2070 SR(R) = 1000* SR(R)
2080 SI(R) = 1000* SI(R)
2090 UN$ = "m"
2100 NEXT R
2110 GOTO 1160
```

EXAMPLE 11.13

Use the BASIC program provided to solve the loop equations obtained in Example 11.8.

Solution

The computer output is:

```
THE LOOP EQUATIONS ARE:
EQUATION 1     (50-j31.85)I1   +(0+j31.85)I2
=(21.21+j0)
EQUATION 2      (0+j31.85)I1   +(0+j30.95)I2
=(-6.13+j3.54)
SOLUTION:
I( 1 )=  241.34E-03 + j 239.08E-03 A
I( 2 )= -133.98E-03 - j 479.75E-04 A
```

EXAMPLE 11.14

Use the BASIC program provided to solve the loop equations obtained in Example 11.9.

Solution

The computer output is:

```
THE LOOP EQUATIONS ARE:
EQUATION 1
     (50-j31.85)I1        +(0+j31.85)I2
+(-50+j0)I3 =(21.21+j0)
EQUATION 2
     (0+j31.85)I1        +(0+j30.95)I2
+(0-j62.8)I3 =(-6.12+j3.54)
EQUATION 3
   (-50+j0)I1            +(0-j62.8)I2
+(100.08+j62.8)I3=(-7.07-j12.25)
SOLUTION:
I( 1 )= 401.90E-03 + j 648.20E-04 A
I( 2 )= 263.90E-04 - j 222.10E-03 A
I( 3 )= 160.30E-03 - j 174.20E-03 A
```

EXAMPLE 11.15

Use the BASIC program provided to solve the node equations obtained in Example 11.10.

Solution

The computer output is:

```
THE NODE EQUATIONS ARE:
EQUATION 1            (20+j15.48)V1
+(0+j15.92)V2 =(424.2+j0)
EQUATION 2             (0-j15.92)V1
+(40+j15.92)V2 =(263.86+j174.21)
SOLUTION:
V( 1 )= 125.50E-01 - j 113.40E-01 V
V( 2 )= 202.00E-02 + j 164.00E-03 V
```

EXAMPLE 11.16

Use the BASIC program provided to solve the node equations obtained in Example 11.12.

Solution

The computer output is:

```
THE NODE EQUATIONS ARE:
EQUATION 1
          (60+j0)V1    +(-20+j0)V2    +(-20+j0)V3
=(546.66-j70.7)
EQUATION 2
          (-20+j0)V1   +(40+j15.48)V2
+(0+j15.92)V3 =(-263.86-j174.21)
EQUATION 3
          (-20+j0)V1   +(0+j15.92)V2
+(20-j15.92)V3 =(0+j0)
SOLUTION:
V( 1 )=  898.13E-02 - j 207.57E-02 V
V( 2 )= -311.93E-02 - j 527.26E-02 V
V( 3 )=  273.03E-02 + j 258.05E-02 V
```

SUMMARY

In this chapter Ohm's Law and Kirchhoff's Voltage and Current Laws as they apply to ac circuits were presented. Series, parallel, and series-parallel ac circuits were analyzed. Methods for obtaining loop and nodal equations for more complex ac circuits were presented. Cramer's Rule was examined as a possible method of solution for the simpler simultaneous equations. A computer program was offered as the means of solving multiple-loop and multiple-node ac circuits.

SUMMARY OF IMPORTANT EQUATIONS

$$\mathbf{I} = \frac{\mathbf{E}}{\mathbf{Z}} \tag{11.1}$$

$$\mathbf{Z}_T = \mathbf{Z}_1 + \mathbf{Z}_2 + \mathbf{Z}_3 + \cdots + \mathbf{Z}_N \tag{11.2}$$

$$\mathbf{Y} = \frac{1}{\mathbf{Z}} \tag{11.3}$$

$$\mathbf{Y}_T = \mathbf{Y}_1 + \mathbf{Y}_2 + \mathbf{Y}_3 + \cdots + \mathbf{Y}_N \tag{11.4a}$$

$$\frac{1}{\mathbf{Z}_T} = \frac{1}{\mathbf{Z}_1} + \frac{1}{\mathbf{Z}_2} + \frac{1}{\mathbf{Z}_3} + \cdots + \frac{1}{\mathbf{Z}_N} \tag{11.4b}$$

$$\mathbf{Z}_{11}\mathbf{I}_1 - \mathbf{Z}_{12}\mathbf{I}_2 - \mathbf{Z}_{13}\mathbf{I}_3 - \cdots - \mathbf{Z}_{1N}\mathbf{I}_N = \mathbf{E}_1 \tag{11.5a}$$

$$-\mathbf{Z}_{21}\mathbf{I}_1 + \mathbf{Z}_{22}\mathbf{I}_2 - \mathbf{Z}_{23}\mathbf{I}_3 - \cdots - \mathbf{Z}_{2N}\mathbf{I}_N = \mathbf{E}_2 \tag{11.5b}$$

$$\vdots$$

$$-\mathbf{Z}_{N1}\mathbf{I}_1 - \mathbf{Z}_{N2}\mathbf{I}_2 - \mathbf{Z}_{N3}\mathbf{I}_3 - \cdots + \mathbf{Z}_{NN}\mathbf{I}_N = \mathbf{E}_N \tag{11.5c}$$

$$\mathbf{Y}_{11}\mathbf{V}_1 - \mathbf{Y}_{12}\mathbf{V}_2 - \cdots - \mathbf{Y}_{1N}\mathbf{V}_N = \mathbf{I}_1 \tag{11.6a}$$

$$-\mathbf{Y}_{21}\mathbf{V}_1 + \mathbf{Y}_{22}\mathbf{V}_2 - \cdots - \mathbf{Y}_{2N}\mathbf{V}_N = \mathbf{I}_2 \tag{11.6b}$$

$$\vdots$$

$$-\mathbf{Y}_{N1}\mathbf{V}_1 - \mathbf{Y}_{N2}\mathbf{V}_2 - \cdots + \mathbf{Y}_{NN}\mathbf{V}_N = \mathbf{I}_N \tag{11.6c}$$

REVIEW QUESTIONS

11.1 State Ohm's Law for ac circuits.

11.2 State Kirchhoff's Voltage Law for ac circuits.

11.3 State Kirchhoff's Current Law for ac circuits.

11.4 Define a series ac circuit.

11.5 Define a parallel ac circuit.

11.6 Define a series-parallel ac circuit.

11.7 State the rules for writing ac circuit loop equations.

11.8 State the rules for writing ac circuit node equations.

11.9 Explain why it is more difficult to analyze ac circuits than dc circuits.

PROBLEMS

11.1 A 50-μF capacitor is in series with a 10-Ω resistor. The applied voltage is $250 \sin(2000t + 70°)$ V. Determine the sinusoidal expression for the current.

11.2 A 10-μF capacitor is in series with a 100-Ω resistor. Find the frequency at which the current leads the voltage by 38.5°. (ans. 199 Hz)

11.3 A 50-Ω resistor is in series with another element. The frequency of the applied voltage is 500 Hz. Determine the element if the current leads the applied voltage by 52°.

11.4 Repeat Problem 11.3 if the current lags the applied voltage by 32°. (ans. 9.95 mH)

11.5 A series *RL* circuit consists of a 10-Ω resistor and a 60-mH coil. The current lags the applied voltage by 60°. Compute the source frequency and the total impedance.

11.6 A series *RL* circuit consists of a 30-Ω resistor and a 30-mH coil. Compute the angle of the impedance and the frequency of the source if the magnitude of the impedance is 40 Ω. (ans. 33.42°, 140 Hz)

11.7 Calculate the impedance of a series *RL* circuit consisting of a 40-Ω resistor and a 15-mH coil at:

a. 100 Hz
b. 500 Hz
c. 1 kHz
d. 10 kHz

11.8 A resistor and a capacitor are in series across a 60-Hz source. The resistor value is 10 kΩ. The voltage across the capacitor is 90 V and the voltage across the resistor is 45 V. Determine:

a. The magnitude of the voltage source
b. The magnitude of the circuit current
c. The reactance of the capacitor
d. The capacitor value
(ans. 100.62 V, 4.5 mA, 20 Ω, 133 μF)

11.9 A 20-Ω resistor and a 10-Ω capacitive reactance are in parallel across a 50-V, 1-kHz source. Compute the magnitude of the source current.

11.10 A series *RLC* circuit ($R = 20\ \Omega$, $C = 2\ \mu F$, $L = 3$ mH) is connected across a 50-V, 2-kHz source. Determine the magnitude and angle of the total current. (ans. 2.49 A, 6°)

11.11 A 5-H coil and a 2-μF capacitor are in series with a resistor. Compute the resistor value if the current is 164 mA from a 120-V, 60-Hz source.

11.12 A 15-Ω resistor, a 30-Ω inductive reactance, and a 15-Ω capacitive reactance are in parallel across a 120-V, 60-Hz source. Compute the magnitude and angle of the total current. (ans. 8.94 A, 27°)

11.13 For the circuit shown in Figure 11.14, determine:

a. The phasor form of the voltage source
b. Frequency in hertz
c. The reactance of the capacitor (or inductor)
d. The polar and rectangular forms of the capacitor (or inductor) impedance
e. The total circuit impedance
f. The phasor circuit current

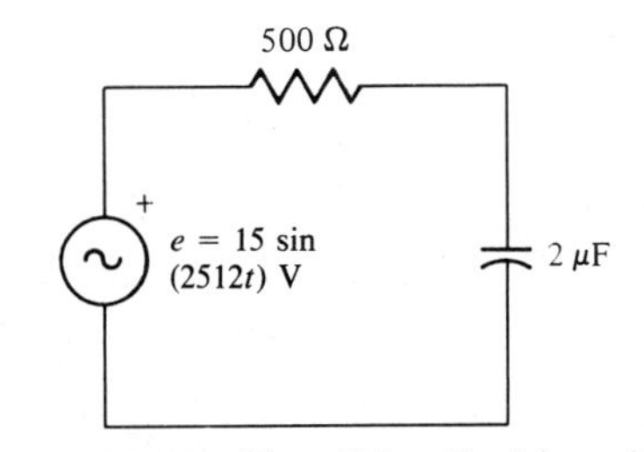

Figure 11.14 Circuit for Problem 11.13.

g. The phasor voltage across each element
h. That KVL holds

11.14 Repeat Problem 11.1 for the circuit shown in Figure 11.15. (ans. $120\underline{/0^\circ}$ V, 60 Hz, 151 Ω, $j151$ Ω, $151\underline{/90^\circ}$ Ω, $181.11\underline{/56.49^\circ}$ Ω, $0.663\underline{/-56.49^\circ}$ A, $66.3\underline{/-56.49^\circ}$ V, $100.11\underline{/33.51^\circ}$ V)

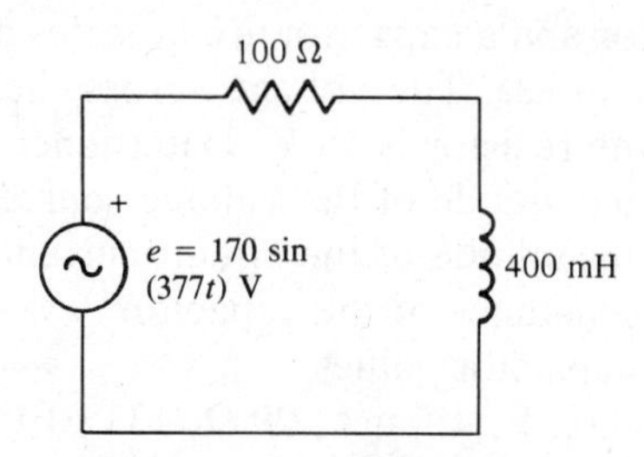

Figure 11.15 Circuit for Problem 11.14.

11.15 For the circuit shown in Figure 11.16, determine:
a. The phasor form of the voltage source
b. The frequency in hertz
c. The reactance of each component
d. The polar and rectangular forms of each impedance
e. The polar and rectangular forms of each admittance
f. The total admittance
g. The total current
h. The current through each element
i. That KCL holds

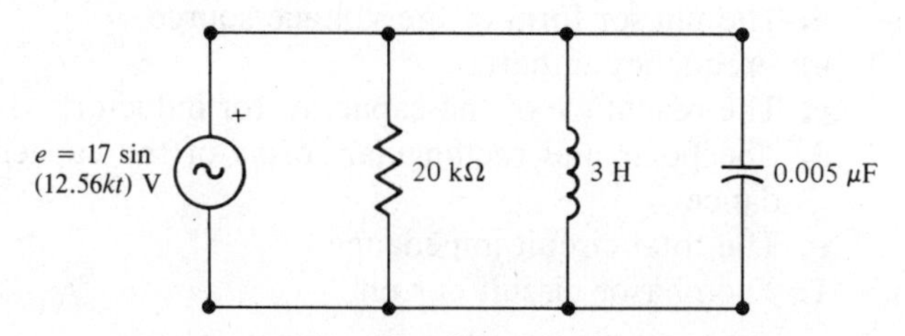

Figure 11.16 Circuit for Problem 11.15.

11.16 For the circuit shown in Figure 11.17, find the current through resistor R_2 both in phasor and in sinusoidal form. [ans. $1.53\underline{/19.44^\circ}$ A, $2.16 \sin(377t + 19.44^\circ)$ A]

11.17 For the circuit in Figure 11.18, use loop equations and Cramer's Rule to find the source current.

11.18 For the circuit shown in Figure 11.18, use nodal analysis and Cramer's Rule to find the voltage at node 1. (ans. $4.32\underline{/30.74^\circ}$ V)

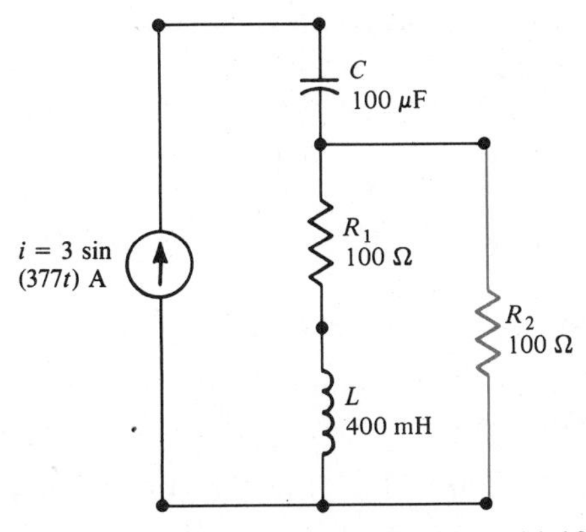

Figure 11.17 Circuit for Problem 11.16.

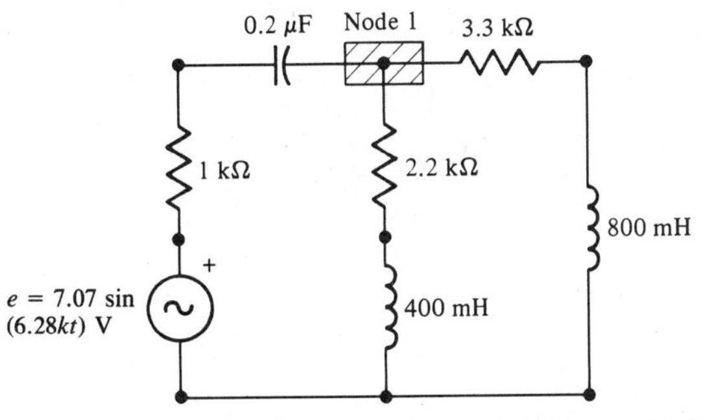

Figure 11.18 Circuit for Problems 11.17 and 11.18.

11.19 For the circuit shown in Figure 11.19, set up the loop equations both in polar and rectangular form.

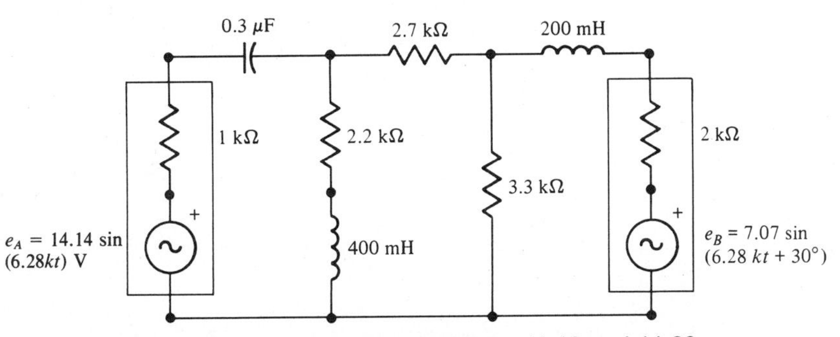

Figure 11.19 Circuit for Problems 11.19 and 11.20.

11.20 For the circuit shown in Figure 11.19, set up the node equations both in polar and rectangular form. [ans. all terms $\times 10^{-3}$; $(1 + j1.883)\mathbf{V}_1 - j1.883$

$\mathbf{V}_2 = 0$, $-j1.883\mathbf{V}_1 + (0.82 + j1.883)\mathbf{V}_2 - 0.45\mathbf{V}_3 - 0.37\mathbf{V}_4 = 0$, $-0.45\mathbf{V}_2 + (0.45 - j0.40)\mathbf{V}_3 = 0$, $-0.37\mathbf{V}_2 + (0.67 - j0.79)\mathbf{V}_4 + j0.79\mathbf{V}_5 = 0$, $j0.79\mathbf{V}_4 + (0.5 - j0.79)\mathbf{V}_5 = 2.17 + j1.25$; $2.13\angle 62.03^\circ\ \mathbf{V}_1 - 1.883\angle 90^\circ\ \mathbf{V}_2 = 0$, $-1.883\angle 90^\circ\ \mathbf{V}_1 + 2.05\angle 66.47^\circ\ \mathbf{V}_2 - 0.45\angle 0^\circ\ \mathbf{V}_3 - 0.37\angle 0^\circ\ \mathbf{V}_4 = 0\angle 0^\circ$, $-0.45\angle 0^\circ\ \mathbf{V}_2 + 0.6\angle -41.63^\circ\ \mathbf{V}_3 = 0\angle 0^\circ$, $-0.37\angle 0^\circ\ \mathbf{V}_2 + 1.04\angle -49.7^\circ\ \mathbf{V}_4 + 0.79\angle 90^\circ\ \mathbf{V}_5 = 0\angle 0^\circ$, $0.79\angle 90^\circ\ \mathbf{V}_4 + 0.93\angle -57.67^\circ\ \mathbf{V}_5 = 2.5\angle 30^\circ$]

PROBLEMS FOR COMPUTER SOLUTION

11.21 Use the BASIC computer program for complex simultaneous equations provided in the chapter to solve the equations obtained in Problem 11.19.

11.22 Repeat Problem 11.21 for the simultaneous equations obtained in Problem 11.20.

CHAPTER 12

Ac Circuit Models

GLOSSARY

admittance parameters (y)—two-port coefficients resulting from independent voltage sources
hybrid parameters (h)—two-port coefficients resulting from an independent current source and an independent voltage source
impedance parameters (z)—two-port coefficients resulting from independent current sources
Norton's Theorem—technique for analyzing electric circuits
port—a pair of electrical terminals
Superposition Theorem—technique for analyzing electric circuits

Thevenin's Theorem—technique for analyzing electric circuits
two-port network—a network with four terminals or two ports

INTRODUCTION

The analysis methods developed earlier in Chapter 8 for dc circuits are applied in this chapter for ac circuits. *Thevenin's, Norton's,* and the *Superposition* theorems are very helpful methods in analyzing ac circuits.

Thevenin's and Norton's equivalent circuits can be used to represent the original circuit and to calculate the unknown quantities in a simplified fashion. The Superposition Theorem is especially useful with ac circuits because it considers the analysis of multisource circuits by considering the effects of each individual source on the circuit. The superposition method of analysis is most valuable when the sources are of different frequencies.

Two-port network models and their parameters are also discussed in this chapter. Standard models are used to represent two-port circuits. The modeling of active devices such as transistors and operational amplifiers is very common in the analysis of electronic circuits. The parameters that are used to define the various models are the characteristics of the network. The following are introduced in this chapter: the *impedance* model, the *admittance* model, and the *hybrid* model.

12.1 THEVENIN'S THEOREM

The use of **Thevenin's Theorem** was shown in Chapter 8 to be a very important technique in the analysis of electric circuits. It allows any load in an ac circuit to be isolated and the rest of the circuit to be replaced by a Thevenin equivalent circuit, which consists of a phasor voltage source and a series impedance.

A procedure for obtaining the Thevenin equivalent of an ac circuit is:

1. Represent each source in its phasor form and each component by its impedance.
2. Remove the load under consideration and label the terminals A and B.
3. Calculate the voltage across the open terminals $\mathbf{V_{AB}}$.
4. Replace each source by its internal impedance and calculate the impedance from A to B, $\mathbf{Z_{AB}}$.
5. Draw the Thevenin equivalent as a phasor voltage source with value $\mathbf{V_{AB}}$ in series with the impedance $\mathbf{Z_{AB}}$.
6. Replace the impedance that was removed in Step 2 and proceed with any required calculations.

EXAMPLE 12.1

For the ac circuit shown in Figure 12.1a, use Thevenin's Theorem to find the capacitor current, I_C.

Solution

The phasor form of the voltage and the impedance and admittance of each component were calculated in Example 11.7.

Figure 12.1b shows the circuit in its phasor form.

Figure 12.1c shows the circuit with $\mathbf{Z_C}$ removed and the terminals labeled A and B.

$\mathbf{Z_1}$ may replace the series combination of $\mathbf{Z_{R_2}}$ and $\mathbf{Z_L}$:

$$\begin{aligned}\mathbf{Z_1} &= \mathbf{Z_{R_2}} + \mathbf{Z_L} \\ &= 50 + j62.8\ \Omega \qquad \text{rectangular form} \\ &= 80.27\,\underline{/51.47^\circ}\ \Omega \qquad \text{polar form}\end{aligned}$$

Figure 12.1d shows the circuit with $\mathbf{Z_1}$ in its proper place. $\mathbf{V_{AB}}$ may be obtained by voltage division:

$$\mathbf{V_{AB}} = \mathbf{E}\frac{\mathbf{Z_1}}{\mathbf{Z_1} + \mathbf{Z_{R_1}}}$$

$$\begin{aligned}\mathbf{Z_1} + \mathbf{Z_{R_1}} &= 50 + 50 + j62.8 \\ &= 100 + j62.8\ \Omega \qquad \text{rectangular form} \\ &= 118.08\,\underline{/32.13^\circ}\ \Omega \qquad \text{polar form}\end{aligned}$$

$$\begin{aligned}\mathbf{V_{AB}} &= 21.21\,\underline{/0^\circ}\,\frac{80.27\,\underline{/51.47^\circ}}{118.08\,\underline{/31.13^\circ}} \\ &= 14.42\,\underline{/19.34^\circ}\ \text{V} \qquad \text{polar form} \\ &= 13.60 + j4.77\ \text{V} \qquad \text{rectangular form}\end{aligned}$$

$$\begin{aligned}\mathbf{Z_{AB}} &= \mathbf{Z_1} \| \mathbf{Z_{R_1}} = \frac{\mathbf{Z_1}\mathbf{Z_{R_1}}}{\mathbf{Z_1} + \mathbf{Z_{R_1}}} \\ &= \frac{(80.27\,\underline{/51.47^\circ})(50)}{118.08\,\underline{/31.13^\circ}} \\ &= 33.99\,\underline{/19.34^\circ}\ \Omega \qquad \text{polar form} \\ &= 32.07 + j11.26\ \Omega \qquad \text{rectangular form}\end{aligned}$$

Figure 12.1e shows Thevenin's equivalent circuit with the capacitor impedance replaced.

The total impedance is:

$$\begin{aligned}\mathbf{Z_T} &= \mathbf{Z_{th}} + \mathbf{Z_C} \\ &= (32.07 + j11.26) - j31.85 \\ &= 32.07 - j20.59\ \Omega \qquad \text{rectangular form} \\ &= 38.11\,\underline{/-32.70^\circ}\ \Omega \qquad \text{polar form}\end{aligned}$$

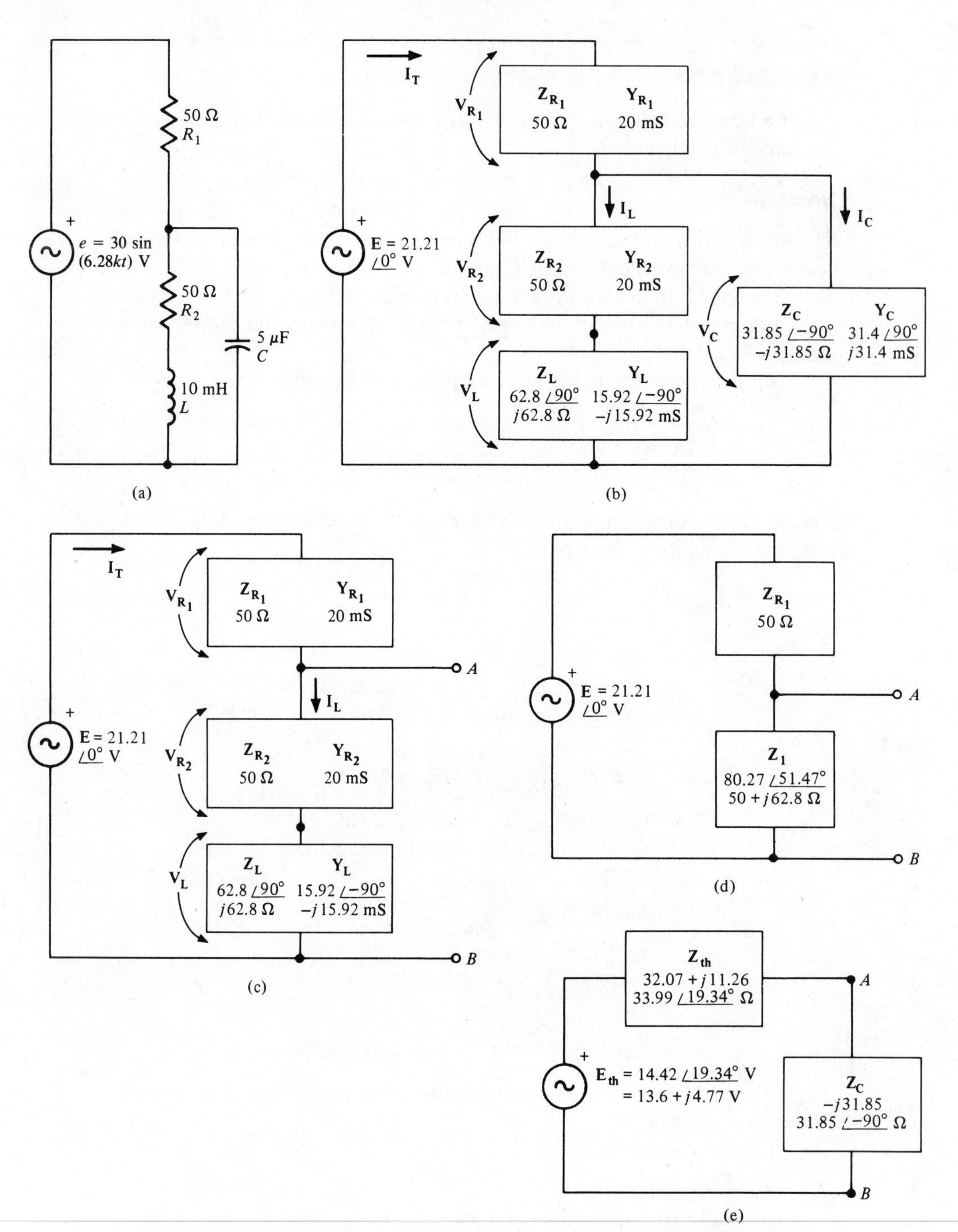

Figure 12.1 Diagrams for Example 12.1. **(a)** Circuit for Example 12.1. **(b)** Circuit in phasor form. **(c)** Circuit with $\mathbf{Z_C}$ removed and terminals labeled *A* and *B*. **(d)** $\mathbf{Z_1}$ replaces $\mathbf{Z_{R_1}}$ and $\mathbf{Z_L}$.

The current is:

$$\mathbf{I_C} = \frac{\mathbf{E_{th}}}{\mathbf{Z_T}} = \frac{14.42\angle 19.34^\circ}{38.11\angle -32.7^\circ}$$
$$= 378.38\angle 52.04^\circ \text{ mA} \qquad \text{polar form}$$
$$= 232.74 + j298.33 \text{ mA} \qquad \text{rectangular form}$$

and

$$i_C = (378.38)(1.414)\sin(6.28kt + 52.04^\circ) \text{ mA}$$
$$= 535.03 \sin(6.28kt + 52.04^\circ) \text{ mA}$$

Note that this matches closely with the answer obtained in Example 11.7.

12.2 NORTON'S THEOREM

The use of **Norton's Theorem** allows, as in the use of Thevenin's Theorem, any impedance in an ac circuit to be isolated. The rest of the circuit is replaced by a Norton equivalent circuit, which consists of a phasor current source and a parallel impedance.

A procedure for obtaining the Norton equivalent of an ac circuit is:

1. Represent each source in its phasor form and each component by its impedance.
2. Remove the load under consideration and label the terminals A and B.
3. Place a short circuit from A to B and calculate the current through it, $\mathbf{I_{AB}}$.
4. Replace each source by its internal impedance and calculate the impedance from A to B, $\mathbf{Z_{AB}}$.
5. Draw the Norton equivalent as a phasor current source with value $\mathbf{I_{AB}}$ in parallel with the impedance $\mathbf{Z_{AB}}$.
6. Replace the impedance that was removed in Step 2 and proceed with any required calculations.

EXAMPLE 12.2

For the ac circuit shown in Figure 12.2a, use Norton's Theorem to find the capacitor current, i_C.

Solution

The phasor form of the voltage and the impedance and admittance of each component were calculated in Example 11.7.

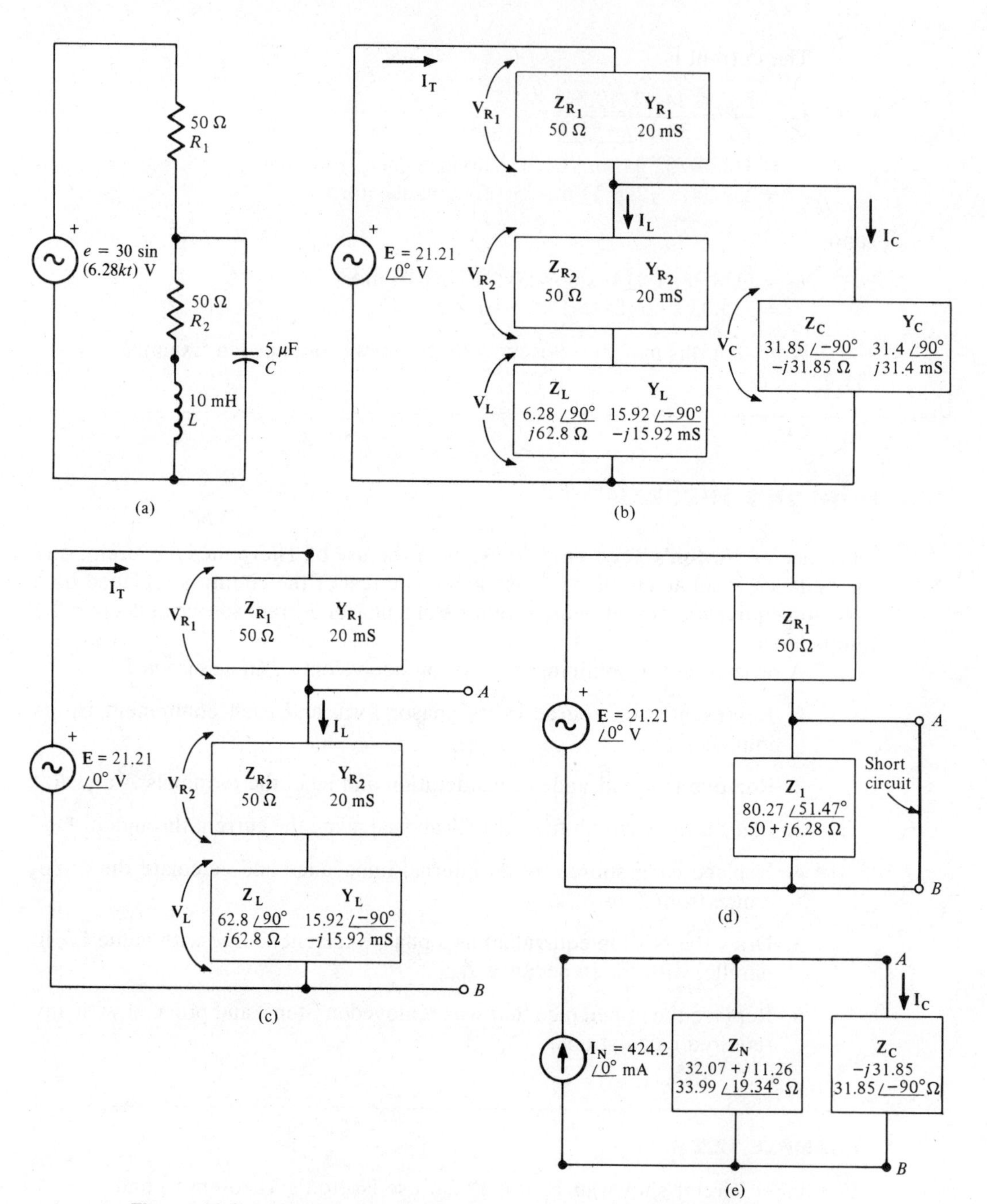

Figure 12.2 Diagrams for Example 12.2. **(a)** Circuit for Example 12.2. **(b)** Circuit in phasor notation. **(c)** Circuit with $\mathbf{Z_C}$ removed and terminals labeled *A* and *B*. **(d)** Short circuit placed from *A* to *B*. **(e)** Norton's equivalent with the load replaced.

Figure 12.2b shows the circuit in its phasor form.

Figure 12.2c shows the circuit with $\mathbf{Z_C}$ removed and the terminals labeled A and B.

Figure 12.2d shows the required short circuit placed from A to B. Note that this shorts out impedance $\mathbf{Z_1}$. The current $\mathbf{I_N}$ is:

$$\mathbf{I_N} = \frac{\mathbf{E}}{\mathbf{Z_{R_1}}} = \frac{21.21\,\underline{/0^\circ}}{50}$$
$$= 424.2\,\underline{/0^\circ}\text{ mA}$$

The Norton impedance is the same as the Thevenin impedance obtained in Example 12.1:

$$\mathbf{Z_N} = \mathbf{Z_{th}} = 32.07 - j20.59 = 38.11\,\underline{/-32.7^\circ}\ \Omega$$

Figure 12.2b shows the Norton equivalent with the capacitor impedance reinserted at terminals A and B.

$\mathbf{I_C}$ may be obtained by current division:

$$\mathbf{I_C} = \mathbf{I_N}\frac{\mathbf{Z_N}}{\mathbf{Z_C} + \mathbf{Z_N}}$$

From Example 12.1:

$$\mathbf{Z_C} + \mathbf{Z_N} = 38.11\,\underline{/-32.7^\circ}\ \Omega$$

$$\mathbf{I_C} = 424.2\,\underline{/0^\circ}\,\frac{33.99\,\underline{/19.34^\circ}}{38.11\,\underline{/-32.7^\circ}}$$
$$= 378.34\,\underline{/52.04^\circ}\text{ mA}$$

and

$$i_C = (378.34)(1.414)\sin(6.28kt + 52.04^\circ)\text{ mA}$$
$$= 534.97\sin(6.28kt + 52.04^\circ)\text{ mA}$$

which checks with Example 12.1.

Note that the preceding Thevenin circuit is related to the Norton circuit by a simple source transformation. If the capacitor was replaced by a short in Thevenin's equivalent, the resulting short-circuit current would be:

$$\mathbf{I_{SC}} = \frac{\mathbf{V_{th}}}{\mathbf{Z_{th}}}$$
$$= \frac{14.42\,\underline{/19.34^\circ}}{33.99\,\underline{/19.34^\circ}}$$
$$= 424.2\,\underline{/0^\circ}\text{ mA}$$

which is the same Norton current value obtained in Example 12.2.

12.3 SUPERPOSITION THEOREM

The **Superposition Theorem** allows an ac circuit with more than one energy source to be reduced to several ac circuits with one energy source each. The superposition theorem for ac circuits may be stated as follows:

In an ac circuit containing more than one source of voltage or current, the current through each circuit component is the phasor sum of the currents generated by each source acting independently. The voltage across each circuit component is the phasor sum of the voltages generated by each source acting independently.

The next example illustrates the use of the Superposition Theorem.

EXAMPLE 12.3

For the ac circuit shown in Figure 12.3a:

a. Use superposition to find each component's current.

b. Show that KCL holds at node 1.

Solution

The impedance and admittance as well as the phasor form of the current source were computed in Example 11.6. The phasor form of the voltage source was computed in Example 11.8.

Figure 12.3b shows the circuit in phasor notation.

Figure 12.3c shows only the current source energizing the circuit. The voltage source was replaced by its internal impedance (a short circuit in this case). The individual component's currents are also labeled with a prime.

This circuit is identical to the one in Example 11.6. Those results are reproduced here:

$$\mathbf{I}'_C = 439\angle 52.26^\circ \text{ mA} \qquad \text{polar form}$$
$$= 268.7 + j347.16 \text{ mA} \qquad \text{rectangular form}$$
$$\mathbf{I}'_R = 279.6\angle -37.74^\circ \text{ mA} \qquad \text{polar form}$$
$$= 221.11 - j171.12 \text{ mA} \qquad \text{rectangular form}$$
$$\mathbf{I}'_L = 222.6\angle -127.74^\circ \text{ mA} \qquad \text{polar form}$$
$$= -136.25 - j176.03 \text{ mA} \qquad \text{rectangular form}$$

Note that the resistor and capacitor currents are labeled downward, whereas the inductor current is labeled left to right.

Figure 12.3d shows only the voltage source energizing the circuit. The current source was replaced by an open circuit. The individual component's currents are labeled with double primes. Note that the resistor and capacitor currents are labeled downward, whereas the inductor current is labeled right to left.

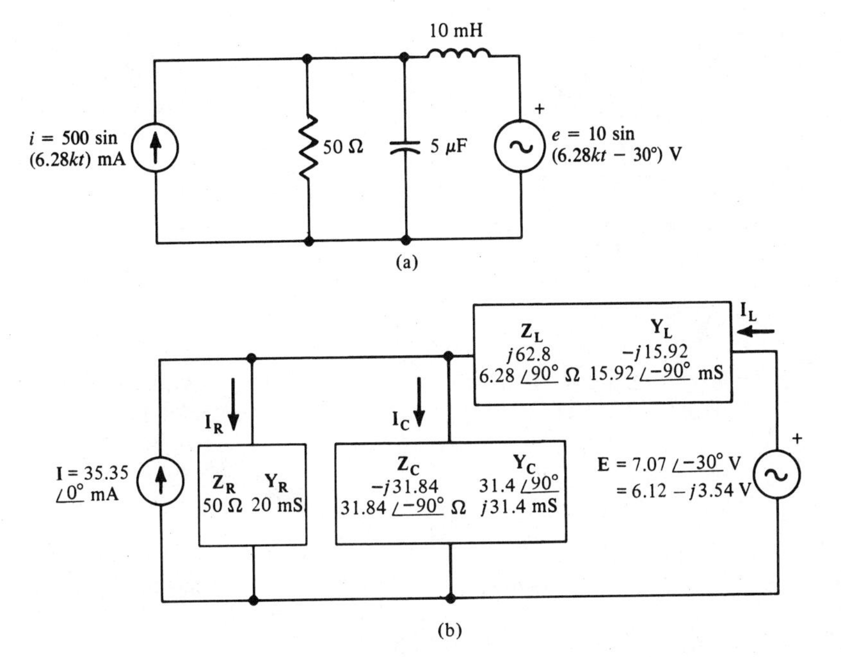

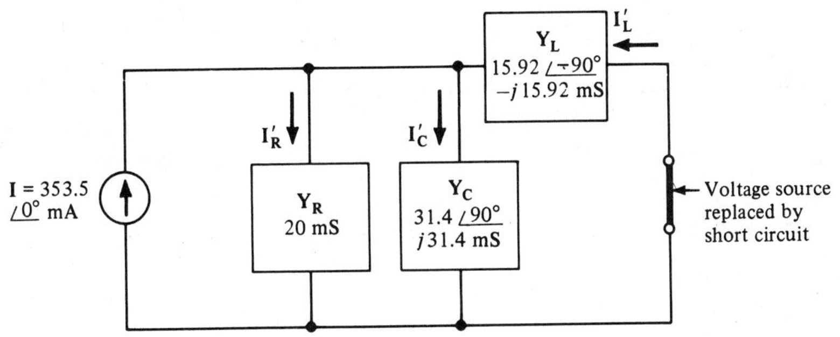

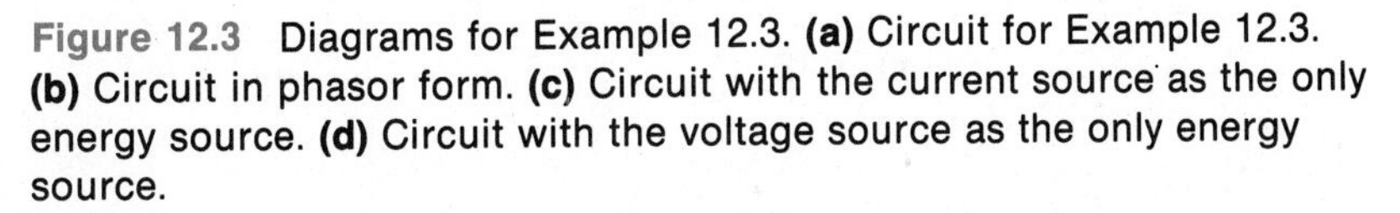

Figure 12.3 Diagrams for Example 12.3. **(a)** Circuit for Example 12.3. **(b)** Circuit in phasor form. **(c)** Circuit with the current source as the only energy source. **(d)** Circuit with the voltage source as the only energy source.

$\mathbf{Y_1} = \mathbf{Y_R} + \mathbf{Y_C}$
$= 20 + j31.4$ mS — rectangular form
$= 31.23\angle 57.5^\circ$ mS — polar form

$\mathbf{Z_1} = \dfrac{1}{\mathbf{Y_1}} = \dfrac{1}{37.23\angle 57.5^\circ \text{ m}}$
$= 26.86\angle -57.5^\circ\ \Omega$ — polar form
$= 14.43 - j22.65\ \Omega$ — rectangular form

$\mathbf{Z_T} = \mathbf{Z_L} + \mathbf{Z_1}$
$= (j62.8) + (14.43 - j22.65)$
$= 14.43 + j40.15\ \Omega$ — rectangular form
$= 42.66\angle 70.23^\circ\ \Omega$ — polar form

$\mathbf{I''_L} = \dfrac{\mathbf{E}}{\mathbf{Z_T}} = \dfrac{7.07\angle -30^\circ}{42.66\angle 70.23^\circ}$
$= 165.73\angle -100.23^\circ$ mA — polar form
$= -29.43 - j163.1$ mA — rectangular form

$\mathbf{V''_1} = \mathbf{I''_L}\mathbf{Z_1} = (0.16573\angle -100.23^\circ)(26.86\angle -57.5^\circ)$
$= 4.45\angle -157.73^\circ$ V — polar form

$\mathbf{I''_C} = \mathbf{V''_1}\mathbf{Y_C} = (4.45\angle -157.73^\circ)(31.4\angle 90^\circ \text{ mA})$
$= 139.73\angle -67.73^\circ$ mA — polar form
$= 52.95 - j129.31$ mA — rectangular form

$\mathbf{I''_R} = \mathbf{V''_1}\mathbf{Y_R} = (4.45\angle -157.73^\circ)(20 \text{ m})$
$= 89\angle -157.73^\circ$ mA — polar form
$= -82.36 - j33.73$ mA — rectangular form

The primed currents due to the current source and the double-primed currents due to the voltage source may be superimposed. The capacitor currents add because they are both in a downward direction:

$\mathbf{I_C} = \mathbf{I'_C} + \mathbf{I''_C}$
$\mathbf{I'_C} \Rightarrow 268.7 + j347.16$
$+\ \mathbf{I''_C} \Rightarrow 52.95 - j129.31$
$\mathbf{I_C} \Rightarrow 321.65 + j217.85$ mA

The resistor currents also add because they are both labeled downward:

$\mathbf{I_R} = \mathbf{I'_R} + \mathbf{I''_R}$
$\mathbf{I'_R} \Rightarrow 221.11 - j171.12$
$+\ \mathbf{I''_R} \Rightarrow -82.36 - j33.73$
$\mathbf{I_R} \Rightarrow 138.75 - j204.85$ mA

The primed inductor current has to be subtracted from the double-primed one to keep the direction indicated in Figure 12.3b:

$$\mathbf{I_L} = \mathbf{I''_L} - \mathbf{I'_L}$$

$$\begin{array}{rl} \mathbf{I''_R} \Rightarrow & -29.43 - j163.1 \\ -\ \mathbf{I'_L} \Rightarrow & -(-136.25 - j176.03) \\ \hline \mathbf{I_L} \Rightarrow & 106.82 + j12.93 \text{ mA} \end{array}$$

Applying KCL at node 1:

$$\mathbf{I} + \mathbf{I_L} = \mathbf{I_R} + \mathbf{I_C}$$

$$\begin{array}{rl} \mathbf{I} \Rightarrow & 353.5 \ + j \\ +\ \mathbf{I_L} \Rightarrow & 106.82 + j12.93 \\ \hline & 460.32 + j12.93 \text{ mA} \end{array} \qquad (1)$$

$$\begin{array}{rl} \mathbf{I_R} \Rightarrow & 138.75 - j204.85 \\ +\ \mathbf{I_C} \Rightarrow & 321.65 + j217.85 \\ \hline & 460.4 \ + j13 \text{ mA} \end{array} \qquad (2)$$

Note that (1) and (2) are approximately the same!

One of the most practical and popular applications of the Superposition Theorem is in the analysis of circuits with sources of different frequencies. Figure 12.4, for example, shows a simple transistor amplifier with a 17-V dc source and a 10-mV, 1000-Hz signal applied. The analysis of this circuit is beyond the scope of

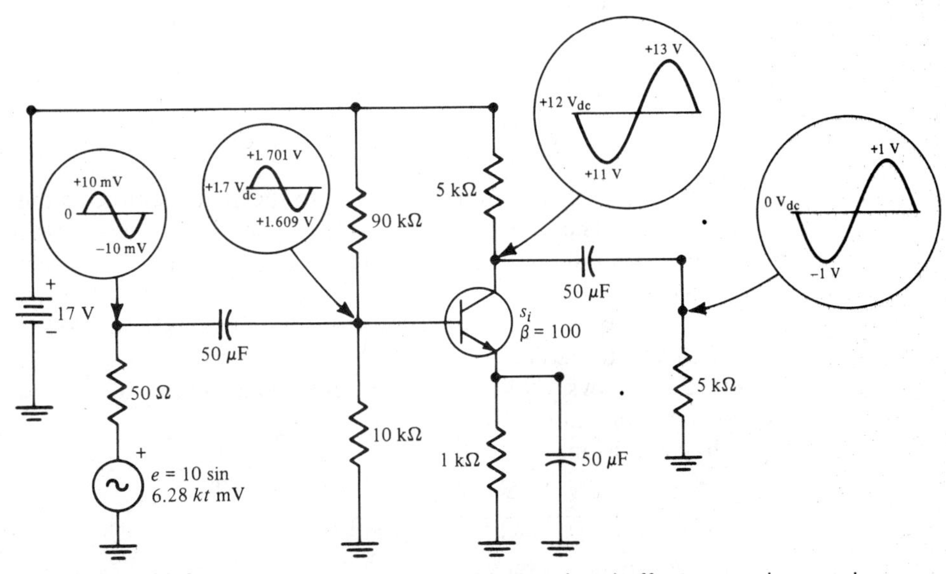

Figure 12.4 A transistor amplifier with the signal effects superimposed on the dc levels at different points in the circuit.

this text. The results, however, are shown, and they indicate that the voltage across each element in the circuit is a superposition of the effects of the dc source and the signal source.

The following example is simpler but illustrates the application of the Superposition Theorem to a circuit with two sources of different frequencies.

EXAMPLE 12.4

For the circuit shown in Figure 12.5a, use superposition to find the output voltage $\mathbf{V}_o$.

Solution

For dc, $f = 0$ and $\omega = 0$, forcing $X_C = \infty$ (open circuit).

Figure 12.5b shows the circuit with the battery as the only energy source and the capacitor replaced by its open-circuit equivalent at dc (zero frequency).

By voltage division:

$$\mathbf{V}_{o_{dc}} = 17\,\frac{10\text{ k}\Omega}{90\text{ k}\Omega + 10\text{ k}\Omega} = 17\,\frac{10\text{ k}\Omega}{100\text{ k}\Omega}$$
$$= 1.7\text{ V}$$

For $\omega = 6.28$ krad/s:

$$X_C = \frac{1}{\omega C} = \frac{1}{(6.28\text{ krad/s})(0.05\ \mu\text{F})}$$
$$= 3.185\text{ k}\Omega$$

$$\mathbf{Z}_C = -j3.185\text{ k}\Omega$$

$$\mathbf{E}_S = 1.0\angle 0^\circ\text{ V}$$

Figure 12.5c shows the circuit subjected to the signal source as the only energy supply. The capacitor is replaced by its impedance at the source frequency, and the effects of the battery are eliminated by replacing it with a short.

Figure 12.5d shows the circuit redrawn so that the 10-kΩ and 90-kΩ resistors are clearly in parallel.

Figure 12.5e shows a 9-kΩ equivalent box replacing the 10-kΩ and 90-kΩ parallel combination.

By voltage division:

$V_{oac} = .467 = Rms$

$$\mathbf{V}_{o_{ac}} = (1\angle 0^\circ)\,\frac{9\text{ k}\Omega}{10\text{ k}\Omega + 9\text{ k}\Omega - j3.185\text{ k}\Omega}$$
$$= (1\angle 0^\circ)(0.467\angle 9.5^\circ)$$
$$= 0.467\angle 9.5^\circ\text{ V}$$

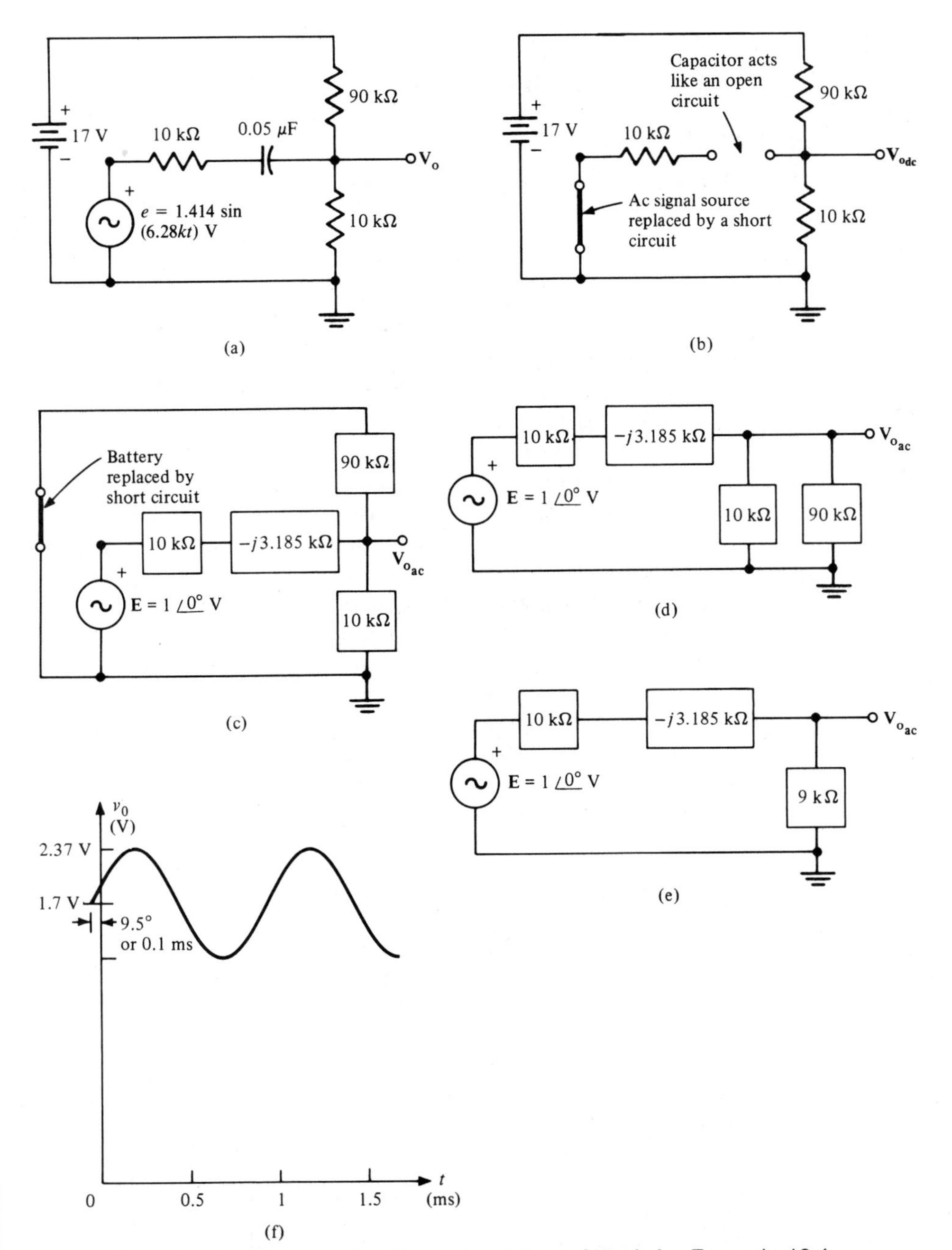

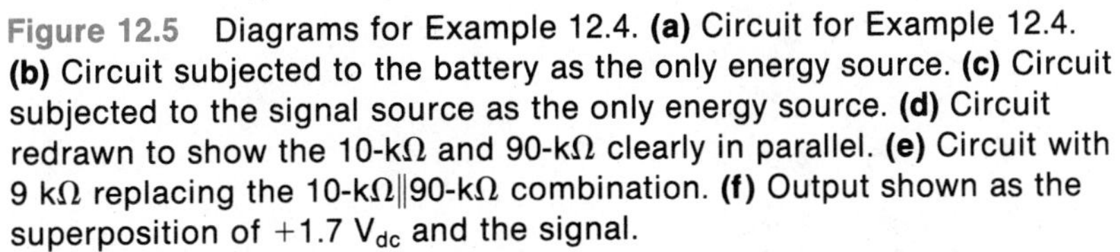

Figure 12.5 Diagrams for Example 12.4. **(a)** Circuit for Example 12.4. **(b)** Circuit subjected to the battery as the only energy source. **(c)** Circuit subjected to the signal source as the only energy source. **(d)** Circuit redrawn to show the 10-kΩ and 90-kΩ clearly in parallel. **(e)** Circuit with 9 kΩ replacing the 10-kΩ‖90-kΩ combination. **(f)** Output shown as the superposition of +1.7 V_{dc} and the signal.

$V_{O_{Pk}} = 0.66$ V

$v_{O_{ac}} = 0.66 \sin(6.28kt + 9.5°)$ V

By superposition:

$$\begin{aligned} v_O &= v_{O_{dc}} + v_{O_{ac}} \\ &= 1.7 + 0.66 \sin(6.28kt + 9.5°) \text{ V} \end{aligned}$$

Figure 12.5f shows a diagram of the output as the superposition of the dc and the signal effects.

12.4 TWO-PORT PARAMETERS

It is often convenient to use a standard model to characterize a four-terminal linear network. In a linear network an increase in input causes a corresponding increase in output. A **port** is defined as a pair of input or output terminals. Hence a four-terminal network is also referred to as a **two-port network.**

The generalized two-port network shown in Figure 12.6a is represented by a box with four terminals. One port, namely, port 1, is considered to be the input side, whereas port 2 is the output side. The currents $\mathbf{I}_1$ and $\mathbf{I}_2$ are shown to enter the box, and the voltages $\mathbf{V}_1$ and $\mathbf{V}_2$ are shown with the positive polarity on top and the negative on the bottom. Any two-port network can be described by a set of two linear equations. The coefficients of the variables in the unknowns, which can be either voltages or currents or both, are called *network parameters*. These parameters clearly define the network within the box.

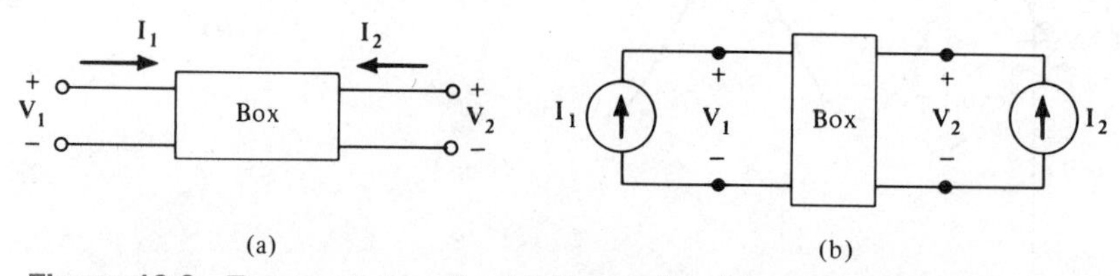

Figure 12.6 Two-port circuits. **(a)** Generalized two-port network. **(b)** Generalized two-port network with current sources at ports 1 and 2.

Connecting two current sources, $\mathbf{I}_1$ and $\mathbf{I}_2$, to port 1 and port 2, respectively, as shown in Figure 12.6b, results in voltage drops, $\mathbf{V}_1$ and $\mathbf{V}_2$, across the ports. Two voltage equations that describe the circuit are:

$$\mathbf{V}_1 = \mathbf{z}_{11}\mathbf{I}_1 + \mathbf{z}_{12}\mathbf{I}_2 \quad (12.1a)$$

$$\mathbf{V}_2 = \mathbf{z}_{21}\mathbf{I}_1 + \mathbf{z}_{22}\mathbf{I}_2 \quad (12.1b)$$

The coefficients are impedances and are therefore called **impedance parameters,** or just *z-parameters* with the units of ohms. The model of the network represented by Eqs. (12.1) can be synthesized by recalling loop equation analysis from Chapter 11. Hence the model shown in Figure 12.7 includes impedance $\mathbf{z}_{11}$ and a dependent voltage source $\mathbf{z}_{12}\mathbf{I}_2$ at the input port and an impedance $\mathbf{z}_{22}$ and a dependent voltage source $\mathbf{z}_{21}\mathbf{I}_1$ at the output port.

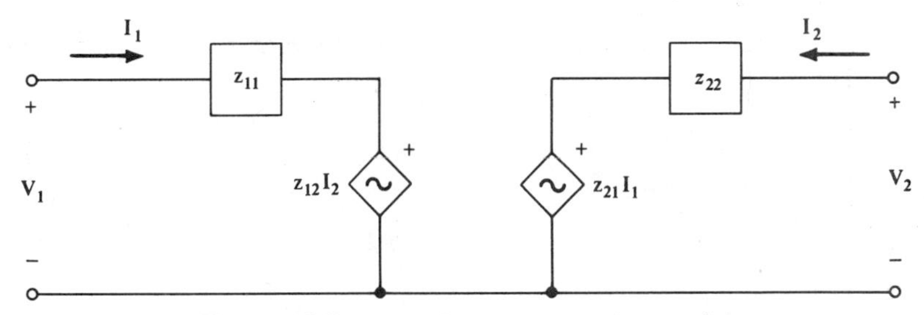

Figure 12.7 Impedance parameter model.

The impedance parameters of a two-port network are defined as follows. For $\mathbf{I}_2 = 0$, from Eq. (12.1a):

$$\mathbf{z}_{11} = \left.\frac{\mathbf{V}_1}{\mathbf{I}_1}\right|_{\mathbf{I}_2=0} \quad \text{input impedance} \tag{12.2}$$

The vertical line denotes the condition, namely, $\mathbf{I}_2 = 0$.
Similarly, from Eq. (12.1b):

$$\mathbf{z}_{21} = \left.\frac{\mathbf{V}_2}{\mathbf{I}_1}\right|_{\mathbf{I}_2=0} \quad \text{forward transfer impedance} \tag{12.3}$$

For $\mathbf{I}_1 = 0$, from Eq. (12.1a):

$$\mathbf{z}_{12} = \left.\frac{\mathbf{V}_1}{\mathbf{I}_2}\right|_{\mathbf{I}_1=0} \quad \text{reverse transfer impedance} \tag{12.4}$$

and from Eq. (12.1b):

$$\mathbf{z}_{22} = \left.\frac{\mathbf{V}_2}{\mathbf{I}_2}\right|_{\mathbf{I}_1=0} \quad \text{output impedance} \tag{12.5}$$

The term *transfer impedance* refers to an impedance which is the ratio between an output and an input quantity. Impedance parameters are also referred to as *open-circuit parameters* because the currents ($\mathbf{I}_1$ and $\mathbf{I}_2$) are set to zero, which in practice means to open-circuit the port.

EXAMPLE 12.5

For the circuit shown in Figure 12.8a, find the impedance parameters at 1 kHz.

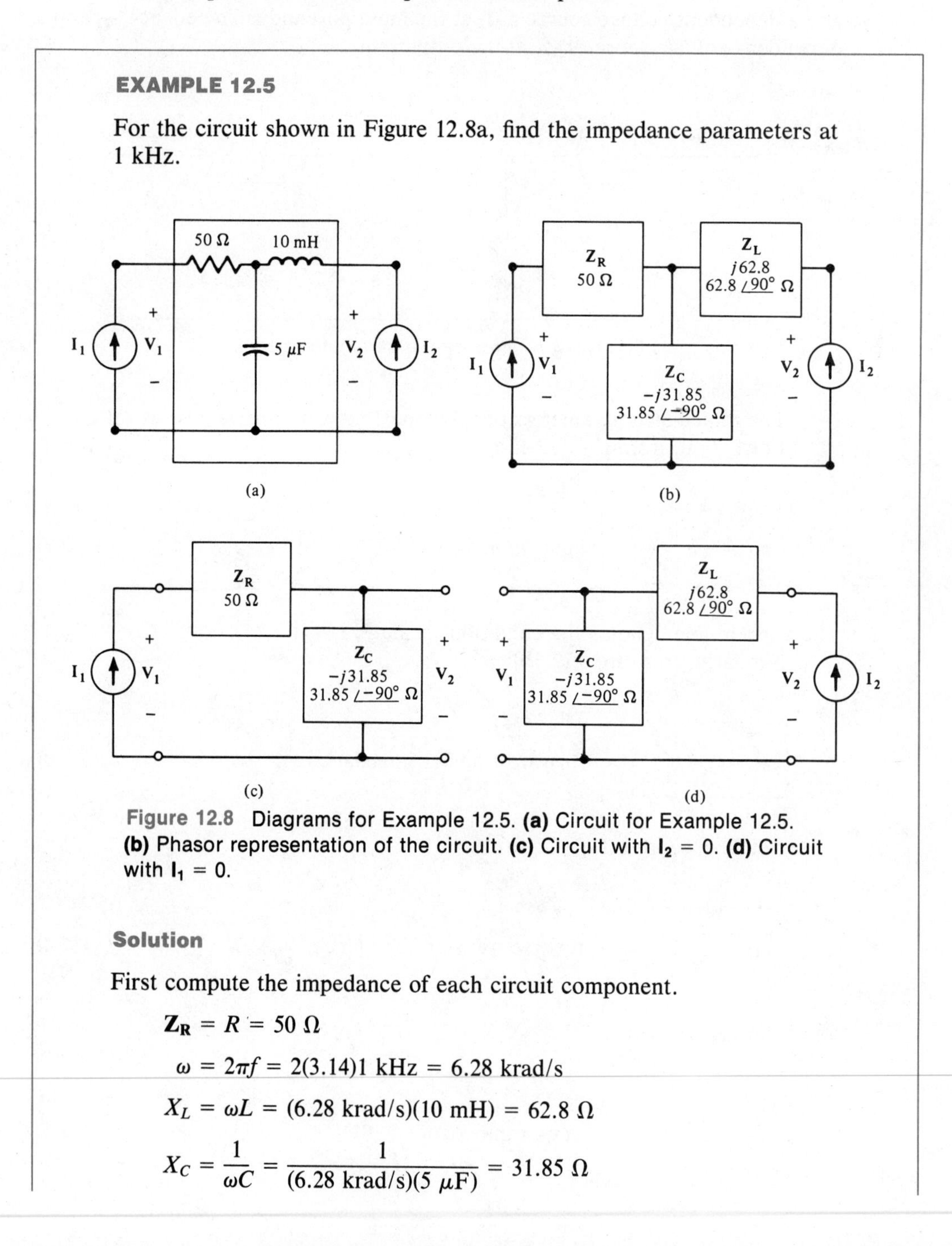

Figure 12.8 Diagrams for Example 12.5. **(a)** Circuit for Example 12.5. **(b)** Phasor representation of the circuit. **(c)** Circuit with $\mathbf{I}_2 = 0$. **(d)** Circuit with $\mathbf{I}_1 = 0$.

Solution

First compute the impedance of each circuit component.

$$\mathbf{Z_R} = R = 50\ \Omega$$

$$\omega = 2\pi f = 2(3.14)1\ \text{kHz} = 6.28\ \text{krad/s}$$

$$X_L = \omega L = (6.28\ \text{krad/s})(10\ \text{mH}) = 62.8\ \Omega$$

$$X_C = \frac{1}{\omega C} = \frac{1}{(6.28\ \text{krad/s})(5\ \mu\text{F})} = 31.85\ \Omega$$

$\mathbf{Z_L} = j62.8\ \Omega = 62.8\underline{/90°}\ \Omega$

$\mathbf{Z_C} = -j31.85\ \Omega = 31.85\underline{/-90°}\ \Omega$

The admittance of the components is:

$\mathbf{Y_R} = 20$ mS

$\mathbf{Y_C} = j31.4\ \text{mS} = 31.4\underline{/90°}\ \text{mS}$

$\mathbf{Y_L} = -j15.92\ \text{mS} = 15.92\underline{/-90°}\ \text{mS}$

Figure 12.8b shows the circuit represented in its phasor notation.

Set $\mathbf{I_2}$ to zero by leaving the output port open. Figure 12.8c shows the circuit with this condition. The following may be computed:

$$\mathbf{z_{11}} = \frac{\mathbf{V_1}}{\mathbf{I_1}} = \mathbf{Z_R} + \mathbf{Z_C}$$

$$= 50 - j62.8\ \Omega \qquad \text{rectangular form}$$

$$= 80.27\underline{/-51.47°}\ \Omega \qquad \text{polar form}$$

$$\mathbf{z_{21}} = \frac{\mathbf{V_2}}{\mathbf{I_1}}$$

$$\mathbf{V_2} = \mathbf{I_1Z_C}$$

$$\frac{\mathbf{V_2}}{\mathbf{I_1}} = \mathbf{Z_C}$$

$$\mathbf{z_{21}} = \mathbf{Z_C} = -j31.85\ \Omega$$

Set $\mathbf{I_1}$ to zero by leaving the input port open as shown in Figure 12.8d. The following may be computed:

$$\mathbf{I_2Z_C} = \mathbf{V_1}$$

$$\frac{\mathbf{V_1}}{\mathbf{I_2}} = \mathbf{Z_C}$$

$$\mathbf{z_{12}} = \mathbf{Z_C} = -j31.85\ \Omega$$

$$\mathbf{V_2} = \mathbf{I_2}(\mathbf{Z_L} + \mathbf{Z_C})$$

$$\frac{\mathbf{V_2}}{\mathbf{I_2}} = \mathbf{Z_L} + \mathbf{Z_C}$$

$$\mathbf{z_{22}} = \mathbf{Z_L} + \mathbf{Z_C}$$

$$= j62.8 - j31.85\ \Omega$$

$$= j30.95\ \Omega$$

The impedance parameter equations are:

$$(50 - j62.8)\mathbf{I_1} + (0 - j31.85)\mathbf{I_2} = \mathbf{V_1}$$

$$(0 - j31.85)\mathbf{I_1} + (0 + j30.95)\mathbf{I_2} = \mathbf{V_2}$$

Note that these resemble loop equations.

Two voltage sources may be connected to the generalized two-port circuit of Figure 12.6 causing the two terminal currents. The two resulting equations in which the terminal voltages are the independent variables and the currents are the dependent variables have admittances as the coefficients:

$$\mathbf{I}_1 = \mathbf{y}_{11}\mathbf{V}_1 + \mathbf{y}_{12}\mathbf{V}_2 \tag{12.6a}$$

$$\mathbf{I}_2 = \mathbf{y}_{21}\mathbf{V}_1 + \mathbf{y}_{22}\mathbf{V}_2 \tag{12.6b}$$

The admittances are called the *y-parameters,* or the **admittance parameters.** The unit for the *y*-parameters is siemens (S). The *y*-parameters are defined by

$$\mathbf{y}_{11} = \left.\frac{\mathbf{I}_1}{\mathbf{V}_1}\right|_{\mathbf{V}_2=0} \quad \text{input admittance} \tag{12.7}$$

$$\mathbf{y}_{12} = \left.\frac{\mathbf{I}_1}{\mathbf{V}_2}\right|_{\mathbf{V}_1=0} \quad \text{reverse transfer admittance (transconductance)} \tag{12.8}$$

$$\mathbf{y}_{21} = \left.\frac{\mathbf{I}_2}{\mathbf{V}_1}\right|_{\mathbf{V}_2=0} \quad \text{forward transfer admittance (transconductance)} \tag{12.9}$$

$$\mathbf{y}_{22} = \left.\frac{\mathbf{I}_2}{\mathbf{V}_2}\right|_{\mathbf{V}_1=0} \quad \text{output admittance} \tag{12.10}$$

The condition imposed on each of the preceding parameter equations is either $\mathbf{V}_1 = 0$ or $\mathbf{V}_2 = 0$. This means that either the input or the output must be shorted. Therefore, the *y*-parameters are also referred to as the *short-circuit parameters*. A *y*-parameter model based on Eqs. (12.6) is shown in Figure 12.9.

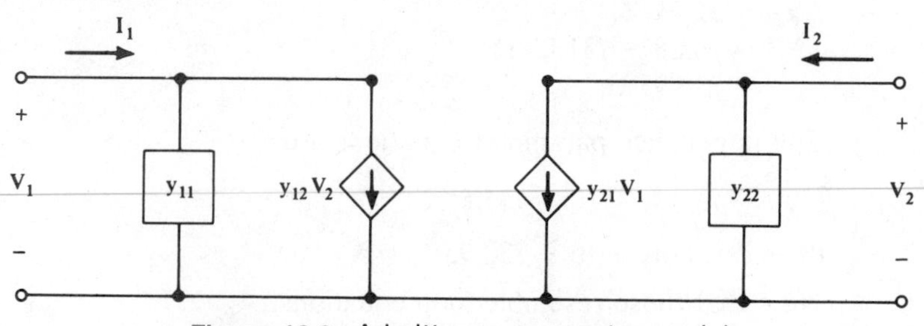

Figure 12.9 Admittance parameter model.

EXAMPLE 12.6

Find the y-parameters at 1 kHz for the circuit shown in Figure 12.10a.

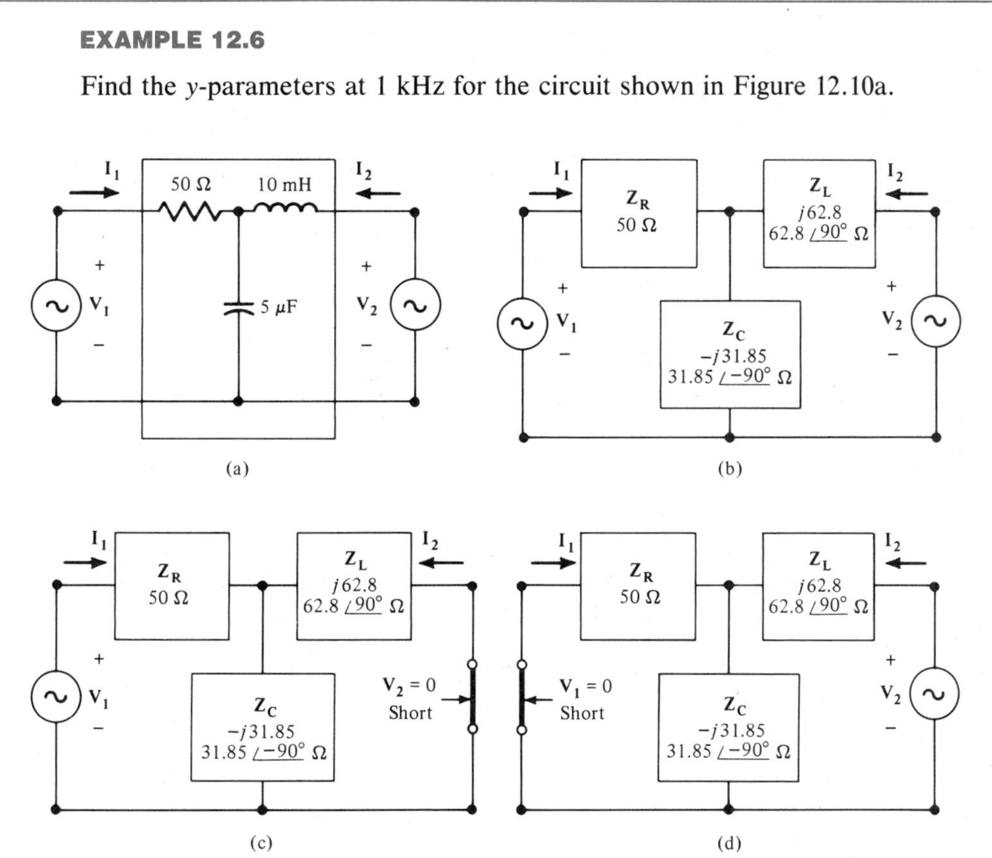

Figure 12.10 Diagrams for Example 12.6. **(a)** Circuit for Example 12.6. **(b)** Circuit in phasor notation. **(c)** Circuit with $\mathbf{V}_2 = 0$. **(d)** Circuit with $\mathbf{V}_1 = 0$.

Solution

Figure 12.10b shows the circuit in its phasor notation. Figure 12.10c shows the circuit with $\mathbf{V}_2 = 0$ (short circuit). From this circuit, the following may be observed:

$$\begin{aligned}\mathbf{Y}_1 &= \mathbf{Y}_C + \mathbf{Y}_L \\ &= j31.4 \text{ mS} - j15.92 \text{ mS} \\ &= j15.48 \text{ mS}\end{aligned}$$

$$\begin{aligned}\mathbf{Z}_1 &= \frac{1}{\mathbf{Y}_1} = \frac{1}{j15.48 \text{ mS}} \\ &= -j64.6\ \Omega\end{aligned}$$

$$\begin{aligned}\mathbf{Z}_T &= \mathbf{Z}_R + \mathbf{Z}_1 \\ &= 50 - j64.6\ \Omega \\ &= 81.7\angle 52.26^\circ\ \Omega\end{aligned}$$

$$\mathbf{Y_T} = \frac{1}{\mathbf{Z_T}} = \frac{1}{81.7\angle 52.26^\circ}\ \text{mS}$$

$$\mathbf{y_{11}} = \mathbf{Y_T} = 12.24\angle{-52.26^\circ}\ \text{mS}$$

$$\mathbf{I_2} = -\mathbf{I_1}\frac{\mathbf{Z_C}}{\mathbf{Z_L} + \mathbf{Z_C}}$$

$$= -\frac{\mathbf{V_1}}{\mathbf{Z_T}}\frac{\mathbf{Z_C}}{\mathbf{Z_L} + \mathbf{Z_C}}$$

$$\frac{\mathbf{I_2}}{\mathbf{V_1}} = -\frac{\mathbf{Z_C}}{\mathbf{Z_T}(\mathbf{Z_L} + \mathbf{Z_C})}$$

$$= -\frac{31.85\angle{-90^\circ}}{(81.7\angle 52.26^\circ)(30.95\angle 90^\circ)}$$

$$\mathbf{y_{21}} = \frac{\mathbf{I_2}}{\mathbf{V_1}} = -12.6\angle{-232.26^\circ}\ \text{mS}$$

Figure 12.10d shows the circuit with $\mathbf{V_1} = 0$ (replaced by a short circuit). The following may be observed:

$$\begin{aligned}\mathbf{Y_2} &= \mathbf{y_R} + \mathbf{Y_C}\\ &= 20 + j31.4\ \text{mS}\\ &= 37.23\angle 57.5^\circ\ \text{mS}\end{aligned}$$

$$\begin{aligned}\mathbf{Z_2} &= \frac{1}{\mathbf{Y_2}} = \frac{1}{37.23\angle 57.5^\circ\ \text{mS}}\\ &= 26.86\angle{-57.5^\circ}\ \Omega\\ &= 14.43 - j22.66\ \Omega\end{aligned}$$

$$\begin{aligned}\mathbf{Z_T} &= \mathbf{Z_L} + \mathbf{Z_2}\\ &= j62.8 + (14.43 - j22.66)\ \Omega\\ &= 14.43 + j40.14\ \Omega\\ &= 42.65\angle 70.23^\circ\ \Omega\end{aligned}$$

$$\begin{aligned}\mathbf{y_{22}} &= \frac{1}{\mathbf{Z_T}} = \frac{1}{42.65\angle 70.23^\circ}\\ &= 23.45\angle{-70.23^\circ}\ \text{mS}\\ &= 7.93 - j22.07\ \text{mS}\end{aligned}$$

$$\mathbf{I_1} = -\mathbf{I_2}\frac{\mathbf{Z_C}}{\mathbf{Z_R} + \mathbf{Z_C}}$$

$$\mathbf{I_2} = \mathbf{V_2}\mathbf{y_{22}}$$

$$\mathbf{I_1} = -\mathbf{V_2}\mathbf{y_{22}}\frac{\mathbf{Z_C}}{\mathbf{Z_R} + \mathbf{Z_C}}$$

$$\mathbf{y}_{12} = -\frac{\mathbf{I}_1}{\mathbf{V}_2} = -\frac{\mathbf{y}_{22}\mathbf{Z}_C}{\mathbf{Z}_R + \mathbf{Z}_C}$$

$$= -\frac{(23.45\underline{/-70.23^\circ})(31.85\underline{/-90^\circ})}{(50 - j31.85)}$$

$$= -7.71 - j9.96 \text{ mS}$$

$$= 12.6\underline{/-127.73^\circ} \text{ mS}$$

The y-parameter equations are:

$$(12.24\underline{/-52.26^\circ} \text{ mS})\mathbf{V}_1 + (12.6\underline{/-127.73^\circ} \text{ mS})\mathbf{V}_2 = \mathbf{I}_1$$

$$(-12.6\underline{/-232.26^\circ} \text{ mS})\mathbf{V}_1 + (23.45\underline{/-70.23^\circ} \text{ mS})\mathbf{V}_2 = \mathbf{I}_2$$

Note that these resemble node equations.

The y-parameter model is used to represent the ac characteristics of the junction field effect transistor (JFET). The transistor is a solid slab of material and can be electrically analyzed only as part of a circuit. The symbol for the JFET is shown in Figure 12.11a and its y-parameter model is given in Figure 12.11b. A practical reduced model is shown in Figure 12.11b. In a practical JFET, y_{11} and y_{12} are approximately zero. Parameter y_{21} is labeled y_{fs} for forward transconductance, and y_{22} is labeled y_d for drain admittance. Notice that the transistor is a three-terminal device, but it is modeled as a two-port network because one of the terminals is always common to both the input and the output.

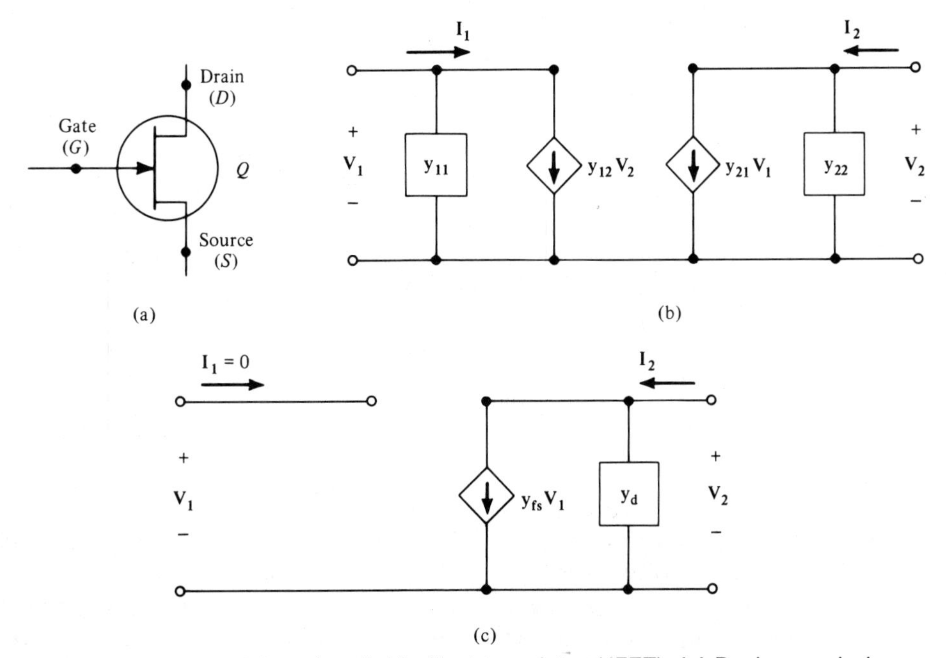

Figure 12.11 A junction field effect transistor (JFET). **(a)** Device symbol. **(b)** y-parameter model.

Hybrid parameters (h) are two-port parameters that are very useful in modeling the bipolar junction transistor (BJT). The general set of h-parameter equations for a two-port network consists of one voltage equation and one current equation:

$$\mathbf{V}_1 = \mathbf{h}_{11}\mathbf{I}_1 + \mathbf{h}_{12}\mathbf{V}_2 \quad (12.11a)$$

$$\mathbf{I}_2 = \mathbf{h}_{21}\mathbf{I}_1 + \mathbf{h}_{22}\mathbf{V}_2 \quad (12.11b)$$

The individual parameters are defined from Eqs. (12.11):

$$\mathbf{h}_{11} = \left.\frac{\mathbf{V}_1}{\mathbf{I}_1}\right|_{\mathbf{V}_2=0} \quad \text{input impedance} \quad (12.12)$$

Parameter $\mathbf{h}_{11}$ is also labeled $\mathbf{h}_i$. The subscript i stands for input.

$$\mathbf{h}_{12} = \left.\frac{\mathbf{V}_1}{\mathbf{V}_2}\right|_{\mathbf{I}_1=0} \quad \text{reverse voltage ratio} \quad (12.13)$$

$\mathbf{h}_{12}$ is also labeled as $\mathbf{h}_r$. The r stands for reverse.

$$\mathbf{h}_{21} = \left.\frac{\mathbf{I}_2}{\mathbf{I}_1}\right|_{\mathbf{V}_2=0} \quad \text{forward current ratio} \quad (12.14)$$

$\mathbf{h}_{21}$ is also labeled $\mathbf{h}_f$. The f stands for forward.

$$\mathbf{h}_{22} = \left.\frac{\mathbf{I}_2}{\mathbf{V}_2}\right|_{\mathbf{I}_1=0} \quad \text{output admittance} \quad (12.15)$$

$\mathbf{h}_{22}$ is also labeled $\mathbf{h}_o$. The o stands for output.

As the name hybrid suggests, these parameters are a mixture of impedance, admittance, current ratios, and voltage ratios. Parameters $\mathbf{h}_{11}$ and $\mathbf{h}_{21}$ are called *short-circuit parameters* because both are obtained with port 2 shorted ($\mathbf{V}_2 = 0$). Parameters $\mathbf{h}_{12}$ and $\mathbf{h}_{22}$ are called *open-circuit parameters* because they are obtained when port 1 is open ($\mathbf{I}_1 = 0$).

The h-parameter two-port circuit model is shown in Figure 12.12.

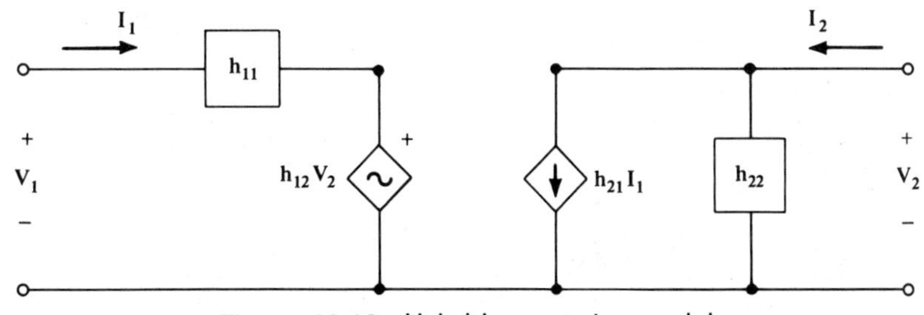

Figure 12.12 Hybrid parameter model.

EXAMPLE 12.7

For the circuit shown in Figure 12.13a, find the hybrid parameters at 1 kHz.

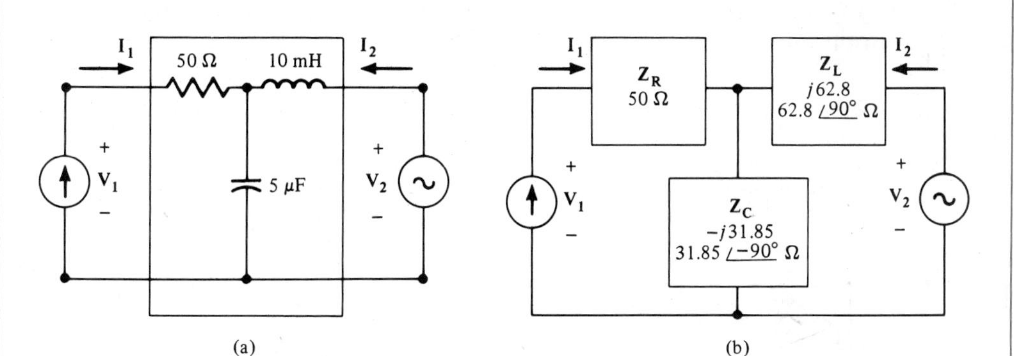

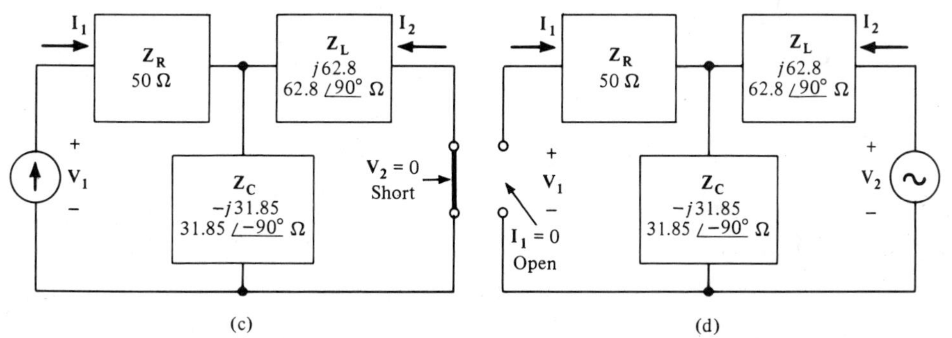

Figure 12.13 Diagrams for Example 12.7. **(a)** Circuit for Example 12.7. **(b)** Circuit in phasor notation. **(c)** Circuit with $\mathbf{V}_2 = 0$. **(d)** Circuit with $\mathbf{I}_1 = 0$.

Solution

Figure 12.13b shows the circuit in phasor notation with $\mathbf{I}_1$ and $\mathbf{V}_2$ as the independent sources to comply with the *h*-parameter specification.

Figure 12.13c shows the circuit with the $\mathbf{V}_2 = 0$ condition. From this circuit, it can be observed that:

$$\mathbf{h}_{11} = \frac{1}{\mathbf{y}_{11}} = 81.7\,\underline{/52.56^\circ}\ \Omega \qquad \text{from Example 12.6}$$

$$\mathbf{I}_2 = -\,\mathbf{I}_1 \frac{\mathbf{Z}_C}{\mathbf{Z}_L + \mathbf{Z}_C}$$

$$\mathbf{h}_{21} = \frac{\mathbf{I}_2}{\mathbf{I}_1} = -\frac{\mathbf{Z}_C}{\mathbf{Z}_L + \mathbf{Z}_C}$$

$$= \frac{-31.85\,\underline{/-90^\circ}}{30.95\,\underline{/90^\circ}}$$

$$= -1.03\,\underline{/-180^\circ}$$

$$= 1.03$$

Figure 12.13d shows the circuit with the $\mathbf{I}_1 = 0$ condition imposed. From this circuit, the following can be observed:

$$\mathbf{h}_{22} = \frac{1}{\mathbf{z}_{22}} = \frac{1}{30.95\,\underline{/90^\circ}\ \Omega} = 32.31\,\underline{/-90^\circ}\ \text{mS} \qquad \text{from Example 12.5}$$

$$\mathbf{V}_1 = \mathbf{V}_2 \frac{\mathbf{Z}_C}{\mathbf{Z}_C + \mathbf{Z}_L}$$

$$\mathbf{h}_{12} = \frac{\mathbf{V}_1}{\mathbf{V}_2} = \frac{\mathbf{Z}_C}{\mathbf{Z}_C + \mathbf{Z}_L}$$

$$= \frac{31.85\,\underline{/90^\circ}}{30.95\,\underline{/90^\circ}}$$

$$= 1.03$$

The h-parameter equations are:

$$(81.7\,\underline{/52.56^\circ}\ \Omega)\mathbf{I}_1 + (1.03)\mathbf{V}_2 = \mathbf{V}_1$$

$$(1.03)\mathbf{I}_1 + (32.31\,\underline{/-90^\circ}\ \text{mS})\mathbf{V}_2 = \mathbf{I}_2$$

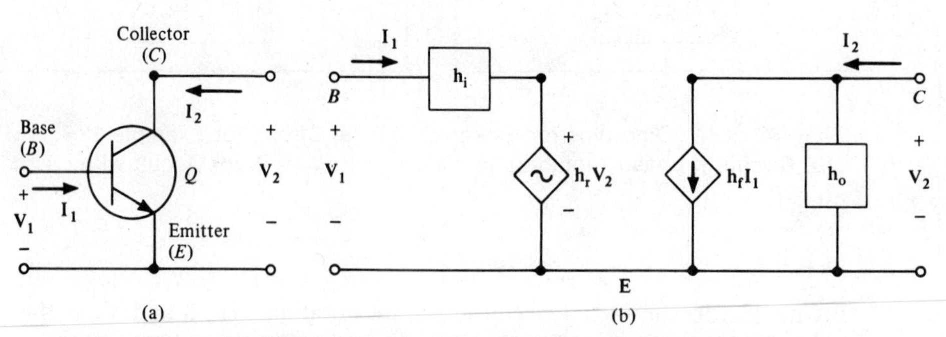

Figure 12.14 An *NPN* bipolar junction transistor (BJT). **(a)** Device symbol. **(b)** h-parameter model.

The symbol of an *NPN* bipolar junction transistor and its hybrid parameter model are shown in Figures 12.14a and 12.14b. The BJT, like the JFET, is also a three-terminal device. The emitter is shown as the common terminal for both port 1 and port 2. Figure 12.14b shows its *h*-parameter model. The next example illustrates the use of a hybrid model to solve a transistor circuit.

EXAMPLE 12.8

A BJT is driven by a signal source $\mathbf{V_B}$, connected between the base and the emitter. A load resistor is connected between the collector and the emitter. The circuit using the hybrid model of the BJT is shown in Figure 12.15a. Calculate the currents $\mathbf{I_1}$ and $\mathbf{I_2}$ and the voltage $\mathbf{V_L}$.

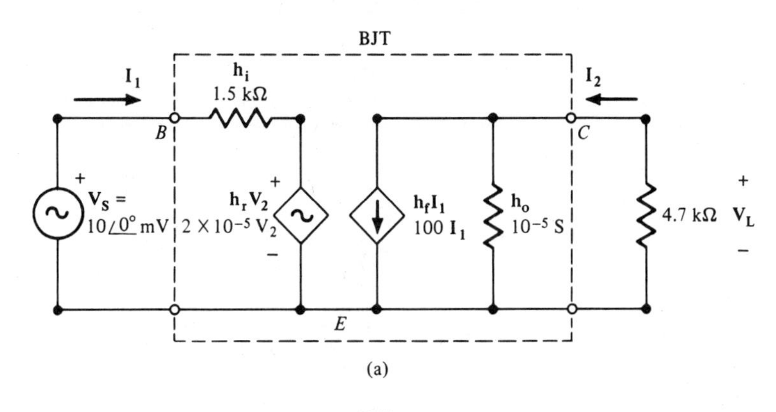

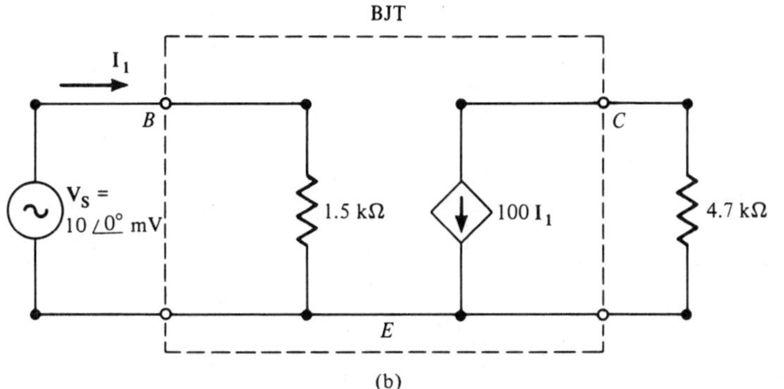

Figure 12.15 Diagrams for Example 12.8. **(a)** Hybrid model with signal source and load connected. **(b)** Simplified **h**-parameter model.

Solution

Since h_r is very small, it may be neglected. Similarly, $h_o = 10^{-5}$ S yields a 100-kΩ impedance and when combined in parallel with the 4.7-kΩ load, it does not significantly influence the result. The simplified circuit with h_r and h_o neglected is shown in Figure 12.15b.

Here it can be observed that:

$$\mathbf{V}_2 = -100\mathbf{I}_1(4.7\ \text{k}\Omega)$$

$$(1.5\ \text{k}\Omega)\mathbf{I}_1 = 10\underline{/0^\circ}\ \text{mV}$$

$$\mathbf{I}_1 = \frac{10\ \text{mV}}{1.5\ \text{k}\Omega} = 6.67\underline{/0^\circ}\ \mu\text{A}$$

$$\mathbf{V}_2 = -100(6.67\ \mu\text{A})(4.7\ \text{k}\Omega) = -3.13\ \text{V}$$

$$\mathbf{I}_2 = 100\mathbf{I}_1 = 100(6.67\ \mu\text{A}) = 667\ \mu\text{A}$$

Other types of two-port parameters are also used. The *z*-, *y*-, and *h*-parameters discussed are perhaps the most common ones. Table 12.1 lists conversions between these three types of parameters.

TABLE 12.1
Two-port parameter conversion table.

From ==>	z		y		h	
To z	z_{11}	z_{12}	$\frac{y_{22}}{D_y}$	$\frac{-y_{12}}{D_y}$	$\frac{D_h}{h_{22}}$	$\frac{h_{12}}{h_{22}}$
	z_{21}	z_{22}	$\frac{-y_{21}}{D_y}$	$\frac{y_{11}}{D_y}$	$\frac{-h_{21}}{h_{22}}$	$\frac{1}{h_{22}}$
y	$\frac{z_{22}}{D_z}$	$\frac{-z_{12}}{D_z}$	y_{11}	y_{12}	$\frac{1}{h_{11}}$	$\frac{-h_{12}}{h_{11}}$
	$\frac{-z_{21}}{D_z}$	$\frac{z_{11}}{D_z}$	y_{21}	y_{22}	$\frac{h_{21}}{h_{11}}$	$\frac{D_h}{h_{11}}$
h	$\frac{D_z}{z_{22}}$	$\frac{z_{12}}{z_{22}}$	$\frac{1}{y_{11}}$	$\frac{-y_{12}}{y_{11}}$	h_{11}	h_{12}
	$\frac{-z_{21}}{z_{22}}$	$\frac{1}{z_{22}}$	$\frac{y_{21}}{y_{11}}$	$\frac{D_y}{y_{11}}$	h_{21}	h_{22}

12.5 CONVERSION BETWEEN PARAMETERS

Conversion between two-port parameters may be easily performed by using Table 12.1 where:

$$D_y = y_{11}y_{22} - y_{12}y_{21}$$

$$D_h = h_{11}h_{22} - h_{12}h_{21}$$

$$D_z = z_{11}z_{22} - z_{12}z_{21}$$

EXAMPLE 12.9

Convert the parameters of the circuit in Example 12.8 to **z**-parameters and draw the **z**-parameter model of the transistor.

Solution

Using Table 12.1:

$$\begin{aligned} \mathbf{D_h} &= \mathbf{h_{11}h_{22}} - \mathbf{h_{12}h_{21}} \\ &= (1.5\ \text{k}\Omega)(10^{-5}) - (100)(2 \times 10^{-5}) \\ &= 0.015 - 0.002 \\ &= 0.013 \end{aligned}$$

$$\begin{aligned} \mathbf{z_{11}} &= \frac{\mathbf{D_h}}{\mathbf{h_{22}}} = \frac{0.013}{10^{-5}} \\ &= 1.3\ \text{k}\Omega \end{aligned}$$

$$\begin{aligned} \mathbf{z_{12}} &= \frac{\mathbf{h_{12}}}{\mathbf{h_{22}}} = \frac{100}{10^{-5}} \\ &= 10\ \text{m} \end{aligned}$$

$$\begin{aligned} \mathbf{z_{21}} &= \frac{\mathbf{h_{21}}}{\mathbf{h_{22}}} = -\frac{2 \times 10^{-5}}{10^{-5}} \\ &= -2\ \Omega \end{aligned}$$

$$\begin{aligned} \mathbf{z_{22}} &= \frac{1}{\mathbf{h_{22}}} = \frac{1}{10^{-5}} \\ &= 100\ \text{k}\Omega \end{aligned}$$

The **z**-parameter model of the transistor is shown in Figure 12.16.

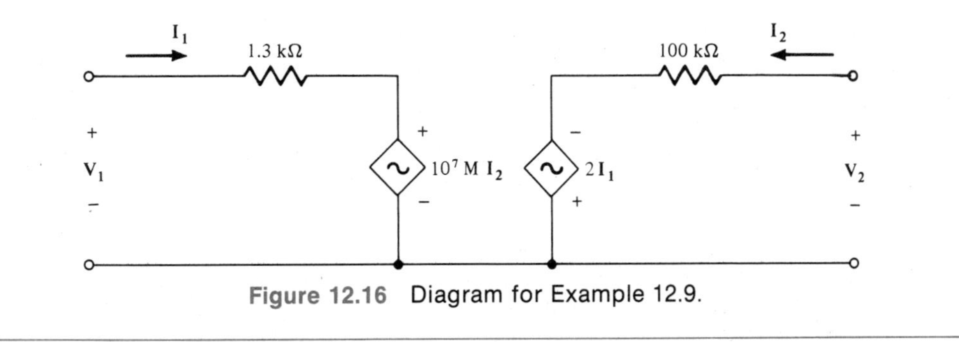

Figure 12.16 Diagram for Example 12.9.

SUMMARY

In this chapter, Thevenin's, Norton's, and the Superposition theorems were examined as they apply to ac circuits. The Superposition Theorem is found to be the only tool available to analyze circuits with sources of different frequencies.

Two-port network parameters were considered in the form of **z**-, **y**-, and **h**-parameters. Examples were given and their application to electronic device modeling was introduced. A conversion table among impedance, admittance, and hybrid parameters was presented.

SUMMARY OF IMPORTANT EQUATIONS

IMPEDANCE PARAMETERS

$$\mathbf{V}_1 = \mathbf{z}_{11}\mathbf{I}_1 + \mathbf{z}_{12}\mathbf{I}_2 \tag{12.1a}$$

$$\mathbf{V}_2 = \mathbf{z}_{21}\mathbf{I}_1 + \mathbf{z}_{22}\mathbf{I}_2 \tag{12.1b}$$

$$\mathbf{z}_{11} = \left.\frac{\mathbf{V}_1}{\mathbf{I}_1}\right|_{\mathbf{I}_2=0} \quad \text{input impedance} \tag{12.2}$$

$$\mathbf{z}_{21} = \left.\frac{\mathbf{V}_2}{\mathbf{I}_1}\right|_{\mathbf{I}_2=0} \quad \text{forward transfer impedance} \tag{12.3}$$

$$\mathbf{z}_{12} = \left.\frac{\mathbf{V}_1}{\mathbf{I}_2}\right|_{\mathbf{I}_1=0} \quad \text{reverse transfer impedance} \tag{12.4}$$

$$\mathbf{z}_{22} = \left.\frac{\mathbf{V}_2}{\mathbf{I}_2}\right|_{\mathbf{I}_1=0} \quad \text{output impedance} \tag{12.5}$$

ADMITTANCE PARAMETERS

$$\mathbf{I}_1 = \mathbf{y}_{11}\mathbf{V}_1 + \mathbf{y}_{12}\mathbf{V}_2 \tag{12.6a}$$

$$\mathbf{I}_2 = \mathbf{y}_{21}\mathbf{V}_1 + \mathbf{y}_{22}\mathbf{V}_2 \tag{12.6b}$$

$$\mathbf{y}_{11} = \left.\frac{\mathbf{I}_1}{\mathbf{V}_1}\right|_{\mathbf{V}_2=0} \quad \text{input admittance} \tag{12.7}$$

$$\mathbf{y}_{12} = \left.\frac{\mathbf{I}_1}{\mathbf{V}_2}\right|_{\mathbf{V}_1=0} \quad \text{reverse transfer admittance (transconductance)} \tag{12.8}$$

$$\mathbf{y}_{21} = \left.\frac{\mathbf{I}_2}{\mathbf{V}_1}\right|_{\mathbf{V}_2=0} \quad \text{forward transfer admittance (transconductance)} \tag{12.9}$$

$$\mathbf{y}_{22} = \left.\frac{\mathbf{I}_2}{\mathbf{V}_2}\right|_{\mathbf{V}_1=0} \quad \text{output admittance} \tag{12.10}$$

HYBRID PARAMETERS

$$\mathbf{V_1} = \mathbf{h_{11}I_1} + \mathbf{h_{12}V_2} \qquad (12.11a)$$

$$\mathbf{I_2} = \mathbf{h_{21}I_1} + \mathbf{h_{22}V_2} \qquad (12.11b)$$

$$\mathbf{h_{11}} = \left.\frac{\mathbf{V_1}}{\mathbf{I_1}}\right|_{V_2=0} \quad \text{input impedance} \qquad (12.12)$$

$$\mathbf{h_{12}} = \left.\frac{\mathbf{V_1}}{\mathbf{V_2}}\right|_{I_1=0} \quad \text{reverse voltage ratio} \qquad (12.13)$$

$$\mathbf{h_{21}} = \left.\frac{\mathbf{I_2}}{\mathbf{I_1}}\right|_{V_2=0} \quad \text{forward current ratio} \qquad (12.14)$$

$$\mathbf{h_{22}} = \left.\frac{\mathbf{I_2}}{\mathbf{V_2}}\right|_{I_1=0} \quad \text{output admittance} \qquad (12.15)$$

REVIEW QUESTIONS

12.1 Describe the process of obtaining Thevenin's equivalent circuit.

12.2 Describe the process of obtaining Norton's equivalent circuit.

12.3 Which method of analysis allows the energy sources to be of different frequencies?

12.4 Define the term electrical port.

12.5 What are impedance parameters?

12.6 What are admittance parameters?

12.7 What are hybrid parameters?

12.8 Define a two-port network.

12.9 Why are **z**-parameters called open-circuit parameters?

12.10 Why are **y**-parameters called short-circuit parameters?

12.11 Why are **h**-parameters named hybrid?

PROBLEMS

12.1 Find the Thevenin equivalent for the circuit shown in Figure 12.17.

12.2 Find the Norton equivalent for the circuit shown in Figure 12.17.
(ans. $12\underline{/0^\circ}$ mA, $3.53\underline{/13.5^\circ}$ kΩ)

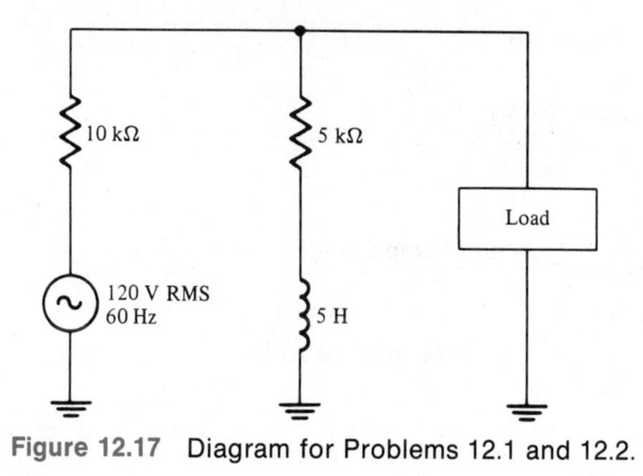

Figure 12.17 Diagram for Problems 12.1 and 12.2.

12.3 Find the Thevenin equivalent for the circuit shown in Figure 12.18.

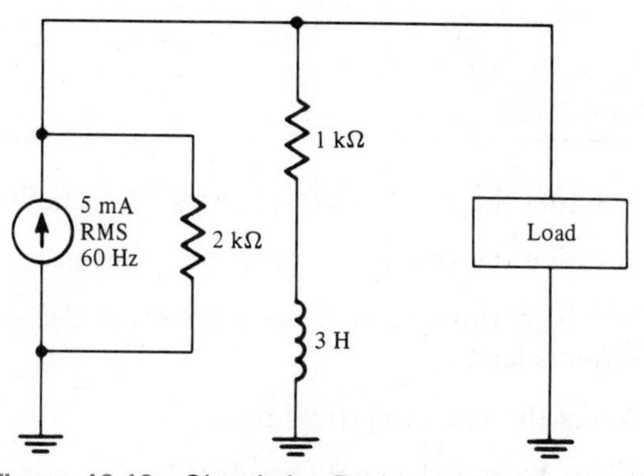

Figure 12.18 Circuit for Problems 12.3 and 12.4.

12.4 Find the Norton equivalent for the circuit shown in Figure 12.18. (ans. $5\angle 0°$ mA, $0.94\angle 27.86°$ kΩ)

12.5 Find the Thevenin equivalent for the circuit shown in Figure 12.19.

12.6 Find the Norton equivalent for the circuit shown in Figure 12.19. (ans. $17.65\angle 0°$ mA, $2.76\angle -24.08°$ kΩ)

12.7 Find the Thevenin equivalent for the circuit shown in Figure 12.20.

12.8 Find the Norton equivalent for the circuit shown in Figure 12.20. (ans. $20\angle 0°$ mA, $7.11\angle -28.43°$ kΩ)

12.9 For the resistor R_L in the circuit shown in Figure 12.21:

a. Find Thevenin's voltage

b. Find Thevenin's impedance

c. Draw Thevenin's equivalent

d. Use the equivalent circuit to find the current through R_L

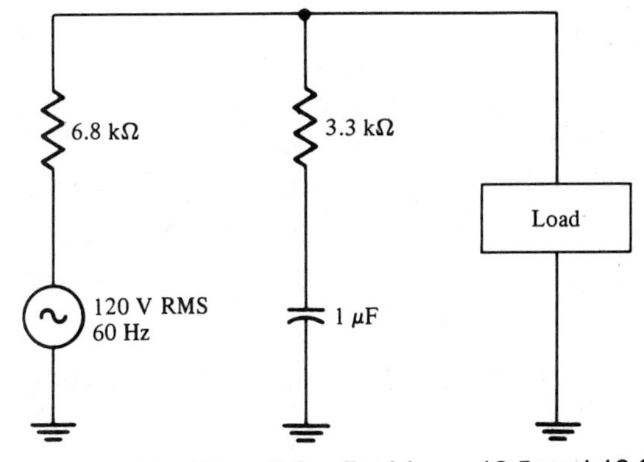

Figure 12.19 Circuit for Problems 12.5 and 12.6.

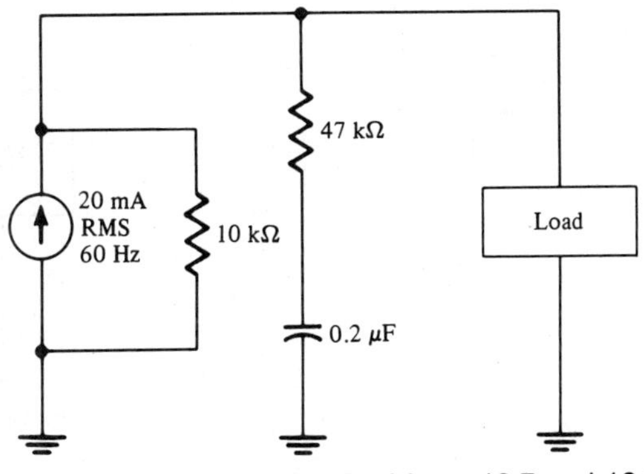

Figure 12.20 Circuit for Problems 12.7 and 12.8.

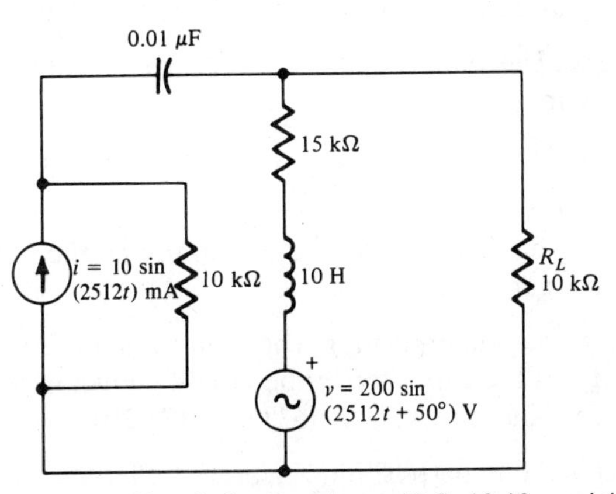

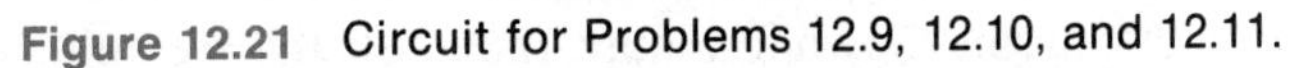
Figure 12.21 Circuit for Problems 12.9, 12.10, and 12.11.

12.10 For the resistor R_L in the circuit shown in Figure 12.21:

a. Find Norton's current

b. Find Norton's impedance

c. Draw Norton's equivalent

d. Use the equivalent circuit to find the current through R_L

(ans. $5.27\,\underline{/9.84°}$ mA, $41.42\,\underline{/13.7°}$ kΩ, $4.26\,\underline{/12.49°}$ mA)

12.11 For the circuit shown in Figure 12.21 use the Superposition Theorem to find the current through R_L.

12.12 Use Thevenin's Theorem to find the current through $\mathbf{Z_L}$ in the circuit shown in Figure 12.22. (ans. $0.43\,\underline{/58.17°}$ mA)

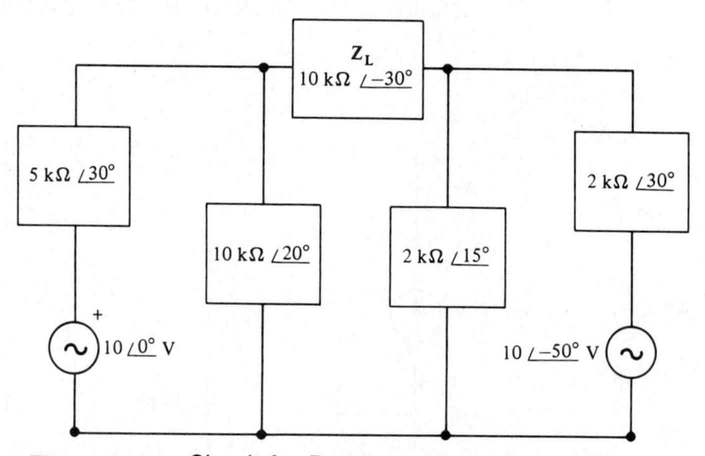

Figure 12.22 Circuit for Problems 12.12, 12.13, and 12.14.

12.13 Use Norton's Theorem to find the current through $\mathbf{Z_L}$ in the circuit of Figure 12.22.

12.14 Use the Superposition Theorem to find the current through $\mathbf{Z_L}$ for the circuit in Figure 12.22. (ans. $0.43\,\underline{/58.17°}$ mA)

12.15 Use Thevenin's Theorem to find the voltage across $\mathbf{Z_L}$ in the circuit of Figure 12.23.

12.16 Use Norton's Theorem to find the voltage across $\mathbf{Z_L}$ in the circuit of Figure 12.23. (ans. $3.86\,\underline{/-15.01°}$ V)

12.17 Use the Superposition Theorem to find the voltage across $\mathbf{Z_L}$ in the circuit of Figure 12.23.

12.18 Use the Superposition Theorem to find the voltage (as a function of time) across the capacitor in the circuit shown in Figure 12.24. Sketch the voltage. [ans. $5 + 4.75 \sin(6280t - 17.72°)$]

12.19 Use the Superposition Theorem to find the voltage across the capacitor in the circuit shown in Figure 12.25. Sketch the voltage as a function of time.

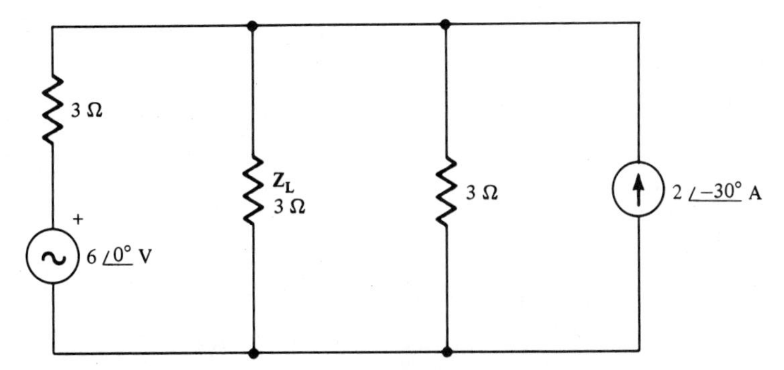

Figure 12.23 Circuit for Problems 12.15, 12.16, and 12.17.

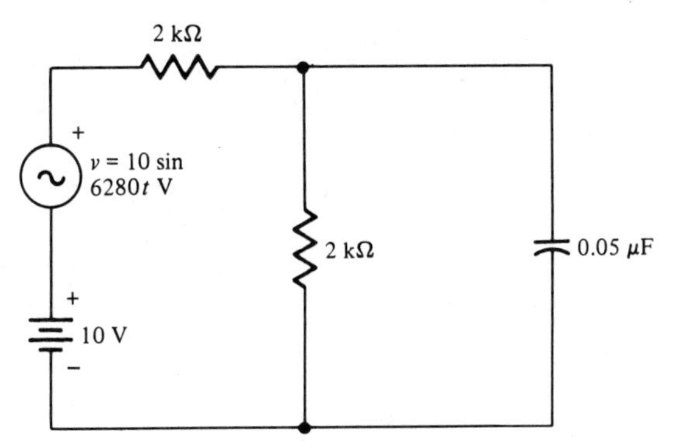

Figure 12.24 Circuit for Problem 12.18.

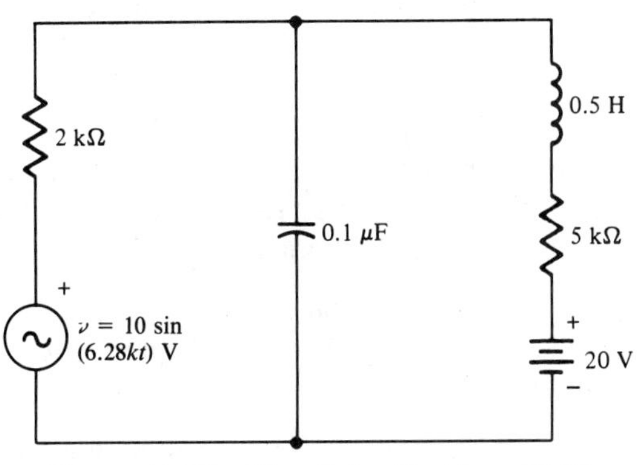

Figure 12.25 Circuit for Problem 12.19.

12.20 Find the **z**-parameter for the two-port network shown in Figure 12.26 at 1 kHz. (ans. $1.88\underline{/-57.83^\circ}$, $1.59\underline{/-90^\circ}$, $1.59\underline{/-90^\circ}$, $2.56\underline{/-38.5^\circ}$, all kΩ)

12.21 Find the **y**-parameters for the two-port network shown in Figure 12.26 at 1 kHz.

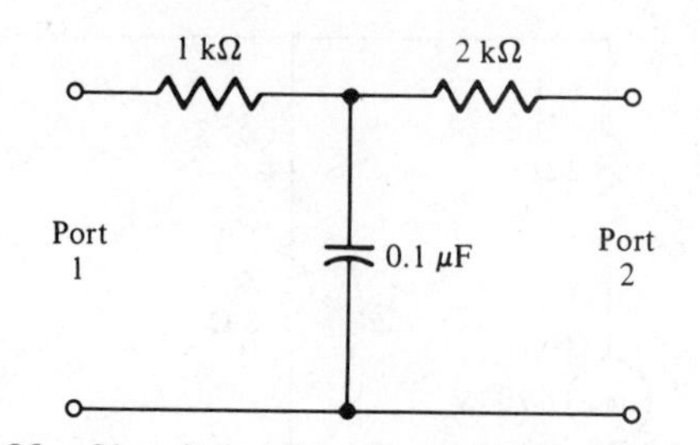

Figure 12.26 Circuit for Problems 12.20, 12.21, and 12.22.

12.22 Find the **h**-parameters for the two-port network shown in Figure 12.26 at 1 kHz. (ans. $2.02\underline{/-28.72^\circ}$ kΩ, $0.62\underline{/-51.52^\circ}$, $-0.62\underline{/-51.52^\circ}$, $0.39\underline{/38.48^\circ}$ mS)

12.23 Repeat Problem 12.20 for the circuit shown in Figure 12.27.

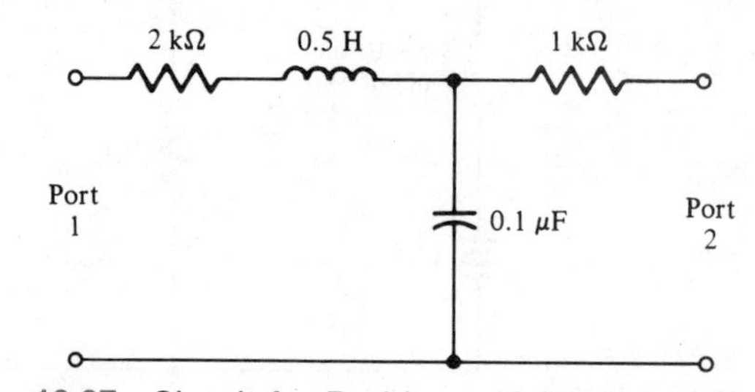

Figure 12.27 Circuit for Problems 12.23, 12.24, and 12.25.

12.24 Repeat Problem 12.21 for the circuit shown in Figure 12.27. (ans. $0.26\underline{/-44.68^\circ}$, $-0.22\underline{/-76.91^\circ}$, $-0.22\underline{/-76.85^\circ}$, $0.35\underline{/50.87^\circ}$, all mS)

12.25 Repeat Problem 12.22 for the circuit shown in Figure 12.27.

12.26 Use the conversion table in this chapter to convert the **z**-parameters obtained in Problem 12.20 to the **y**-parameter equivalent (check your results with answers to Problem 12.21).

12.27 Use the conversion table in this chapter to convert the **y**-parameters obtained in Problem 12.21 to their equivalent **h**-parameters (check your results with answers to Problem 12.22).

12.28 Use the conversion table in this chapter to convert the **z**-parameters obtained in Problem 12.23 to the corresponding **y**-parameters (check your results with answers to Problem 12.24).

12.29 Use the conversion table in this chapter to convert the **y**-parameters obtained in Problem 12.24 to the corresponding **h**-parameters (check your results with answers to Problem 12.25).

CHAPTER 13

Resonance and Filter Circuits

GLOSSARY

active circuit—circuit that exhibits a power gain
active filter—filter circuit that contains an electronic amplifier
attenuator—circuit with less output than input
bandwidth (BW)—the difference between the upper and lower cutoff frequencies
decibel—a unit of the logarithmic ratio of two power levels
filter—electric network that passes signals at some frequencies and attenuates signals at other frequencies

frequency response—a characteristic of the circuit, which describes behavior of the circuit with changes of the source voltage or current
half-power frequencies—frequencies at which the power delivered to the circuit is one-half the power delivered at resonance
passive circuit—circuit that exhibits a power loss
passive filter—filter circuit constructed from passive components (R, L, C)
quality factor (Q)—the ratio of the total energy stored at resonance to the energy dissipated per cycle. Also, a measure of the selectivity of a resonant circuit
resonance—the condition for which the total current and voltage in an *RLC* circuit are in phase
resonant frequency—the frequency at which resonance occurs
selectivity—the circuit's ability to discriminate between signals of different frequencies

INTRODUCTION

Resonance is a phenomenon that is common to all physical systems including electric circuits. **Resonance** occurs when energy is applied to a system at a specific frequency f_r, which causes the system to respond in such a way that it can become unstable and even destroy itself if its response is not limited in some way. Several common examples of resonance are well known: the crystal glass that can be shattered if the singer emits a certain note or the bridge that collapses because of soldiers walking on it "in-step." In both of these examples the systems (glass, bridge) were forced at their natural resonant frequencies. The glass resonates at an audio frequency, whereas the bridge has a very low resonant frequency.

Electric circuits that contain energy storage elements, such as inductors and capacitors, will also resonate and exhibit a maximum response at a particular frequency, f_r, proportional to the circuit component values. Because of this type of behavior, many electronic systems employ inductors, capacitors, and resistors to construct tuned circuits, which are more sensitive to the signal at one particular frequency. Tuned circuits are used in radio and television in both transmission and reception of radio frequency signals.

Electric filters are circuits used to pass signals at some frequencies and reject or *attenuate* signals at other frequencies. Different characteristics of filter circuits are obtained through the use of combinations of reactive and resistive elements.

13.1 SERIES RESONANT CIRCUITS

The circuit shown in Figure 13.1 contains a variable frequency sinusoidal voltage source, a resistor, inductor, and a capacitor, all connected in series. Such a circuit

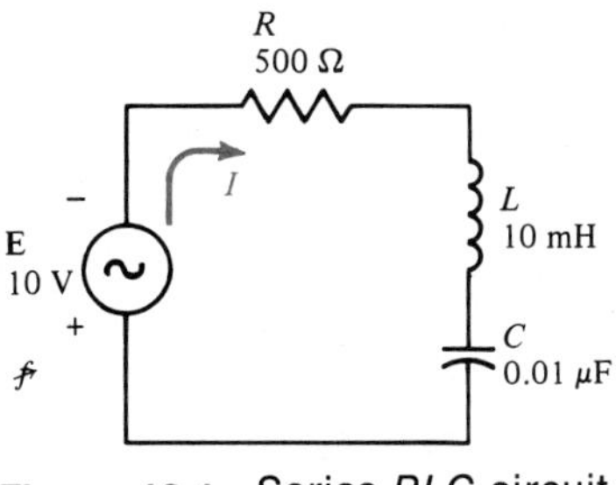

Figure 13.1 Series *RLC* circuit.

exhibits an impedance, **Z,** which varies with frequency; consequently, the current in the circuit is a function of frequency.

For specific values of R, L, and C there exists one and only one frequency at which the total impedance of the circuit is purely resistive. Furthermore, the current is maximum and in phase with the applied voltage. This condition is called ***resonance,*** *and the frequency,* f_r*, at which this condition occurs is called the* ***resonant frequency.***

The total impedance of the circuit in Figure 13.1 is given by

$$\mathbf{Z_T} = R + j(X_L - X_C)$$

The resonance condition will occur if:

$$\boxed{X_L = X_C} \tag{13.1}$$

or

$$\omega L = \frac{1}{\omega C}$$

Solving for ω,

$$\omega^2 = \frac{1}{LC}$$

and

$$\boxed{\omega_{rs} = \frac{1}{\sqrt{LC}}} \tag{13.2}$$

where subscript r denotes *resonant* frequency and s stands for *series* circuit. Hence, ω_{rs} is the resonant-radian frequency for a series *RLC* circuit. Furthermore, because $\omega_{rs} = 2\pi f_{rs}$,

$$f_{rs} = \frac{1}{2\pi\sqrt{LC}} = \frac{0.159}{\sqrt{LC}} \quad (13.3)$$

At resonance, since the inductive and capacitive reactances are equal, the total impedance is:

$$\mathbf{Z_T} = R \quad (13.4)$$

The current I in the circuit is given by

$$\mathbf{I} = \frac{\mathbf{E}}{\mathbf{Z_T}} = \frac{E}{R} \quad (13.5)$$

EXAMPLE 13.1

In the circuit of Figure 13.1, $R = 500\ \Omega$, $C = 0.01\ \mu\text{F}$, $L = 10$ mH, and $\mathbf{E} = 10\angle 0^\circ$ V.

Determine:

a. The resonant frequency, f_{rs}

b. The current I at resonance

c. The voltage drop $\mathbf{V_R}$, $\mathbf{V_L}$, and $\mathbf{V_C}$ at resonance

Solution

a. By Eq. (13.3),

$$f_{rs} = \frac{0.159}{\sqrt{(1 \times 10^{-2})(1 \times 10^{-8})}} = 15.9 \text{ kHz}$$

b. By Eq. (13.5),

$$\mathbf{I} = \frac{10\angle 0^\circ}{5 \times 10^2} = 20\angle 0^\circ \text{ mA}$$

c. The voltage $\mathbf{V_R}$ dropped across the resistor is the voltage of the source:

$$\mathbf{V_R} = \mathbf{E} = 10\angle 0^\circ \text{ V}$$

The voltage $\mathbf{V_L}$ is given by

$$\mathbf{V_L} = (jX_L)(\mathbf{I}) = j2\pi fL\mathbf{I}$$
$$= 2\pi(15{,}900)(1 \times 10^{-2})(20 \times 10^{-3})\angle 90^\circ = 20\angle 90^\circ \text{ V}$$

Similarly

$$\mathbf{V_C} = (-jX_C)\mathbf{I} = -j\frac{1}{2\pi fC}\mathbf{I}$$
$$= \frac{(1)(20 \times 10^{-3})\angle -90^\circ}{2\pi(15{,}900)(1 \times 10^{-8})} = 20\angle -90^\circ \text{ V}$$

Notice that the magnitude of the voltage across the inductor and across the capacitor is, in this case, the same as the voltage applied.

EXAMPLE 13.2

Assume that the capacitance in Example 13.1 is variable. Calculate the capacitance value necessary for the circuit to have a resonant frequency of 10 kHz.

Solution

Solving Eq. (13.3),

$$f_{rs} = \frac{0.159}{\sqrt{LC}}$$

$$f_{rs}^2 = \frac{(0.159)^2}{LC}$$

and

$$C = \frac{(0.159)^2}{Lf_{rs}^2}$$

Substituting and solving for C gives

$$C = \frac{(0.159)^2}{(1 \times 10^{-2})(1 \times 10^4)^2} = 0.0253 \times 10^{-6}$$
$$= 0.0253\ \mu\text{F}$$

13.2 FREQUENCY RESPONSE

The **frequency response** is a characteristic of a circuit that describes the behavior of the circuit parameters with changes in frequency of the source, voltage, or current. The frequency response can be represented by a correspondence table or a plot of a factor such as current, voltage, impedance, or phase angle versus frequency. The frequency response plot is the more commonly used method, and the frequency axis is usually calibrated on a logarithmic scale.

The frequency response plot of impedance versus frequency of the circuit in Figure 13.1 is given in Figure 13.2. The current versus frequency plot is shown in Figure 13.3, and the phase angle of the impedance versus frequency is plotted in Figure 13.4. Notice the vertical axis for each plot is a linear scale and the frequency is plotted on a logarithmic scale. Such a plot is considered to be a *semilog plot*. A tabulation of values is also provided (see Table 13.1). The magnitude of the impedance as a function of frequency is calculated by:

$$|\mathbf{Z_T}| = \sqrt{R^2 + \left(2\pi fL - \frac{1}{2\pi fC}\right)^2} \tag{13.6}$$

and the phase angle of the impedance is calculated by

$$\phi_Z = \tan^{-1}\frac{X_L - X_C}{R}$$

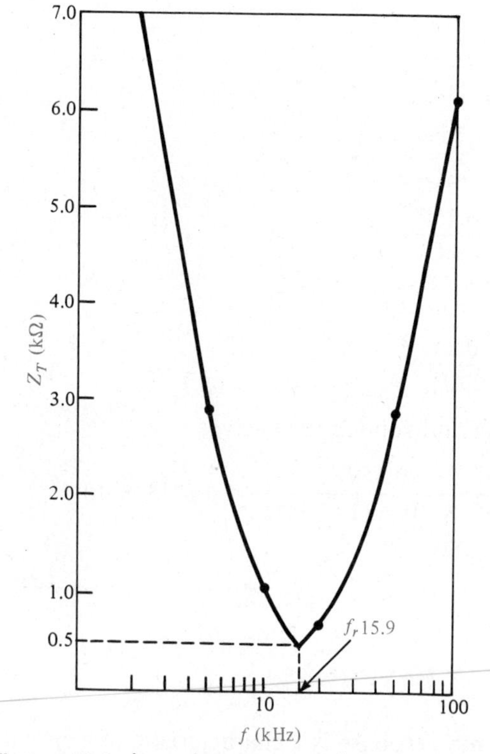

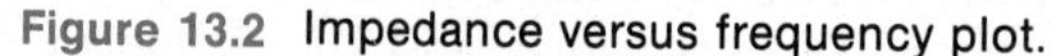

Figure 13.2 Impedance versus frequency plot.

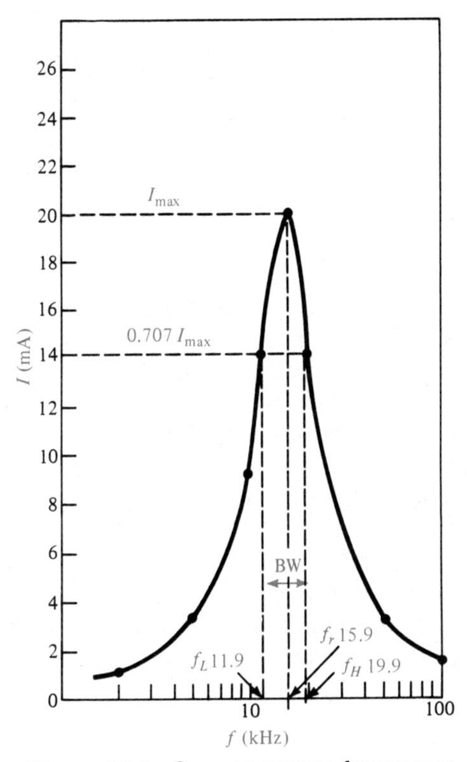

Figure 13.3 Current versus frequency plot.

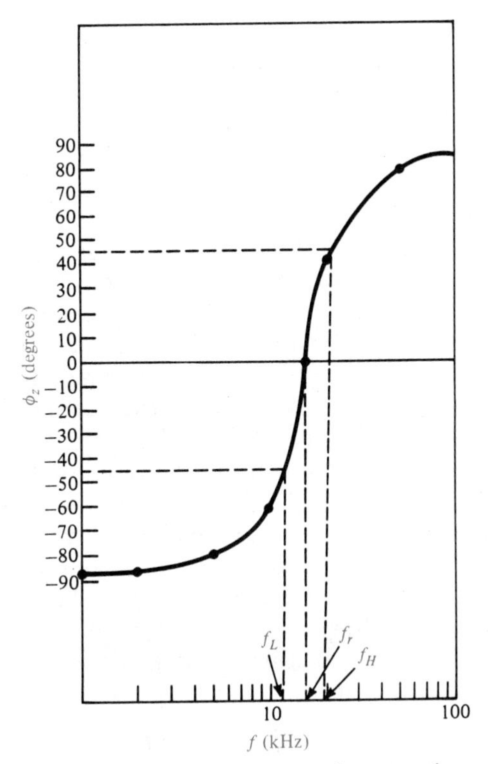

Figure 13.4 Impedance phase angle versus frequency plot.

TABLE 13.1
Frequency response values

f (kHz)	Z_T (kΩ)	ϕ_Z (deg.)	I (mA)
1.0	15.8	−88.2	0.63
2.0	7.84	−86.4	1.28
5.0	2.91	−80.1	3.44
10.0	1.08	−62.5	9.26
15.9*	0.5	0	20.0
20.0	0.68	42.7	14.71
50.0	2.87	80.0	3.48
100.0	6.14	85.3	1.63

* Resonance

or

$$\phi_Z = \tan^{-1}\left[\frac{2\pi fL - 1/(2\pi fC)}{R}\right] \tag{13.7}$$

The impedance is minimum at resonance, and the current is at its maximum at the resonant frequency. The phase difference between voltage and current is zero at resonance, as it can be seen on the plot of phase angle of the impedance. When the impedance angle is negative, it is an indication that the circuit is capacitive, and the positive phase angle indicates that the circuit is inductive.

Both the impedance and current frequency response plots show that the circuit responds differently to signals (input voltages) at different frequencies. To better define the behavior of the *RLC* circuit, we define two *critical frequencies,* f_L and f_H, at which the power delivered to the circuit is one-half of the power delivered to the circuit at resonance, which is maximum. Those two frequencies are indicated on the current plot of Figure 13.3 and are also referred to as: *lower cutoff* frequency for f_L and *upper cutoff* frequency for f_H, as well as the **half-power frequencies.** Hence, at resonance the power delivered to the circuit is P_{max}, and at the half-power frequencies the power delivered is $0.5P_{max}$. Consequently, at resonance,

$$P_{max} = (I_{max})^2R$$

where I_{max} is the maximum effective value of the current and at the half-power frequencies,

$$0.5P_{max} = \tfrac{1}{2}(I_{max})^2R$$

or

$$0.5P_{max} = (0.707I_{max})^2R$$

Hence, we conclude that the current in the circuit at the half-power frequencies is 0.707 of I_{max}.

The **bandwidth (BW)** of the circuit is defined as the difference between the upper and lower cutoff frequencies.

$$\text{BW} = f_H - f_L \tag{13.8}$$

The **quality factor (*Q*)** of the *RLC* circuit is a measure of how well the circuit can select between signals at different frequencies. Mathematically, it is given by:

$$Q = \frac{f_r}{\text{BW}} \tag{13.9}$$

EXAMPLE 13.3

From the plot in Figure 13.3 determine the bandwidth, BW, and the quality factor, Q, for the RLC circuit of Figure 13.1.

Solution

From Figure 13.3,

$$0.707I_{max} = 0.707 \times 20 \times 10^{-3} = 14.14 \text{ mA}$$

Therefore, from the graph,

$$f_L \simeq 11.9 \text{ kHz} \quad \text{and} \quad f_H \simeq 19.9 \text{ kHz}$$

and by Eq. (13.8),

$$\text{BW} = 19.9 \times 10^3 - 11.9 \times 10^3 = 8 \text{ kHz}$$

The quality factor, Q, is calculated by Eq. (13.9):

$$Q = \frac{15.9 \times 10^3}{8 \times 10^3} \simeq 2$$

A circuit with a $Q < 10$ is considered to be a *wide bandwidth,* or low Q circuit, and one with a $Q > 10$ is a *narrow bandwidth,* or high Q circuit. The higher the Q value, the more selective the circuit because it has a narrower bandwidth. **Selectivity** is the circuit's ability to discriminate between signals of different frequencies.

If the Q of an RLC circuit is high ($Q > 10$), then the critical frequencies can be estimated by assuming that the response is symmetrical about the resonant frequency value. Therefore,

$$f_L = f_r - \frac{\text{BW}}{2} \tag{13.10}$$

and

$$f_H = f_r + \frac{\text{BW}}{2} \tag{13.11}$$

This concept is illustrated in Figure 13.5.

The quality factor, Q, is also the ratio of the total energy stored at resonar to the energy dissipated per cycle. In a series RLC circuit it is the ratio of reactive power of the inductor or capacitor to the real power dissipated by resistor

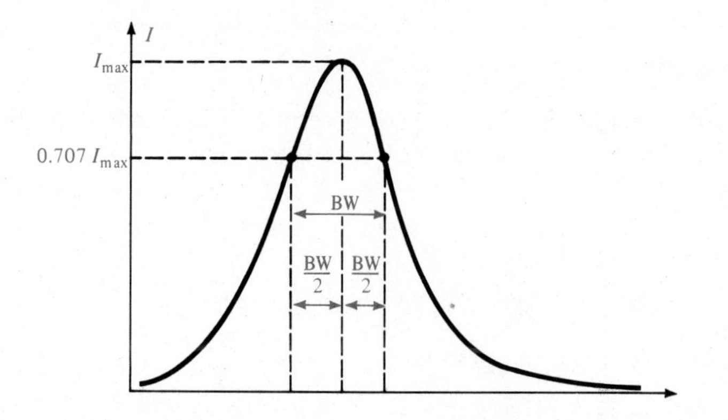

Figure 13.5 Symmetrical frequency response plot of a high Q circuit.

$$Q_{rs} = \frac{I^2X_L}{I^2R} = \frac{I^2X_C}{I^2R}$$

or

$$\boxed{Q_{rs} = \frac{X_L}{R} = \frac{X_C}{R}} \qquad (13.12)$$

where R is the total resistance including the resistance of the coil. If the total resistance, R, of the circuit is the resistance of the coil only, then $Q_{rs} = Q$, and Q is the quality factor of the coil. From Eq. (13.2),

$$Q_{rs} = \frac{2\pi f_{rs}L}{R} = \frac{1}{2\pi f_{rs}CR}$$

Substituting Eq. (13.3) for f_r,

$$\boxed{Q_{rs} = \frac{1}{R}\sqrt{\frac{L}{C}}} \qquad (13.13)$$

Substituting Eqs. (13.3) and (13.13) into Eq. (13.9) and solving for bandwidth gives

$$\text{BW} = \frac{\dfrac{0.159}{\sqrt{LC}}}{\dfrac{1}{R}\sqrt{\dfrac{L}{C}}} = \boxed{\frac{0.159R}{L}} \qquad (13.14)$$

EXAMPLE 13.4

Using the component values of the circuit in Figure 13.1, determine Q, BW, f_L, and f_H.

Solution

By Eq. (13.13),

$$Q_{rs} = \frac{1}{5 \times 10^2} \sqrt{\frac{1 \times 10^{-2}}{1 \times 10^{-8}}} = 2$$

The BW is calculated by (13.9) with the value for f_{rs} obtained in Example 13.1:

$$\text{BW} = \frac{f_r}{Q}$$

$$= \frac{15.9 \times 10^3}{2} \simeq 8 \text{ kHz}$$

or, by Eq. (13.14),

$$\text{BW} = \frac{(0.159)(5 \times 10^2)}{1 \times 10^{-2}} \simeq 8 \text{ kHz}$$

From Eqs. (13.10) and (13.11),

$$f_L = 15.91 - \frac{8 \text{ k}\Omega}{2} = 11.9 \text{ kHz}$$

and

$$f_H = 15.9\text{K} + \frac{8 \text{ k}\Omega}{2} = 19.9 \text{ kHz}$$

EXAMPLE 13.5

A series RLC circuit has the following component values: $L = 50\ \mu\text{H}$, $C = 0.05\ \mu\text{F}$, and $R = 10\ \Omega$. Calculate the resonant frequency, f_{rs}, the quality factor, Q_{rs}, and the bandwidth, BW.

Solution

By Eq. (13.3),

$$f_{rs} = \frac{0.159}{\sqrt{(50 \times 10^{-6})(5 \times 10^{-8})}} = 100.56 \text{ kHz}$$

The Q of the circuit is calculated from Eq. (13.13):

$$Q_{rs} = \frac{1}{10}\sqrt{\frac{50 \times 10^{-6}}{5 \times 10^{-8}}} = 3.16$$

The bandwidth is given by Eq. (13.9):

$$\text{BW} = \frac{100{,}560}{3.16} \simeq 31.82 \text{ kHz}$$

The cutoff frequencies, f_L and f_H, for any RLC series circuit are determined next from the fact that the phase shift at the critical frequencies is $-45°$ and $+45°$ for f_L and f_H, respectively, as shown in Figure 13.4. To determine the lower frequency cutoff value:

$$\tan(-45°) = \frac{(X_L - X_C)}{R} = -1$$

Hence

$$X_C - X_L = R$$

or

$$\frac{1}{2\pi f_L C} - 2\pi f_L L = R$$

and

$$1 - 4\pi^2 f_L^2 LC = 2\pi f_L RC$$

Rearranging terms gives

$$(4\pi^2 LC)f_L^2 + 2\pi RCf_L - 1 = 0$$

The preceding equation is a quadratic equation that results in the following when solved for f_L:

$$f_L = \frac{1}{2\pi}\left(-\frac{R}{2L} + \frac{1}{2}\sqrt{\left(\frac{R}{L}\right)^2 + \frac{4}{LC}}\right) \quad \textbf{(13.15)}$$

The upper cutoff frequency, f_H, can be derived in a similar manner, by setting

$$\tan 45° = \frac{X_L - X_C}{R} = 1$$

Solving for f_H yields

$$f_H = \frac{1}{2\pi}\left(\frac{R}{2L} + \frac{1}{2}\sqrt{\left(\frac{R}{L}\right)^2 + \frac{4}{LC}}\right) \quad \textbf{(13.16)}$$

EXAMPLE 13.6

For the circuit in Example 13.5, determine f_L and f_H.

Solution

By Eq. (13.15),

$$f_L = \frac{1}{2\pi}\left(-\frac{10}{2(50\times10^{-6})} + \frac{1}{2}\sqrt{\left(\frac{10}{50\times10^{-6}}\right)^2 + \frac{4}{(50\times10^{-6})(5\times10^{-8})}}\right)$$
$$= 86.017 \text{ kHz}$$

f_H is calculated by Eq. (13.16):

$$f_H = \frac{1}{2\pi}\left(\frac{10}{2(50\times10^{-6})} + \frac{1}{2}\sqrt{\left(\frac{10}{50\times10^{-6}}\right)^2 + \frac{4}{(50\times10^{-6})(5\times10^{-8})}}\right)$$
$$= 117.824 \text{ kHz}$$

The effects that different components in an RLC resonant circuit have on the frequency response are examined in the plots of Figures 13.6 and 13.7. In Figure 13.6 the curves represent the change in the response for different values of resistance, R. As the value of R increases, the circuit becomes less selective and the BW increases because of the drop in the value of Q. This is consistent with Eqs. (13.13) and (13.14). The resonant frequency remains the same as long as L and C are fixed.

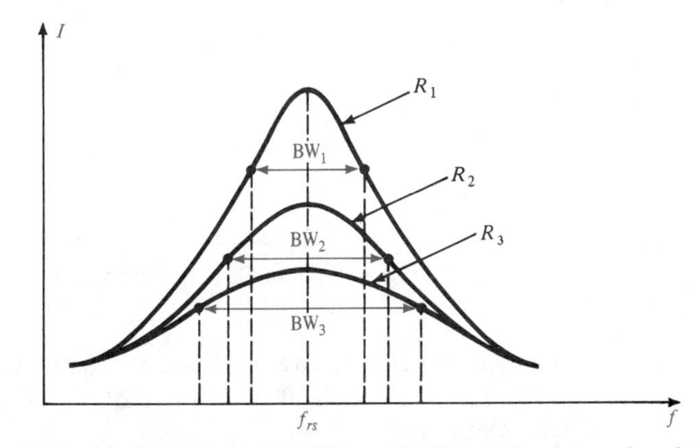

Figure 13.6 Changes in bandwidth with different values of resistance R but the same values of L and C (resistors $R_1 < R_2 < R_3$).

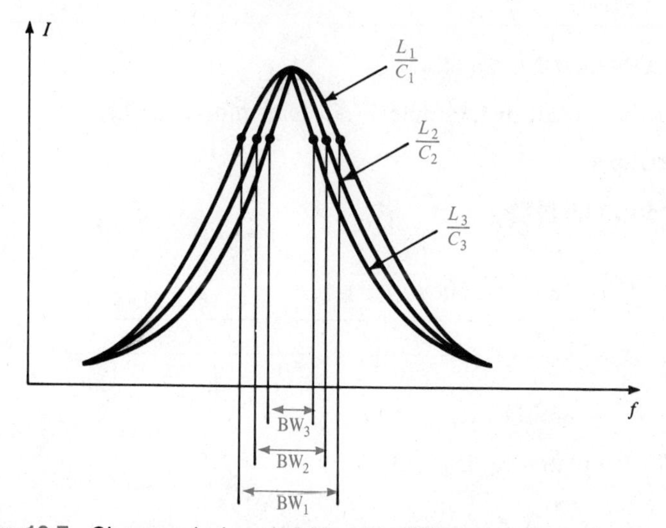

Figure 13.7 Changes in bandwidth with different values of L/C and a fixed value of R ($L_1/C_1 < L_2/C_2 < L_3/C_3$).

The curves of Figure 13.7 show, according to Eq. (13.3), that the selectivity increases if the ratio of L to C increases, which means that the bandwidth "narrows" and the Q of the circuit increases. Notice that the maximum value of the current remains constant as long as the value of R does not change.

13.3 PARALLEL RESONANT CIRCUITS

Resonance was defined earlier in the chapter as the condition for which the total voltage and current in the circuit are in phase. The circuit in Figure 13.8 is a parallel RLC circuit driven by a current source.

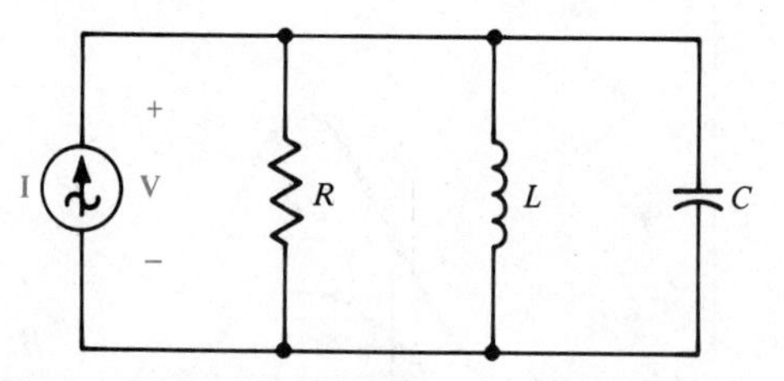

Figure 13.8 Parallel RLC resonant circuit.

When the voltage and current are in phase, at resonance, the impedance of the circuit is purely resistive. This also means that the admittance, **Y**, of the circuit is real. The admittance is:

$$\mathbf{Y} = G + j(B_C - B_L)$$

At resonance

$$\boxed{B_C = B_L} \tag{13.17}$$

or

$$\omega_{rp} C = \frac{1}{\omega_{rp} L}$$

where ω_{rp} is the radiant resonant frequency for a parallel circuit. Solving for ω_{rp},

$$\omega_{rp}^2 = \frac{1}{LC}$$

and

$$\omega_{rp} = \frac{1}{\sqrt{LC}}$$

Furthermore,

$$\boxed{f_{rp} = \frac{1}{2\pi\sqrt{LC}}} \tag{13.18}$$

Notice that the resonant frequency in a parallel circuit is also a function of L and C only. The circuit voltage, V, is a function of frequency and is given by

$$\mathbf{V} = \frac{\mathbf{I}}{\mathbf{Y}}$$

At resonance, $\mathbf{Y} = G = 1/R$; then

$$V_{rp} = \frac{I}{G} = IR$$

At resonance the admittance is minimum and the impedance is maximum. Since the current is constant, the voltage is also maximum at the resonant frequency, as shown in Figures 13.9 and 13.10.

The circuit shown in Figure 13.8 is not very practical, because a practical inductor has a resistance in series due to the resistance of the windings. The circuit in Figure 13.11a shows a more practical parallel resonant circuit, with R_S as source resistance. This circuit is also referred to as a "tank" circuit and is used as a tuned circuit in electronic applications. A variable capacitor is used to change the resonant frequency, making the circuit selective at different frequencies.

The practical circuit of Figure 13.11a can be converted to an equivalent circuit in the form of the basic *RLC* parallel circuit shown earlier in Figure 13.8. This requires the conversion of the inductive branch containing L and R to two

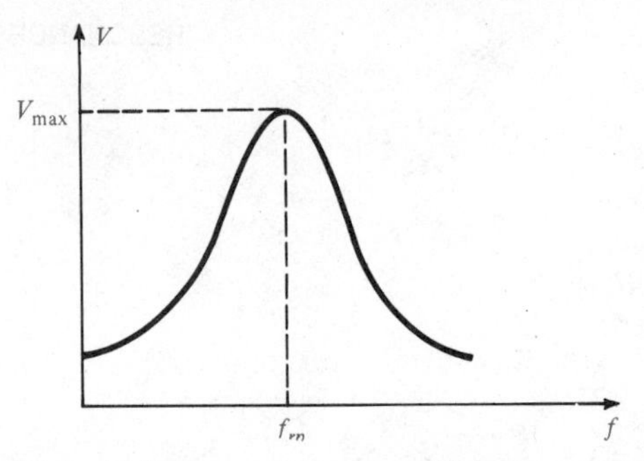

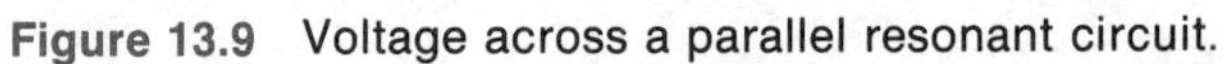
Figure 13.9 Voltage across a parallel resonant circuit.

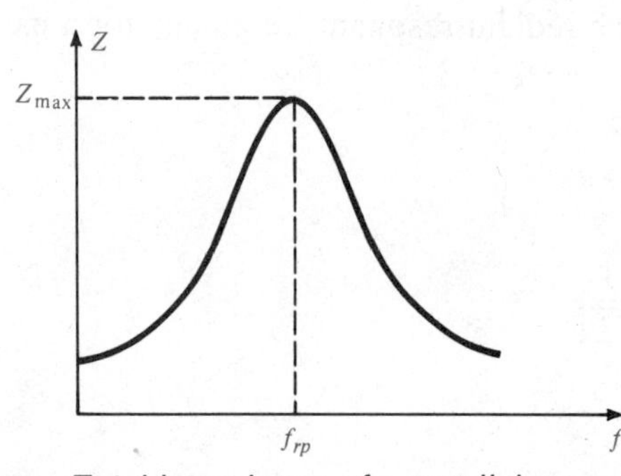

Figure 13.10 Total impedance of a parallel resonant circuit.

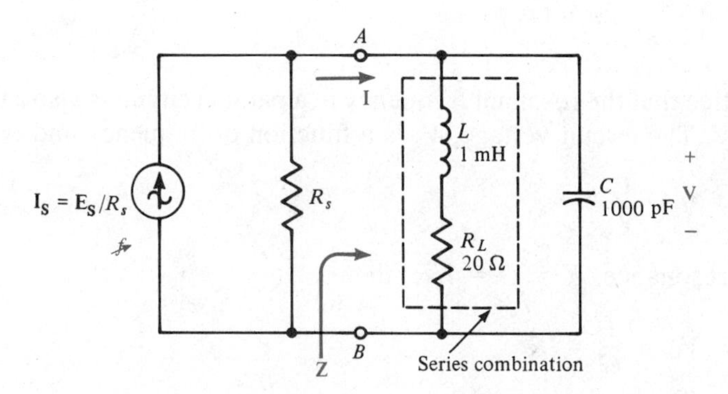

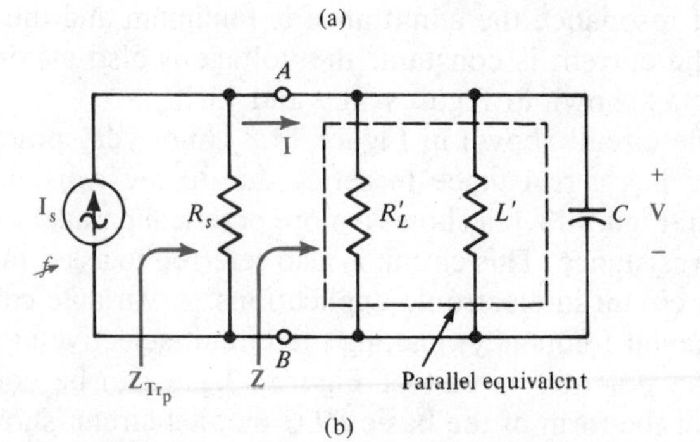

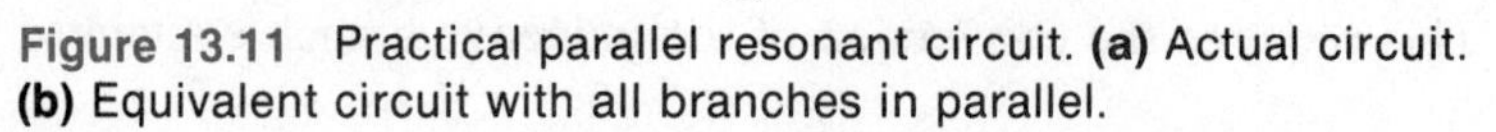
Figure 13.11 Practical parallel resonant circuit. **(a)** Actual circuit. **(b)** Equivalent circuit with all branches in parallel.

parallel branches. The resulting circuit is shown in Figure 13.11b. The equivalent parallel inductor winding resistance is designated by R'_L, and L' is the parallel equivalent inductance. The admittance of the two branches must be equal; hence,

$$\mathbf{Y} = \frac{1}{R'_L} + \frac{1}{j\omega L'} = \frac{1}{R_L + j\omega L}$$

Rationalizing the denominator of the expression on the right,

$$\frac{1}{R'_L} + \frac{1}{j\omega L'} = \frac{R_L - j\omega L}{R_L^2 + \omega^2 L^2}$$

Equating the real parts,

$$\frac{1}{R'_L} = \frac{R_L}{R_L^2 + \omega^2 L^2}$$

and

$$\boxed{R'_L = \frac{R_L^2 + \omega^2 L^2}{R_L}} \tag{13.19}$$

Equating the reactive parts gives

$$\frac{1}{j\omega L'} = \frac{-j\omega L}{R_L^2 + \omega^2 L^2}$$

resulting in:

$$\omega L' = \frac{R_L^2 + \omega^2 L^2}{\omega L}$$

Dividing both sides by ω gives

$$\boxed{L' = \frac{R_L^2 + \omega^2 L^2}{\omega^2 L}} \tag{13.20}$$

Substituting L' from Eq. (13.20) for L in Eq. (13.18), at resonance,

$$f_{rp} = \frac{1}{2\pi \sqrt{\left[\frac{R_L^2 + 4\pi^2 f_{rp}^2 L^2}{4\pi^2 f_{rp}^2 L}\right] C}}$$

Solving for f_{rp} results in:

$$\boxed{f_{rp} = \frac{0.159}{\sqrt{LC}} \sqrt{1 - \frac{R_L^2 C}{L}}} \tag{13.21}$$

Notice that for $(R_L'^2 C/L) > 1$, the quantity under the radical is a negative number; hence f_{rp} is imaginary. In practical terms, this means that the circuit will not resonate.

EXAMPLE 13.7

In the circuit of Figure 13.11a, $L = 1$ mH, $R_L = 20\ \Omega$, and $C = 1000$ pF. Calculate the resonant frequency, f_{rp}, of the circuit.

Solution

By Eq. (13.21),

$$f_{rp} = \frac{0.15}{\sqrt{(1 \times 10^{-3} \times 1 \times 10^{-3})}} \sqrt{1 - \frac{(20)^2 \times 1 \times 10^{-3}}{1 \times 10^{-3}}}$$
$$\simeq 159 \text{ kHz}$$

The impedance of the practical parallel circuit, $\mathbf{Z_T}$, is calculated from the equivalent circuit in Figure 13.11b.

$$\mathbf{Y} = \frac{1}{\mathbf{Z}} = \frac{1}{R_L'} + \frac{1}{j\omega L'} + j\omega C = \frac{j'\omega L' + R_L' - \omega^2 R_L' L' C}{j'\omega L' R_L'}$$

and

$$\mathbf{Y} = \frac{1}{R_L'} + \frac{j'(\omega^2 L' C - 1)}{\omega L'} \qquad (13.22)$$

Since at resonance the admittance must be purely resistive,

$$\mathbf{Y_{rp}} = \frac{1}{R_L'}$$

Therefore,

$$\mathbf{Z_{rp}} = \frac{1}{\mathbf{Y_{rp}}} = R_L'$$

From Eq. (13.19),

$$\mathbf{Z_{rp}} = \frac{R_L^2 + \omega_{rp}^2 L^2}{R_L} \qquad (13.23)$$

Substituting Eq. (13.21) into Eq. (13.23) simplifies the expression for $\mathbf{Z_{rp}}$:

$$\mathbf{Z_{rp}} = R'_{L_{rp}} = \frac{L}{R_L C} \tag{13.24}$$

The total impedance at resonance as seen by the current source, $\mathbf{Z_{T_{rp}}}$, is the parallel combination of the source resistance and $R'_{L_{rp}}$.

$$\mathbf{Z_{T_{rp}}} = R_s \| R'_{L_{rp}} \tag{13.25}$$

The voltage $\mathbf{V}$ across the parallel resonant circuit is given by

$$\mathbf{V} = (\mathbf{I_s})(\mathbf{Z_{T_{rp}}}) \tag{13.26a}$$

or

$$\mathbf{V} = (\mathbf{I})(\mathbf{Z_{T_{rp}}}) \tag{13.26b}$$

The current I_s is constant and $Z_{T_{rp}}$ is maximum at resonance; hence, the voltage V is maximum at resonance, as previously shown in Figure 13.9.

EXAMPLE 13.8

Determine the total impedance of the parallel circuit of Figure 13.11a at resonance if (a) R_s = 100 kΩ, (b) R_s = 1 kΩ.

Solution

a. From Eq. (13.24),

$$\mathbf{Z_{rp}} = \frac{1 \times 10^{-3}}{(2 \times 10^{1})(1 \times 10^{-3})} = 50\ \text{k}\Omega$$

$$\mathbf{Z_{T_{rp}}} = 100\ \text{k}\Omega \| 50\ \text{k}\Omega = 33.33\ \text{k}\Omega$$

b. $\mathbf{Z_{rp}}$ is the same as in (a), and

$$\mathbf{Z_{T_{rp}}} = 1\ \text{k}\Omega \| 50\ \text{k}\Omega = 980\ \Omega$$

The quality factor of a parallel resonant circuit is defined as before for the series circuit to be the ratio of the reactive power of the inductor or the capacitor to the real power dissipated by the resistance of the circuit. Hence

$$Q_{rp} = \frac{V_{rp}^2/X'_{L_{rp}}}{V_{rp}^2/Z_{T_{rp}}} = \frac{V_{rp}^2/X_{C_{rp}}}{V_{rp}^2/Z_{T_{rp}}}$$

or

$$Q_{rp} = \frac{R'_L \| R_s}{X'_{L_{rp}}} = \frac{R'_L \| R_s}{\omega_{rp} L'} \tag{13.27}$$

Similarly, from the capacitor,

$$Q_{rp} = \frac{R'_L \| R_s}{X_{C_{rp}}} = (R'_L \| R_s)\omega_{rp} C \tag{13.28}$$

For $R_s \gg R'_L$, Eq. (13.28) can be reduced to

$$Q_{rp} \simeq \frac{R'_L}{\omega_{rp} L'} = \frac{\omega_{rp} L}{R_L} \tag{13.29}$$

Notice that this result is the same as for the series resonance circuit, $Q_{rp} = Q$, the quality factor of the coil, and $Q = \frac{X_L}{R}$.

Finally

$$Q_{rp} = \frac{2\pi f_{rp} L}{R_L} = Q \tag{13.30}$$

The bandwidth of the parallel resonant circuit is given by the same equation as that for a series circuit:

$$\text{BW} = \frac{f_{rp}}{Q_{rp}} \tag{13.31}$$

EXAMPLE 13.9

For the circuit in Figure 13.11a calculate Q and the bandwidth. (Assume $R_s \simeq \infty$.)

Solution

It was calculated in Example 13.7 that $f_{rp} \cong 159$ kHz, and by Eq. (13.30),

$$Q_{rp} = \frac{2\pi(159 \times 10^3)(1 \times 10^{-2})}{20}$$

$$\simeq 50$$

The BW is calculated from Eq. (13.31):

$$\text{BW} = \frac{159 \text{ k}\Omega}{50} = 3.18 \text{ kHz}$$

Assuming that the parallel resonant circuit is driven from a current source with a very high value of R_S, then $Q_{rp} = Q$; let us examine the relation between R_L', L', and Q from Eq. (13.19):

$$R_L' = R_L + \frac{\omega^2 L^2}{R_L}$$

Substituting for Q_{rp} from Eq. (13.29) and recalling $Q_{rp} = Q$,

$$\boxed{R_L' = R_L + Q^2 R_L} \tag{13.32}$$

and for a large value of Q ($Q > 10$),

$$\boxed{R_L' \simeq Q^2 R_L} \tag{13.33}$$

Similarly, it can be shown that

$$\boxed{L' = \frac{L}{Q^2} + L} \tag{13.34}$$

For a large value of Q,

$$\boxed{L' \simeq L} \tag{13.35}$$

For $Q > 10$, the resonant frequency is given, as in Eq. (13.18), by

$$f_{rp} = \frac{0.159}{\sqrt{LC}}$$

EXAMPLE 13.10

For the circuit in Figure 13.12, calculate:

a. $Z_{T_{rp}}$

b. f_{rp}

c. Q_{rp}

d. BW

e. f_H

f. 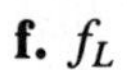 f_L

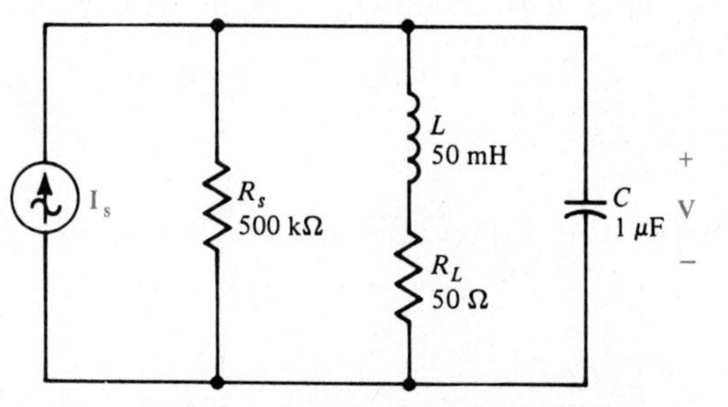

Figure 13.12 Circuit for Example 13.9.

Solution

a. Z_{rp} is calculated first by Eq. (13.24):

$$Z_{rp} = R'_L = \frac{50 \times 10^{-3}}{50 \times 1 \times 10^{-7}} = 10 \text{ k}\Omega$$

Then $Z_{T_{rp}}$, from Eq. (13.25), is:

$$Z_{T_{rp}} = 500 \text{ k}\Omega \| 10 \text{ k}\Omega = 9.8 \text{ k}\Omega$$

Notice that the value of R_s is very large compared to the value of R'_L and $Z_{T_{rp}} \simeq Z_{rp}$.

b. By Eq. (13.21),

$$f_{rp} = \frac{0.159}{\sqrt{(50 \times 10^{-3})(1 \times 10^{-7})}} \sqrt{1 - \frac{(50)(1 \times 10^{-7})}{50 \times 10^{-3}}}$$
$$\simeq 2.25 \text{ kHz}$$

c. The Q of the circuit is solved by Eq. (13.30):

$$Q_{rp} = \frac{2\pi(2.25 \times 10^3)(50 \times 10^{-3})}{50} = 14.13$$

d. Finally, the bandwidth is obtained from Eq. (13.31):

$$\text{BW} = \frac{2.25 \times 10^3}{14.13} = 159 \text{ Hz}$$

e. Since Q is relatively high, it is assumed that the response plot is symmetrical about f_{rp}, and by Eq. (13.11),

$$f_H = 2.25 \times 10^3 + \frac{159}{2} \simeq 2.33 \text{ kHz}$$

f. By Eq. (13.10),

$$f_L = 2.25 \times 10^3 - \frac{159}{2} \simeq 2.17 \text{ kHz}$$

Parallel resonant circuits are driven from current sources rather than voltage sources. Voltage sources have low internal resistances, causing $Z_{T_{rp}}$ to become small. Electronic transistor circuits simulate current sources with high internal resistances.

EXAMPLE 13.11

A practical parallel resonant circuit is connected to an electronic transistor circuit, as shown in Figure 13.13. The capacitor is variable from 10 pF to 200 pF. Determine the range of frequencies for which the circuit can be resonant if the capacitor is varied from minimum to maximum capacitance.

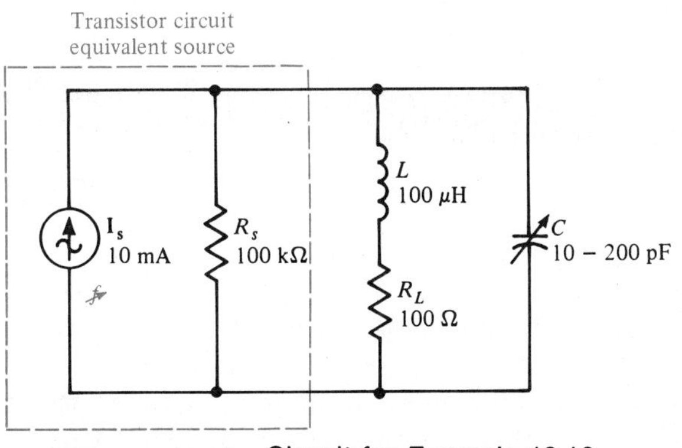

Figure 13.13 Circuit for Example 13.10.

Solution

The resonant frequency, f_{rp}, with a capacitance value of 10 pF, is given by Eq. (13.21):

$$f_{rp1} = \frac{0.153}{\sqrt{(1 \times 10^{-4})(10 \times 10^{-12})}} \sqrt{1 - \frac{(100^2)(10 \times 10^{-12})}{1 \times 10^{-4}}}$$
$$\simeq 5.028 \times 10^5 = 5.028 \text{ MHz}$$

Similarly, for $C = 200$ pF,

$$f_{rp2} = \frac{0.159}{\sqrt{(1 \times 10^{-4})(2 \times 10^{-10})}} \sqrt{1 - \frac{(100^2)(2 \times 10^{-10})}{1 \times 10^{-4}}}$$
$$\simeq 1.124 \times 10^6 = 1.124 \text{ MHz}$$

Therefore, varying the capacitor fully will change the resonant frequency of the circuit from 1.124 MHz to 5.028 MHz.

The effect of R_L, L/C, and Q on the frequency response of the circuit is shown in Figures 13.14a, 13.14b, and 13.14c, respectively. The higher the value of R_L, the wider the bandwidth and the less selective the circuit is. The larger the ratio of L/C or the Q value, the more selective the circuit becomes.

13.4 DECIBELS

Alexander Graham Bell, who invented the telephone, also defined a logarithmic ratio of power, and the unit of that ratio is called the *bel.*

$$\text{bel} = \log_{10} \frac{P_2}{P_1} \tag{13.36}$$

A smaller unit, the **decibel,** proved to be more useful:

$$\text{dB} = \text{decibel} = 10 \log \frac{P_2}{P_1} \tag{13.37}$$

Hence

The decibel is the unit of the logarithmic ratio of two power levels.

In many instances it is more meaningful to know the ratio of two electrical quantities in a system than to know the magnitude of a quantity. An example of

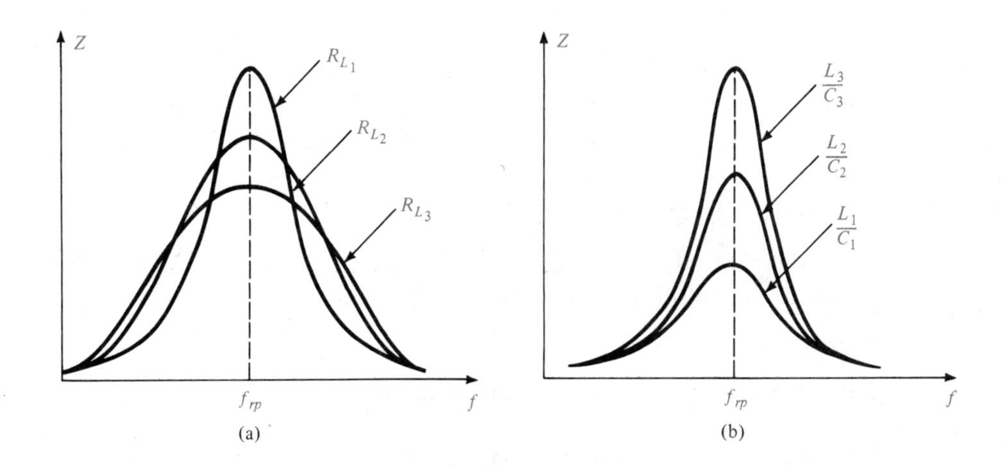

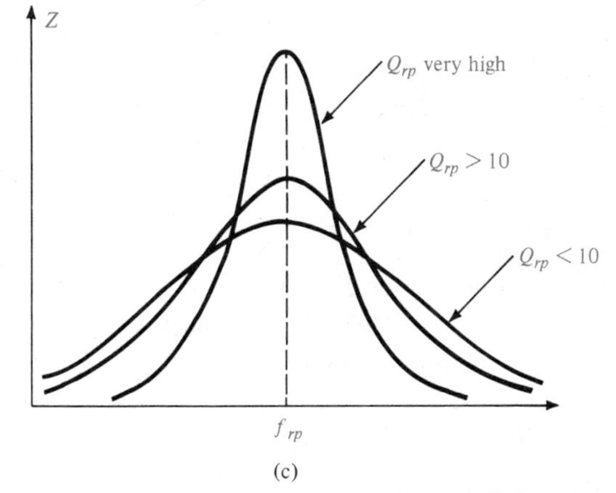

Figure 13.14 Frequency response curves of a parallel resonant circuit. **(a)** Effect of inductor resistance on impedance ($R_{L_1} < R_{L_2} < R_{L_3}$). **(b)** Effect of ratio L/C on impedance ($L_1/C_1 < L_2/C_2 < L_3/C_3$). **(c)** Effect of Q on frequency response.

this was shown earlier when we defined the cutoff frequencies to occur when the power delivered to the circuit is one-half of the maximum power delivered. In this case we are not interested in the magnitude of the power but rather in the ratio.

The reason why the logarithmic ratio is a popular measure has to do with the fact that it had its beginning in the telephone industry, which related power levels to sound intensity. The human ear interprets changes in sound intensity in a logarithmic manner; therefore, it makes sense to measure acoustical power changes as logarithmic ratios. However, the advantage of using logarithmic ratios in other than acoustical systems has to do with the fact that when systems made up of a number of subsystems are analyzed, it is easier to compute overall system ratios when they are expressed in decibels. This is demonstrated in this section.

In electronics, where *amplifiers* are used to magnify voltage, current, or power of the applied signal, it is very common to express the ratios of signal output power to input signal in decibels.

EXAMPLE 13.12

Determine the ratio in decibels of the power delivered to a resonant circuit at the cutoff frequencies to the maximum power.

Solution

Since the power delivered to a circuit at "cutoff" is $0.5P_{\max}$, then

$$\text{dB} = 10 \log \frac{0.5P_{\max}}{P_{\max}} = 10 \log 0.5 \simeq -3 \text{ dB}$$

The cutoff frequencies are also called the −3-dB frequencies because at frequencies f_L and f_H, the power is *down* 3 dB from the maximum.

EXAMPLE 13.13

The symbol for an amplifier circuit is shown in Figure 13.15. If the input power to the amplifier is 50 mW and the output power is 1.0 W, what is the power gain of the amplifier in decibels?

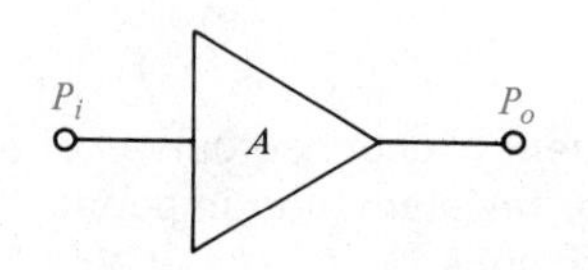

Figure 13.15 Symbol of amplifier circuit.

Solution

The gain of the amplifier is given by

$$\text{dB} = 10 \log \frac{P_0}{P_i} = 10 \log \frac{1}{50 \times 10^{-3}} \simeq 13 \text{ dB}$$

An amplifier circuit that yields power gain is an **active circuit.** Circuits that are not capable of having a gain are called **passive circuits.** The diagram in Figure 13.16 shows a *cascaded* system (output of one circuit connected to input of another) made up of active and passive circuits. *Passive circuits are* **attenuators.** Attenuators provide an output that is smaller than the input—hence, a loss. "Loss" is the opposite of "gain."

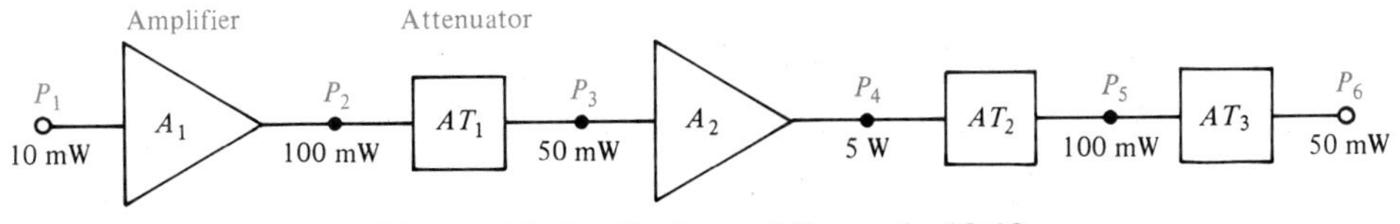

Figure 13.16 System of Example 13.13.

EXAMPLE 13.14

From Figure 13.16, determine:

a. The individual power gains both as a straight ratio and in decibels

b. The total power gain of the system as a straight ratio and in decibels

Solution

a. The gain of the first amplifier is:

$$A_1 = \frac{P_2}{P_1} = \frac{100 \times 10^{-3}}{10 \times 10^{-3}} = 10$$

The dB gain of the first amplifier is:

$$\text{dB}_{A_1} = 10 \log 10 = 10 \text{ dB}$$

The loss or attenuation of AT_1 is:

$$AT_1 = \frac{P_3}{P_2} = \frac{50 \times 10^{-3}}{100 \times 10^{-3}} = 0.5$$

In dB,

$$\text{dB}_{AT_1} = 10 \log 0.5 \simeq -3 \text{ dB}$$

For A_2:

$$A_2 = \frac{P_4}{P_3} = \frac{5}{50 \times 10^{-3}} = 100$$

and

$$\text{dB}_{A_2} = 10 \log 100 = 20 \text{ dB}$$

Continuing in the same manner,

$$AT_2 = \frac{P_5}{P_4} = \frac{100 \times 10^{-3}}{5} = 0.02$$

and

$$\text{dB}_{AT_2} = 10 \log 0.02 \simeq -17 \text{ dB}$$

Finally,

$$AT_3 = \frac{P_6}{P_5} = \frac{50 \times 10^{-3}}{100 \times 10^{-3}} = 0.5$$

resulting in

$$dB_{AT_3} = 10 \log 0.5 \simeq -3 \text{ dB}$$

b. The overall gain of the system can be obtained by

$$A_{TOT} = \frac{P_6}{P_1} = \frac{50 \times 10^{-3}}{10 \times 10^{-3}} = 5$$

or by finding the product of each of the stages:

$$\begin{aligned} A_{TOT} &= A_1 \times AT_1 \times A_2 \times AT_2 \times AT_3 \\ &= 10 \times 0.5 \times 100 \times 0.02 \times 0.5 = 5 \end{aligned}$$

The total gain in decibels is:

$$\begin{aligned} dB_{A_{TOT}} &= 10 \log \frac{P_6}{P_1} \\ &= 10 \log 5 \simeq 7 \text{ dB} \end{aligned}$$

A_{TOT} in decibels can be found by algebraically summing the individual decibel ratios. Recall that the log of the product of different values is equivalent to the sum of their logarithms. Therefore,

$$\begin{aligned} dB_{TOT} &= dB_{A_i} + dB_{AT_1} + dB_{A_2} + dB_{AT_2} = dB_{AT_3} \\ &= 10 - 3 + 20 - 17 - 3 = 7 \text{ dB} \end{aligned}$$

The power ratios of a circuit can be calculated using voltage and current ratios.

$$\boxed{A = \frac{P_2}{P_i} = \frac{V_2^2/R_2}{V_1^2/R_1} = \frac{I_2^2R_2}{I_1^2R_1}} \tag{13.38}$$

In decibels,

$$\begin{aligned} dB_A &= 10 \log \frac{P_2}{P_1} \\ &= 10 \log \frac{V_2^2/R_2}{V_1^2/R_1} = 10 \log \left[\left(\frac{V_2}{V_1}\right)^2 \left(\frac{R_1}{R_2}\right)\right] \end{aligned}$$

$$\boxed{dB_A = 20 \log \frac{V_2}{V_1} + 10 \log \frac{R_1}{R_2}} \tag{13.39}$$

In many practical systems the impedances at the input and output are matched; therefore, $R_1 = R_2$ and Eq. (13.39) reduces to:

$$\boxed{dB_A = 20 \log \frac{V_2}{V_1}} \tag{13.40}$$

Similarly, it can be shown that for $R_1 = R_2$,

$$\boxed{dB_A = 20 \log \frac{I_2}{I_1}} \tag{13.41}$$

EXAMPLE 13.15

An amplifier with a load resistance, R_1, of 100 Ω and a gain of 20 dB is driven from a 0.5-V source, as shown in Figure 13.17. Determine the power and voltage across the load if (a) the input resistance R_1 = 10 kΩ and (b) R_i = 100 Ω.

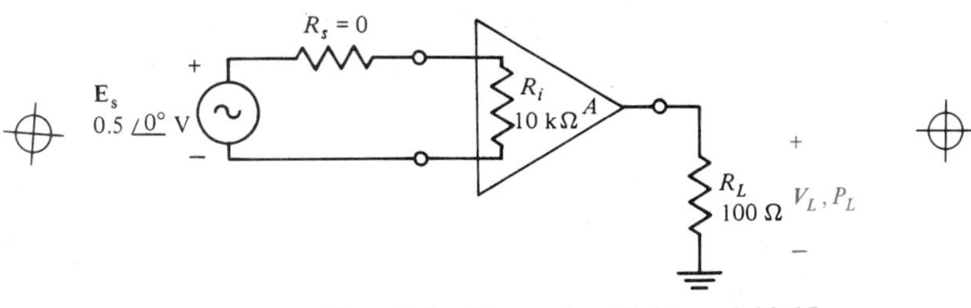

Figure 13.17 Circuit for Examples 13.14 and 13.15.

Solution

a. First, the input power is determined as

$$P_i = \frac{(0.5)^2}{10 \times 10^3} = 25\ \mu W$$

The output power is determined from the decibel gain relationship of the amplifier:

$$20\ dB = 10 \log \frac{P_L}{P_i}$$

$$= 10 \log \frac{P_L}{25 \times 10^{-6}}$$

Dividing both sides by 10 gives

$$2 = \log \frac{P_L}{25 \times 10^{-6}}$$

Taking the antilog of both sides yields

$$100 = \frac{P_L}{25 \times 10^{-6}}$$

and the load power, P_L, is:

$$P_L = 100(25 \times 10^{-6}) = 2.5 \text{ mW}$$

The voltage across the load is calculated from the power formula

$$P_L = \frac{V_L^2}{R_L}$$

Rearranging terms gives

$$V_L = \sqrt{P_L R_L} = \sqrt{2.5 \times 10^{-3} 100}$$
$$= 0.5 \text{ V}$$

b. Since the load resistance and the input resistance are both 100 Ω, the output voltage can be calculated by Eq. (13.40).

$$20 = 20 \log \frac{V_L}{0.5}$$

Dividing both sides by 20,

$$1 = \log \frac{V_L}{0.5}$$

Taking the antilog of both sides,

$$10 = \frac{V_L}{0.5}$$

Solving for V_L yields

$$V_L = 5 \text{ V}$$

and the power at the load is computed from the power-voltage relationship:

$$P_L = \frac{(5)^2}{100} = 0.25 \text{ W}$$

At times it is convenient to express power as a decibel ratio with reference to a known value of power. One such unit is dBW:

$$\text{dBW} = 10 \log \frac{P}{1 \text{ W}} \tag{13.42}$$

Hence, the reference power is 1 W. Another commonly used unit is dBm:

$$\boxed{\text{dBm} = 10 \log \frac{P}{1 \text{ mW}}} \tag{13.43}$$

with 1 mW as the reference.

EXAMPLE 13.16

Express 2 W in dBW and dBm.

Solution

By Eq. (13.42),

$$10 \log \frac{2}{1} \simeq 3 \text{ dBW}$$

and by Eq. (13.43),

$$10 \log \frac{2}{1 \times 10^{-3}} \simeq 33 \text{ dBm}$$

13.5 FILTERS

***Filters** are electric or electronic circuits that pass signals of some frequencies and attenuate signals at other frequencies.*

Filters constructed of passive components (R, L, C) are called **passive filters.** Those filters that contain amplifier circuits are called **active filters.** In this section we examine various types of passive filters. Active filters exhibit superior characteristics to passive filters and are studied as part of electronic circuits.

The simplest filter construction includes a resistor and a capacitor (RC circuit) or a resistor and an inductor (RL circuit). Using one of these combinations, we can build filters that will result in either *low-pass* or *high-pass* characteristics.

A low-pass filter passes signals at low frequencies and attenuates signals at high frequencies.

The frequency response characteristic of a low-pass filter is shown in Figure 13.18c, and the actual circuits, RC and RL, are shown in Figures 13.18a and 13.18b, respectively. An ideal low-pass filter passes signals at all frequencies up to the cutoff, or 3-dB frequency, f_c, and rejects all signals above that frequency. A practical passive filter attenuates all frequencies, as shown in Figure 13.18c; i.e.,

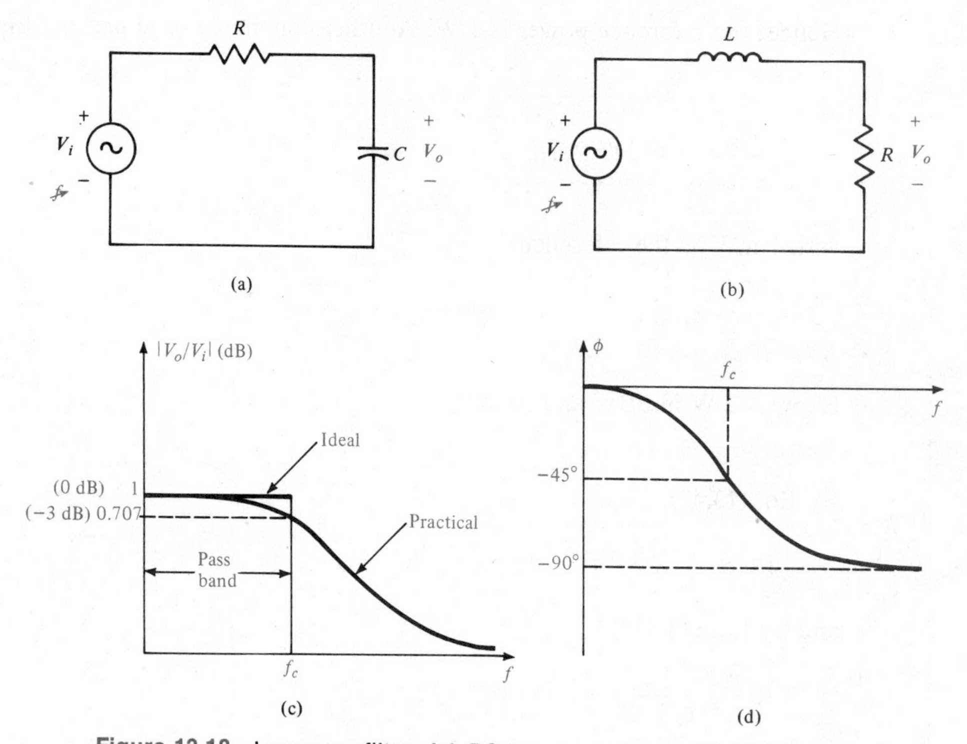

Figure 13.18 Low-pass filter. **(a)** *RC* low-pass circuit. **(b)** *RL* low-pass circuit. **(c)** Frequency response characteristic. **(d)** Phase angle versus frequency plot.

there is only a slight drop in magnitude (3 dB from maximum level) up to f_c and more significant attenuation above cutoff.

From the *RC* circuit in Figure 13.18a, we can conclude that at low frequencies the reactance of the capacitor, X_C, is high, causing most of the voltage V_i to be dropped across it. This results in a high-level V_o. At high frequencies X_C is low, and V_o decreases with an increase in the frequency of the signal. This filter is also referred to as a *shunt capacitance* filter because the load is connected in parallel with the capacitor.

In the *RL* network in Figure 13.18b, the inductor has a low reactive value, X_C at low frequencies; hence, signals at low frequencies are readily passed through, but signals at high frequencies are attenuated because the value of X_L increases with an increase in frequency.

The output voltage of the low-pass *RC* network is:

$$\mathbf{V_o} = \frac{\dfrac{\mathbf{V_i}}{j\omega C}}{R + \dfrac{1}{j\omega C}}$$

$$= \frac{\mathbf{V_i}}{1 + j\omega RC}$$

Solving for $\mathbf{V}_o/\mathbf{V}_i$,

$$\frac{\mathbf{V}_o}{\mathbf{V}_i} = \frac{1}{1 + j\omega RC} \tag{13.44}$$

Letting

$$\omega_c = \frac{1}{RC} \tag{13.45}$$

Substituting Eq. (13.45) into Eq. (13.44) yields

$$\frac{\mathbf{V}_o}{\mathbf{V}_i} = \frac{1}{1 + j\omega/\omega_c} \tag{13.46}$$

The magnitude of Eq. (13.46) becomes

$$\left|\frac{\mathbf{V}_o}{\mathbf{V}_i}\right| = \frac{1}{\sqrt{1 + (\omega/\omega_c)^2}} \tag{13.47}$$

and the phase angle ϕ, obtained from Eq. (13.46), is:

$$\phi = -\tan^{-1}\left(\frac{\omega}{\omega_c}\right) \tag{13.48}$$

From Eq. (13.45), the cutoff frequency, f_c, is:

$$f_c = \frac{1}{2\pi RC} = \frac{0.159}{RC} \tag{13.49}$$

The plot of the phase angle ϕ versus frequency is shown in Figure 13.18d. Notice that the maximum phase shift between input and output is $-90°$ and is exactly $-45°$ at the cutoff frequency.

EXAMPLE 13.17

Design a low-pass *RC* filter to have a cutoff frequency f_c of 20 kHz and a resistance R of 10 kΩ.

Solution

From Figure 13.18a, the only unknown is the capacitor, C. Rearrange Eq. (13.49):

$$C = \frac{0.159}{Rf_c} = \frac{0.159}{10 \times 10^3 \times 20 \times 10^3}$$
$$= 7.95 \times 10^{-10} = 795 \text{ pF}$$

The response of the *RL* low-pass filter is explained next. From the circuit in Figure 13.18b,

$$\mathbf{V_o} = \frac{\mathbf{V_i}R}{R + j\omega L}$$
$$= \frac{\mathbf{V_i}}{1 + j\omega L/R}$$

Solving for $\mathbf{V_o}/\mathbf{V_i}$ yields

$$\boxed{\frac{\mathbf{V_o}}{\mathbf{V_i}} = \frac{1}{1 + j\omega L/R}} \tag{13.50}$$

Letting

$$\boxed{\omega_c = \frac{R}{L}} \tag{13.51}$$

and substituting in Eq. (13.50) results in

$$\boxed{\frac{\mathbf{V_o}}{\mathbf{V_i}} = \frac{1}{1 + j\omega/\omega_c}} \tag{13.52}$$

Notice that Eqs. (13.46) and (13.52) are identical; thus, Eqs. (13.47) and (13.48) can be used for the *RL* circuit. However, for the *RL* circuit,

$$\boxed{f_c = \frac{R}{2\pi L}} \tag{13.53}$$

A straight-line approximation of the magnitude and phase angle response to frequency is shown in Figure 13.19. The plot is drawn on a semilog scale, with the frequency on the logarithmic axis. This type of plot was first proposed by H. W. Bode; hence, it is called a *Bode plot.* With the cutoff frequency known, it is possible to draw an approximation to the actual response curves. With a simple *RC* and *RL* low-pass filter, the magnitude plot is "flat" at 0 dB up to f_c, and then the response attenuates, with a slope of -20 dB per decade. A decade means a tenfold increase in frequency. The phase plot is flat at 0° up to $\frac{1}{10}$ of f_c; then it drops with a slope of $-45°$ per decade up until it reads $10f_c$ and stays flat at $-90°$ for all remaining values of f.

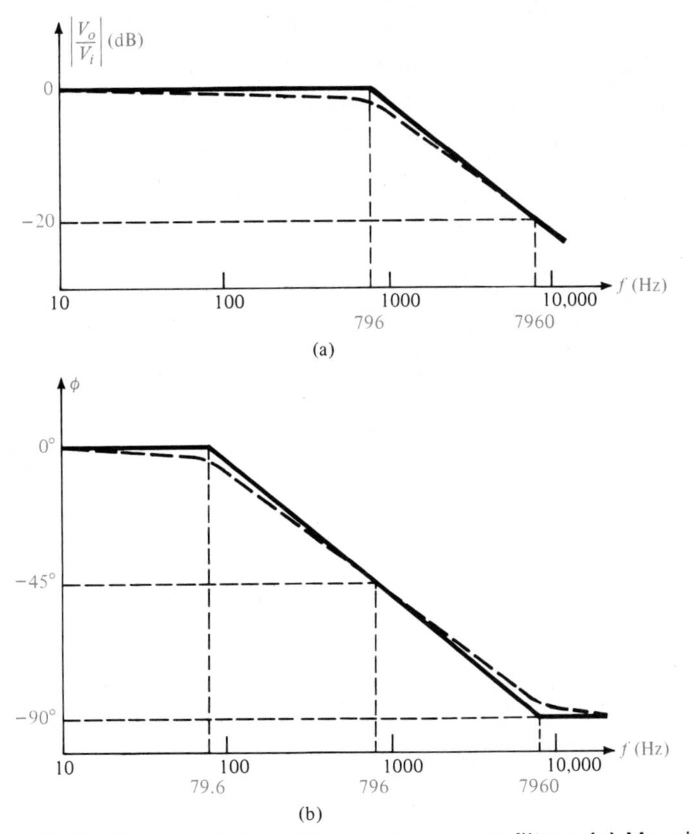

Figure 13.19 Bode plots for *RC* and *RL* low-pass filters. **(a)** Magnitude plot. **(b)** Phase plot.

EXAMPLE 13.18

An *RL* low-pass filter consists of a resistor $R = 1$ kΩ and an inductor $L = 0.2$ H. Determine the cutoff frequency f_c, and plot the Bode magnitude and phase angle versus frequency characteristics.

Solution

By Eq. (13.53),

$$f_c = \frac{10^3}{2\pi \times 0.2} = 796 \text{ H}_2$$

The Bode plots of magnitude and phase are shown in Figure 13.19. The dashed lines represent the actual plot.

High-pass filters attenuate signals up to the cutoff frequency and pass signals above the cutoff frequency.

Two simple high-pass filters are shown in Figure 13.20. Notice that the circuits are identical to the low-pass configurations. By taking the output across the resistor in the *RC* network and across the inductor in the *RL* network, the circuits become high-pass filters.

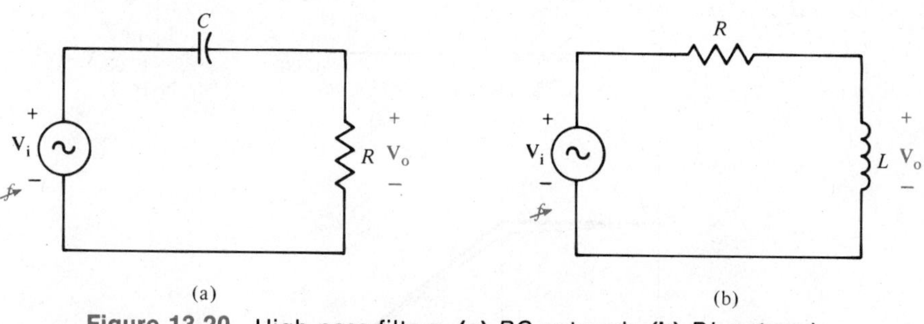

Figure 13.20 High-pass filters. **(a)** *RC* network. **(b)** *RL* network.

The equation for the output voltage $\mathbf{V_o}$ in the *RC* network is obtained from:

$$\mathbf{V_o} = \frac{\mathbf{V_i}(R)}{R + \dfrac{1}{j\omega C}}$$

$$= \frac{\mathbf{V_i}}{1 - \dfrac{j}{\omega RC}}$$

Furthermore,

$$\boxed{\frac{\mathbf{V_o}}{\mathbf{V_i}} = \frac{1}{1 - \dfrac{1}{\omega RC}}} \tag{13.54}$$

Letting

$$\omega_c = \frac{1}{RC} \tag{13.55}$$

The preceding result shows that the cutoff frequency is the same regardless if the circuit is used as a low-pass or high-pass filter:

$$f_c = \frac{0.159}{RC} \tag{13.56}$$

Equation (13.54) becomes

$$\frac{\mathbf{V_o}}{\mathbf{V_i}} = \frac{1}{1 - \dfrac{j\omega C}{\omega}} \tag{13.57}$$

The magnitude of $\mathbf{V_o}/\mathbf{V_i}$ is:

$$\left|\frac{\mathbf{V_o}}{\mathbf{V_i}}\right| = \frac{1}{\sqrt{1 + \left(\dfrac{\omega_c}{\omega}\right)^2}} \tag{13.58}$$

and the phase angle of $\mathbf{V_o}/\mathbf{V_i}$ is given by

$$\phi = \tan^{-1}\left(\frac{\omega_c}{\omega}\right) \tag{13.59}$$

Similarly, it can be shown that for the *RL* high-pass circuit,

$$f_c = \frac{R}{2\pi L} \tag{13.60}$$

the magnitude of $\mathbf{V_o}/\mathbf{V_i}$ is given by Eq. (13.58), and the phase relation is given by Eq. (13.59). The Bode approximation and the actual plots of magnitude and phase versus frequency are shown in Figure 13.21.

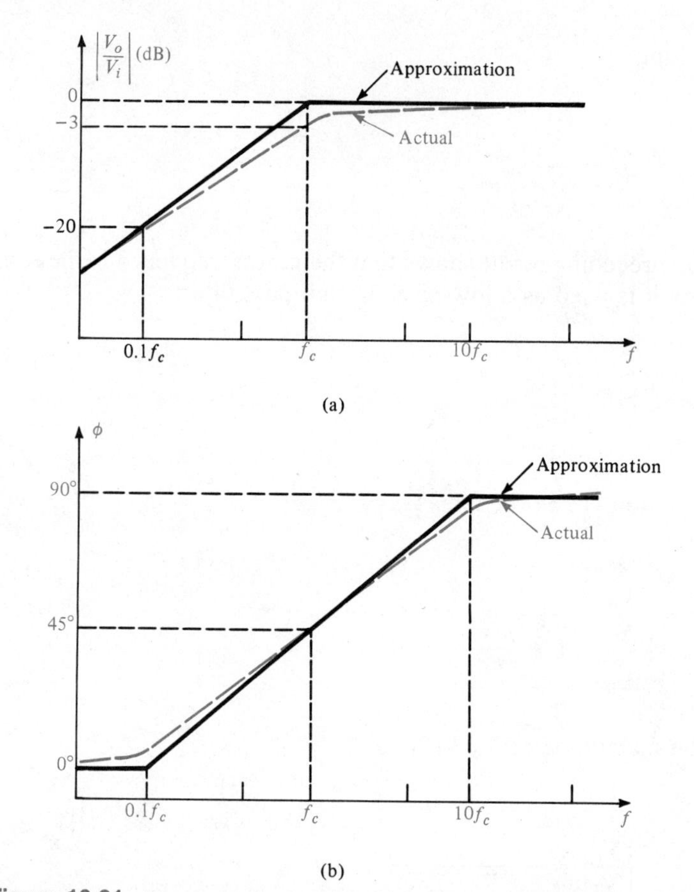

Figure 13.21 Responses of a high-pass filter. **(a)** Magnitude plot. **(b)** Phase angle plot.

A *band-pass* filter passes signals of frequencies within a particular band and attenuates signals at frequencies outside this band. Recall from previous discussions that *RLC* resonant circuits were selective to a band of frequencies; hence, they are band-pass filters. Band-pass filters are very useful in many applications. Some of the common uses are as follows:

1. Communications—to "tune in" a narrow band of radio frequencies
2. Audio—to equalize the sound level ("graphic equalizers")
3. Audio—speaker crossover networks

The magnitude characteristic of a band-pass filter is shown in Figure 13.22. Band-pass filters are classified as *narrow-band* or *wideband* filters. Generally, it is accepted that if the ratio of f_H to f_L is greater than 1.5, the filter is a wideband filter.

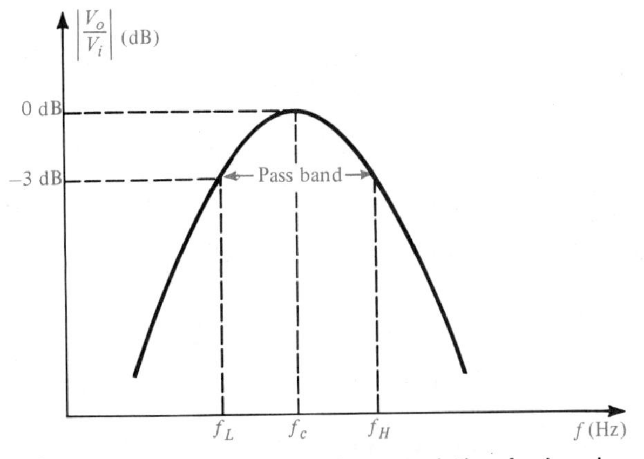

Figure 13.22 Frequency response characteristic of a band-pass filter.

EXAMPLE 13.19

A series *RLC* circuit is connected to a load R_L and an audio source, as shown in Figure 13.23a. Determine:

a. The center frequency, f_{rs}

b. f_L and f_H

c. BW

d. If the filter is a narrow-band or a wideband filter

e. The Bode plot of the magnitude, $|V_o/V_i|$

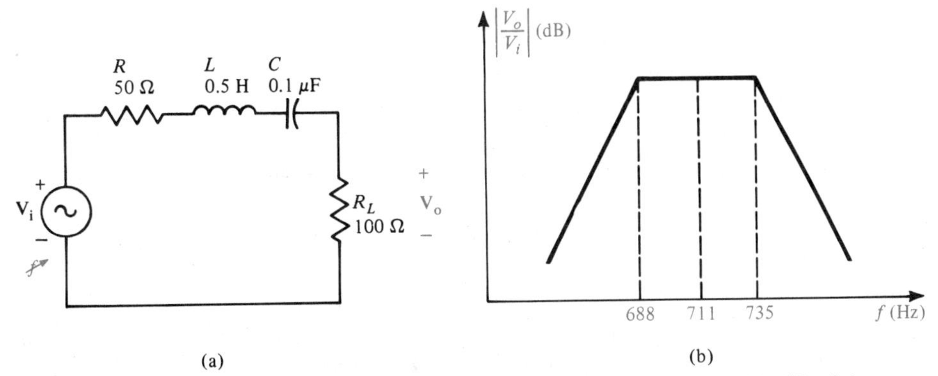

Figure 13.23 Band-pass series *RLC* circuit for Example 13.18. (Resistance *R* includes the source resistance and the resistance of the coil.) **(a)** Circuit diagram. **(b)** Bode plot.

Solution

a. The center frequency is determined from Eq. (13.3):

$$f_{rs} = \frac{0.159}{\sqrt{LC}} = \frac{0.159}{\sqrt{0.5 \times 10^{-7}}} = 711\ \text{Hz}$$

b. The low cutoff frequency is calculated by Eq. (13.15):

$$f_L = \frac{1}{2\pi}\left(-\frac{150}{2(0.5)} + \frac{1}{2}\sqrt{\left(\frac{150}{0.5}\right)^2 + \frac{4}{(0.5)(10^{-7})}}\right) = 688\ \text{Hz}$$

and f_H is calculated by Eq. (13.16):

$$f_H = \frac{1}{2\pi}\left(\frac{150}{2(0.5)} + \frac{1}{2}\sqrt{\left(\frac{150}{0.5}\right)^2 + \frac{4}{(0.5)(10^{-7})}}\right) = 735\ \text{Hz}$$

c. The bandwidth is simply

$$\text{BW} = f_H - f_L = 735 - 688 = 47\ \text{Hz}$$

d. From the preceding discussion, the value of the ratio f_H to f_L determines the type of band-pass filter:

$$\frac{f_H}{f_L} = \frac{735}{688} = 1.068$$

Since $1.068 < 1.5$, it is a *narrow-band* filter.

e. The Bode plot is shown in Figure 13.23b.

The circuit in Figure 13.23a can be viewed as two filters, a low-pass *RL* filter and a high-pass *RC* filter with the output across the resistor in both cases. If the cutoff frequency of the low-pass filter (*RL*) is greater than the cutoff frequency of the high-pass filter, the result is a band-pass filter with f_L and f_H determined by the cutoff frequencies.

The last type of filter that we will consider in this chapter is the *band-reject* filter.

Band-reject filters reject signals at frequencies within a specified band and pass signals at all other frequencies.

The frequency characteristic of a band-reject filter is shown in Figure 13.24. It is apparent that if two filters are used—a low-pass filter with a cutoff frequency, f_L,

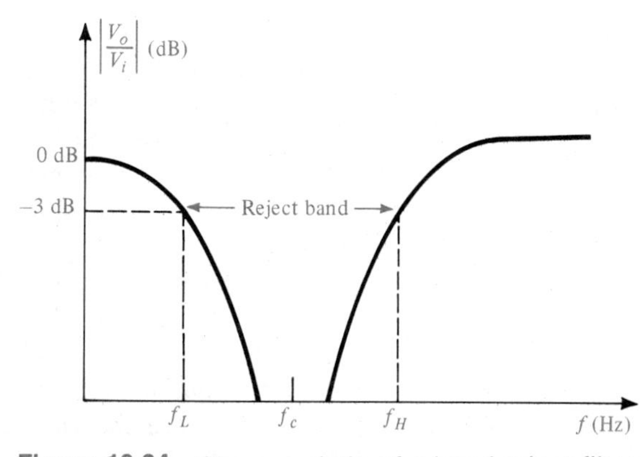

Figure 13.24 Characteristic of a band-reject filter.

lower than the cutoff frequency, f_H, of the high-pass filter—a band-reject filter will result.

In practice various *LC* circuit combinations are used to implement passive filters of any of the types discussed in this chapter. Examples of some *LC* filters are shown in Figure 13.25. Sections of *LC* filters can be interconnected to form multistage filters, resulting in various frequency response characteristics. Filter

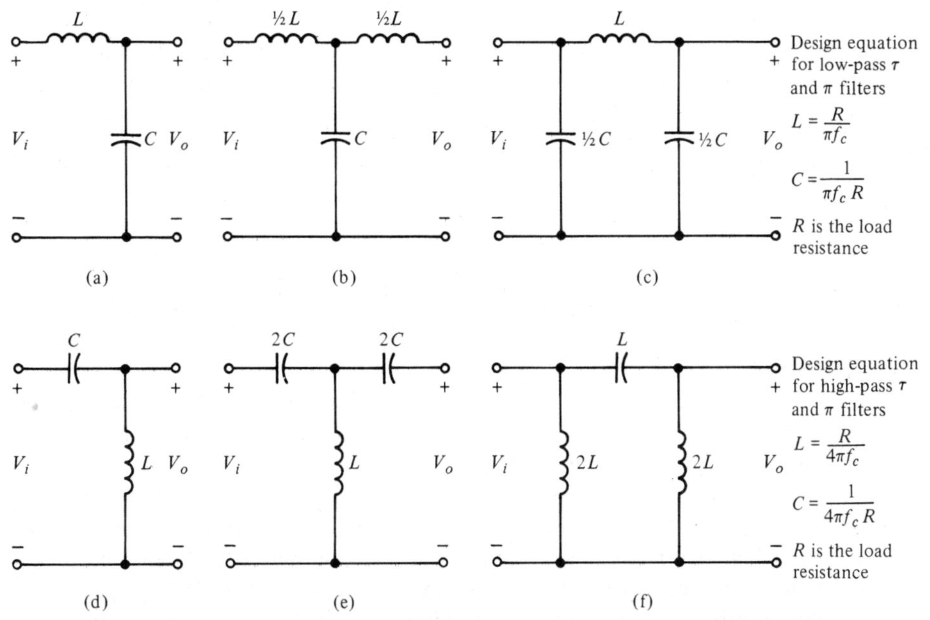

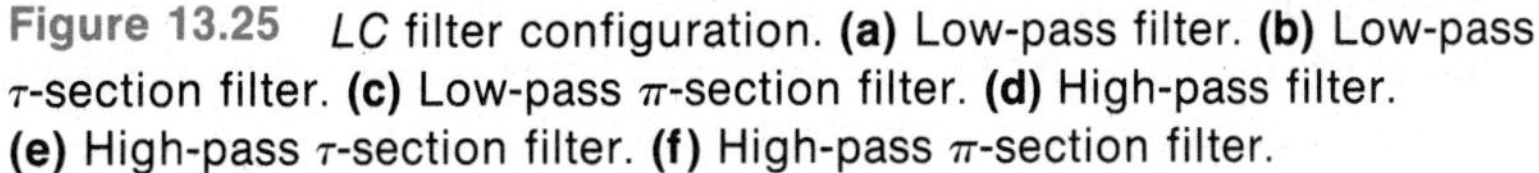

Figure 13.25 *LC* filter configuration. **(a)** Low-pass filter. **(b)** Low-pass τ-section filter. **(c)** Low-pass π-section filter. **(d)** High-pass filter. **(e)** High-pass τ-section filter. **(f)** High-pass π-section filter.

designers usually consult handbooks that feature standard filter designs, their corresponding equations, and frequency response characteristics.

Active filters, as mentioned earlier, are filters constructed of electronic amplifier circuits, resistors, and capacitors. The advantage of such filters lies in the fact that no inductors, which can be bulky and expensive, are needed; a voltage gain can be achieved through the amplifier; and input and output impedances independent of frequency are provided.

Quartz crystals are used as filters because they exhibit both series and parallel resonant characteristics with Q's as high as 1 million. These devices are perfect for making highly selective narrow-band filters. Crystal filters are constantly replacing LC filter elements in electronic communications. The symbol of a quartz crystal and its electrical equivalent circuit are shown in Figure 13.26.

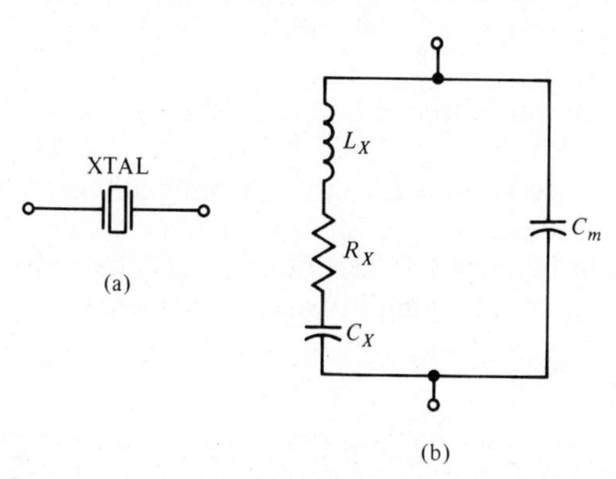

Figure 13.26 Quartz crystal. **(a)** Electrical symbol. **(b)** Electrical equivalent circuit.

SUMMARY

Resonance is the condition where, in an RLC circuit, the current and the voltage are in phase. This condition occurs if the impedance or admittance is purely resistive. The impedance is purely resistive if the capacitive reactance is equal to the inductive reactance.

In a series resonant circuit the impedance is minimum and the current is maximum when the circuit is driven at the resonant frequency. In a parallel resonant circuit, the total impedance is maximum and the voltage across the parallel branches is maximum. Series resonant circuits are usually supplied by voltage sources, and parallel resonant circuits are driven by current sources.

A measure of the selectivity of a resonant circuit is the quality factor, Q. The higher the value of Q, the more selective the circuit is, which also means that its bandwidth is narrow. The bandwidth is the difference between the high and low cutoff frequencies. At the cutoff frequencies, power delivered to the load is one-half of the maximum power delivered to the circuit.

Electrical filters are circuits that pass signals at a specific range of frequencies and attenuate signals at all other frequencies. The following types of filters were examined: *low-pass, high-pass, band-pass,* and *band-reject*. Only passive filters were discussed. Sections of R, L, and C networks can be interconnected to form filters with a variety of frequency response characteristics. Bode plots are a convenient method for presenting an approximation of frequency responses.

SUMMARY OF IMPORTANT EQUATIONS

$$\omega_{rs} = \frac{1}{\sqrt{LC}} \tag{13.2}$$

$$f_{rs} = \frac{1}{2\pi\sqrt{LC}} = \frac{0.159}{\sqrt{LC}} \tag{13.3}$$

$$\text{BW} = f_H - f_L \tag{13.8}$$

$$Q = \frac{f_r}{\text{BW}} \tag{13.9}$$

$$f_L = f_r - \frac{\text{BW}}{2} \tag{13.10}$$

$$f_H = f_r + \frac{\text{BW}}{2} \tag{13.11}$$

$$Q_{rs} = \frac{1}{R}\sqrt{\frac{L}{C}} \tag{13.13}$$

$$\text{BW} = \frac{\dfrac{0.159}{\sqrt{LC}}}{\dfrac{1}{R}\sqrt{\dfrac{L}{C}}} = \frac{0.159R}{L} \tag{13.14}$$

$$f_L = \frac{1}{2\pi}\left(-\frac{R}{2L} + \frac{1}{2}\sqrt{\left(\frac{R}{L}\right)^2 + \frac{4}{LC}}\right) \tag{13.15}$$

$$f_H = \frac{1}{2\pi}\left(\frac{R}{2L} + \frac{1}{2}\sqrt{\left(\frac{R}{L}\right)^2 + \frac{4}{LC}}\right) \tag{13.16}$$

$$f_{rp} = \frac{1}{2\pi\sqrt{LC}} \tag{13.18}$$

$$R'_L = \frac{R_L^2 + \omega^2 L^2}{R_L} \tag{13.19}$$

$$L' = \frac{R_L^2 + \omega^2 L^2}{\omega^2 L} \tag{13.20}$$

$$f_{rp} = \frac{0.159}{\sqrt{LC}} \sqrt{1 - \frac{R_L^2 C}{L}} \tag{13.21}$$

$$\mathbf{Z_{rp}} = \frac{R_L^2 + \omega_{rp}^2 L^2}{R_L} \tag{13.23}$$

$$\mathbf{Z_{rp}} = R'_{L_{rp}} = \frac{L}{R_L C} \tag{13.24}$$

$$\mathbf{Z_{T_{rp}}} = R_s \| R'_{L_{rp}} \tag{13.25}$$

$$Q_{rp} = \frac{2\pi f_{rp} L}{R_L} = Q \tag{13.30}$$

$$R'_L = R_L + Q^2 R_L \tag{13.32}$$

$$\text{dB} = \text{decibel} = 10 \log \frac{P_2}{P_1} \tag{13.37}$$

$$\text{dB}_A = 20 \log \frac{V_2}{V_1} \tag{13.40}$$

$$\text{dB}_A = 20 \log \frac{I_2}{I_1} \tag{13.41}$$

$$\text{dBW} = 10 \log \frac{P}{1\ \text{W}} \tag{13.42}$$

$$\text{dBm} = 10 \log \frac{P}{1\ \text{mW}} \tag{13.43}$$

$$\omega_c = \frac{1}{RC} \tag{13.45}$$

$$\left| \frac{\mathbf{V_o}}{\mathbf{V_i}} \right| = \frac{1}{\sqrt{1 + (\omega/\omega_c)^2}} \tag{13.47}$$

$$\phi = -\tan^{-1}\left(\frac{\omega}{\omega_c}\right) \tag{13.48}$$

$$f_c = \frac{R}{2\pi L} \tag{13.53}$$

$$\phi = \tan^{-1}\left(\frac{\omega_c}{\omega}\right) \tag{13.59}$$

REVIEW QUESTIONS

13.1 Define resonance.

13.2 What circuit components determine the resonant frequency in a series *RLC* circuit?

13.3 What is the value of the impedance in a series *RLC* circuit at resonance?

13.4 Is it possible for the voltage across the capacitor or inductor to be higher than the applied voltage in a series resonant circuit? Under what condition? Why?

13.5 What is the significance of the frequency response plot?

13.6 Define bandwidth.

13.7 Define quality factor, Q.

13.8 How is the bandwidth related to the quality factor?

13.9 What determines if a resonant circuit is a high Q or low Q circuit?

13.10 How does the resistance value affect the bandwidth of a series resonant circuit?

13.11 What is the difference between an ideal and a practical parallel resonant circuit?

13.12 Under what condition is $Q_{rp} = Q$ in a parallel resonant circuit?

13.13 Under what condition is the resonant frequency of an ideal parallel *RLC* circuit the same as that of a practical parallel circuit?

13.14 At what frequency is the voltage maximum across the capacitor in a parallel resonant circuit?

13.15 Define the unit decibel.

13.16 What does cascading of stages refer to?

13.17 Explain the difference between a passive filter and an active filter.

13.18 List the different configurations for a low-pass filter.

13.19 What are Bode plots?

13.20 Explain the characteristics of a band-reject filter.

13.21 List three applications of electrical filters.

13.22 List the advantages of an active filter over a passive filter.

PROBLEMS

13.1 In the series circuit of Figure 13.27, $E_s = 20$ V, $R = 10\ \Omega$, $L = 1$ mH, and $C = 500$ pF. Calculate: (a) f_{rs} and (b) the current I at resonance.

13.2 Repeat Problem 13.1 for $C = 0.05\ \mu$F. (ans. 22.5 kHz, 2 A)

13.3 Determine the voltage dropped $\mathbf{V_R}$, $\mathbf{V_L}$, and $\mathbf{V_C}$ at resonance for the values in Problem 13.1.

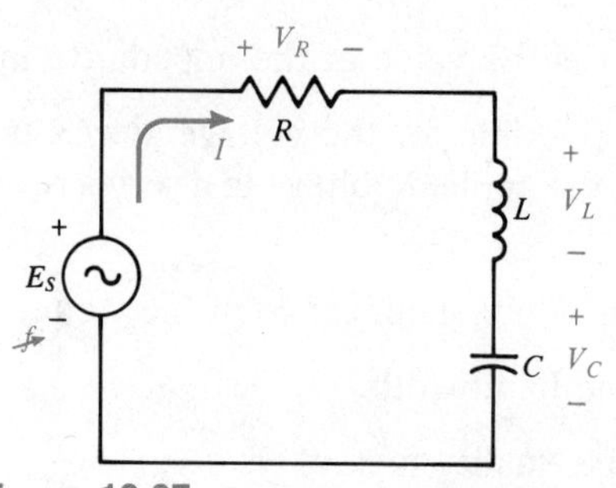

Figure 13.27 Series resonant circuit.

13.4 For the circuit of Figure 13.27, $R = 100\ \Omega$, $C = 200$ pF, determine the value of L if the resonant frequency is 50 kHz. (ans. 0.05 H)

13.5 Repeat Problem 13.4 for $f_{rs} = 100$ kHz.

13.6 Determine the quality factor, Q_{rs}, of the circuit in Problem 13.1. (ans. 140)

13.7 Determine the bandwidth of a series circuit with $R = 1$ kΩ, $L = 10\ \mu$H, and $C = 100$ pF. Repeat for $R = 100\ \Omega$.

13.8 Determine f_L and f_H in the circuit of Problem 13.7. (ans. 1.46 MHz, 17.36 MHz)

13.9 A series resonant circuit with $R = 60\ \Omega$, $L = 75$ mH, and $C = 500$ pF is energized by a 10-V source.
a. Calculate the bandwidth and the resonance frequency.
b. Repeat (a) for $R = 200\ \Omega$.

13.10 Calculate the value of R in a series resonant circuit with $L = 680\ \mu$H and $C = 150$ pF to yield a bandwidth of 10 kHz. (ans. 42.77 Ω)

13.11 Determine the cutoff frequencies, f_H and f_L, in the circuit of Problem 13.10.

13.12 In the practical tank circuit of Figure 13.28, $R_s = 100$ kΩ, $L = 100\ \mu$H, $R_L = 100\ \Omega$, and $C = 1000$ pF. Calculate the resonant frequency, f_{rp}. (ans. 477.19 kHz)

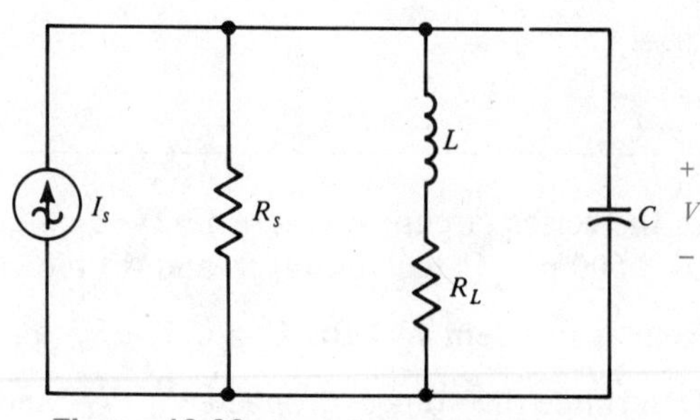

Figure 13.28 Parallel resonant circuit.

13.13 Repeat Problem 13.12 for $C = 100$ pF.

13.14 For Problem 13.12 calculate the total impedance $\mathbf{Z}_{\mathbf{T}_{rp}}$. (ans. $\approx$1 kΩ)

13.15 The circuit in Figure 13.28 has the following component values: $R_s = 1$ MΩ, $L = 220$ μH, and $R_L = 120$ Ω. Determine the value of C for a resonant frequency $f_{rp} = 100$ kHz.

13.16 Determine the quality factor Q_{rp} and the bandwidth for the circuit in Problem 13.12. (ans. 3, 159.06 kHz)

13.17 A practical parallel resonant circuit with a high Q has a lower cutoff frequency of 18 kHz and a resonant frequency of 21 kHz. What is the bandwidth?

13.18 A tank circuit with $L = 6.8$ mH, $R_L = 80$ Ω, and $C = 200$ pF is driven by a 10-mA source with a 50-kΩ source resistance. Determine the voltage across the capacitor at resonance. (ans. 447.4 V)

13.19 Design a parallel resonant circuit to have $f_{rp} = 500$ kHz when driven from a very high ($\simeq \infty$) resistance current source. The Q of the circuit should be 100, and R_L of the coil is 5 Ω.

13.20 An amplifier delivers 2 W of power to a load with an input power of 25 mW. Calculate the power gain in decibels. (ans. 23 dB)

13.21 An amplifier has an input of 50 mW and an output of 10 mW. What is its attenuation in decibels?

13.22 In the system of Figure 13.29, what is the overall gain in decibels of the system? (ans. 12 dB)

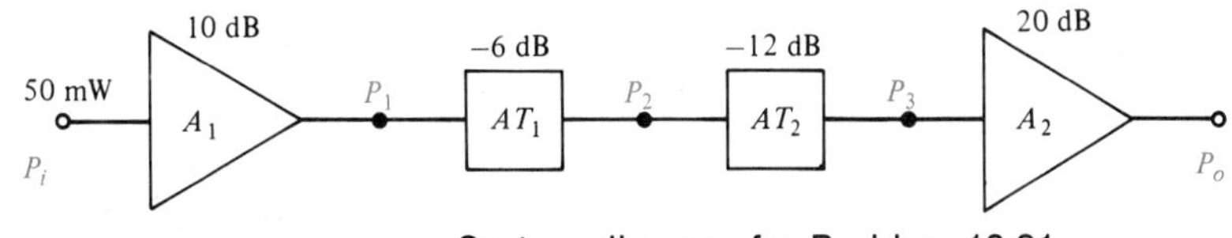

Figure 13.29 System diagram for Problem 13.21.

13.23 The system shown in Figure 13.29 is driven by a 50-mW source. Determine the output power.

13.24 Calculate the dBm value of 20 mW. (ans. 13 dBm)

13.25 Calculate the dBW and dBm values of 3.5 W.

13.26 A voltage amplifier has the same input and output resistance. If the input voltage is 20 mV and the gain of the amplifier is 23 dB, what is the output voltage of the amplifier? (ans. 0.285 V)

13.27 A voltage amplifier with an input resistance of 10 kΩ and an output resistance of 200 Ω is driven by a 100-mV source. Calculate the gain of the amplifier in decibels if the output voltage is 5 V.

13.28 An *RC* low-pass filter shown in Figure 13.30 has the following values: $R = 750\ \Omega$ and $C = 0.1\ \mu F$. Calculate the cutoff frequency, f_c, and sketch the Bode magnitude $|\mathbf{V_o}/\mathbf{V_i}|$ and phase angle responses. (ans. 2.12 kHz)

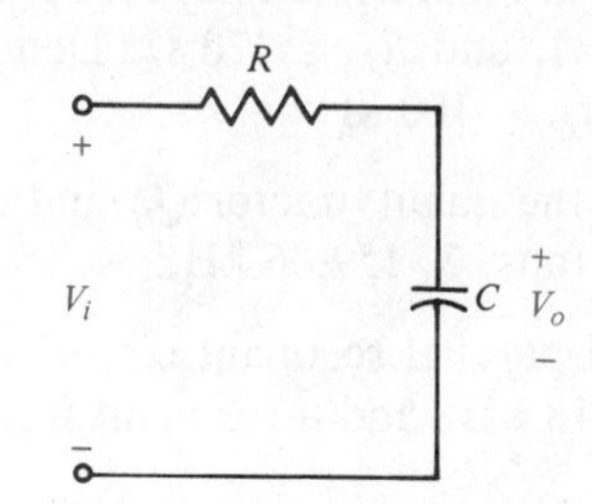

Figure 13.30 Low-pass filter for Problems 13.27 and 13.28.

13.29 Repeat Problem 13.28 for $R = 100\ \Omega$ and $C = 1\ \mu F$.

13.30 Sketch the Bode magnitude and phase angle responses for the *RC* high-pass filter, with $R = 750\ \Omega$ and $C = 0.1\ \mu F$, shown in Figure 13.31. (ans. 2.12 kHz)

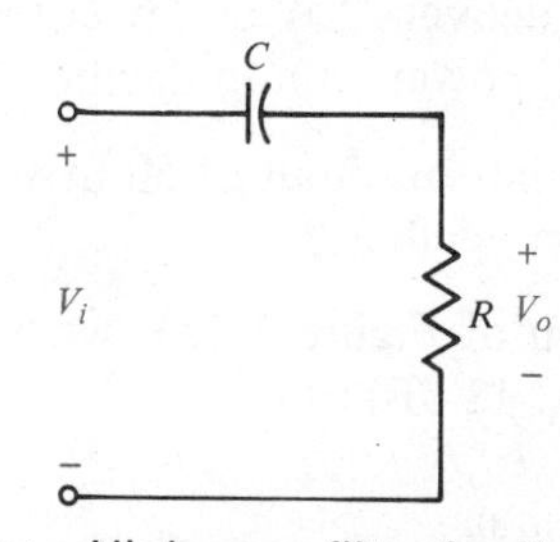

Figure 13.31 High-pass filter for Problem 13.28.

13.31 The *RL* circuit in Figure 13.32 can be used as a low-pass and a high-pass filter. The output voltage across the inductor is the high-pass output (V_{oH}). The voltage across the resistor is the low-pass output (V_{oL}). If $L = 5$ H and $R = 100\ \Omega$, calculate the cutoff frequency of the circuit, and plot the Bode magnitude plots, $|\mathbf{V_{oH}}/\mathbf{V_i}|$ and $|\mathbf{V_{oL}}/\mathbf{V_i}|$, both on the same set of axes.

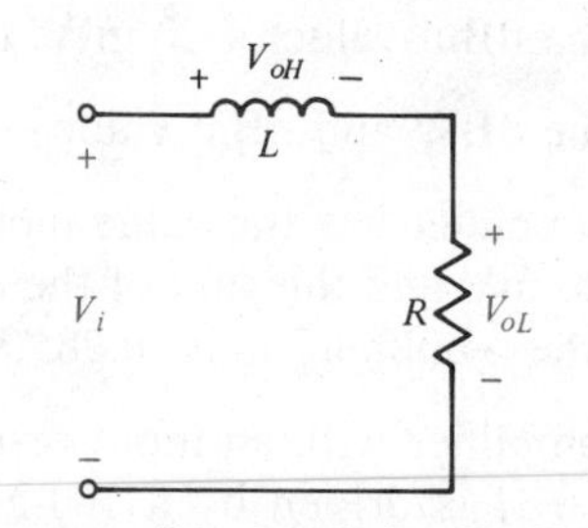

Figure 13.32 *RL* circuit for Problem 13.29.

13.32 A series RLC circuit is shown in Figure 13.33. Calculate the resonant (center) frequency of the filter, f_L and f_H. Sketch the Bode plot of $|\mathbf{V_o}/\mathbf{V_i}|$. (ans. 21,171 Hz, 20,988 Hz, 21,306 Hz)

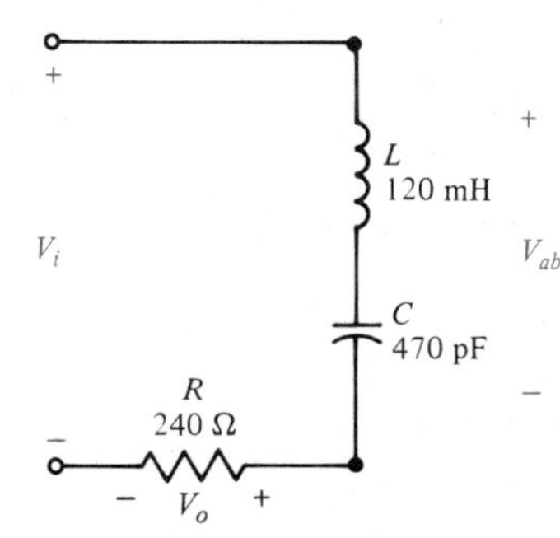

Figure 13.33 *RLC* circuit for Problems 13.30, 13.31, and 13.32.

13.33 Repeat Problem 13.32 for $L = 500$ mH.

13.34 What type of filter will result if the output of the circuit in Figure 13.33 is taken across points $ab(V_{ab})$? What is the center frequency? (ans. 21,171 Hz)

13.35 For the tank circuit of Figure 13.34, which is used as a band-pass filter, determine the center frequency, f_H and f_L.

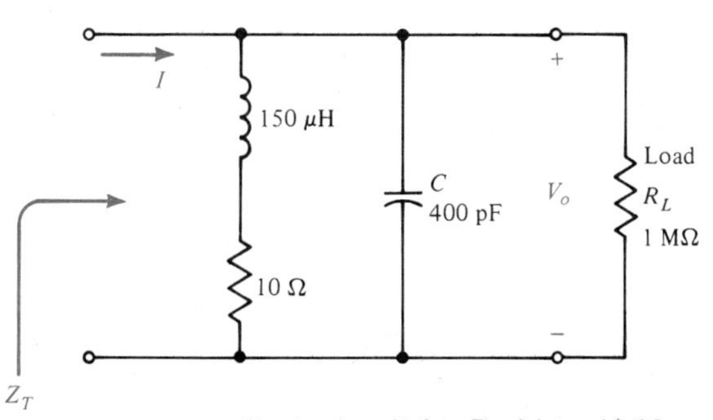

Figure 13.34 Tank circuit for Problem 13.33.

13.36 Sketch the Bode magnitude plot of $|\mathbf{Z_T}|$ of the circuit in Figure 13.34. (ans. 1 MΩ)

13.37 Using the low-pass section filter circuit in Figure 13.25b and the corresponding equations, design a filter with a cutoff frequency of 1 kHz if the load is a resistor, $R = 10$ kΩ.

13.38 Repeat Problem 13.35 for the high-pass filter in Figure 13.25f. (ans. 1.59 H, 0.00796 μF)

SECTION IV

FREQUENCY, POWER, AND INSTRUMENTATION

CHAPTER 14

Power (Single-Phase and Polyphase Systems)

GLOSSARY

active (average, real) power—power to resistive components
apparent power—product of the effective voltage and current
balanced 3-ϕ load—three equal independent loads powered from a 3-ϕ source

instantaneous power—product of voltage and current at an instant in time
phase sequence—the order in which the phases of voltage or current follow each other
polyphase circuit—an alternating current circuit powered by sources that generate a number of voltages of the same magnitude and frequency and are related in time
power factor—the ratio of active power to the apparent power
reactive power—power supplied by the source for energy storage in reactive components
transformer—a device used in electrical energy distribution to transfer energy inductively from one circuit to another by stepping the voltage up or down

INTRODUCTION

Direct current electric power considerations were discussed in Chapters 4 and 6. In this chapter the primary type of electric power, ac power, will be examined. Electric energy is supplied by utility companies worldwide primarily as alternating sinusoidal voltage and current. In the United States, the power companies operate and distribute electric energy in the form of *three-phase,* sinusoidal, 60-Hz power. Three-phase power systems are the "backbone" of industry. Only residential and some commercial users utilize *single-phase* power, which is derived from a *polyphase* system, of which three-phase is most commonly used.

Sinusoidal electric power is generated either by *rotating machines,* called generators, or by oscillator-type electronic circuits. The distribution of electric power is via electrical wires and with the help of transformers. A power system also contains switches, fuses, breakers, and other control devices through which a load is powered. In this chapter we examine and analyze power systems.

14.1 POWER IN SINUSOIDAL CIRCUITS

When connected to a sinusoidal source, a network consisting of resistors, inductors, and capacitors dissipates energy in the form of heat due to the resistive element only. Capacitors store energy during one-quarter of the sinusoidal cycle and deliver the stored energy to the circuit during the next quarter of the cycle. Inductors have a similar effect on the circuit. Hence, the resistors are the only components that dissipate energy. There is a constant exchange of shared energy between the magnetic field of the inductor and the electric field of the capacitor, which influences the current in the circuit but does not affect the average power.

*The power to the resistances in a circuit is called the **active, real,** or **average power.***

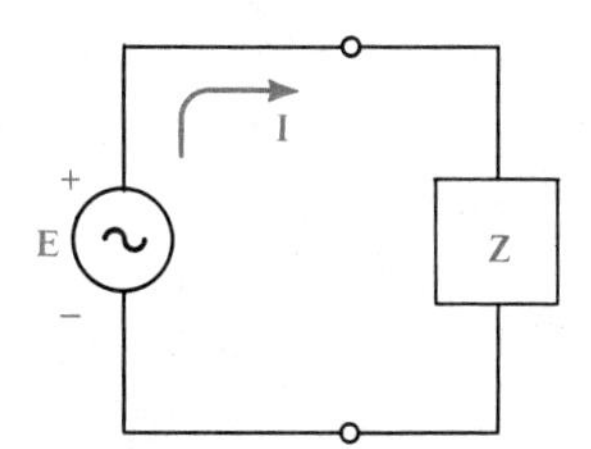

Figure 14.1 Generalized ac load connected to a sinusoidal source.

Figure 14.1 depicts a general impedance of a load connected to a sinusoidal source. The load consists of resistors, inductors, and capacitors. The instantaneous power, p, is given by

$$\boxed{p = vi} \qquad (14.1)$$

where the instantaneous voltage, v, is:

$$v = V_m \cos(\omega t + \theta)$$

and the resulting instantaneous current is:

$$i = I_m \cos\omega t$$

θ is the phase angle between the *voltage* and the *current* ($\theta = \theta_v - \theta_i$). Equation (14.1) can be rewritten as

$$\begin{aligned} p &= V_m I_m \cos(\omega t) \cos(\omega t + \theta) \\ &= 2VI \cos(\omega t) \cos(\omega t + \theta) \end{aligned}$$

where $V = V_{\text{eff}}$. Using the trigonometric identity

$$\cos A \cos B = \tfrac{1}{2}[\cos(A - B) + \cos(A + B)]$$

the instantaneous power equation becomes

$$\boxed{p = VI \cos\theta + VI \cos(2\omega t + \theta)} \qquad (14.2a)$$

If the circuit is purely resistive, then there is no phase difference between the voltage and the current, $\theta = 0°$, $\cos 0° = 1$, and Eq. (14.2a) reduces to

$$\boxed{p_R = VI(1 + \cos 2\omega t)} \qquad (14.2b)$$

and if the circuit is purely reactive (capacitive or inductive, $\cos \pm 90° = 0$), Eq. (14.2a) becomes

$$p_X = VI \cos(2\omega t \pm 90°)$$

$$= \boxed{\pm VI \sin(2\omega t)} \qquad \begin{array}{l} + \text{ inductive} \\ - \text{ capacitive} \end{array} \qquad (14.2c)$$

The plots of current, voltage, and power for the general circuit [Eq. (14.2a)] are shown in Figure 14.2. The power waveform is at double the frequency of the voltage and current waves. The first term in Eq. (14.2a), $VI \cos\theta$, is a constant and represents the average value of the power waveform. Since the average power is the power to the resistive part only, then

$$\boxed{P_{av} = P = VI \cos\theta} \qquad \text{W} \qquad (14.3)$$

Notice that the product of effective voltage and current is not the real power supplied or average power supplied to the circuit. The product of voltage and current supplied by the source is called the **apparent power** and is measured in *volt-ampere* (VA) units.

$$\boxed{P_{app} = S = VI} \qquad \text{VA} \qquad (14.4)$$

Equation (14.3) is now rewritten as

$$\boxed{P = S \cos\theta} \qquad \text{W} \qquad (14.5)$$

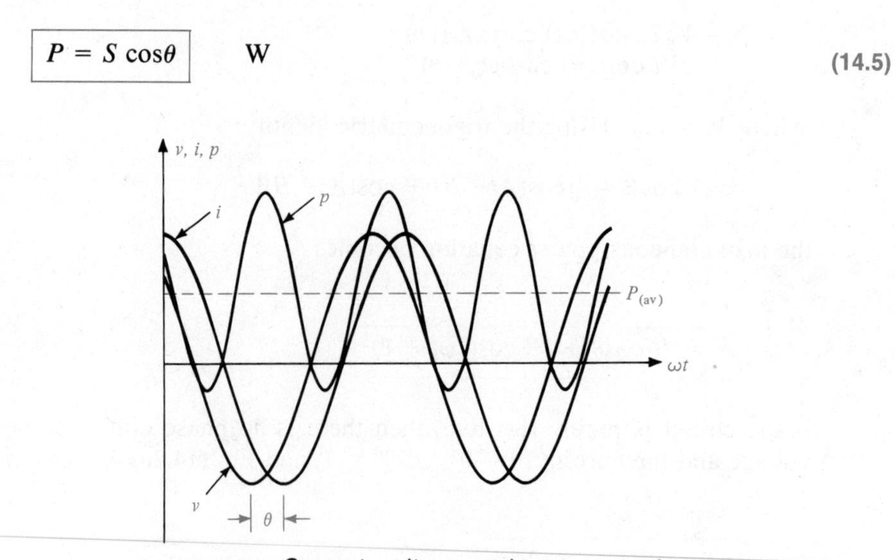

Figure 14.2 Current, voltage, and power waveforms.

The term "apparent" power stems from the fact that the source is apparently supplying that power, which is $V \times I$. For a *purely* resistive circuit, $\theta = 0°$ and

$$\boxed{P_R = VI = S} \qquad \text{W} \tag{14.6}$$

Since the real power loss in a general circuit is not the same as the product VI, there must be a quantity that accounts for the difference between the two. This quantity is called **reactive power,** Q or P_q. It is the power supplied by the source for energy storage in the reactive components. (The symbol Q here is not to be confused with the quality factor discussed in Chapter 13.) It is mathematically defined as:

$$\boxed{Q = VI \sin\theta} \qquad \text{VAR} \tag{14.7}$$

VAR stands for *volt-ampere reactive*.

The power triangle is a vectorial representation of average power, reactive power, and apparent power. The power triangle in Figure 14.3 illustrates these relationships. This diagram depicts the power relationship for a circuit that has reactive power, Q_L, due to circuit inductance, and reactive capacitance, Q_c. However, the resultant reactive power, Q, is the difference between the two. In this case $Q_L > Q_c$. The Pythagorean relationship from the triangle yields the magnitude of S,

$$\boxed{S = \sqrt{P^2 + Q^2}} \qquad \text{VA} \tag{14.8}$$

The *complex power* S can be defined as

$$\boxed{\mathbf{S} = P + jQ = S\underline{/\theta}} \qquad \text{VA} \tag{14.9}$$

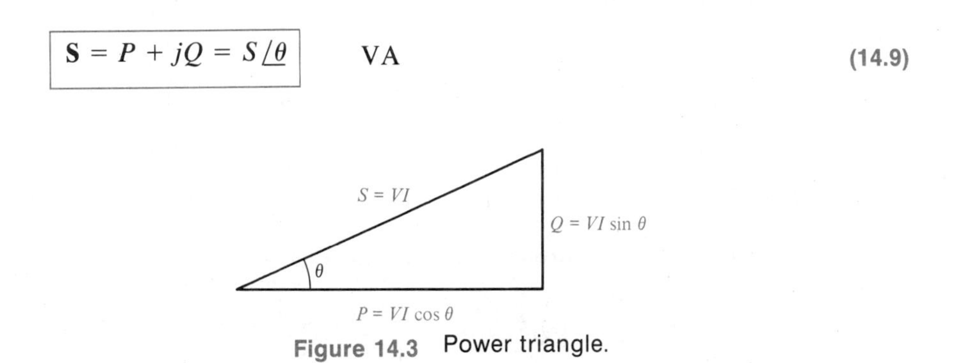

Figure 14.3 Power triangle.

The phase angle, θ, was defined earlier as the difference in phase between the voltage and the current waves. Hence, if the angle θ is the phase angle of the impedance **Z**, then the complex power is:

$$\boxed{\mathbf{S} = \mathbf{I}^2\mathbf{Z}} \qquad \text{VA} \tag{14.10}$$

The product **IV** is not equal to **S** because the angle of **S** would be the sum of the current and voltage angles. Since the angle of the complex power is the difference between the voltage and current angles, we write the complex power **S** as

$$\boxed{\mathbf{S} = \mathbf{V}\mathbf{I}^*} \qquad \text{VA} \tag{14.11}$$

where $\mathbf{I}^*$ stands for the conjugate of **I.** If $\mathbf{I} = \mathbf{I}\underline{/\theta}$, then $I^* = I\underline{/-\theta}$ and the angle of **S** is $\theta_v - \theta_i$.

The active power in a branch is given by

$$\boxed{P = I^2R} \qquad \text{W} \tag{14.12}$$

and the total reactive power of a branch is:

$$\boxed{Q = I^2X} \qquad \text{VAR} \tag{14.13}$$

EXAMPLE 14.1

A 60-Hz voltage source of 117 V is connected to an impedance $\mathbf{Z} = 100\underline{/-60^\circ}\ \Omega$. Calculate:

a. The apparent power, S

b. The active power, P

c. The reactive power, Q

d. Draw the power triangle.

Solution

a. The current through the impedance is:

$$\mathbf{I} = \frac{\mathbf{V}}{\mathbf{Z}} = \frac{117\underline{/0^\circ}}{100\underline{/-60^\circ}} = 1.17\underline{/60^\circ}\ \text{A}$$

The apparent power is calculated by Eq. (14.4):

$$S = 117 \times 1.17 = 136.9 \text{ VA}$$

b. The active power is calculated by Eq. (14.3), since $\theta = (0° - 60°) = -60°$:

$$P = 117 \times 1.17 \cos(-60°) = 68.45 \text{ W}$$

The active power can also be determined by using Eq. (14.12) if the value of R is calculated first. Converting the impedance value to rectangular form,

$$\mathbf{Z} = 100\underline{/-60°} = 50 - j86.6$$

The resistance value is $R = 50\ \Omega$. Consequently, by Eq. (14.12),

$$P = (1.17)^2\, 50 \simeq 68.45 \text{ W}$$

c. The reactive power is calculated by Eq. (14.7):

$$Q = 117 \times 1.17 \sin(-60°) = 118.55 \text{ VAR} \qquad \text{(capacitive)}$$

The negative sign is replaced by the indication that the network is capacitive. The same result can be obtained using Eq. (14.13), where X is the reactance of 86.6 Ω capacitive obtained before.

$$Q = (1.17)^2(86.6) \simeq 118.55 \text{ VAR} \qquad \text{(capacitive)}$$

d. The power triangle is shown in Figure 14.4.

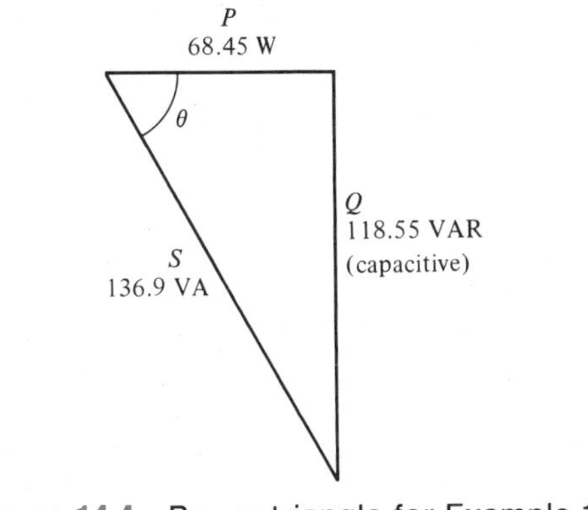

Figure 14.4 Power triangle for Example 14.1.

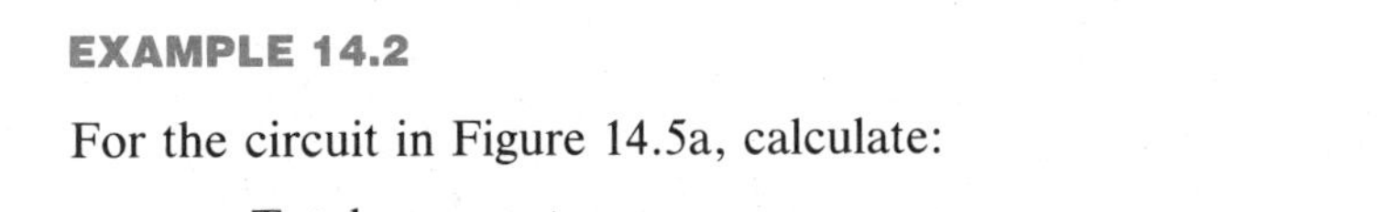

EXAMPLE 14.2

For the circuit in Figure 14.5a, calculate:

a. Total apparent power

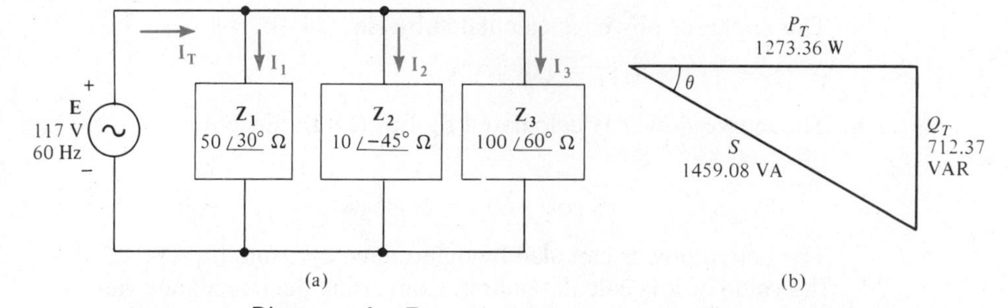

Figure 14.5 Diagrams for Example 14.2. **(a)** Circuit diagram. **(b)** Power triangle.

b. Total reactive power

c. Total active power

d. Draw the power triangle.

Solution

The voltage is the same across all the branches; currents are different.

$$\mathbf{I}_1 = \frac{117\,\underline{/0^\circ}}{50\,\underline{/30^\circ}} = 2.34\,\underline{/-30^\circ}\text{ A}$$

$$\mathbf{I}_2 = \frac{117\,\underline{/0^\circ}}{10\,\underline{/-45^\circ}} = 11.7\,\underline{/45^\circ}\text{ A}$$

$$\mathbf{I}_3 = \frac{117}{100\,\underline{/60^\circ}} = 1.17\,\underline{/-60^\circ}\text{ A}$$

The complex power of each branch is given by Eq. (14.11):

$$\mathbf{S}_1 = \mathbf{V}_1\mathbf{I}_1^* = (117)(2.34\,\underline{/30^\circ}) = 273.78\,\underline{/30^\circ} = 237.1 + j136.89\text{ VA}$$

Similarly,

$$\mathbf{S}_2 = (117)(11.7\,\underline{/-45^\circ}) = 1368.9\,\underline{/-45^\circ} = 967.81 - j967.81\text{ VA}$$

and

$$\mathbf{S}_3 = (117)(1.17\,\underline{/60^\circ}) = 136.89\,\underline{/60^\circ} = 68.45 + j118.55\text{ VA}$$

a. The total active power is the sum of the average powers of the individual branches. The branch active powers are the real parts of the complex powers:

$$\begin{aligned} P_T &= P_1 + P_2 + P_3 \\ &= 237.1 + 967.81 + 68.45 = 1273.36\text{ W} \end{aligned}$$

b. The total reactive power is the algebraic sum of the individual branch reactive power components,

$$\begin{aligned} Q_T &= Q_1 + Q_2 + Q_3 \\ &= 136.89 + (-967.81) + 118.55 \\ &= 712.37 \text{ VAR} \quad \text{(capacitive)} \end{aligned}$$

c. The apparent power is calculated by Eq. (14.8):

$$\begin{aligned} S &= \sqrt{(1173.15)^2 + (712.37)^2} \\ &= 1459.08 \text{ VA} \end{aligned}$$

d. The power triangle for the circuit is shown in Figure 14.5b.

EXAMPLE 14.3

For the circuit of Figure 14.6 calculate the following:

a. Total apparent power

b. Total active power

c. Total reactive power

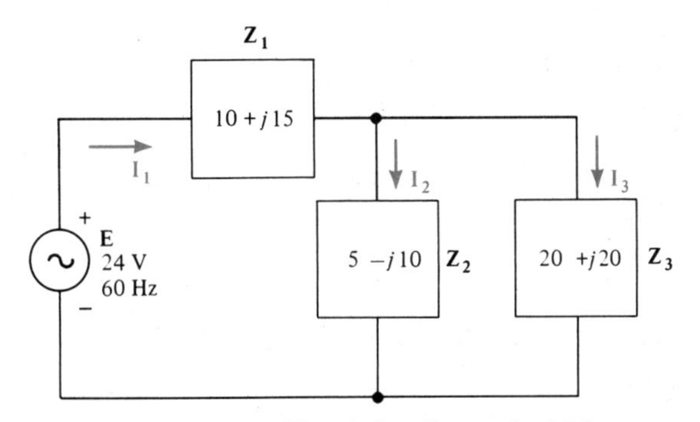

Figure 14.6 Circuit for Example 14.3.

Solution

The total impedance is calculated first:

$$\begin{aligned} \mathbf{Z_T} &= \mathbf{Z_1} + (\mathbf{Z_2} \parallel \mathbf{Z_3}) \\ &= 10 + j5 + \frac{(5 - j10)(20 + j20)}{(5 - j10) + (20 + j20)} \\ &= 10 + j5 + 8.96 - j7.58 \\ &= 18.96 - j2.58 = 19.13\angle{-7.75^\circ}\ \Omega \end{aligned}$$

The total current is:

$$\mathbf{I_T} = \frac{24\angle 0^\circ}{19.13\angle -7.75^\circ} = 1.25\angle +7.75^\circ \text{ A}$$

The complex power **S** is calculated by Eq. (14.11):

$$\mathbf{S} = (24\angle 0^\circ)(1.25\angle -7.75^\circ) = 30\angle -7.75^\circ \simeq 29.73 - j4.05$$

a. The apparent power is the magnitude of **S**; therefore, from the preceding,

$$S = 30 \text{ VA}$$

b. The active power is the real part of **S**; hence

$$P = 29.73 \text{ W}$$

c. The reactive power is the imaginary component of **S**; hence

$$Q = 4.05 \text{ VAR} \qquad \text{(capacitive)}$$

EXAMPLE 14.4

The following loads are connected in parallel across a 440-Vac source:

Motor 1 using 2 kVA and 1.5 kW

Motor 2 using 1.2 kVA and 1.0 kW

Heater dissipating energy at 5 kW

Determine the total active, reactive, and apparent powers delivered to the circuit.

Solution

Motor 1 has the following specifications:

$$S_1 = 2 \text{ kVA} = V_1 I_1$$

$$P_1 = 1.5 \text{ kW}$$

The reactive power can be calculated by first solving for the phase angle from Eq. (14.3):

$$\cos\theta_1 = \frac{P}{VI} = \frac{1.5 \text{ kW}}{2 \text{ kVA}} = 0.75$$

$$\theta_1 = \cos^{-1} 0.75 = 41.4^\circ$$

The reactive power for motor 1, Q, is calculated by Eq. (14.7):

$$Q_1 = 2 \times 10^3 \sin 41.4^\circ = 1322.9 \text{ VAR}$$

The specifications for motor 2 are calculated the same way as for motor 1.

$$S_2 = 1.2 \text{ kVA}$$

$$P_2 = 1 \text{ kW}$$

$$\cos\theta_2 = \frac{1 \text{ kW}}{1.2 \text{ kVA}} = 0.833$$

$$\theta_2 = \cos^{-1} 0.833 = 33.56°$$

$$Q_2 = 1.2 \times 10^3 \sin 33.56° = 663 \text{ VAR}$$

The heater specifications are:

$$S_3 = 5 \text{ kVA}$$

$$P_3 = 5 \text{ kW}$$

$$Q_3 = 0 \text{ VAR}$$

The total active power is:

$$P_T = P_1 + P_2 + P_3 = 1.5 \text{ kW} + 1 \text{ kW} + 5 \text{ kW} = 7.5 \text{ kW}$$

The reactive power is:

$$\begin{aligned} Q_T &= Q_1 + Q_2 + Q_3 \\ &= 1322.9 + 663 + 0 = 1985.9 \text{ VAR} \end{aligned}$$

and the apparent power is given by Eq. (14.8):

$$S_T = \sqrt{(7.5 \times 10^3)^2 + (1985.9)^2} = 7758.5 \text{ VA}$$

14.2 POWER FACTOR AND POWER FACTOR CORRECTION

The **power factor,** F_p, of a system is defined as the ratio of the real power to the apparent power, or the cosine of the angle θ between voltage and current.

$$F_p = \frac{P}{S} = \cos\theta \tag{14.14}$$

The power factor, then, has values between 0 and 1. When the system is purely resistive, $\theta = 0°$ and $F_p = 1$. For a purely reactive load, the power factor is 0. For a power factor between 0 and 1 it is customary to state the value and to indicate if the system is inductive or capacitive. If the load is inductive, it has a *lagging* power factor; the current lags the voltage in an inductive circuit. If the load is capacitive, it is a *leading* power factor. The power factor is also expressed as a percent value when multiplied by 100%.

The power factor value of the system is very important to the plant engineer. The kilowatt-hour meter, which records the energy used, measures the active (average) power multiplied by the length of time power was maintained. However, as far as the utility company is concerned, it delivers a voltage and a current to power the customer's system; hence, $I \cdot V$, or apparent power. Only when the load is purely resistive will the $I \cdot V$ supplied by the utility company equal the energy recorded by the kilowatt-hour meter at the customer's site. In all other cases the kilowatt-hour meter will indicate a smaller value than the energy that the generators "apparently" supply ($\cos\theta < 1$). For that reason, electric companies monitor the power factor of individual consumers and charge higher rates if the power factor of the load is not near 100% ($F_p = 1$). Usually, the higher rates apply to power factors below 95%.

The power factor of an industrial plant is usually inductive and less than 100% due to the large use of induction motors. The power factor of such systems is usually "corrected" by increasing the F_p value to 1 or close to 1. This is accomplished by adding capacitors across the line. The next few examples illustrate this technique.

EXAMPLE 14.5

In Example 14.4, what are the power factors, F_p, of the motors, and what is the total power factor of the system?

Solution

From Example 14.4 the power factor of motor 1 is:

$$F_{p1} = \cos\theta_1 = 0.75 \qquad \text{lagging}$$

The motor is an inductive load, and in an inductor the current lags the voltage; hence, the power factor is lagging.

For motor 2,

$$F_{p2} = \cos\theta_2 \simeq 0.83 \qquad \text{lagging}$$

The total power factor of the system is given by Eq. (14.14):

$$F_p = \frac{7.5 \times 10^3}{7758.5} \simeq 0.97 \qquad \text{lagging,} \quad \text{or} \quad 97\% \qquad \text{lagging}$$

EXAMPLE 14.6

An inductive load of 20 kW exists, and a power factor of 0.8 lagging is to be corrected, so that the total power factor is unity. Design the value of the parallel capacitor.

Solution

The system is shown in Figure 14.7. For the power factor to be unity, the total reactive power must be zero. Therefore, the reactive power due to the capacitor must be equal to the reactive power of the load, which is inductive (lagging power factor).

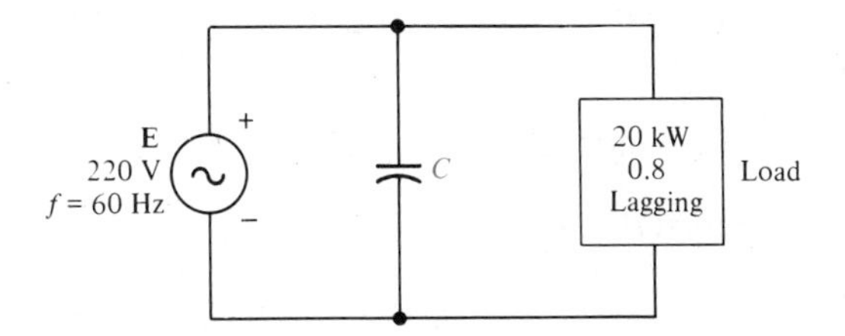

Figure 14.7 Circuit for Example 14.6.

The reactive power at: 0.8 lagging power factor is calculated next. The phase angle is:

$$\theta = \cos^{-1}0.8 = 36.87°$$

The apparent power is calculated from Eq. (14.5):

$$S = \frac{P}{\cos\theta} = \frac{20\ \text{k}\Omega}{0.8} = 25\ \text{kVA}$$

The reactive power is given by Eq. (14.7):

$$Q = 25 \times 10^3 \sin 36.87° = 15\ \text{kVAR}$$

The reactive power due to the capacitor must also be 15 kVAR. Since the capacitor is connected directly across the source, the voltage across it is the same as the source voltage, and

$$Q_c = \frac{V_c^2}{X_c}$$

Solving for X_c,

$$X_c = \frac{V_c^2}{Q} = \frac{(220)^2}{15 \times 10^3} = 3.23\ \Omega$$

and

$$c = \frac{0.159}{fX_c} = \frac{0.159}{60 \times 3.23} = 820.4\ \mu\text{F}$$

Notice that this is a fairly large capacitance value of a high operating voltage. Oil-filled capacitors are generally used for this application.

14.3 MAXIMUM POWER TRANSFER TO AN AC LOAD

The Maximum Power Transfer Theorem was introduced in Chapter 6. It was shown that maximum power is transferred from a source to the load if the load resistance is equal to the internal resistance of the source. An ac-type source, usually a *rotating machine generator,* has an internal impedance that is inductive and supplies loads that are resistive or reactive, or both.

The total load with impedance $\mathbf{Z_L}$ is connected to a source with internal impedance $\mathbf{Z_i}$ in Figure 14.8. If

$$\mathbf{Z_L} = R_L \pm jX_L \quad \text{and} \quad \mathbf{Z_i} = R_i \pm jX_i$$

the current is maximized if $jX_L = -jX_i$, yielding a total circuit impedance of

$$\boxed{Z_T = R_L + R_i} \tag{14.15}$$

The active power delivered to the load is:

$$P_L = I^2R_L = \left(\frac{E}{Z_T}\right)^2 R_L$$

resulting in

$$P_L = \left(\frac{E}{R_L + R_i}\right)^2 R_L \qquad \text{if } jX_L = -jX_i$$

Recall from Chapter 6 that maximum power is transferred to the load when $R_L = R_i$; then

$$\boxed{P_L = \frac{E^2}{4R_i}} \tag{14.16}$$

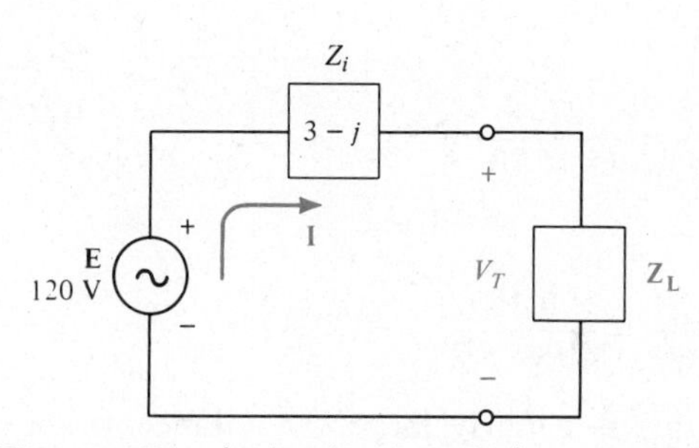

Figure 14.8 Ac load connected to a source.

Let us now recap the conditions imposed on the circuit for maximum power transfer to the load:

1. $jX_L = -jX_i$, or the reactance of the source and load must be equal in magnitude but opposite in sign.
2. The load resistance, R_L, must be equal to the source resistance, R_i.

This can be summarized as follows:

Maximum transfer of power from the source to the load occurs when the impedance of the load is the complex conjugate of the impedance at the source.

Mathematically, this relationship is given as

$$\mathbf{Z_L} = \mathbf{Z_i^*} \tag{14.17}$$

In some instances, the internal impedance of the source is not purely resistive but the load is a pure resistance. In such cases the maximum power is delivered to the load when the resistance is equal to the magnitude of the source impedance.

$$Z_L = R_L = \sqrt{R_i^2 + X_i^2} \tag{14.18}$$

EXAMPLE 14.7

A load is to be matched to the source of Figure 14.8.

a. Determine the impedance of the load for maximum power transfer and the power to the load.
b. Find the load resistor value if the load is purely resistive; also the power to the load.

Solution

a. From Eq. (14.17), with $\mathbf{Z_i} = (3 - j)\ \Omega$,

$$\mathbf{Z_L} = (3 + j)\ \Omega$$

By Eq. (14.16),

$$P_{L_{max}} = \frac{(120)^2}{4(3)} = 1200\ \text{W}$$

b. By Eq. (14.18),

$$R_L = \sqrt{3^2 + 1^2} = 3.16\ \Omega$$

The current **I** is:

$$\mathbf{I} = \frac{\mathbf{E}}{\mathbf{Z_T}} = \frac{120}{3 - j + 3.16} = 19.23\,\underline{/9.22^\circ}\ \text{A}$$

The power to the load is given by

$$P_L = |I|^2 R_L = (19.23)^2\,(3.16) = 1168.55\ \text{W}$$

14.4 POLYPHASE CIRCUITS AND SYSTEMS

The study of ac circuits up until this point was restricted to *single-phase* circuits, meaning circuits driven by sources that supply a single voltage.

Polyphase circuits *are powered by sources that generate a number of voltages of the same magnitude and frequency and are related in time.*

The most commonly used polyphase source is the three-phase (3-ϕ) generator. It is more efficient and economical to generate, transmit, and utilize three-phase power than single-phase power. Most of the power is generated and transmitted in the "three-phase" mode. Industrial loads are mainly of the three-phase type.

A three-phase generator is a rotating machine that essentially holds three generators in one. This machine is physically smaller and lighter and, as a result, more efficient than three single-phase generators. In the transmission of power, it requires less copper to transmit the same amount of power in a three-phase system than in a single-phase system. The voltage regulation of 3-ϕ power lines is better than that of single-phase lines. Three-phase systems can power both single- and three-phase loads. Even automobile designers replaced the dc generator with a three-phase alternator because of its advantages of size and efficiency. However, the generated ac power must be converted to dc to charge an automobile battery.

The basic construction of a three-phase generator is shown in Figure 14.9a. A magnet M is rotated mechanically. Three sets of windings, Aa, Bb, and Cc, positioned physically 120° from each other, make up the stator part of the alternator. An ac voltage is induced in the stator windings as the magnet rotates. Since the windings are exactly the same, the induced voltage in each winding is of the same magnitude and frequency but displaced 120° in phase. The induced voltages due to each winding are shown in Figure 14.9b, and the phase diagram is given in Figure 14.9c. It is apparent that the three-phase single machine generates three separate voltages. Hence, these can be used as three separate sources with six independent lines or, as will be shown next, as *four-wire* or *three-wire* systems.

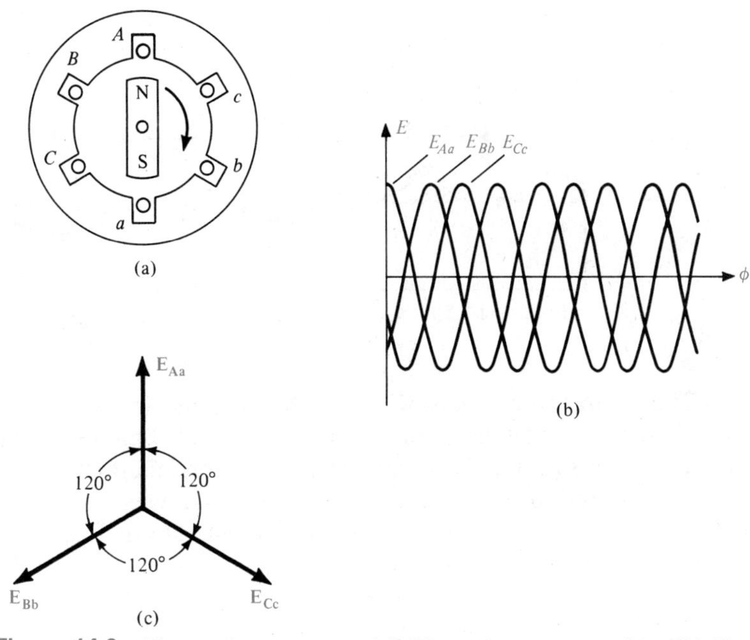

Figure 14.9 Three-phase power. **(a)** Alternator construction. **(b)** Phase relation. **(c)** Phasor diagram.

A *balanced* 3-ϕ generator supplies three voltages equal in magnitude. If a three-phase balanced generator supplies power to three equal independent loads **(balanced 3-ϕ loads),** the total instantaneous power is given by

$$p = 3VI \cos\theta \tag{14.19}$$

where θ is the phase angle between the voltage, V, and current, I, in each load. In resistive loads the voltage and current are in phase, and the *active* power is:

$$P = 3VI \tag{14.20}$$

The three-phase alternator can be connected as a four-wire system if the return terminals of each independent generator are connected together to form one *neutral* terminal. This allows the loads to be connected to the alternator through four wires referred to as four lines. The diagram in Figure 14.10 shows a four-wire wye-connected system. The coils represent the balanced windings of the generator, and the three loads are also wye-configured. The particular voltages and currents are as indicated in the diagram. At this point it is necessary to become familiar with the terminology of three-phase systems.

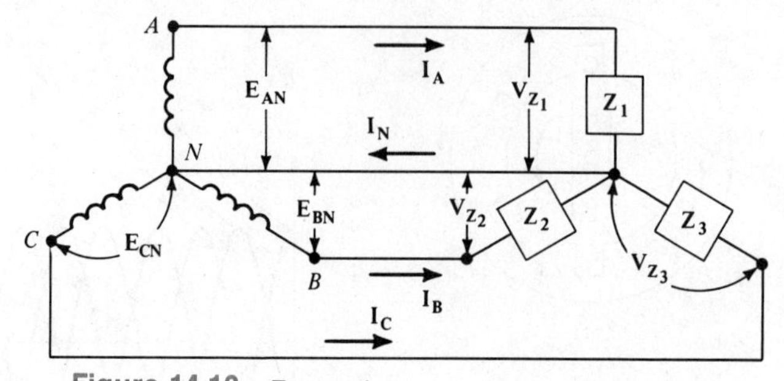

Figure 14.10 Four-wire, wye-connected system.

There are three "phases" and a neutral line in a four-wire system. The voltages between each line and the neutral are called phase voltages, E_ϕ. Hence, $\mathbf{E_{AN}}$, $\mathbf{E_{BN}}$, and $\mathbf{E_{CN}}$ are phase voltages. The potential differences, $\mathbf{E_{AB}}$, $\mathbf{E_{BC}}$, and $\mathbf{E_{CA}}$, are referred to as line-to-line, or just line, voltages, E_L. The currents $\mathbf{I_A}$, $\mathbf{I_B}$, and $\mathbf{I_C}$ are the phase currents. The phasor diagram in Figure 14.11 shows the relationships between phase and line voltages. The line voltage $\mathbf{E_{AB}}$ is given by

$$\mathbf{E_{AB}} = \mathbf{E_{AN}} + \mathbf{E_{NB}} = \mathbf{E_{AN}} - \mathbf{E_{BN}} \quad (14.21)$$

Similarly,

$$\mathbf{E_{BC}} = \mathbf{E_{BN}} + \mathbf{E_{NC}} = \mathbf{E_{BN}} - \mathbf{E_{CN}} \quad (14.22)$$

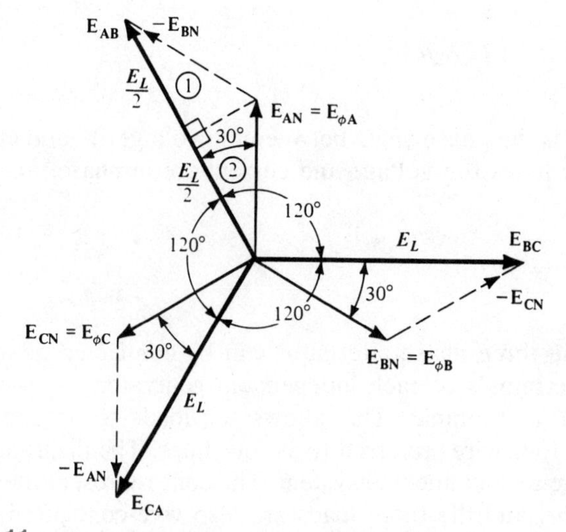

Figure 14.11 Phase and line voltage relationships in a wye-connected generator.

and

$$\mathbf{E_{CA}} = \mathbf{E_{CN}} + \mathbf{E_{NA}} = \mathbf{E_{CN}} - \mathbf{E_{AN}} \tag{14.23}$$

From the phasor diagram in Figure 14.11, it is apparent that each of the line-voltage phasors forms an isosceles triangle with two phase-voltage phasors. The altitude bisects the base and forms two right triangles, as shown in the diagram. Hence

$$\frac{E_L}{2} = E_\phi \cos 30° \tag{14.24}$$

Solving for the line voltage,

$$E_L = \sqrt{3}\, E_\phi \tag{14.25}$$

This result indicates that in a four-wire, 3-ϕ balanced system, there are two sets of voltages available: three identical but 120° out-of-phase voltages and three line voltages also equal in magnitude and 120° out of phase. However, the magnitude of each line voltage is larger than the magnitude of each phase voltage by a factor of $\sqrt{3}$, and there is a 30° shift between the corresponding line and phase voltages. The voltage across each of the loads in the wye connection of Figure 14.10 is the same as the phase voltage.

The phase currents $\mathbf{I_A}$, $\mathbf{I_B}$, and $\mathbf{I_C}$ are also the line currents in the 3-ϕ, four-wire wye-connected systems.

$$I_L = I_\phi \tag{14.26}$$

If the load is balanced, meaning $\mathbf{Z_1} = \mathbf{Z_2} = \mathbf{Z_3}$, then the phase or line currents are equal in magnitude, I_L, and also displaced by 120°, as shown in Figure 14.12. The angle θ is the phase difference between voltage and current. The current in the neutral wire is equal to the sum of the currents in the three lines:

$$\mathbf{I_N} = \mathbf{I_A} + \mathbf{I_B} + \mathbf{I_C} \tag{14.27}$$

The time relationship between line currents is shown in Figure 14.13. Notice that the sum of the three line currents at any position in time is equal to zero. Therefore, it can be concluded that the current in the neutral line is zero for a *balanced*

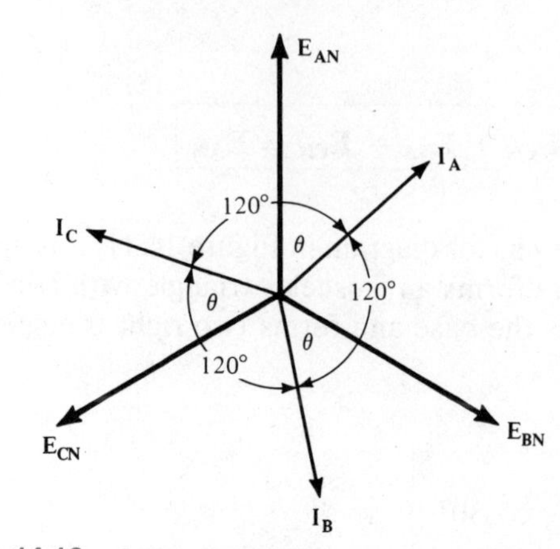

Figure 14.12 Current relationships in a 3-ϕ wye system.

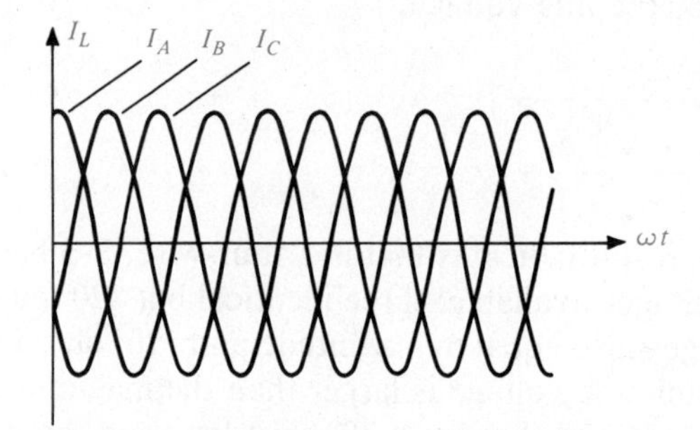

Figure 14.13 Relationship between line currents in a 3-ϕ, four-wire system.

load. If no current flows in the neutral wire, then it can be removed without affecting the system. This results in a 3-ϕ, *three-wire, wye-* or *star*-connected system. Notice that this applies only to a balanced load, which means all three load branches cannot be identical. Otherwise an imbalance exists, which causes the voltage drops across the loads to be different. A 3-ϕ, three-wire, *wye*-connected system is shown in Figure 14.14.

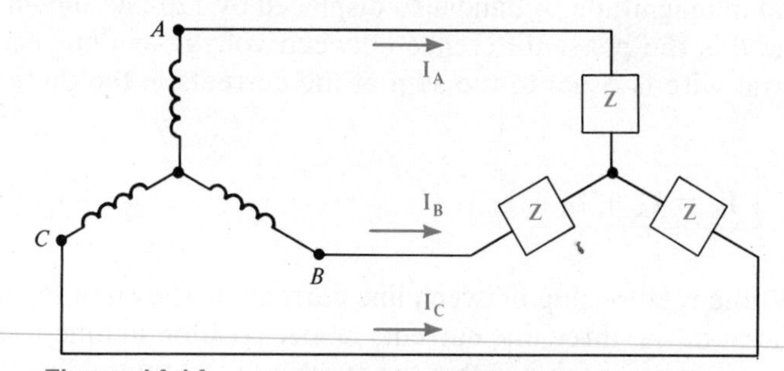

Figure 14.14 3-ϕ, three-wire, wye-connected system.

EXAMPLE 14.8

A 3-ϕ wye-connected alternator generates line voltages of 440 V. Calculate the phase voltages.

Solution

Rearranging Eq. (14.25) and solving for E_ϕ,

$$E_\phi = \frac{440}{\sqrt{3}} \simeq 254 \text{ V}$$

EXAMPLE 14.9

The line voltages of a 3-ϕ, wye-connected system are 208 V. The loads $\mathbf{Z}_1$, $\mathbf{Z}_2$, and $\mathbf{Z}_3$ are $20 + j10\ \Omega$. Calculate the magnitudes of the line currents.

Solution

By rearranging Eq. (14.25),

$$E_\phi = \frac{208}{\sqrt{3}} \simeq 120 \text{ V}$$

Since

$$\mathbf{Z}_1 = \mathbf{Z}_2 = \mathbf{Z}_3 = \mathbf{Z} = 20 + j10 = 22.36\underline{/26.56^\circ}\ \Omega$$

then

$$I_L = \frac{E_\phi}{|Z|} = \frac{120}{22.36} = 5.38 \text{ A}$$

EXAMPLE 14.10

The phase voltage of a 3-ϕ, wye-connected system is 240 V. The load is made up of three 10-Ω resistors. Calculate: (a) the current in each load resistor, and (b) the power delivered to each load.

Solution

a. The current in each load resistor is:

$$I_L = \frac{E_\phi}{R} = \frac{240}{10} = 24 \text{ A}$$

b. The total power delivered to the load is given by Eq. (14.20):

$$P = 3(240)(24) = 17.28\ \text{kW}$$

The system shown in Figure 14.15 has a delta-connected generator and a delta-connected load, resulting in a *delta*-connected system. A delta system is inherently a three-wire system with the phase voltages being the same as the line voltages.

$$\mathbf{E}_{\phi} = \mathbf{E}_{L} = \mathbf{V}_{Z} \tag{14.28}$$

It is important to notice that when the load is balanced, as is usually the case in three-phase systems, the delta load can be connected to three lines, and it does not matter if the alternator is connected in wye or delta. The phasor diagram for the currents in the delta-connected load of Figure 14.15 is given in Figure 14.16.

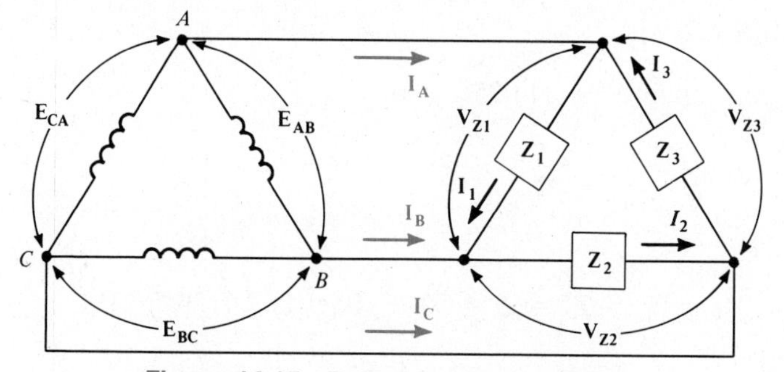

Figure 14.15 Delta-connected system.

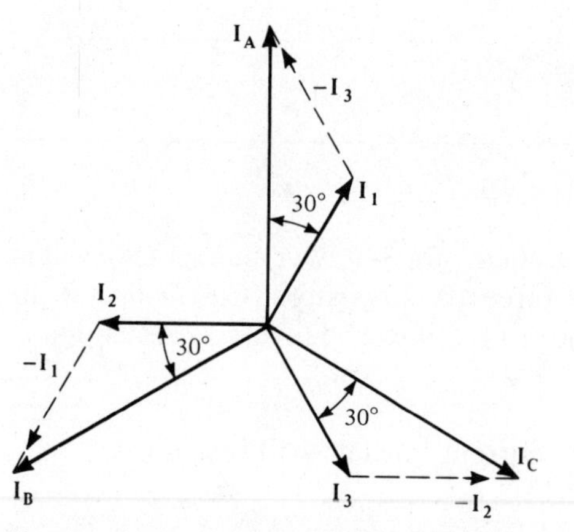

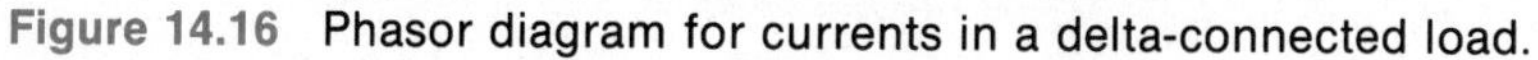
Figure 14.16 Phasor diagram for currents in a delta-connected load.

The line current is given by Kirchhoff's Current Law:

$$\boxed{\mathbf{I_A} = \mathbf{I_1} - \mathbf{I_3}} \tag{14.29}$$

Similarly

$$\boxed{\mathbf{I_B} = \mathbf{I_2} - \mathbf{I_1}} \tag{14.30}$$

and

$$\boxed{\mathbf{I_C} = \mathbf{I_3} - \mathbf{I_2}} \tag{14.31}$$

The triangles formed by the addition of those phasors can be solved for the line currents by trigonometric relations. If we consider currents I_1, I_2, and I_3 to be the phase currents for each load, I_ϕ, and I_A, I_B, and I_C the line currents, I_L, then

$$\boxed{I_L = \sqrt{3}\, I_\phi} \tag{14.32}$$

EXAMPLE 14.11

A three-phase, delta-connected balanced load is energized from 440-V lines and draws line currents of 20 A each. Calculate the phase voltage, phase currents, and the magnitude of the impedance of each load branch.

Solution

Since the phase voltages and line voltages are the same in a balanced delta-connected load, then

$$V_\phi = V_Z = 440 \text{ V}$$

The phase currents are calculated by rearranging Eq. (14.32):

$$I_\phi = I_1 = I_2 = I_3 \simeq \frac{20}{3} = 11.56 \text{ A}$$

and the magnitude of impedances Z_1, Z_2, and Z_3 is:

$$Z = \frac{V_Z}{I_Z} = \frac{440}{11.56} = 38.06\ \Omega$$

A *delta-wye*-connected system is shown in Figure 14.17a. The circuit contains a balanced delta-connected source and a balanced wye-connected load. Line currents $\mathbf{I}_{L_1}$, $\mathbf{I}_{L_2}$, and $\mathbf{I}_{L_3}$ are the corresponding load currents $\mathbf{I}_{Z_1}$, $\mathbf{I}_{Z_2}$, and $\mathbf{I}_{Z_3}$. Notice that the delta-wye system is a three-wire system because in a balanced system the sum of the load currents at the common node is zero and there is no need for a neutral. The phase voltages generated are the same as the line voltages. The phase voltages of the load are related to the line voltages as follows:

$$\mathbf{E}_{\phi_1} = \mathbf{E}_{AB} = \mathbf{V}_{Z_1} - \mathbf{V}_{Z_2}$$

$$\mathbf{E}_{\phi_2} = \mathbf{E}_{BC} = \mathbf{V}_{Z_2} - \mathbf{V}_{Z_3}$$

$$\mathbf{E}_{\phi_3} = \mathbf{E}_{CA} = \mathbf{V}_{Z_1} - \mathbf{V}_{Z_3}$$

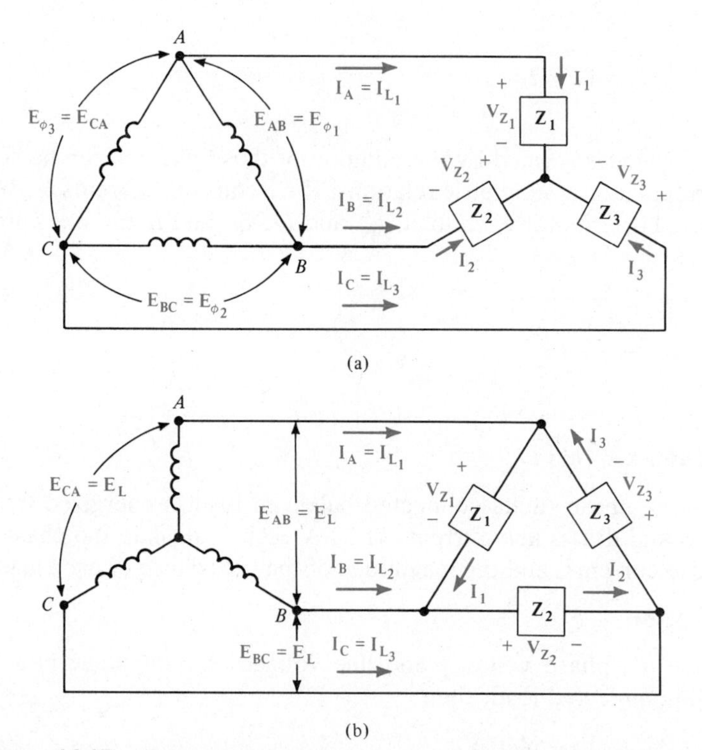

Figure 14.17 Three-phase systems. **(a)** Delta-wye-connected system. **(b)** Wye-delta-connected system.

Since the load is balanced $\mathbf{Z}_1 = \mathbf{Z}_2 = \mathbf{Z}_3 = \mathbf{Z}$ the magnitude of each phase voltage of the load is:

$$\boxed{V_Z = \frac{E_\phi}{\sqrt{3}}} \qquad \textbf{(14.33)}$$

The circuit of a balanced wye-delta-connected system is shown in Figure 14.17b. In this type of connection the line voltages and the phase voltages of the load are the same. The relation between load currents and line voltages is as follows:

$$\mathbf{I_{L_1}} = \mathbf{I_A} = \mathbf{I_1} - \mathbf{I_3}$$

$$\mathbf{I_{L_2}} = \mathbf{I_B} = \mathbf{I_2} - \mathbf{I_1}$$

$$\mathbf{I_{L_3}} = \mathbf{I_C} = \mathbf{I_3} - \mathbf{I_2}$$

Since the system is balanced, $\mathbf{Z_1} = \mathbf{Z_2} = \mathbf{Z_3}$, $I_1 = I_2 = I_3 = I_Z$, and $I_{L_1} = I_{L_2} = I_{L_3} = I_L$.

Then

$$\boxed{I_L = \sqrt{3}\, I_Z} \qquad (14.34)$$

EXAMPLE 14.12

The circuit in Figure 14.17a has a balanced load: $\mathbf{Z_1} = \mathbf{Z_2} = \mathbf{Z_3} = 5 + j10\ \Omega$. The line currents are: $\mathbf{I_A} = 10\underline{/0^\circ}$ A, $\mathbf{I_B} = 10\underline{/120^\circ}$ A, $\mathbf{I_C} = 10\underline{/-120^\circ}$ A.

Determine:

a. Load currents

b. Load voltages

c. Magnitude of line voltages

Solution

a. Since in a delta-wye-connected system, the load currents are the same as the line currents,

$$\mathbf{I_1} = \mathbf{I_A} = 10\underline{/0^\circ}\text{ A}$$

$$\mathbf{I_2} = \mathbf{I_B} = 10\underline{/120^\circ}\text{ A}$$

$$\mathbf{I_3} = \mathbf{I_C} = 10\underline{/-120^\circ}\text{ A}$$

b. The load voltages are calculated by

$$\mathbf{V_{Z_1}} = \mathbf{I_1Z_1} = 10\underline{/0^\circ} \times (5 + j10) = 111.8\underline{/63.43^\circ}\text{ V}$$

$$\mathbf{V_{Z_2}} = \mathbf{I_2Z_2} = 10\underline{/120^\circ} \times (5 + j10) = 111.8\underline{/183.42^\circ}\text{ V}$$

$$\mathbf{V_{Z_3}} = \mathbf{I_2Z_2} = 10\underline{/-120^\circ} \times (5 + j10) = 111.8\underline{/-56.57^\circ}\text{ V}$$

c. Rearranging Eq. (14.33):

$$E_\phi = \sqrt{3}\, V_Z$$
$$= \sqrt{3}\ 111.8 = 193.65\text{ V}$$

EXAMPLE 14.13

For the circuit in Figure 14.17b, the line voltages are as follows: $\mathbf{E_{AB}} = 120\angle 0^\circ$ V, $\mathbf{E_{CA}} = 120\angle 120^\circ$ V, $\mathbf{E_{BC}} = 120\angle -120^\circ$ V. The balanced load values are $Z_1 = Z_2 = Z_3 = 5 + j10\ \Omega$. Calculate:

a. Voltage across each phase of the load

b. Currents in each phase of the load

c. Magnitude of the line currents

Solution

a. Since the load voltages are equal to the line voltages,

$$\mathbf{V_{Z_1}} = \mathbf{E_{AB}} = 120\angle 0^\circ \text{ V}$$

$$\mathbf{V_{Z_2}} = \mathbf{E_{BC}} = 120\angle -120^\circ \text{ V}$$

$$\mathbf{V_{Z_3}} = \mathbf{E_{CA}} = 120\angle 120^\circ \text{ V}$$

b. The load currents are calculated next:

$$\mathbf{I_1} = \frac{\mathbf{V_{Z_1}}}{\mathbf{Z_1}} = \frac{120\angle 0^\circ}{5 + j10} = 10.73\angle -63.43^\circ \text{ A}$$

$$\mathbf{I_2} = \frac{\mathbf{V_{Z_2}}}{\mathbf{Z_2}} = \frac{120\angle -120^\circ}{5 + j10} = 10.73\angle -183.43^\circ \text{ A}$$

$$\mathbf{I_3} = \frac{\mathbf{V_{Z_3}}}{\mathbf{Z_3}} = \frac{120\angle 120^\circ}{5 + j10} = 10.73\angle 56.57^\circ \text{ A}$$

c. By Eq. (14.34),

$$I_A = I_B = I_C = I_L = \sqrt{3} \times 10.73 = 18.58 \text{ A}$$

The apparent power, S_T, delivered to a balanced wye load is equal to three times the effective voltages across the individual load branches and the effective current through each branch.

$$S_T = 3E_\phi I_\phi = 3E_L/\sqrt{3}\, I_L$$

$$\boxed{S_T = \sqrt{3}\, E_L I_L} \qquad \text{VA} \tag{14.35}$$

For a delta load,

$$S_T = 3E_\phi I_\phi = \frac{3E_L \times I_L}{\sqrt{3}} = \sqrt{3}\, E_L I_L \qquad \text{VA}$$

which yields the same result as for the wye load. Since, for a balanced load, the power factor, $F_p = \cos\theta$, is the same for each branch, then the total power delivered to the load is:

$$P_T = 3\, E_\phi I_\phi \cos\theta \qquad \text{W} \tag{14.36a}$$

or

$$P_T = \sqrt{3}\, E_L I_L \cos\theta \qquad \text{W} \tag{14.36b}$$

The equations above apply to both wye- and delta-connected balanced loads.
The total reactive power is given by

$$Q_T = 3E_\phi I_\phi \sin\theta = \sqrt{3}\, E_L I_L \sin\theta \qquad \text{VAR} \tag{14.37}$$

Notice that the angle θ is the phase difference between the voltage and current of a single branch, and the power factor of the total balanced load is the same as the power factor of an individual branch. The complex power is given by

$$\mathbf{S_T} = P_T + jQ_T \qquad \text{VA} \tag{14.38}$$

and the magnitude of $\mathbf{S_T}$ is:

$$S_T = \sqrt{P_T^2 + Q_T^2} \qquad \text{VA} \tag{14.39}$$

EXAMPLE 14.14

A three-phase balanced load is powered from 440-V lines and 50-A current in each line. If the load exhibits a power factor of 0.9, calculate the active power consumed by the load.

Solution

The total active power is given by Eq. (14.36b):

$$P_T = \sqrt{3}\,(440)(50)(0.9) = 34.295 \text{ kW}$$

EXAMPLE 14.15

An industrial plant powered from a 440-V, 3-ϕ line consumes 150 kVA at a power factor of 0.9 lagging. Calculate:

a. The current in each line

b. The total active power delivered

c. The total reactive power in the load

Solution

a. The current in each line is calculated from Eq. (14.35):

$$I_L = \frac{S_T}{\sqrt{3}\,E_L} = \frac{150 \times 10^3}{\sqrt{3}\,(440)} = 196.82 \text{ A}$$

b. By Eq. (14.5),

$$P_T = (150 \times 10^3)(0.9) = 135 \text{ kW}$$

c. To calculate the reactive power, the angle θ between the voltage and current of an individual branch must be calculated from the power factor value:

$$\theta = \cos^{-1} 0.9 = 25.84°$$

Substituting S for VI in Eq. (14.7),

$$Q_T = S \sin \theta$$
$$= 150 \times 10^3 \times 0.43 = 65.28 \text{ kVAR}$$

EXAMPLE 14.16

The following balanced loads are connected to an 11,600-V, 3-ϕ source: a wye-connected 1000-HP motor operating at 90% efficiency with a 0.92 lagging power factor, 20-kW heaters, and a 200-kVAR bank of capacitors. Determine:

a. The total active power delivered to the load

b. The total reactive power supplied by the source

c. The apparent power delivered by the source

Solution

a. Active power is consumed by the motor and the heaters. The heaters consume only active power, but the motor's active power must be determined from its efficiency. The power to the motor is:

$$P_i = P_o/\eta = \frac{1000 \times 746}{0.9} = 828.9 \text{ kW}$$

The total active power to the load is:

$$P_T = P_M + P_H = 828.9 \times 10^3 + 20 \times 10^3 = 848.9 \text{ kW}$$

b. Total reactive power is due to the capacitors and the inductance of the motor. The angle θ between the phase voltage and the phase current of the motor is determined from the power factor of the motor:

$$\theta = \cos^{-1} 0.92 = 23.07°$$

The apparent power of the motor is calculated by solving for S in Eq. (14.14):

$$S_M = \frac{P_M}{F_p} = \frac{849.9 \times 10^3}{0.92} = 923.8 \text{ kVA}$$

By Eq. (14.7), substituting $S = VI$,

$$Q_M = (923.8 \times 10^3) \sin 23.07° = 362 \text{ kVAR} \qquad \text{(inductive)}$$

Finally,

$$Q_T = Q_M - Q_C$$
$$= 362 \times 10^3 - 200 \times 10^3 = 162 \text{ kVAR}$$

c. The total apparent power is given by Eq. (14.39):

$$S_T = \sqrt{(848.9 \times 10^3)^2 + (162 \times 10^3)^2} = 864.2 \text{ kVA}$$

The **phase sequence** is an important factor in describing a three-phase system. It determines the direction of rotation of 3-ϕ motors and whether or not two 3-ϕ systems can be connected in parallel. Phase sequence refers to the order in which the phasors of voltage or current follow each other. Figure 14.18a depicts a

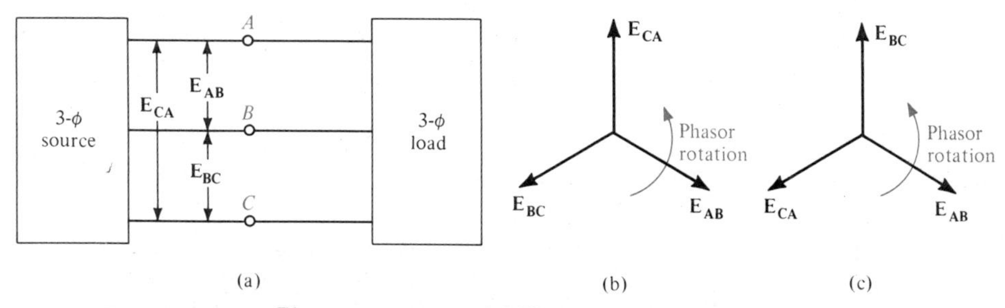

Figure 14.18 Phase sequence. **(a)** Three-phase system. **(b)** Phase sequence *ABC*. **(c)** Phase sequence *ACB*.

three-phase system, and the possible phase sequences are shown in the phasor diagrams, with the maximum line voltage E_{AB} selccted as the reference point. It is assumed that the phasors rotate in a counterclockwise direction. In Figure 14.18b the phasors rotate through the reference in the following order: $\mathbf{E_{AB}}$, $\mathbf{E_{BC}}$, and $\mathbf{E_{CA}}$. Considering the first subscript of each, we say that the sequence is *ABC*. If phasors $\mathbf{E_{BC}}$ are interchanged with $\mathbf{E_{CA}}$, which amounts to interchanging the connections of the two lines to the load, this results in the phasor diagram in Figure 14.18c. This phase *reversal* results in the sequence $\mathbf{E_{AB}}$, $\mathbf{E_{CA}}$, $\mathbf{E_{BC}}$, or *ACB*. If *ABC* is referred to as the *positive* sequence, *ACB* is called the *negative* sequence.

Because it is important to know the phase sequence, several methods are used to determine the sequence using special instruments, such as a two-lamp with capacitor circuit shown in Figure 14.19. One of the lamps lights up brightly and the other one is dimmer. The sequence is: *bright, dim,* capacitor. If L_1 is bright and L_2 is on, but dimmer, then the phase sequence is *ABC*.

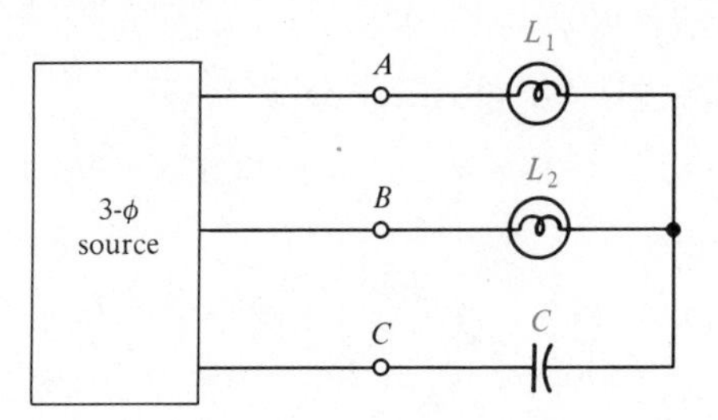

Figure 14.19 Circuit used to determine the phase sequence of a 3-ϕ source.

A polyphase system in general is considered to be *unbalanced* if the voltages and currents on each branch of the load are unequal and the phase angle between the voltages and currents in each phase differs. Generally, the systems are balanced by design, but occasionally a system becomes unbalanced due to the varied loading in each phase. The following examples illustrate the analysis of such systems.

EXAMPLE 14.17

For the system in Figure 14.20 determine the line currents and the total active power dissipated in the load. The phase sequence is *ABC*.

Solution

With a phase sequence of *ABC*, the voltage phasors are as follows:

$$\mathbf{E_{AB}} = 440\,\underline{/0^\circ}\text{ V}$$

$$\mathbf{E_{BC}} = 440\,\underline{/-120^\circ}\text{ V}$$

$$\mathbf{E_{CA}} = 440\,\underline{/120^\circ}\text{ V}$$

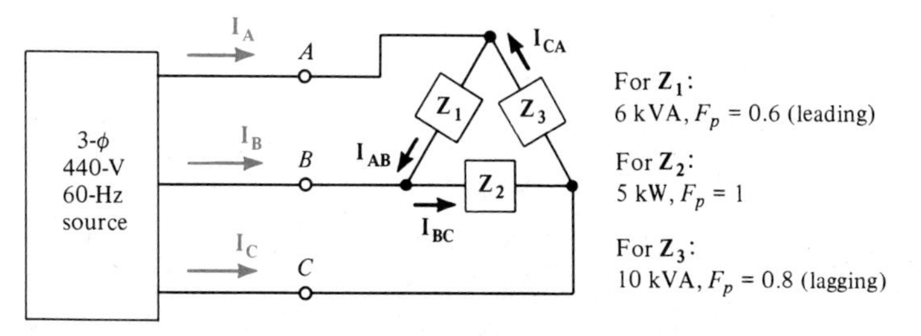

Figure 14.20 System diagram for Example 14.14.

The currents in each of the loads are determined next. For $\mathbf{Z_1}$, $F_p = 0.6$; hence

$$\theta = \cos^{-1} 0.6 = 53.13°$$

θ is also the impedance angle, and because it is a leading power factor, then $\mathbf{Z_1}$ is capacitive. Therefore, the current leads the voltage by 53.13° in Z_1. Hence

$$I_{AB} = \frac{6 \times 10^3}{440} = 13.64 \text{ A}$$

and

$$\mathbf{I_{AB}} = 13.64\underline{/53.13°} \text{ A}$$

Load $\mathbf{Z_2}$ is purely resistive ($F_p = 1$) and the voltage and current in this branch are in phase, so $\theta_2 = 0°$:

$$I_{BC} = \frac{5 \times 10^3}{440} = 11.36 \text{ A}$$

Since E_{BC} is at an angle of $\underline{/-120°}$ (with reference to E_{AB}),

$$\mathbf{I_{BC}} = 11.36\underline{/-120°} \text{ A}$$

For $\mathbf{Z_3}$,

$$\theta_3 = \cos^{-1} 0.8 = 36.78°$$

(current lagging the voltage)

$$I_{CA} = \frac{10 \times 10^3}{440} = 22.73 \text{ A}$$

and

$$\mathbf{I_{CA}} = 22.73\underline{/120° - 36.87°} = 22.73\underline{/83.13°} \text{ A}$$

To find the line currents we apply KCL:

$$\mathbf{I_A} = \mathbf{I_{AB}} - \mathbf{I_{CA}}$$

$$\mathbf{I_B} = \mathbf{I_{BC}} - \mathbf{I_{AB}}$$

$$\mathbf{I_C} = \mathbf{I_{CA}} - \mathbf{I_{BC}}$$

Solving for $\mathbf{I_A}$ gives

$$\mathbf{I_A} = 13.64\underline{/53.13^\circ} - 22.73\underline{/83.13^\circ}$$
$$= 12.87\underline{/-64.9^\circ}\text{ A}$$

Similarly, for $\mathbf{I_B}$,

$$\mathbf{I_B} = 11.36\underline{/-120^\circ} - 13.64\underline{/53.13^\circ}$$
$$= 24.95\underline{/236^\circ}\text{ A}$$

and for $\mathbf{I_C}$,

$$\mathbf{I_C} = 22.73\underline{/83.13^\circ} - 11.36\underline{/-120^\circ}$$
$$= 33.48\underline{/75.45^\circ}\text{ A}$$

The total active power is the sum of the individual active powers consumed by the load:

$$P_1 = S_1 \cos\theta = 6 \times 10^3 \times 0.6$$
$$= 3.6\text{ kW}$$

$$P_2 = 5\text{ kW}$$

and

$$P_3 = 10 \times 10^3 \times 0.8 = 8\text{ kW}$$

Hence,

$$P_T = P_1 + P_2 + P_3$$
$$= 3.6 \times 10^3 + 5 \times 10^3 + 8 \times 10^3 = 16.6\text{ kW}$$

14.5 TRANSFORMERS

The study of power in ac systems would not be complete without considering the transformer, because it is probably the transformer that is responsible for us having ac power in the outlet instead of dc.

*A **transformer** is a device used in electrical energy distribution to transfer energy inductively from one circuit to another by stepping the voltage up or down.*

Transformers are used to increase or reduce the voltage or the current in a circuit. They can also be used to isolate one circuit from another electrically and to match impedances between two circuits.

The transformer is generally constructed of two magnetically coupled coils, as shown in Figure 14.21a. The two coils are electrically isolated but magnetically coupled through the iron core that serves as the frame for the structure. If an ac source is connected to the *primary* coil, it develops a changing flux, ϕ, which links the secondary coil through the magnetic core. An individual voltage will result in both coils. Recall from Faraday's Law (Chapter 10) that the voltage induced in a

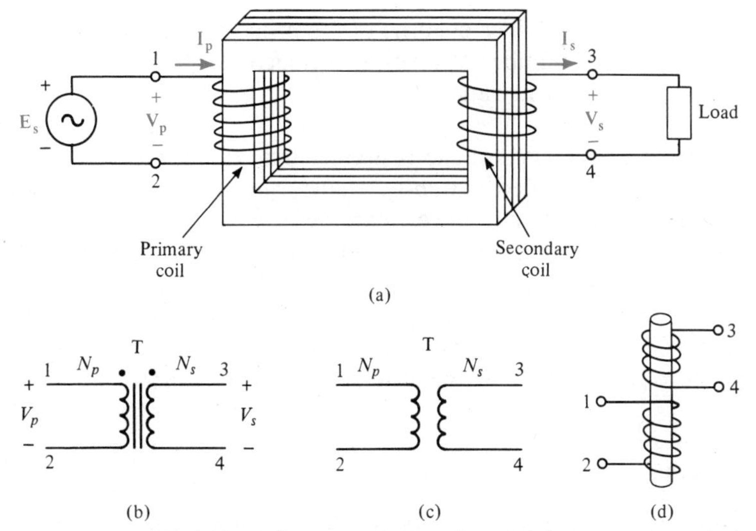

Figure 14.21 Basic transformer construction. **(a)** Iron-core transformer structure with source and load connected to it. **(b)** Electrical symbol of iron-core transformer. **(c)** Electrical symbol of air-core transformer. **(d)** Structure of air-core transformer.

coil is proportional to the number of turns of the coil and the rate of change of flux linking the coil. Therefore, the alternating current supplied by the source to the primary coil produces a changing flux at the same rate as the frequency of the applied energy, which in turn induces a changing voltage in the secondary coil. The iron core is used to enhance the coupling properties between the coils, thus increasing the mutual flux. Two coils wound on a paper or plastic cylinder with air in the middle make up an air-core transformer, as shown in Figure 14.21d.

An ideal transformer transfers energy from the primary to the secondary circuit without losses.

In an ideal transformer the dc resistance of the windings is neglected, the coupling coefficient between the windings is unity, and the magnetic losses, such as hysteresis and eddy currents, are zero. In the analysis of transformers in this chapter, we assume that the transformers are ideal. Practical power transformers have efficiencies of 90% or better.

In an ideal transformer the voltages across the primary and secondary coils are proportional to the ratio between the turns of the coils:

$$\boxed{\frac{V_p}{V_s} = \frac{N_p}{N_s} = a} \qquad (14.40)$$

where: N_p = number of turns of the primary coil
N_s = number of turns of the secondary coil
a = turns ratio

If the secondary voltage is greater than the primary voltage, which means that $a < 1$, it is called a *step-up* transformer. A *step-down* transformer has a secondary voltage that is lower than the primary voltage ($a > 1$).

The two dots on top of the transformer symbol in Figure 14.21b are called *polarity marks*. The polarity marks in Figure 14.21b indicate that a current entering terminal 1 will be at maximum at the same time as the current through terminal 3; hence, the voltage polarities are assumed to be positive at terminal 1 with respect to 2 and also positive at terminal 3 with respect to 4.

EXAMPLE 14.18

An ideal transformer that has 600 turns in the primary and 120 turns in the secondary is connected to a 120-V line. Determine:

a. The turns ratio

b. The secondary voltage

Solution

a. By Eq. (14.40),

$$a = \frac{600}{120} = 5$$

b. The secondary voltage is:

$$V_s = \left(\frac{1}{5}\right)120 = 24 \text{ V}$$

The transformer is a step-down.

By definition, the power to the primary of an ideal transformer must be dissipated in the resistive load connected across the secondary. Therefore, for the ideal transformer

$$P_p = P_s$$

or

$$V_pI_p = V_sI_s$$

Solving for the voltage ratio,

$$\frac{V_p}{V_s} = \frac{I_s}{I_p} = a \qquad (14.41)$$

Hence, in an ideal transformer the voltage ratio of the primary to the secondary is directly proportional to the turns ratio and inversely proportional to the current ratio.

EXAMPLE 14.19

For the transformer connection in Figure 14.22, determine the output (secondary) voltage and current.

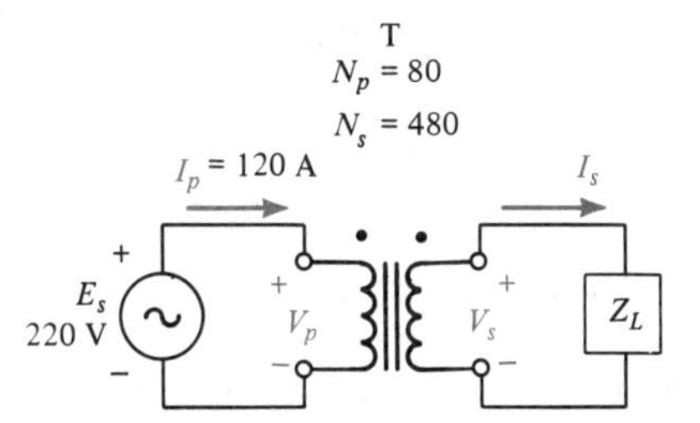

Figure 14.22 Transformer circuit used in Example 14.16.

Solution

The turns ratio, a, is given by Eq. (14.40):

$$a = \frac{80}{480} = \frac{1}{6}$$

Therefore, the secondary voltage is:

$$V_s = 220 \times 6 = 1320 \text{ V}$$

and the output current is:

$$I_s = 120 \times \frac{1}{6} = 20 \text{ A}$$

The transformer in Example 14.19 is a step-up transformer. The voltage at the output is six times larger than the input voltage, but the output current is six times smaller than the input current.

The secondary current, I_s, in the ideal transformer of Figure 14.23a is a function of the load resistor, Z_L:

$$\boxed{I_s = \frac{V_s}{Z_L}} \tag{14.42}$$

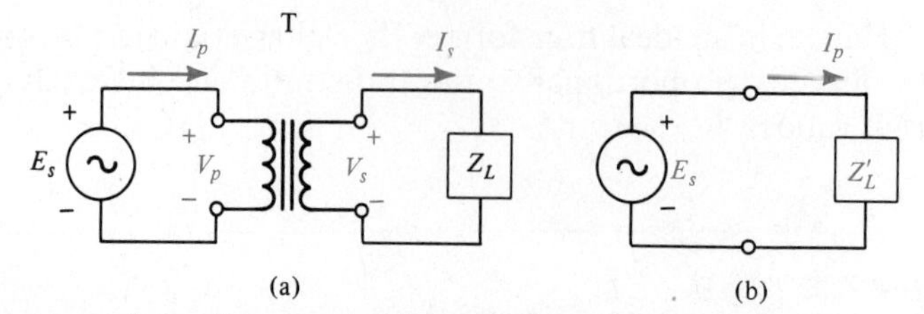

Figure 14.23 Impedance matching. **(a)** Transformer with load connected to secondary. **(b)** Load impedance reflected to secondary.

The primary current, I_p, is then given by Eq. (14.41) as

$$I_p = \frac{V_s}{aZ_L} \tag{14.43}$$

Taking the reciprocal of Eq. (14.43), and multiplying both sides by V_p yields

$$\frac{V_p}{I_p} = aZ_L\left(\frac{V_p}{V_s}\right)$$

Recognizing from Eq. (14.39) that $V_p/V_s = a$, then

$$\frac{V_p}{I_p} = a^2Z_L = Z'_L \tag{14.44}$$

Since the ratio V_p/I_p represents an equivalent impedance in the primary circuit, we call this impedance the reflected load impedance, Z'_L. Therefore, an impedance in the secondary is transformed to the primary as a larger or smaller impedance relative to the square of the turns ratio. This makes the transformer an ideal device for changing (transforming) impedance values for the purpose of *matching impedances*. One such application is the matching of source and load impedances for maximum power transfer.

The loudspeaker often has to be matched to the power amplifier of the sound system. In order to transfer maximum power from the amplifier to the loudspeaker, the impedance of the load (loudspeaker) must be the same as the internal impedance of the amplifier. However, this is not always the case; the speaker impedance is sometimes lower than the internal impedance of the amplifier. Hence, a transformer is used to match the two impedances.

EXAMPLE 14.20

The internal impedance of an audio power amplifier is 3200 Ω. Design a transformer to match an 8-Ω loudspeaker to the amplifier.

Solution

The load impedance, Z_L, is 8 Ω, and the impedance of the source is 3200 Ω. Hence, the reflected load impedance into the primary of the transformer, Z'_L, must also be 3200 Ω. Therefore, by Eq. (14.44),

$$a^2 = \frac{Z'_L}{Z_L} = \frac{3200}{8} = 400$$

and

$$a = \sqrt{400} = 20$$

Therefore, a step-down transformer with a turns ratio of 20 should be used.

Step-down transformers reflect the load impedance to the primary as a larger impedance value. The opposite is true for step-up transformers.

EXAMPLE 14.21

What is the reflected load impedance Z'_L at the primary of a transformer with a turns ratio $a = \frac{1}{10}$ (step-up) if the load is a capacitor of 10 μF operating at a frequency of 60 Hz? What is the "effective" capacitance in the primary?

Solution

The magnitude of the load impedance is given by

$$Z_L = \frac{1}{2\pi fC} = \frac{1}{2\pi(60)(10 \times 10^{-6})} = 265\ \Omega$$

and by Eq. (14.44),

$$Z'_L = \left(\frac{1}{10}\right)^2 265 = 2.65\ \Omega$$

The effective capacitance in the primary of the transformer is calculated from the reactance equation for a capacitor:

$$Z'_L = X'_C = \frac{1}{2\pi fC'}$$

Solving for C', the effective capacitance in the primary gives

$$C' = \frac{1}{2\pi f X'_C}$$

$$= \frac{1}{2\pi(60)(2.65)} = 1000\ \mu\text{F}$$

The preceding example illustrates that a transformer can also be used to increase or decrease the capacitance value in a circuit.

The practical iron-core transformer differs from the ideal transformer in that the ideal transformer is considered lossless and has an unlimited frequency response. However, the practical transformer has losses due to the resistance of the windings of the coils, called copper losses, flux leakage at the windings, and core losses. Core losses are called hysteresis and eddy current losses. The frequency response of a transformer is limited due to the sensitivity of the inductors to frequency and due to the effective shunt capacitances between the primary and secondary coil windings.

The efficiency of iron-core transformers is relatively high, especially in large-power applications, where the frequency of operation is constant and the efficiency is in the high ninetieth percentile. To achieve such high efficiencies, it is critical to make the transformer core of high-resistivity, silicon steel laminations. In addition, the windings of the primary and secondary coils are tightly wound on top of each other to reduce flux leakage, and, finally, the resistance of the coil wires is kept at a minimum to reduce copper losses.

Transformers are rated by the power they can handle, their nominal primary and secondary voltages, and the nominal operating frequency. The power-handling capacity is specified in volt-amperes, VA, and not in watts because the power factor varies depending on the type of load. Hence, the power-handling capacity is simply the product of the voltage and current.

EXAMPLE 14.22

A power transformer is rated at 10 kVA, 440 V/120 V (440 V primary, 120 V secondary). Calculate the nominal current in the primary and in the secondary.

Solution

Assuming that the transformer is very efficient ($\approx$100%), $S_p = S_s$:

$$I_p = \frac{10 \times 10^3}{440} = 22.73\ \text{A}$$

and

$$I_s = \frac{10 \times 10^3}{120} = 83.33\ \text{A}$$

14.6 SPECIAL-PURPOSE TRANSFORMERS

Iron-core transformers can be made to deliver several different voltages from a single source by using a number of secondary coils. A power transformer with multiple secondary windings is shown in Figure 14.24. The transformer has a "center-tapped" winding, which means that a terminal is available at the center of one winding making available two separate 120-V connections or one 240-V connection. The second, isolated winding provides 6.3 V.

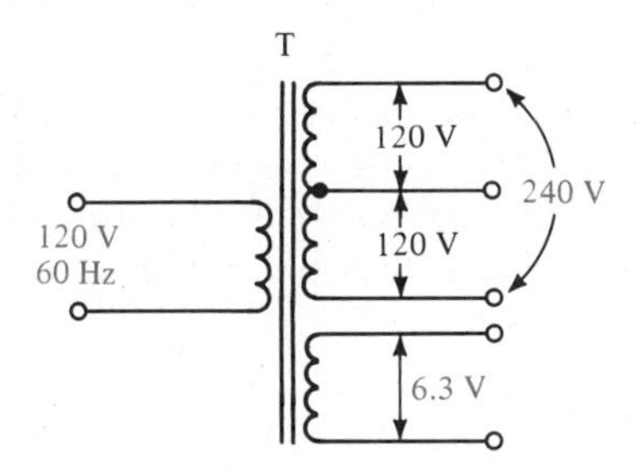

Figure 14.24 Transformer with multiple secondaries.

A transformer constructed of a single winding is called an autotransformer and is shown in Figure 14.25. As in the case of the transformers discussed until now, in this transformer the voltage ratio of the primary to the secondary is directly proportional to the turns ratio between the primary and the secondary. If the input is applied to the full length of the coil (terminals 1, 3) the output of 240 V is present across the "secondary," terminals 2, 3. If the source is connected between terminals 2 and 3 and the load between terminals 1 and 3, it becomes a step-up transformer, and the entire length of the coil is the "secondary." The disadvantage of the autotransformer is that there is no electrical isolation between the primary and secondary windings. A conventional two-winding transformer can be connected as an autotransformer by connecting one leg of the primary to one leg of the secondary winding. Such a hookup results in an increase in power-handling capacity of the transformer.

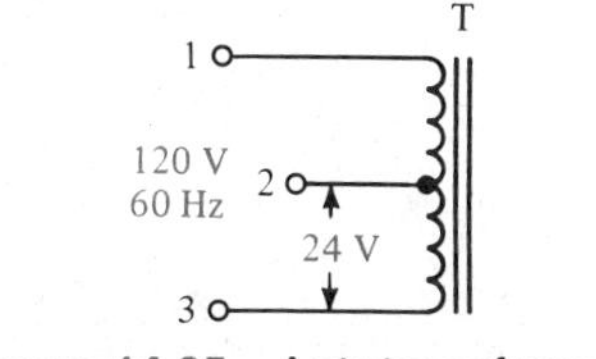

Figure 14.25 Autotransformer.

Potential transformers (PTs) are precision transformers providing a secondary voltage, which is exactly proportional to the primary voltage regardless of the variations in load. The secondary voltage is also in phase with the primary voltage.

Current transformers (CTs) are precision transformers that deliver a secondary current at a precise ratio to the primary current, regardless of variations in load. The phase angle is zero between the primary and secondary currents. Both the PT and the CT are used with measuring instruments to measure currents and voltages in power circuits accurately.

For current-measurement purposes we sometimes use a *toroidal* current transformer. As shown in Figure 14.26, a donut-shaped core with many windings serves as the secondary of the current transformer, and the conductor whose current is to be measured is the primary. By Eq. (14.41), the current measured with the ammeter (secondary current) is the current in the conductor (primary) multiplied by a, where a is the reciprocal of the number of turns in the toroidal coil. Since this is a CT, current sensed by the meter multiplied by the number of turns of the toroidal coil is the exact value of the current in the conductor.

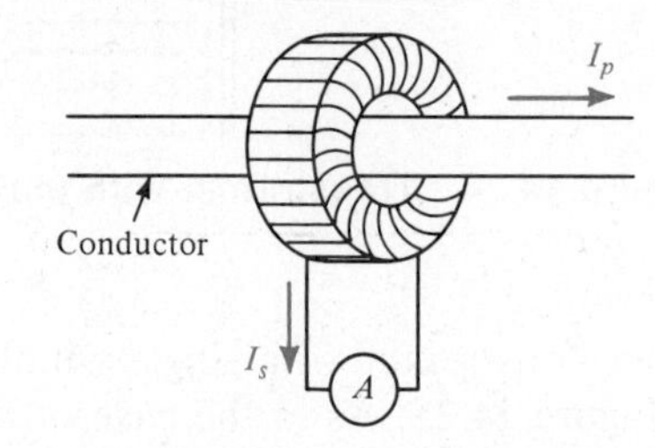

Figure 14.26 Toroidal current transformer (CT).

Air-core transformers are used primarily in high-frequency communication circuits. They are much smaller and lighter than power transformers and are usually tunable to act as a pass-band filter for signals at different frequencies. Various types of transformers are shown in Figure 14.27.

(a)

(a) Isolation transformer (Reprinted with permission of the Control Systems Division of United Technologies Corp.).

Figure 14.27 Assortment of transformers. (Courtesy Stancor Corp.)

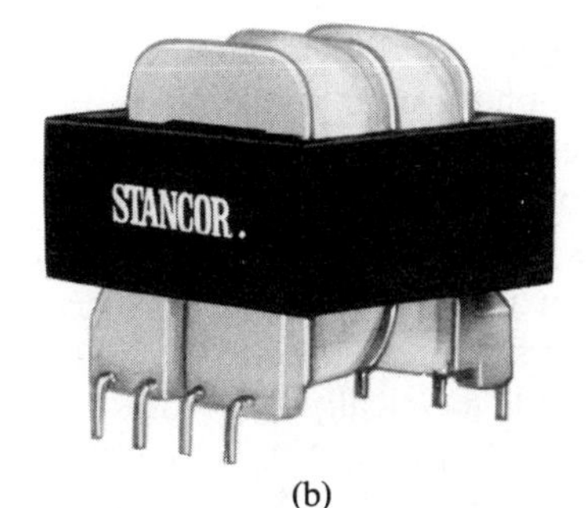

(b)

(b) Printed circuit board transformer (STANCOR® is the registered trademark of Hamilton Standard Controls, Inc. - reprinted with permission of the Control Systems Division of United Technologies Corp.)

(c)

(c) Low profile transformer (STANCOR® is the registered trademark of Hamilton Standard Controls, Inc. - reprinted with permission of the Control Systems Division of United Technologies Corp.)

Figure 14.27 (continued)

SUMMARY

Power is delivered to ac circuits in single-phase or three-phase form. Three-phase power is commonly used in industry, and single-phase power is used in residential and commercial applications. Ac loads can be purely resistive but often contain reactances as well as resistances. Industrial-type equipment is usually inductive due to the use of motors.

The product of voltage and current delivered by an ac generator is the apparent power measured in volt-amperes. The real or active power consumed by a load is the apparent power multiplied by the cosine of the phase angle between the voltage and current waves, and it is measured in watts. The reactive power measured in VARs is the apparent power multiplied by the sine of the phase angle between the voltage and current waves.

The power factor is the ratio of the active power to the apparent power. A purely resistive load has a power factor of unity. A purely reactive load has a zero power factor. Most industrial loads have an inductive power factor (also called a lagging power factor) of less than one. Utility companies charge a penalty for power consumption at a power factor less than 0.95, or 95%. The power factor can be *corrected,* or increased to a value higher than 0.95, by connecting a capacitor across the line.

Polyphase circuits, and especially three-phase circuits, were studied. The advantages of three-phase over single-phase power were discussed. Three-phase loads are connected in a wye or delta configuration to absorb three times the power of a single-phase load.

Transformers were shown to be important electrical devices that are used to transfer energy from a source to a load by increasing or decreasing the voltage or current. Transformers are also used to match the impedance of a load to a source and to isolate the load electrically from the source. Various types of transformers were discussed.

SUMMARY OF IMPORTANT EQUATIONS

$$p = vi \tag{14.1}$$

$$P_{\text{ave}} = P = VI\cos\theta \tag{14.3}$$

$$P_{\text{app}} = S = VI \tag{14.4}$$

$$P = S\cos\theta \tag{14.5}$$

$$Q = VI\sin\theta \tag{14.7}$$

$$S = \sqrt{P^2 + Q^2} \tag{14.8}$$

$$\mathbf{S} = P + jQ = S\underline{/\theta} \tag{14.9}$$

$$\mathbf{S} = \mathbf{VI^*} \tag{14.11}$$

$$P = I^2R \tag{14.12}$$

$$Q = I^2X \tag{14.13}$$

$$F_p = \cos\theta \tag{14.14}$$

$$\mathbf{Z_L} = \mathbf{Z_i^*} \tag{14.17}$$

$$S_T = \sqrt{3}\,E_L I_L \tag{14.35}$$

$$P_T = \sqrt{3}E_\phi I_\phi \cos\theta \tag{14.36a}$$

$$P_T = \sqrt{3}\,E_L I_L \cos\theta \tag{14.36b}$$

$$Q_T = 3E_\phi I_\phi \sin\theta\ \sqrt{3}\,E_L I_L \sin\theta \tag{14.37}$$

$$\frac{V_p}{V_s} = \frac{N_p}{N_s} = a \tag{14.40}$$

$$\frac{V_p}{V_s} = \frac{I_s}{I_p} = a \tag{14.41}$$

$$\frac{V_p}{I_p} = a^2Z_L = Z_L' \tag{14.44}$$

REVIEW QUESTIONS

14.1 Define the term active power.

14.2 Define the term apparent power and state its units.

14.3 Explain the term power factor. What is the difference between a leading and lagging power factor?

14.4 State the maximum power transfer condition in an ac circuit.

14.5 Explain why 3-ϕ generators are more desirable than single-phase generators.

14.6 Explain what a balanced 3-ϕ load means.

14.7 State the advantages and disadvantages of a wye-connected system over a delta-connected system.

14.8 To what does the phase sequence in a 3-ϕ system refer?

14.9 Describe the construction of a transformer.

14.10 Describe the turns ratio of a step-up transformer.

14.11 Explain how a transformer can be used to match the impedance of load to a source.

14.12 Explain the isolation property of a transformer.

14.13 Describe the function of a current transformer.

PROBLEMS

14.1 A 60-Hz voltage source forces a current of $2\angle 45^\circ$ A in a load, and $\mathbf{Z} = 40\angle 30^\circ\ \Omega$. Determine:
a. The apparent power, S
b. The active power, P
c. The reactive power, Q
d. Draw the power triangle.

14.2 A 60-Hz, $177\angle 0^\circ$ V source supplies a current of $0.5\angle 30^\circ$ A to a load. Calculate:
a. The load impedance, $\mathbf{Z}$
b. The active power, P, delivered to the load
(ans. $234\angle -30^\circ\ \Omega$, 50.72 W)

14.3 For the circuit in Figure 14.28, determine:
a. The total apparent power delivered by the source
b. The total active power loss in the circuit
c. Sketch the power triangle.

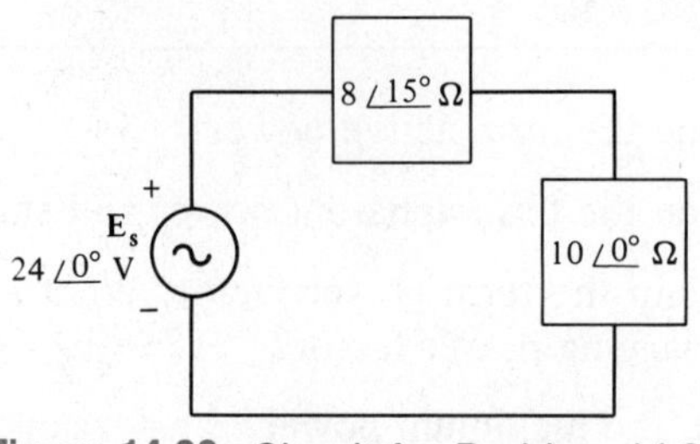

Figure 14.28 Circuit for Problem 14.3.

14.4 Repeat Problem 14.3 for the circuit in Figure 14.29. (ans. 0.194 VA, 0.188 W, 0.0048 VAR)

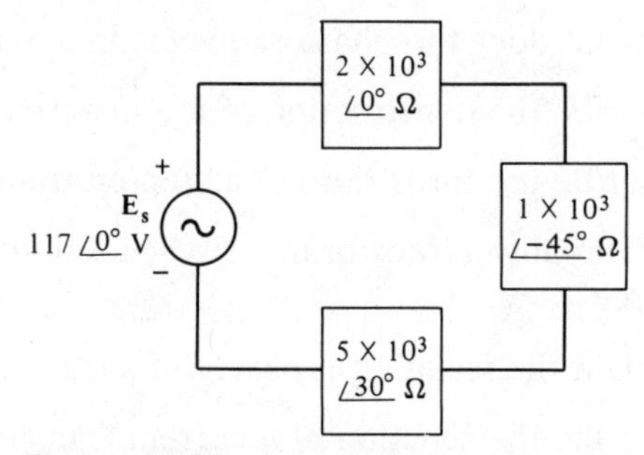

Figure 14.29 Circuit for Problem 14.4.

14.5 For the circuit in Figure 14.30, determine:

a. Total active power
b. Total reactive power
c. Sketch the power triangle.

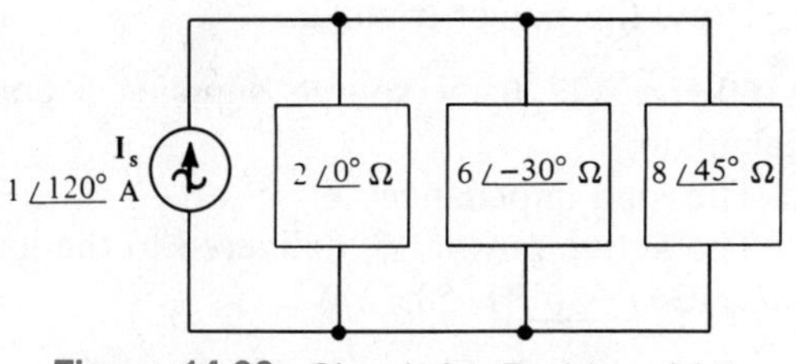

Figure 14.30 Circuit for Problem 14.5.

14.6 Repeat Problem 14.5 for the circuit in Figure 14.31. (ans. 8.61 kW, 3.24 kVAR)

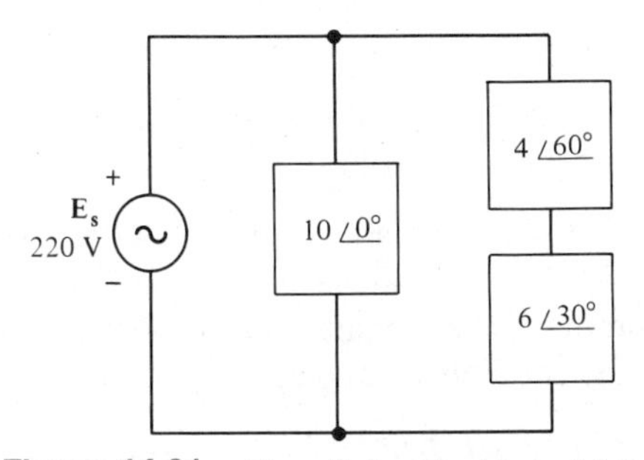

Figure 14.31 Circuit for Problem 14.6.

14.7 Calculate the power factor for the circuit in Figure 14.29.

14.8 A load with a leading power factor of 0.82 consumes 600 W. What is the apparent power and the reactive power in the load? (ans. 731.7 VA, 418.64 VAR)

14.9 An inductive 100-kVA load, with 440 V (60 Hz) across it, has a power factor of 0.72 lagging. Determine the value of a parallel capacitor to correct the power factor to 0.98 lagging.

14.10 An ac voltage source with an internal impedance of $0.2 + j0.6\ \Omega$ and 220 V is to transfer maximum power to a load. Determine the value of the load and the maximum power to the load. (ans. $0.2 - j0.6\ \Omega$, 60.5 kW)

14.11 For the circuit in Figure 14.32, determine the source impedance and the terminal voltage, $\mathbf{V_T}$, if the load is matched to the source for maximum power transfer.

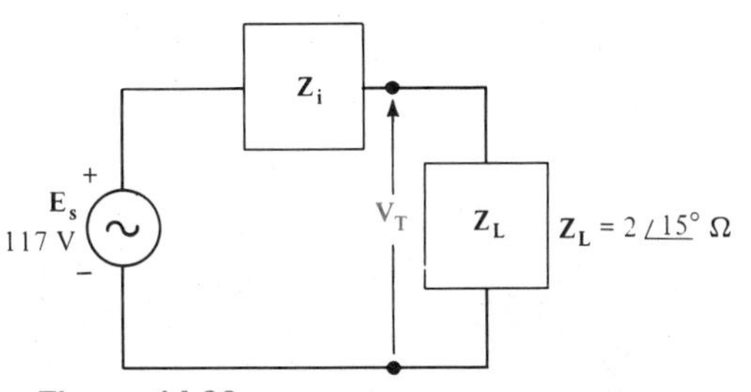

Figure 14.32 Circuit for Problem 14.11.

14.12 Calculate the phase voltages of a 3-ϕ wye-connected generator with line voltages of 220 V. (ans. 127 V)

14.13 Repeat Problem 14.12 for a delta-connected generator.

14.14 A 3-ϕ wye-connected system with line voltages of 440 V drives a balanced wye load with $\mathbf{Z_1} = \mathbf{Z_2} = \mathbf{Z_3} = 8 + j6\ \Omega$. Calculate the magnitude of the line and phase currents. Calculate the total active power to the load. (ans. 25.4 A, 15.48 kW)

14.15 Repeat Problem 14.14 for a balanced load of $3 - j2\ \Omega$.

14.16 A 3-ϕ delta balanced load of $0.5 + j\ \Omega$ is supplied from 220-V lines. Determine the line and phase currents. (ans. 176 A, 101.16 A)

14.17 A three-phase balanced 60-Hz delta source supplies power to a balanced wye load. Each phase of the load contains a 10-Ω resistor and a 50-mH inductor. If the magnitude of the load currents is 12 A, calculate the magnitude of the line voltages.

14.18 A three-phase balanced wye source supplies power to a balanced delta load. The load phase currents have a magnitude of 6 A each. Determine the magnitude of the line currents. (ans. 10.39 A)

14.19 An industrial delta-connected load is powered from 4400-V, 3-ϕ lines. Calculate the active power to the load and the reactive power in the load if the load consumes 250 kVA at a 0.95 lagging power factor.

14.20 Determine the secondary voltage and current for a transformer with a primary voltage of 220 V, a primary current of 1.2 A, and a turns ratio, $a = 4$. (ans. 55 V, 4.8 A)

14.21 Determine the turns ratio of an ideal transformer with a primary voltage of 117 V and a secondary voltage of 9 V.

14.22 A step-up transformer with a turns ratio of $a = \frac{1}{3}$ is used to match a load to a source. If the source impedance is 90 Ω, what should the load impedance be for a matched condition? (ans. 810 Ω)

14.23 It is necessary to match a 10-Ω load to a 2500-Ω source. Determine the transformer turns ratio necessary to accomplish this task.

CHAPTER 15

Measurements and Instrumentation

GLOSSARY

ammeter—instrument used to measure current
analog instrument—measuring instrument that displays results in a continuous manner by using a pointer to indicate a change in distance on a scale
automatic test equipment (ATE)—programmable test systems used in repetitive high-volume testing

clamp-on ammeter—instrument used to measure ac currents by external sensing of the flow
dc power supply—equipment that supplies dc power at a constant voltage
digital frequency counter—instrument used to measure frequency and period of periodic waves
digital instrument—measuring instrument that provides a numerical display of the measured quantity
digital multimeter (DMM)—digital instrument used to measure current, voltage, and resistance
digital voltmeter (DVM)—digital instrument used to measure voltage
electrodynamometer—a movement used in the construction of ammeters, voltmeters, and wattmeters
error—difference between the actual and measured value
function generator—laboratory instrument that produces various types of "standard" ac signals at variable amplitude and frequency
galvanometer—instrument used to sense and display very low values of current
impedance bridge—instrument used to measure capacitance, inductance, and resistance
light-emitting diode (LED)—indicating device used with digital instruments
liquid crystal display (LCD)—indicating device used with digital instruments
loading effect—the effect that the measuring instrument has on the circuit
megohmmeter (megger)—instrument used to measure very high resistance values
meter—measuring instrument
moving-iron movement—mechanism for measuring current
ohmmeter—instrument used to measure resistance values
oscilloscope—instrument used to display time-varying signals
rectification—process of converting ac to pulsating dc
sampling oscilloscope—instrument used to display high-frequency signals by using the sampling technique
sensitivity—factor describing the loading effect of a meter, given in units of ohms per volt
spectrum analyzer—instrument used to display signal amplitudes versus frequency
storage oscilloscope—instrument used to display single events and low-frequency ac signals
voltmeter—instrument used to measure voltage
volt-ohm-milliammeter (VOM)—instrument used to measure current, voltage, and resistance
wattmeter—instrument used to measure the power delivered to a device or circuit
Wheatstone bridge—instrument used to measure resistance accurately

INTRODUCTION

Measuring instruments called **meters** are the eyes and ears of electrical technicians and engineers. It is through these instruments that an operational knowledge of the devices and circuits is gained. There are instruments available to measure every electrical quantity; however, the ease and accuracy with which measurements can be made vary from instrument to instrument.

Both **analog** and **digital instruments** are available. The terms analog and digital refer to the method of display of the electrical quantity. Analog instruments provide a continuously changing display by using a pointer to indicate a change in distance on a scale. The change in the quantity measured causes the pointer to change its position relative to the scale. A digital instrument displays the value of the measured quantity in numerical form. Since the measured electrical quantities are present in analog form, every change represents a change in value. Therefore, the sensing portions of the instruments are the same, but the circuitry needed to cause the display of the result is substantially different between analog and digital instruments.

In this chapter, both analog and digital instruments used to measure dc and ac quantities are discussed. Various measurement techniques are illustrated.

15.1 ANALOG METERS

The most commonly used indicating device used with analog instruments is the *permanent-magnet moving-coil* (PMMC) meter movement. The basic movement is also referred to as the d'Arsonval movement, after its inventor. The construction of such a device is shown in Figure 15.1. In this device, a coil wound on a soft iron cylindrical core is suspended between two poles of a permanent magnet. A pointer is attached to the coil. A current through the coil causes the coil to rotate in the permanent magnetic field and the pointer to move with it. The distance traveled by the pointer is a function of the magnitude of the current in the coil. The

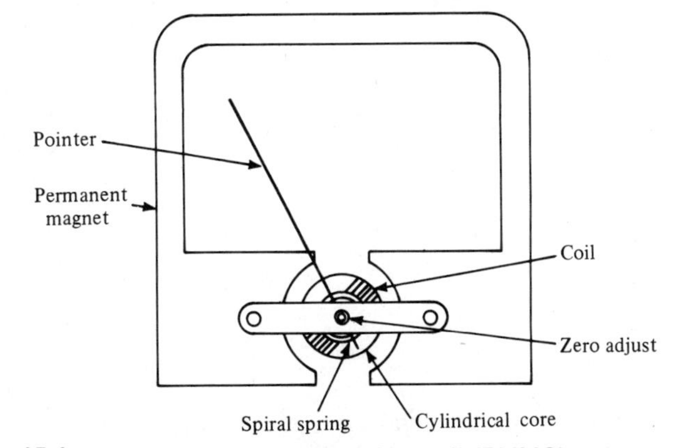

Figure 15.1 Permanent-magnet moving-coil (PMMC) meter movement.

direction of coil rotation and, subsequently, the direction of pointer deflection depend on the direction of the current. There are two spiral springs on each side of the coil cylinder. One end of each spring is attached to the coil and the other to a fixed point or a zero-adjust screw. The springs serve as conductors of current to the coil and to return the pointer to zero position with no current applied. The zero-adjust screw is used to calibrate the pointer to a zero indication on the scale. The coil cylinder is held in place by means of *jewel bearings* at each end or by a *taut-band suspension* system, which consists of metal bands holding up the cylinder. The taut-band system is less rugged than the jewel bearings but results in less friction. The PMMC movement is usually designed to have a *full-scale deflection* of micro- or milliamps of current.

The **galvanometer** is a very sensitive PMMC meter. Galvanometers deflect for minute currents as low as picoamps (pA). The zero is usually at the center of the scale to detect currents in either direction. They are very delicate instruments and must be handled with a great deal of care.

The *electrodynamic* movement, or the **electrodynamometer,** is a device that also has a moving coil, but the magnetic field is not generated through permanent magnets but with permanent field coils. The electrodynamometer is capable of measuring dc and ac currents, whereas the PMMC device measures only dc currents. The electrodynamometer is used to measure the true RMS value of ac signals.

Moving-iron instruments operate on the principle that when a current passes through a coil that surrounds two adjacent pieces of soft iron, the soft iron pieces will become magnetized and will repel each other. Hence, if one soft iron piece, called a *vane,* is stationary, and the other is movable with a pointer attached to it, then the deflection of the pointer is proportional to the current through the coil. Two types of moving-iron mechanisms are illustrated in Figures 15.2a and 15.2b.

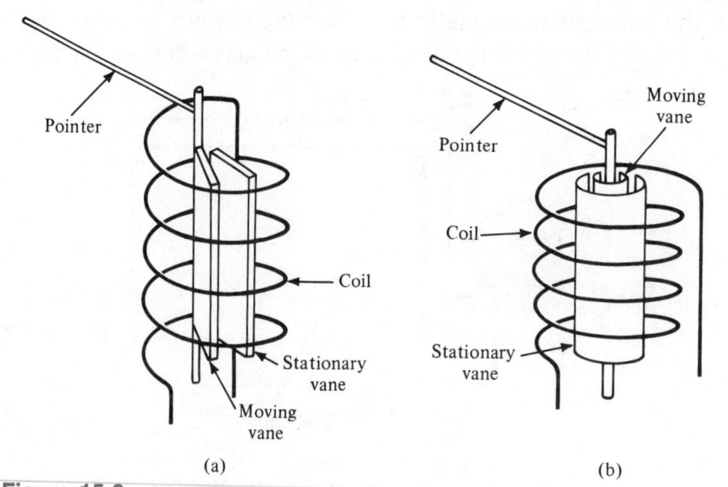

Figure 15.2 Moving-iron mechanisms. **(a)** Radial iron vane type (book-type iron vane). **(b)** Concentric iron vane.

The most basic analog measuring instrument is the **ammeter.** The ammeter measures the current in a circuit. Instruments that measure low current values, such as milliamperes and microamperes, are referred to respectively as *milliammeters* or *microammeters*. The movements used in ammeters are essentially those discussed earlier in this section. A basic PMMC can be used to measure currents at different levels. The *full-scale deflection* (FSD) of the movement, which is given in microamperes, milliamperes, or amperes, determines the maximum current that can be measured by the instrument. The FSD of the ammeter is usually different than the FSD of the movement. It is through the use of additional shunt resistors that the instrument is made to have a different FSD than the movement itself.

The ammeter is inserted in series with the device or circuit; the current is measured as illustrated in Figure 15.3. The current that flows in the meter is the current in the load. The meter has polarity marked on its terminals, and the connection showing conventional current entering the positive terminal of the ammeter will assure positive deflection on the scale.

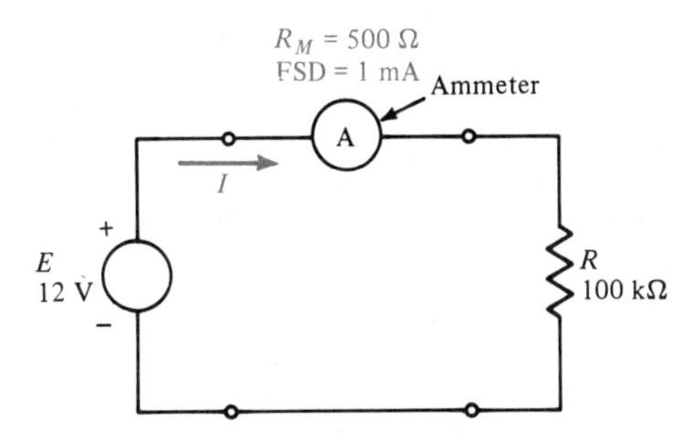

Figure 15.3 Ammeter inserted in the circuit to measure current.

The ammeter has an internal resistance, R_M, due to the coil windings. Therefore, inserting the instrument in series with the load will result in a change of the total circuit resistance and also a change in the current. Hence, the internal resistance of the ammeter is an important factor in determining the accuracy of the measurement.

The effect of the measuring instrument on the circuit is called the ***loading effect.***

An ammeter with a very low resistance as compared to the load resistance has a "negligible" loading effect. Ideally, an ammeter should have zero resistance. Since in a practical meter the internal resistance is not zero, there is also a voltage drop, V_M, across the instrument and power consumed by it.

EXAMPLE 15.1

The ammeter in Figure 15.3 has an FSD of 1 mA and an internal resistance $R_M = 500\ \Omega$. Calculate (a) the theoretical current in the circuit without the ammeter connected, (b) the current measured by the ammeter, and (c) the voltage dropped across the ammeter.

Solution

a. The current in the circuit is given by Ohm's Law:

$$I = \frac{12}{100 \times 10^3} = 0.12 \text{ mA}$$

b. The ammeter reads:

$$I = \frac{12}{100 \times 10^3 + 500} = 0.1194 \text{ mA}$$

c. By Ohm's Law,

$$V_M = (500)(0.1194 \times 10^{-3}) = 0.0597 \text{ V}$$

It becomes apparent that the meter introduces an error in the current measurement due to the loading effect.

Assume that the ammeter of Figure 15.3 is to be used to measure currents up to 1 A. If a current of 1 A were to flow through this meter, which can handle currents only up to 1 mA, it would be destroyed. However, by designing a shunt resistor to be placed across the terminals of the movement, it is possible to increase the FSD capacity and to operate the meter safely. This is illustrated in the next example.

EXAMPLE 15.2

Convert the ammeter circuit of Figure 15.3 to an ammeter with an FSD of 1 A.

Solution

A shunt resistor connected across the original movement will act to divert most of the current and to assure that the current through the movement is 1 mA at maximum. The meter and the shunt resistor are shown in Figure 15.4. By Kirchhoff's Current Law, current I_S through the short is:

$$\begin{aligned} I_S &= I_{FS} - I_M \\ &= 1 \text{ A} - 1 \text{ mA} = 999 \text{ mA} \end{aligned}$$

The voltage across the shunt resistor is the same as the voltage across the meter; hence,

$$\begin{aligned} V_S &= V_{MFS} = R_M \times I_M \\ &= (500)(1 \text{ mA}) = 0.5 \text{ V} \end{aligned}$$

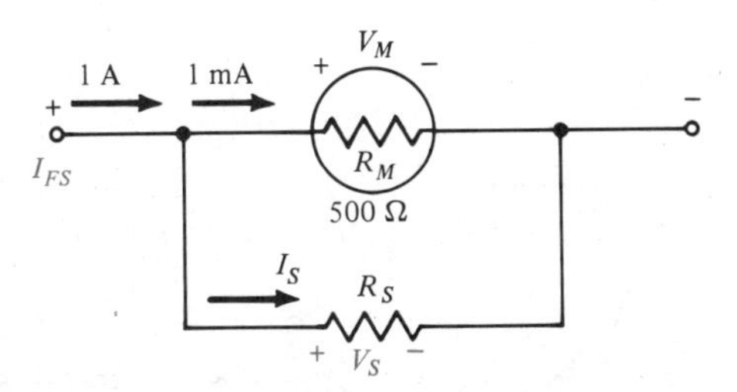

Figure 15.4 Shunt resistor for Example 15.3.

Therefore, the shunt resistance is:

$$R_S = \frac{V_S}{I_S} = \frac{0.5}{999 \times 10^{-3}} = 0.5005\ \Omega$$

(The result is carried out to four decimal places because the accuracy of the instrument depends on the precision of the shunt resistor. Thus, shunt resistors used in instrumentation are precision-type resistors.)

Inserting and connecting internally the shunt resistor of Example 15.2 yields an ammeter with an FSD of 1 A. Of course, the original meter scale will have to be recalibrated for 0- to 1-A indications.

EXAMPLE 15.3

A multirange milliammeter is shown in Figure 15.5a. Determine the FSD at different switch settings.

Solution

The voltage across the meter at FSD must always be the same; hence,

$$V_{MFS} = (500)(1\text{ mA}) = 0.5\text{ V}$$

The FSD current with the switch in *position A* is 1 mA, the FSD of the basic movement, because no shunt resistance is present.

The current through shunt resistor R_{S_1}, when the switch is in position B, is:

$$I_{S_1} = \frac{V_{MFS}}{R_{S_1}} = \frac{0.5}{55.56} = 9\text{ mA}$$

Therefore, in position B,

$$\begin{aligned} I_{FS} &= I_{S_1} + I_M \\ &= 9\text{ mA} + 1\text{ mA} = 10\text{ mA} \end{aligned}$$

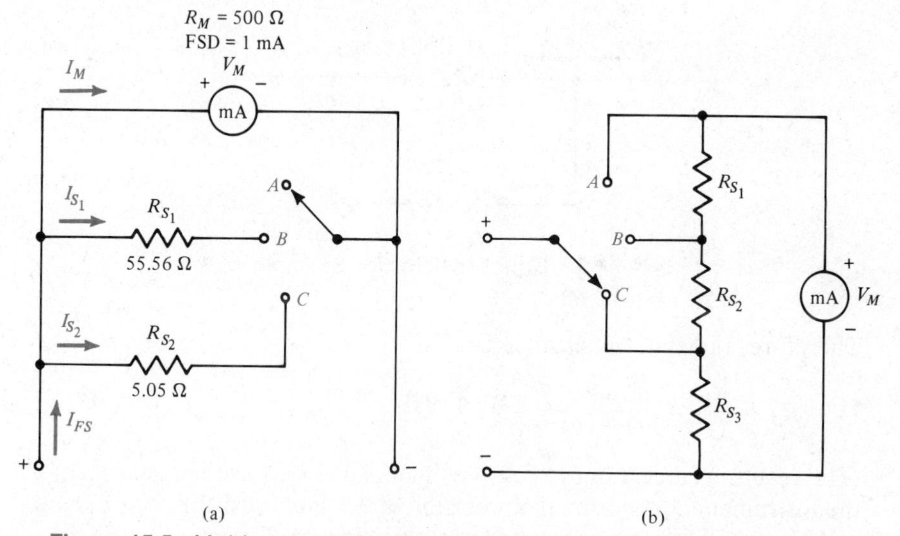

Figure 15.5 Multirange milliammeters. **(a)** Simple multirange. **(b)** Ayrton shunt.

Similarly, in position C,

$$I_{S_2} = \frac{0.5}{6.05} = 99 \text{ mA}$$

and the full-scale current in position C is:

$$\begin{aligned} I_{FS} &= I_{SZ} + I_M \\ &= 99 \text{ mA} + 1 \text{ mA} = 100 \text{ mA} \end{aligned}$$

Therefore, the milliammeter has three ranges, 1 mA, 10 mA, and 100 mA.

The designer of the meter circuit in Example 15.3 must be careful to specify that the switch must be of the *make-before-break* type rather than a *break-before-make* switch. A make-before-break switch will connect to the next position before disconnecting from the previous one. In one case the switch will connect, for example, to position C before disengaging from position B. This is necessary to avoid a situation such as the following: If the scale is switched from the 100-mA range to the 10-mA range because the current is, for example, 5 mA and the B scale will give a better reading, then the meter will not be overloaded during the instant when the switch is transferred from C to B. Without the make-before-break feature, during the instant of switching the current would be applied directly to the movement without a shunt.

The *Ayrton,* or *universal,* shunt design shown in Figure 15.5b is frequently used to construct a multirange ammeter. In this circuit there is always a shunt resistor across the meter movement, hence eliminating the need for a make-before-break selector switch.

The basic ammeter can be used to construct a **voltmeter.** A voltmeter is an instrument that measures voltage across devices and circuits. It was shown earlier that a voltage is dropped across a PMMC movement due to the current through it and the internal resistance of the instrument. Since the resistance is fairly constant, the voltage across the instrument is directly proportional to the deflection of the meter. The voltage across a current meter is usually small due to the low internal resistance; hence, the voltage range of the meter is extended by adding series resistance, R_V, to the instrument.

The PMMC movement used earlier in this section can be converted to a voltmeter. The design procedure is illustrated in the next example.

EXAMPLE 15.4

The PMMC movement in Figure 15.6 is to be converted to a voltmeter with an FSD of 50 V. Design the value of the series resistor, R_V, also called the *multiplier*.

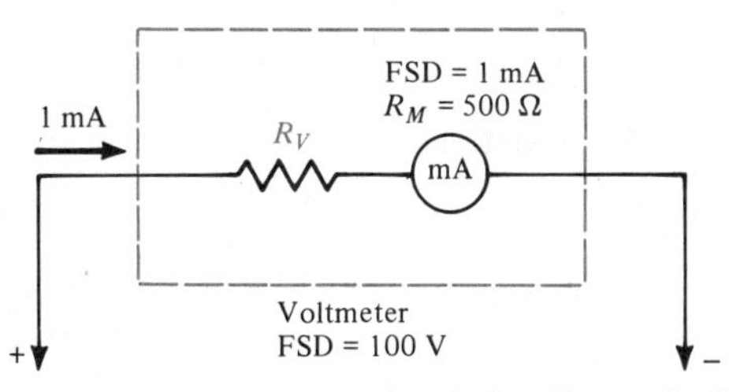

Figure 15.6 Voltmeter circuit for Example 15.4.

Solution

Since the FSD of the movement is 1 mA and an FSD of the voltmeter is to be 50 V, then the total voltmeter resistance must be:

$$R_T = \frac{50}{1 \times 10^{-3}} = 50 \text{ k}\Omega$$

Resistor, R_V, and meter resistance, R_M, are connected in series and form the total resistance; hence,

$$\begin{aligned} R_V &= R_T - R_M \\ &= 50 \times 10^{-3} - 500 = 45.5 \text{ k}\Omega \end{aligned}$$

A multirange voltmeter is shown in Figure 15.7. A combination of three series multiplier resistors and a switch are used to convert a single PMMC movement to a three-range voltmeter.

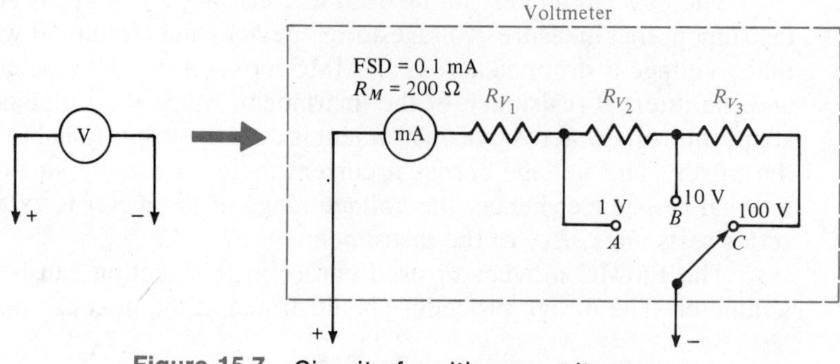

Figure 15.7 Circuit of multirange voltmeter.

EXAMPLE 15.5

Design the values of R_V, R_{V_2}, and R_{V_3} in the circuit of Figure 15.7 for the voltmeter to have 1-V, 10-V, and a 100-V ranges when the switch is in positions *A*, *B*, and *C*, respectively.

Solution

1-V RANGE. The total resistance for the switch in position *A* is:

$$R_{T_1} = R_{V_1} + R_M$$

Since the total current is the FSD current of the PMMC, the total resistance is also:

$$R_{T_1} = \frac{1}{0.1 \times 10^{-3}} = 10\ \text{k}\Omega$$

Therefore

$$R_{V_1} = 10\ \text{k}\Omega - 200\ \Omega = 9.8\ \text{k}\Omega$$

10-V RANGE. When the switch is in position *B*,

$$R_{T_2} = R_{V_1} + R_{V_2} + R_M$$

The total resistance is also:

$$R_{T_2} = \frac{10}{0.1 \times 10^{-3}} = 100\ \text{k}\Omega$$

Solving for R_{V_2} in the total resistance equals

$$R_{V_2} = R_{T_2} - R_{V_1} - R_M$$
$$= 100\ \text{k}\Omega - 9.8\ \text{k}\Omega - 0.2\ \text{k}\Omega = 90\ \text{k}\Omega$$

100-V RANGE. Switch is in position *C*:

$$R_{T_3} = R_{V_1} + R_{V_2} + R_{V_3} + R_M$$

and also

$$R_{T_3} = \frac{100}{0.1 \times 10^{-3}} = 1\ \text{M}\Omega$$

Solving for R_{V_3} from the total resistance equals

$$R_{V_3} = R_{T_3} - R_{V_1} - R_{V_2} - R_M$$
$$= 1\ \text{M}\Omega - 90\ \text{k}\Omega - 9.8\ \text{k}\Omega - 0.2\ \text{k}\Omega = 900\ \text{k}\Omega$$

A voltmeter connected *across* the resistor is shown in Figure 15.8. Since voltage is measured between two points, the voltmeter is always connected across the device whose voltage is being measured. A voltmeter also has an internal resistance R_M; hence, it also changes the original circuit conditions when connected. Therefore, a voltmeter can also have a loading effect on a circuit. Since the voltmeter is a parallel resistance, it affects the circuit if its internal resistance is not much higher than the component it is connected to. Therefore, ideally a voltmeter should have infinite resistance.

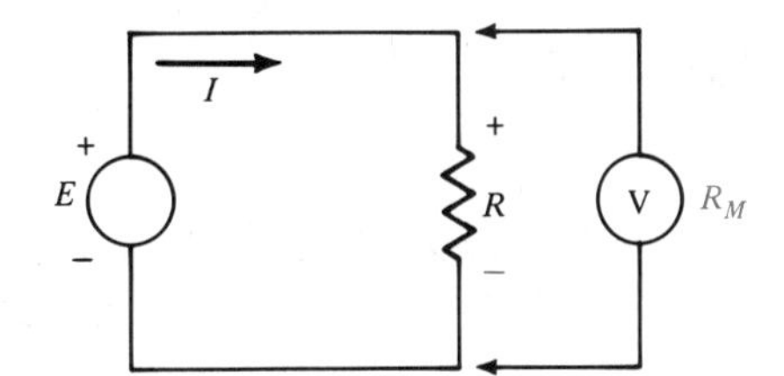

Figure 15.8 Voltmeter connected across a resistor.

The meter ***sensitivity*** *is a factor describing the loading effect of the meter.*

Sensitivity is given mathematically as the reciprocal of the full-scale deflection current of the movement,

$$\boxed{S_M = \frac{1}{I_{MFS}}} \qquad \Omega/\text{V} \tag{15.1}$$

where S_M stands for meter sensitivity and I_{MFS} is the full-scale current. The loading effect of a multirange voltmeter, which is the resistance of the voltmeter at a particular scale, is simply the product of the sensitivity value and the full-scale deflection of the particular voltage range.

EXAMPLE 15.6

For the voltmeter designed in Example 15.5, determine the sensitivity and the meter resistance of each range.

Solution

The meter sensitivity is given by Eq. (15.1):

$$S_M = \frac{1}{0.1 \times 10^{-3}} = 10\ \text{k}\Omega/\text{V}$$

The resistance on the 1-V range is:

$$R_{M_1} = 10 \times 10^3 \times 1 = 10\ \text{k}\Omega$$

Similarly, on the 10-V range,

$$R_{M_2} = 10 \times 10^3 \times 10 = 100\ \text{k}\Omega$$

Finally, on the 100-V range,

$$R_{M_3} = 10 \times 10^3 \times 100 = 1\ \text{M}\Omega$$

The information obtained in Example 15.6 is utilized in Example 15.7 to illustrate the loading effect of the meter.

EXAMPLE 15.7

The voltmeter from Examples 15.5 and 15.6 is used to measure the voltage of the circuit in Figure 15.9. Calculate the voltage measured by the voltmeter on the 10-V and the 100-V ranges. What is the actual calculated voltage across R_2?

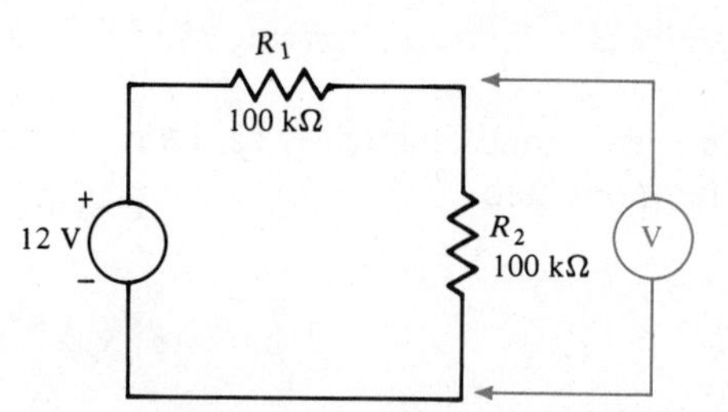

Figure 15.9 Circuit for Example 15.7.

Solution

The internal resistance of the voltmeter is connected in parallel with R_2.

On the *10-V range,* the meter resistance is 100 kΩ; therefore, the parallel resistance is:

$$R_{p_2} = R_2 \| R_{M_2} = 100\ \text{k}\Omega \| 100\ \text{k}\Omega = 50\ \text{k}\Omega$$

and by the Voltage Divider Rule, the voltmeter voltage is:

$$V_{M_2} = (12)\frac{50\ \text{k}\Omega}{100\ \text{k}\Omega + 50\ \text{k}\Omega} = 4\ \text{V}$$

On the *100-V range,* the meter resistance was found to be 1 MΩ; therefore, the parallel resistance is:

$$R_{p_3} = R_2 \| R_{M_3} = 100\ \text{k}\Omega \| 1\ \text{M}\Omega = 90.91\ \text{k}\Omega$$

and the voltmeter voltage is:

$$V_{M_3} = (12)\frac{90.9\ \text{k}\Omega}{100\ \text{k}\Omega + 90.9\ \text{k}\Omega} = 5.71\ \text{V}$$

The actual voltage across R_2 without the voltmeter is:

$$V_{R_2} = (12)\frac{100\ \text{k}\Omega}{100\ \text{k}\Omega + 100\ \text{k}\Omega} = 6\ \text{V}$$

This result shows that the voltage measured with the voltmeter is not accurate, and the lower the internal resistance of the meter, the larger the error.

An **ohmmeter** is used to measure resistance values. Once again, the basic PMMC movement is used to construct an ohmmeter. The circuit of an ohmmeter is shown in Figure 15.10a. A battery is used as a constant voltage source, and resistors R_1 and R_2 are used to limit the current through the current meter. When the leads are left open, there is no current in the circuit and the ohmmeter scale indicates an infinite resistance, as shown in Figure 15.10b at the extreme left of the scale. When the two leads are shorted together, a current limited by R_1 and R_2 flows through the meter to deflect fully. Resistor R_2 is a potentiometer, which is

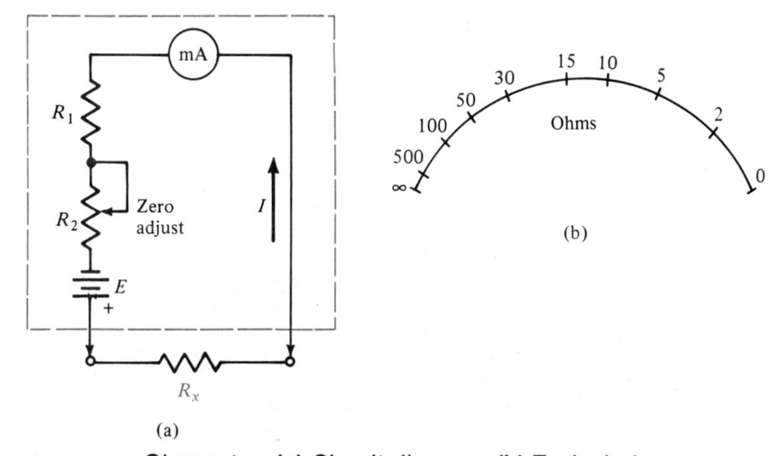

Figure 15.10 Ohmmeter. **(a)** Circuit diagram. **(b)** Typical ohmmeter scale.

varied to calibrate the ohmmeter for full-scale deflection (0 Ω) when the leads are shorted together. An unknown resistor, R_x, connected to the leads of the instrument completes the circuit and causes the pointer to deflect. The ohmmeter scale is nonlinear. The current through the movement is the ratio of the battery voltage and the sum of the resistors:

$$\boxed{I = \frac{E}{R_1 + R_2 + R_M + R_x}} \tag{15.2}$$

Resistance measurements of devices must be performed with the devices isolated from the circuit and no power other than the ohmmeter internal battery connected to it. Otherwise, the measurement is not accurate, and the instrument might be damaged if there is external power applied.

All three types of meters—ammeter, voltmeter, and ohmmeter—utilize the same display type described, the PMMC movement. The only difference is in the method in which the measured quantity is converted to feed the basic ammeter movement. Hence, it is possible to combine all functions into one and to use a single instrument called a *multimeter*. The multimeter shown in Figure 15.11 is a very popular multimeter also called a *volt-ohm-milliammeter* (VOM). This particular VOM is capable of measuring dc voltages and currents as well as resistance, and it also has a decibel scale for relative decibel measurements. The meter has a sensitivity of 20,000 Ω/V on dc ranges and 1000 Ω/V on ac ranges. Hence, the resistance of the meter varies with each scale.

Figure 15.11 Simpson multimeter. (Courtesy Simpson Electric Company.)

The VOM measures ac quantities by *rectifying* the ac signal first and then sensing its dc component.

***Rectification** is the process of converting ac to pulsating dc.*

Since the meter mechanism detects dc only, it is necessary to convert ac to some form of dc. The rectifier circuits of Figures 15.12a and 15.12b are used for this purpose. The coil of the movement will respond to the dc level (average) of the pulsating wave. The diodes in the circuit are the key elements in the rectification process. The ac scales are calibrated so that when the meter responds to the dc component of the current, the pointer will indicate RMS, or peak-to-peak, values. Because of that type of calibration, such instruments' scales are calibrated for the 0.707 RMS factor of a sinusoidal wave and will not measure accurately ac waves other than sinusoids.

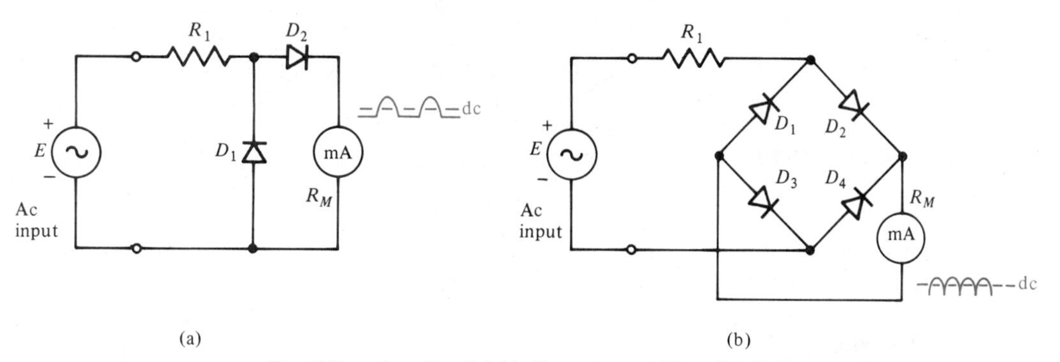

Figure 15.12 Rectifier circuits. **(a)** Half-wave rectifier. **(b)** Full-wave bridge rectifier.

The meter measuring ac has a limited frequency response. The reactance of the PMMC varies with frequency, and so will the amount of current through it. There will be less current at higher frequencies. Hence, PMMC-type instruments are effective up to approximately 10 kHz.

The **clamp-on ammeter** is used extensively to measure ac currents in industrial applications, where it is not practical to open the circuit and insert an ammeter to perform the measurement. The clamp-on ammeter shown in Figure 15.13 wraps its jaws around the current-carrying conductor and, through transformer action with the conductor as the primary and the windings in the clamp portion of the ammeter as the secondary, senses and displays the current value. Clamp-on ammeters are designed to operate at about a 60-Hz frequency value and measure high values of currents, up to 1000 A.

Electronic-type multimeters are either of the analog or digital type and generally provide better loading characteristics due to a constant resistance value on all scales. Such instruments are referred to as *vacuum tube voltmeters* (VTVM), *transistorized voltmeters* (TVM), and *digital multimeters* (DMM). These instru-

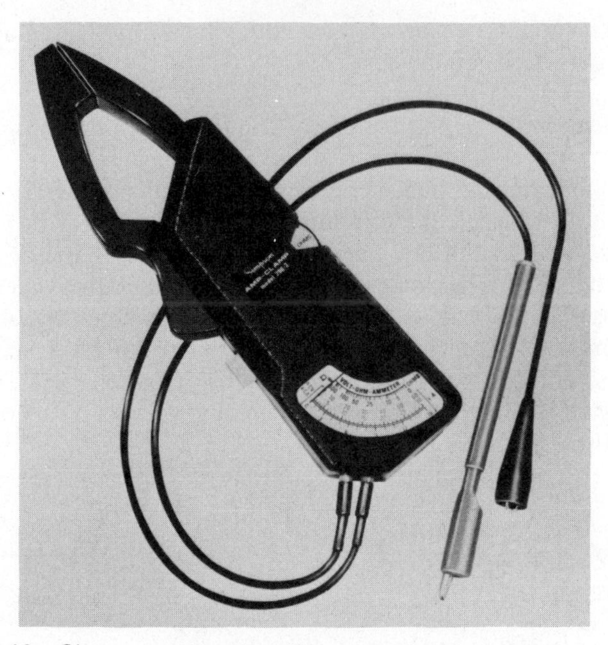

Figure 15.13 Clamp-on ammeter. (Courtesy Simpson Electric Company.)

ments require an internal or external source to power the electronic circuits. Such instruments are further discussed in the next section.

A commonly used instrument for accurate measurements of resistance is the **Wheatstone bridge.** A circuit that is constructed of components in four arms arranged in a diamond shape, as shown in Figure 15.14, is called a bridge. The unknown resistance value, R_x, is not read from the face of the meter but from the setting of the potentiometer, R_4. The bridge is said to be *balanced,* which means no current through the meter, if the ratios of the resistors are as follows:

$$\frac{R_2}{R_3} = \frac{R_4}{R_x} \tag{15.3}$$

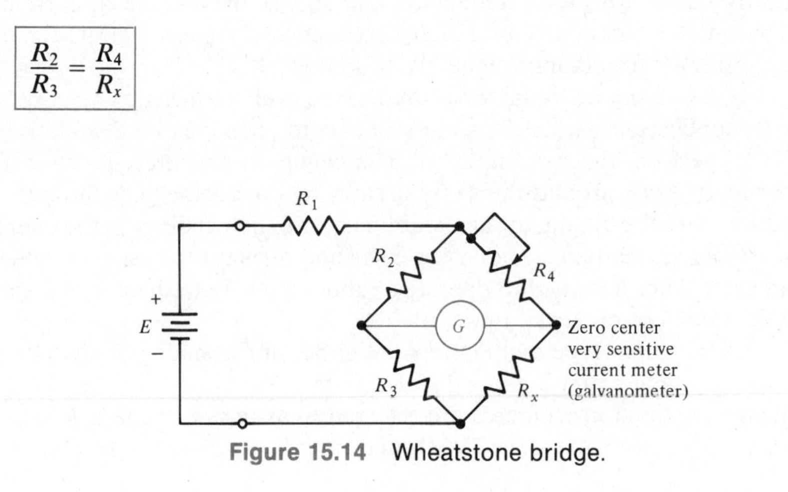

Figure 15.14 Wheatstone bridge.

or

$$R_x = R_4 \frac{R_3}{R_2} \tag{15.4}$$

If resistors R_3 and R_2 are precision resistors of constant values, then the ratio of R_3/R_2 is always constant. After adjusting the potentiometer R_4 for the galvanometer to read zero (balanced condition), it is necessary only to know the value of the potentiometer at that setting and multiply by the known ratio of resistors R_3/R_2.

A photograph of a commercial Wheatstone bridge instrument is shown in Figure 15.15a, and that of a *Kelvin bridge,* which is used for measuring low-

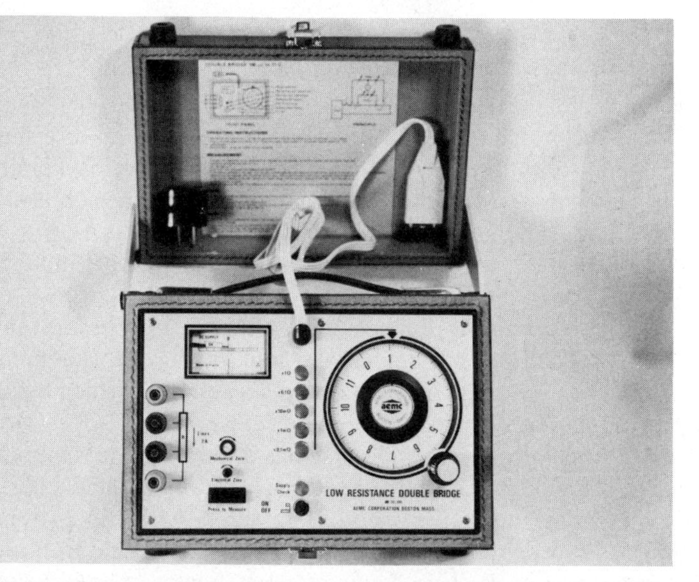

Figure 15.15 Commercial resistance bridges. **(a)** Wheatstone bridge. (Courtesy AEMC Corporation-Boston, MA) **(b)** Kelvin bridge. (Courtesy AEMC Corporation-Boston, MA)

resistance values, is shown in Figure 15.15b. The Wheatstone bridge is not as accurate as the Kelvin bridge for measuring low-resistance values. Kelvin bridges typically measure accurately resistances from 1 $\mu\Omega$ to 1 Ω.

A **megohmmeter,** commonly known as a **megger,** is used to measure very high resistances. The instrument is widely used to measure insulation resistance of wires, insulation between motor windings, and windings to frame resistance. A photograph of a megger is shown in Figure 15.16. An internal generator, powered by a hand crank, is used to produce the high voltage (a few hundred volts) necessary to force a current through the high resistance measured.

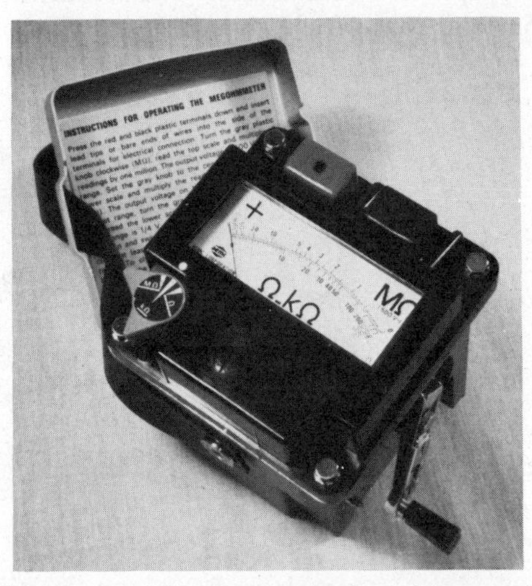

Figure 15.16 Megger used to measure high-resistance values. (Courtesy AEMC Corporation-Boston, MA)

Ac bridge-type instruments are used to measure inductance, capacitance, impedance, and the Q factor. Those bridges are ac versions of the Wheatstone bridge with resistive and reactive components in each arm of the bridge. Some of the more commonly used ac bridges are *Maxwell, Hay, Owen, Schering,* and *Wien* bridges. An **impedance bridge** used to measure inductance, capacitance, and resistance is shown in Figure 15.17.

Power can be determined by measuring the current through and voltage across a circuit or a device and multiplying the two quantities. However, it is simpler to use an instrument called a **wattmeter,** which displays the power directly in watts. The electrodynamometer is well suited for power measurements.

Recall from earlier discussions of the electrodynamometer that there are two sets of coils in the instrument: two stationary field coils and one moving coil. Connecting the stationary coils in series with the current and the moving coil across in order to measure voltage will result in a deflection of the coil corresponding to the power—hence, a wattmeter. Figure 15.18 illustrates the connection of a wattmeter to a resistor to measure power. Wattmeters are used with ac and dc power measurements. Notice that a wattmeter is a four-terminal device.

Figure 15.17 Impedance (*RLC*) bridge. (Copyright 1988 Hewlett-Packard Company. Reproduced with permission.)

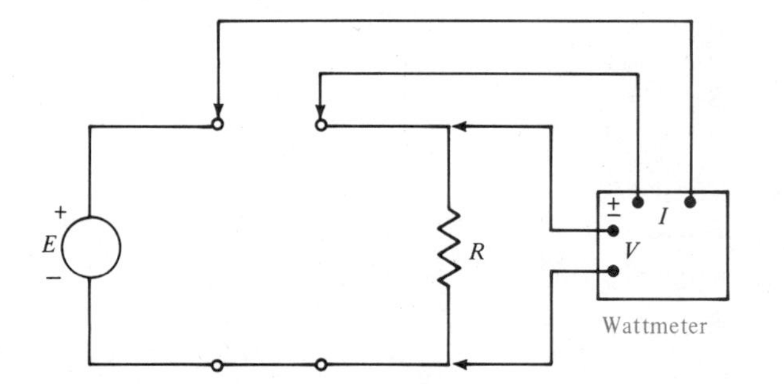

Figure 15.18 Wattmeter connection.

15.2 DIGITAL MEASURING INSTRUMENTS

Digital measuring instruments display the measured electrical quantity as a number. Digital displays are much easier to read than analog displays, as discussed earlier. They are also more accurate and more durable and can provide more functions than analog meters. However, digital instruments have more complex electronic circuitry than analog meters, constructed of electronic integrated circuits.

The block diagram shown in Figure 15.19 depicts a general digital measuring instrument. The quantity measured is always analog (continuous), such as voltage, current, power, and so on. The analog quantity is first ''conditioned'' to appear as a voltage at proper voltage levels, corresponding to the actual input. Next, the voltage level is converted to a digital signal represented by pulses (highs

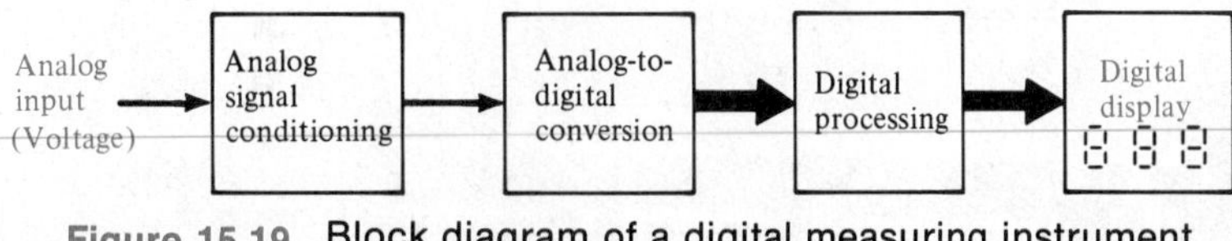

Figure 15.19 Block diagram of a digital measuring instrument.

and lows), which stand for binary numbers. Those highs and lows representing numbers are further converted, decoded, and delivered to a display unit, which displays the value of the measured quantities as decimal numbers. The type of display that is usually used is either the **light-emitting diode (LED)** type or the **liquid crystal display (LCD)** type. Each numeral is displayed by up to seven segments.

All digital measuring instruments are essentially **digital voltmeters (DVM)** because the measured quantity is converted to an equivalent voltage and the instrument actually senses and measures that voltage. Essentially all the instruments discussed earlier in this chapter are available as digital meters. Two types of **digital multimeters (DMMs)** are shown in Figure 15.20. The hand-held meter is

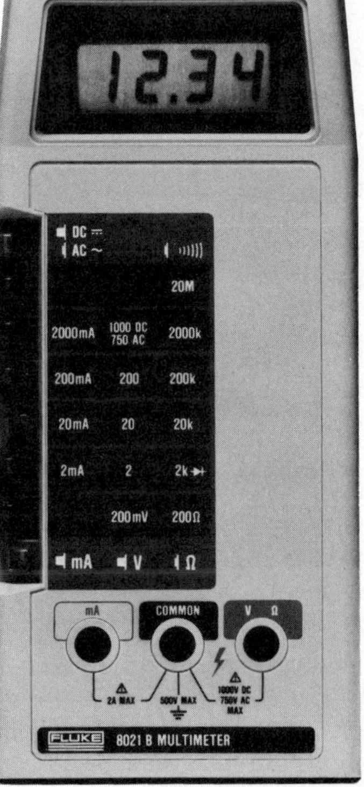

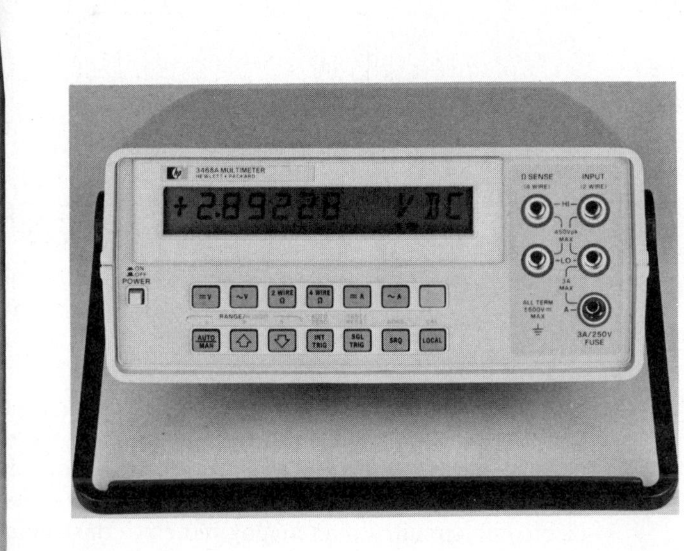

Figure 15.20 Digital multimeters. **(a)** Hand-held type. (Reproduced with permission from the John Fluke Mfg. Co., Inc.) **(b)** Bench model. (Copyright 1988 Hewlett-Packard Company. Reproduced with permission.)

powered by batteries and has an LCD-type display. The laboratory-type meter has an LED display. The laboratory-type DMM is capable of measuring dc and ac voltages and currents as well as resistance.

15.3 FREQUENCY-MEASURING INSTRUMENTS

Most commonly used frequency-measuring instruments are **digital frequency counters.** Frequency counters measure the frequency by counting the number of events and displaying in digital form the count for a specific time interval. Modern counters display the number of cycles per unit time, number of pulses, and events. A *universal counter-timer* is shown in Figure 15.21. This instrument is capable of displaying the period of a repetitive waveform in addition to its frequency and of counting pulses. It can measure frequencies up to 100 MHz with an accuracy of 1 part per million (ppm).

Analog frequency meters such as the one shown in Figure 15.22 are used at low frequencies in the power industry. **Oscilloscopes** are also used to determine

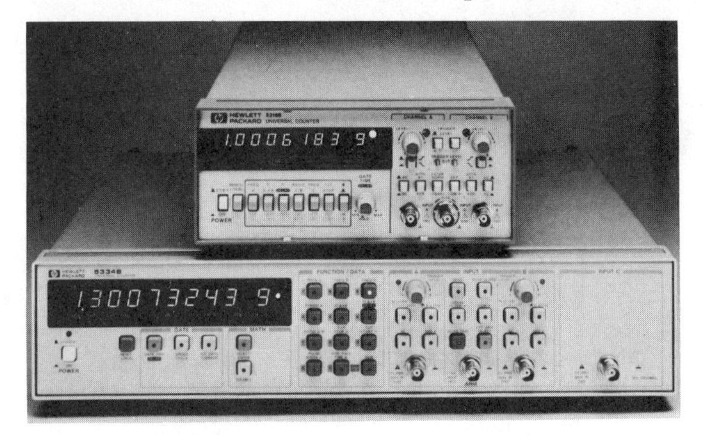

Figure 15.21 Universal counter. (Copyright 1988 Hewlett-Packard Company. Reproduced with permission.)

Figure 15.22 Analog frequency meter. (Courtesy Yokogawa Corporation of America, Newnan, GA)

frequency by measuring time intervals of the wave and then calculating the frequency.

15.4 OSCILLOSCOPES

The oscilloscope, since its invention, has been regarded as the most versatile electrical/electronics instrument. It is a very powerful tool in laboratory measurements of dc and ac quantities. However, its main advantage over other instruments is that it displays an accurate graph of the wave. There is no substitute for a visual display of the waveform measured. A photograph of a general-purpose oscilloscope is shown in Figure 15.23.

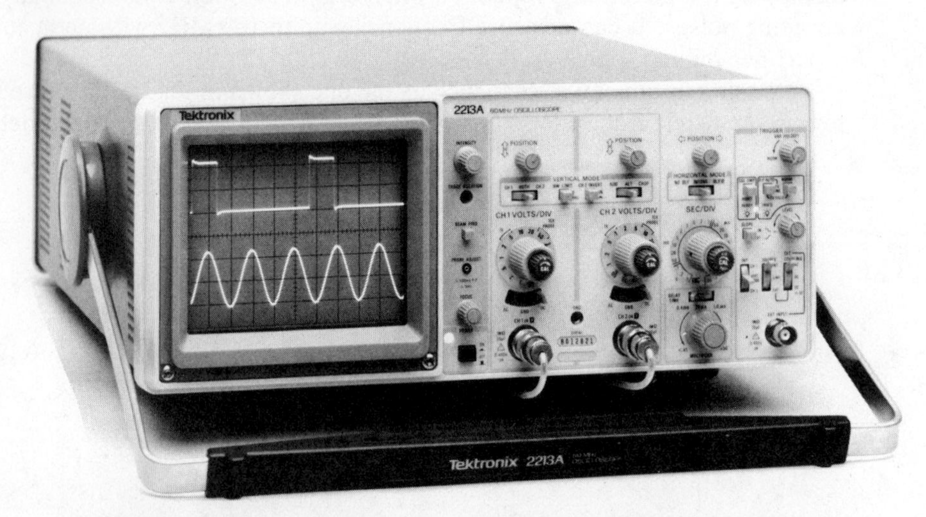

Figure 15.23 General-purpose oscilloscope. (Courtesy Tek Photo, Tektronix, Inc.)

All oscilloscopes contain three major parts: (1) cathode-ray tube (CRT), (2) vertical amplifier section, and (3) horizontal time-control section. The CRT is used to display the wave. The vertical amplifier section is used to control and calibrate the display in the vertical direction. This particular instrument has two identical vertical amplifier controls because this is a *dual-trace* oscilloscope. It is capable of displaying two signals simultaneously. Each signal display can be controlled independently in the vertical direction. Oscilloscopes are available with single-trace, dual-trace, and four-trace capability. Some specialized oscilloscopes can display as many as 24 channels of signal information. Finally, the horizontal section consists of the controls for the time-base calibration and the *trigger controls*. The trigger circuit assures proper synchronization of the measured wave with an internal signal in order for the display to start at a particular point and appear stationary on the screen. There is also a control for synchronizing the measured wave to an external signal.

15.16 Why are LCDs used in portable measuring instruments?

15.17 Explain and give an example of an observer error.

15.18 What is meant by a well-regulated power supply?

15.19 What is the purpose of a function generator?

PROBLEMS

15.1 Using a PMMC mechanism with FSD of 20 μA and an internal resistance of 400 Ω, design a 1-A ammeter.

15.2 Using the mechanism in Problem 15.1, design a multirange milliammeter with 100-μA, 1-mA, and 5-mA scales. Use the circuit of Figure 15.5a. Repeat for Figure 15.5b. (ans. 100 Ω, 8.16 Ω, 1.606 Ω)

15.3 Using the movement of Problem 15.1, design a 1-V voltmeter.

15.4 A 50-μA, 1-kΩ movement is used to design a multirange voltmeter. What are the multiplier resistor values if the desired scales are 10 V, 20 V, and 100 V? (ans. 199 kΩ, 200 kΩ, 1.6 MΩ)

15.5 Repeat Problem 15.4 for a 0.5-mA, 100-Ω movement.

15.6 Determine the actual current without the milliammeter in the circuit and the milliammeter reading in the circuit of Figure 15.33. (ans. 2.31 mA)

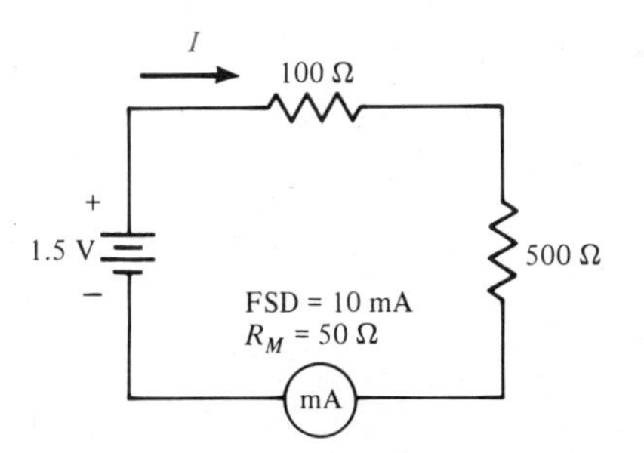

Figure 15.33 Circuit for Problem 15.6.

15.7 A voltmeter has a sensitivity of 10 kΩ/V.
a. What is its internal resistance on the 100-V scale?
b. What is its internal resistance on the 1-V scale?

15.8 What is the sensitivity of a meter with an FSD of 0.5 mA? (ans. 2 kΩ/V)

15.9 The meter from Problem 15.7 is used to measure the voltage across R_2 in Figure 15.34. Determine the actual voltage across R_2 and the voltage measured with the voltmeter on the 10-V scale.

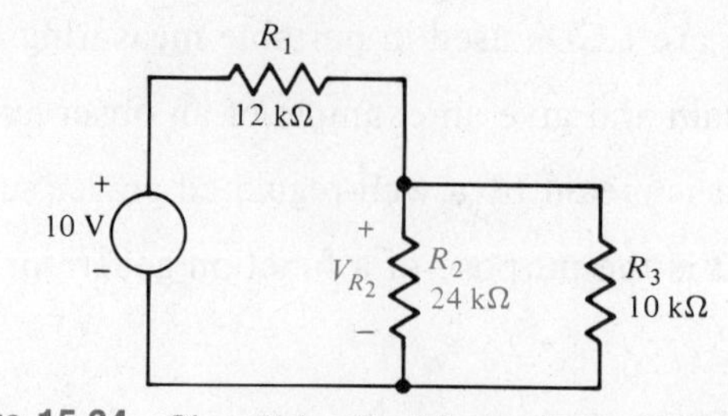

Figure 15.34 Circuit for Problems 15.7, 15.8, and 15.9.

15.10 Repeat Problem 15.9 for the movement in Problem 15.8. (ans. 3.03 V)

15.11 Repeat Problem 15.9 for the voltage across resistor R_1.

15.12 The Wheatstone bridge of Figure 15.14 is balanced with R_2 = 100 kΩ, R_3 = 25 kΩ, and R_4 set at 12 kΩ. What is the value of the unknown resistor R_x? (ans. 3 kΩ)

15.13 The peak-to-peak value of a sinusoidal wave is displayed over three vertical divisions and four horizontal divisions. The vertical and horizontal sensitivity knobs are set at 0.5 V/division and 10 μs/division. Calculate the effective value of the wave and its frequency.

15.14 Repeat Problem 15.13 for a vertical display of 5.2 divisions and 1.8 horizontal divisions. (ans. 55.55 kHz)

15.15 A voltmeter measures 11.7 V across a battery when the actual voltage is 12 V. Calculate the error in percent.

APPENDICES

APPENDIX A

Glossary

ac—acronym for alternating current
ac current—same as alternating current
ac voltage—same as alternating voltage
acceleration—the time rate at which velocity changes
active (average, real) power—power to resistive components
active circuit—circuit that exhibits a power gain
active filter—filter circuit that contains an amplifier
admittance (Y)—the ability of ac circuits to allow ac current and the phase between the voltage and the current; measured in siemens (S)
admittance parameters (y)—two-port coefficients resulting from independent voltage sources
American Wire Gauge (AWG) numbers—classification of wires according to their diameters (the higher the number, the smaller the diameter)
ammeter—instrument used to measure current
ampere (A)—unit of measurement of electric current
amplifier—an electronic circuit used to magnify electronic signals
analog instrument—measuring instrument that displays results in a continuous manner by using a pointer to indicate a change in angle on a scale
angular velocity (ω)—the rate at which an angle changes; measured in radians per second (rad/s); same as radian frequency
apparent power—product of the effective voltage and current
aspect ratio—ratio of length to width of a material
atom—smallest sample of an element
atomic number—the number of electrons (and therefore of protons) in an atom
attenuator—circuit with less output than input

automatic test equipment (ATE)—programmable test systems used in repetitive high-volume testing
average value—the mean value over a period of time
balanced 3-ϕ load—three equal independent loads powered from a 3-ϕ source
bandwidth (BW)—the difference between the upper and lower cutoff frequencies
BASIC—a very popular higher-order computer programming language used primarily with microcomputers
battery—a device made up of many electric cells
B_C—symbol for capacitive susceptance; measured in siemens (S)
B_L—inductive susceptance; measured in siemens (S)
capacitance—property of matter that describes the ability to store separate particles of opposite charge; also opposition to changes in voltage; measured in farads (F)
capacitor—a circuit component built to show the property of capacitance
charge (Q)—electrical property of certain particles of matter
clamp-on ammeter—instrument used to measure ac currents by external sensing
complex numbers—numbers having a real and an imaginary part
complex plane—a plane described by a real and an imaginary axis
compound—material formed by the chemical combination of elements
computer—electronic system used to carry out calculations
computer program—a set of instructions that directs a computer to solve a problem
computer run—the actual running of a computer program
conductance—ability to conduct charge and allow an electric current; measured in siemens (S)
conductor—material with low resistivity
conventional current—the assumed flow of positive charge
coulomb (C)—unit of electrical charge
Cramer's Rule—a method of solving simultaneous equations
current—the uniform motion of charged particles
Current Divider Rule (CDR)—a rule that allows the direct computation of a particular current when the total current to a parallel combination is known
cycle—one repetition of a repeating waveform
cycles per second (cps)—an older unit for frequency
dc power supply—equipment that supplies dc power at constant voltage
decaying exponential—the time representation of a capacitor's discharge or the collapse of a current in an inductor
decibel—a unit of the logarithm of the ratio of two power levels
degree—angular measurement ($360° = 1$ circle)
dependent sources—devices that generate energy as functions of circuit parameters (voltage or current)
determinant—an orderly arrangement of coefficients used in Cramer's Rule
dielectric—the insulating material between the plates of a capacitor
dielectric constant—the ratio of capacitance of a capacitor with the given dielectric to the capacitance of a capacitor having air as the dielectric
diffused resistor—integrated circuit resistor, part of a silicon chip

digital frequency counter—instrument used to measure frequency and period of periodic waveforms
digital instrument—measuring instrument that provides a numerical display of the quantity measured
digital multimeter (DMM)—digital instrument used to measure current, voltage, and resistance
digital voltmeter (DVM)—digital instrument used to measure voltage
effective value—value indicating the energy content of a waveform; same as RMS value
electric circuit—interconnection of a number of electrical components
electric current—same as current
electric potential—same as voltage
electrical engineering—a branch of engineering dealing with electricity and magnetism
electrical machines—motors and generators
electricity—study of resulting forces when two charged bodies are placed in the vicinity of each other
electrodynamometer—a movement used in the construction of ammeters, voltmeters, and wattmeters
electrolyte—salt solution used in battery cells
electrolytic capacitor—large-value capacitors that are usually polarized
electron—an atomic particle attributed with negative charge
electronic distribution—the way electrons are distributed in an atom
electronics—branch of electrical engineering dealing with the processing and control of electric signals
electromotive force (EMF)—same as voltage
energy—the ability to do work
equivalent circuit—a circuit with identical terminal characteristics as another circuit
error—difference between the actual and the measured values
farad (F)—unit of measurement for capacitance
filter—electric network that passes signals at some frequencies and attenuates signals at other frequencies
force—external influence on a body that tends to affect the motion of the body
frequency—the number of repetitions per second
frequency response—a characteristic of the circuit that describes behavior of the circuit with changes of the input frequencies
function generator—laboratory instrument that produces various types of "standard" ac signals at various amplitudes and frequencies
galvanometer—instrument used to sense and display very low values of current
Gauss' Elimination Method—a mathematical technique used to solve simultaneous equations
gradian—angular measurement (400 grad = 1 circle)
half-power frequencies—frequencies at which the power to the circuit is half of the power at resonance
henry (H)—unit of measurement for inductance

hertz (Hz)—unit of frequency
hybrid parameters (h)—two-port coefficients resulting from an independent current source and an independent voltage source
ideal current source—device that delivers electrical energy to a source at a constant value of current
ideal source—device used to model the behavior of practical sources
ideal voltage source—device that delivers electrical energy to a load at a constant voltage
impedance (Z)—opposition of ac circuits to ac current and the angle between the voltage and the current
impedance bridge—instrument used to measure capacitance, inductance, and resistance
impedance parameters (z)—two-port coefficients resulting from independent current sources
in phase—not having any time or angular difference
independent source—device that generates energy independent of other circuit components
inductance (*L*)—opposition to changes in voltage; measured in henrys (H)
inductor—a circuit component built to show the property of inductance
inertia—the opposition of an object to changes in velocity
instantaneous power—product of voltage and current at any instant in time
instantaneous value—the value of a varying waveform at a particular instant of time
insulator—material with high resistivity
integrated circuit (IC)—a self-contained complete electronic circuit that occupies a small area
joule (J)—unit of energy
kilogram (kg)—unit of mass
Kirchhoff's Current Law (KCL)—the sum of the currents entering a node equals the sum of the currents leaving the node
Kirchhoff's Voltage Law (KVL)—in a closed loop, the sum of the voltage rises equals the sum of the voltage drops
Kirchhoff, Gustav Robert (1824–1887)—German physicist to whom the formulation of many electrical laws is attributed
lag—when a waveform starts after the reference
Law of Charges—like charges repel, unlike charges attract
lead—when a waveform starts before a reference
light-emitting diode (LED)—indicating device used with digital instruments
liquid crystal display (LCD)—indicating device used with digital instruments
load—device that transforms electrical energy into other forms of energy, such as light and heat
loading effect—the effect that a measuring instrument has on the circuit
loop—a closed path in an electric circuit
loop equations—a set of simultaneous equations that results from applying KVL to each loop in an electric circuit

magnetic field—the region of "altered space" that will interact with the poles of a magnet
magnetism—the property of being a magnet; having the power to attract
mass—quantity of matter
matter—one of the two basic entities making up the physical world
megohmmeter (megger)—instrument used to measure very high resistance values
mesh—same as loop
meter—measuring instrument
meter (m)—unit of distance
microfarad (μF)—one millionth farad
microhenry (μH)—one millionth henry
millihenry (mH)—one one-thousandth henry
Millman's Theorem—a theorem applicable to multisource circuits containing exclusively practical voltage sources connected in parallel, which allows the substitution by a single simple equivalent voltage source
molecule—smallest sample of a compound
moving-iron movement—mechanism for measuring current
negative charge—electrical property attributed to electrons
neutron—atomic particle with no charge
newton (N)—unit of force
node—electrical connection for two or more components
node equations—a set of simultaneous equations that results from applying KCL to each node of an electric circuit
Norton's equivalent circuit—an equivalent circuit consisting of an ideal current source and a parallel impedance
Norton's Theorem—technique for analyzing electric circuits
nucleus—center of an atom containing protons and neutrons
ohmmeter—instrument used to measure resistance values
open circuit—circuit with a disconnected or interrupted path
orbits—paths around the nucleus followed by electrons
oscilloscope—instrument used to display time-varying signals
parallel ac circuit—ac circuit in which all the impedances are connected between the same two nodes
parallel circuit—circuit in which all the components are connected between the same two nodes
passive circuit—circuit that exhibits a power loss
passive filter—filter circuit from passive components (R, L, C)
peak-to-peak value—the value between the negative and the positive peaks
peak value—the maximum value a varying waveform achieves
period—the time in seconds needed for one repetition of a repeating waveform
phase angle—the angular difference between a waveform and its reference
phase sequence—the order in which the phases of voltage or current follow each other
phasor—a line rotating at a constant speed; a number that represents a quantity that can be described by a phasor

phasor notation—representation of electrical components by their describing phasors
photoconductor—a device whose resistance is a function of illumination
picofarad (pF)—a millionth of a millionth farad
polar form—the form of a complex number having a magnitude and an angle
polyphase circuit—an ac circuit powered by sources that generate a number of voltages of the same magnitude and frequency and are related in time
port—a pair of electrical terminals
positive charge—electrical property attributed to protons
potentiometer—variable resistor
power—work done per unit time; rate of delivering energy
power factor—ratio of active power to apparent power
primary batteries—sources of energy that cannot be recharged
proton—atomic particle attributed with positive charge
quality factor (*Q*)—the ratio of the total energy stored at resonance to the energy dissipated per cycle. Also, a measure of the selectivity of a resonance circuit
radian (rad)—angular measurement (2π rad = 1 circle)
radian frequency (ω)—same as angular velocity ($\omega = 2\pi f$)
reactance (*X*)—opposition of circuits to ac current
reactive power—power supplied by the source for energy storage in reactive components
rectangular form—the form of a complex number having a real and an imaginary part ($a + jb$)
rectification—process of converting ac to pulsating dc
reference waveform—a waveform to which others are compared
resistance—measure of the opposition to electric current
resistor—current-opposing device
resonance—the condition for which the total current and voltage in an *RLC* circuit are in phase
resonant frequency—the frequency at which resonance occurs
rheostat—variable resistor with two of its terminals utilized
rising exponential—the time representation of the charging of a capacitor or the current buildup in an inductor
RMS value—root mean square value; same as effective value
sampling oscilloscope—instrument used to display high-frequency signals by using the sampling technique
scientific notation—method of representing numbers
second (s)—unit of time
secondary batteries—rechargeable sources of energy
selectivity—the circuit's ability to discriminate between signals of different frequencies
self-inductance—same as inductance
semiconductor—material having conductivity between a conductor and an insulator
semiconductor diode—a device that allows current in one direction only

sensitivity—factor describing the loading effects of a meter; given in units of ohms per volt (Ω/V)
series ac circuit—ac circuit in which all the impedances are connected end to end in string fashion
series circuit—circuit in which all the components are connected end to end in string fashion
series-parallel ac circuit—ac circuit containing both series and parallel combinations of impedances
series-parallel circuit—circuit containing both series and parallel connections of components
sheet resistance—resistance of a square of material
shell—same as orbit
short circuit—a path of zero resistance
solar cell (photovoltaic cell)—device that converts light energy to electrical energy
spectrum analyzer—instrument used to display signal amplitudes versus frequency
speed—time rate of change of position
steady-state response—response of a system after all the initial disturbances and variations have subsided
storage oscilloscope—instrument used to display single events and low-frequency ac signals
Superposition Theorem—technique for analyzing electric circuits
superposition theorem—a theorem that states that the current in any element or voltage across any element in a linear independent multisource circuit is equivalent to the algebraic sum of the voltage or current due to each source acting alone
susceptance (B)—ability of inductors or capacitors to allow ac current; measured in siemens (S); $B = 1/X$
T—symbol for a period; measured in seconds (s)
telegraph—communication system used to transmit messages by electric impulses
telephone—system used for transmitting and receiving speech and data
thermistor—device whose resistance changes with changes in temperature
Thevenin's equivalent circuit—an equivalent circuit represented by an ideal voltage source and a series impedance
Thevenin's Theorem—technique for analyzing electric circuits
time (t)—sequence of events
time constant—RC for capacitors; L/R for inductors; the time needed for a capacitor to charge to 63.2% of its final value; the time needed for an inductor's current to build up to 63.2% of its final value
transducer—device used to convert one form of energy to another
transformer—a device used in electrical energy distribution to transfer energy inductively from one circuit to another by stepping the voltage up or down
transient response—the behavior of a system from the time it is disturbed to the time it achieves its steady-state response
transistor—semiconductor electronic device that controls current

trimmer capacitor—a variable capacitor used to make small capacitance adjustments in electronic equipment

tuning capacitor—a capacitor used in radio receivers to tune in a particular radio station

tuning slug—an iron screw that changes the value of an inductor when screwed into or out of an inductor's core

two-port network—a network with four terminals or two ports

vacuum tube—device used as the main element in electronic circuits and systems

variable capacitor—capacitor whose value may be changed

velocity (v)—speed and direction

volt (V)—unit of voltage

voltage (V)—entity that will cause electrically charged particles to move uniformly. Also called electric potential, electromotive force (EMF), and electric tension

volt-ohm-milliammeter (VOM)—instrument used to measure current, voltage, and resistance

Voltage Divider Rule (VDR)—method of calculating the voltage across each series resistor without first calculating the total series current

voltaic cell—source of electric energy

voltmeter—instrument used to measure voltage

watt (W)—unit of power

wattmeter—instrument used to measure electric power

waveshape—the form of a varying voltage or current

Wheatstone bridge—instrument used to measure resistance accurately

work—the manifestation of the expenditure of energy

working voltage (WVDC)—a capacitor rating that describes the maximum voltage that can safely be applied across a capacitor

APPENDIX B

Electrical Symbols

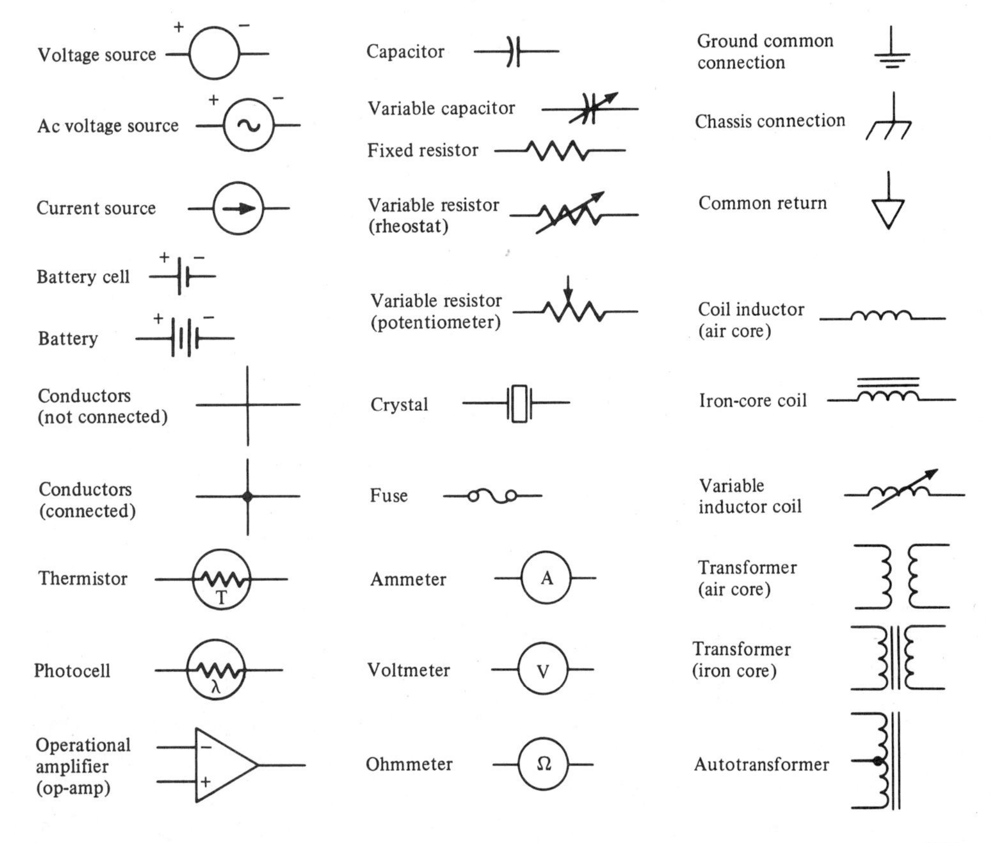

APPENDIX C

Color-Code Chart and Standard Resistance Values

Color-code chart

Color	Significant figure	Multiplier	Tolerance %
Black	0	10^0	—
Brown	1	10^1	—
Red	2	10^2	—
Orange	3	10^3	—
Yellow	4	10^4	—
Green	5	10^5	—
Blue	6	10^6	—
Violet	7	10^7	—
Gray	8	10^8	—
White	9	10^9	—
Gold	—	10^{-1}	5
Silver	—	10^{-2}	10
No band	—	—	20

Standard resistance values for ±5% resistors

To obtain the ohmic value, multiply the numbers in the table by the appropriate multiplier.

Values		Multipliers
1.0	3.3	10^0
1.1	3.6	10^1
1.2	3.9	10^2
1.3	4.3	10^3
1.5	4.7	10^4
1.6	5.1	10^5
1.8	5.6	10^6
2.0	6.2	*10^7
2.2	6.8	
2.4	7.5	
2.7	8.2	
3.0	9.1	

*The highest standard resistance value is 100 MΩ.

APPENDIX D

Complex Number Conversions

Complex numbers may be stated in either *rectangular* or *polar* form. For addition and subtraction, the rectangular form is preferred, whereas for multiplication and division the polar form is easier to handle. The two forms of a complex number are as follows:

$$\mathbf{P} = a + jb \qquad \text{rectangular form}$$
$$= r\angle\theta \qquad \text{polar form}$$

Conversion from one form to the other is a very common task. The conversion may be implemented by using trigonometry and the Pythagorean Theorem or simply by using one of the many calculators available that do the conversion.

Rectangular-to-Polar Conversion

Using trigonometry and the Pythagorean Theorem (refer to Figure D.1).

In this case, a and b are known and r and θ are to be found according to the following relationships:

$$r = \sqrt{a^2 + b^2} \qquad \text{magnitude}$$

$$\theta = \tan^{-1}\frac{b}{a} \qquad \text{angle}$$

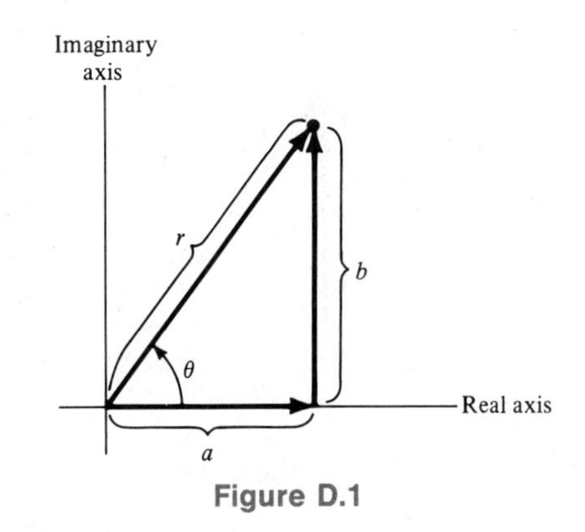

Figure D.1

EXAMPLE D.1

First quadrant (*a* positive, *b* positive); refer to Figure D.2.

$$\mathbf{P} = 3 + j4 \qquad \text{rectangular form}$$

$$r = \sqrt{3^2 + 4^2} = \sqrt{9 + 16} = \sqrt{25} = 5$$

$$\theta = \tan^{-1}\frac{4}{3} = \tan^{-1}(1.33) = 53.13°$$

$$\mathbf{P} = 5\underline{/53.13°} \qquad \text{polar form}$$

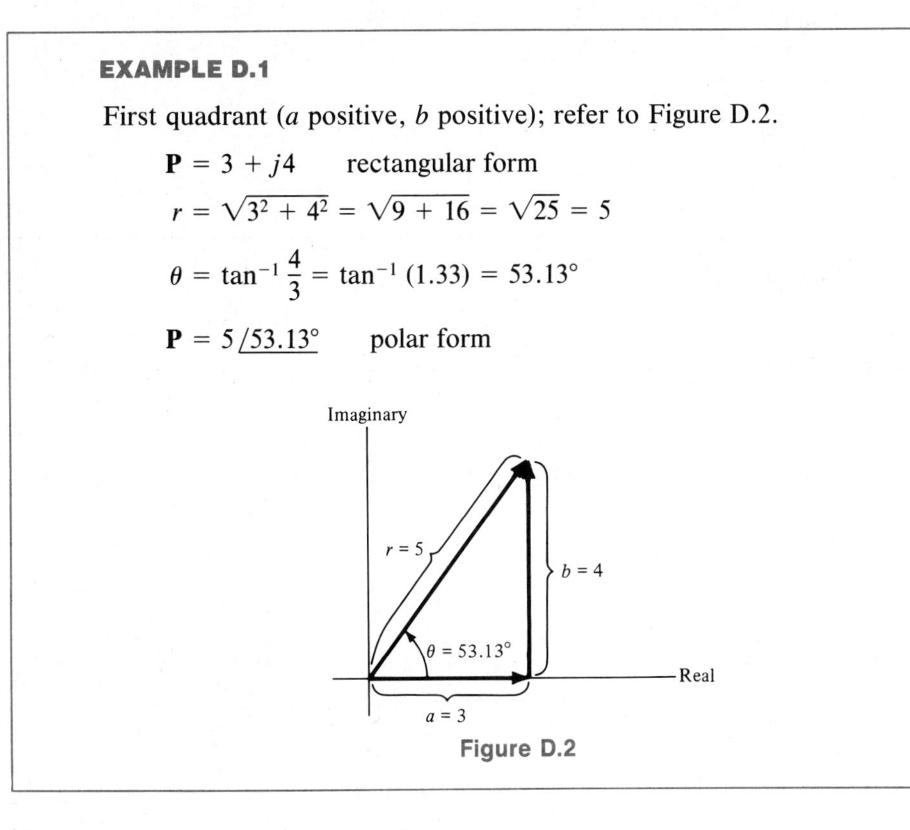

Figure D.2

EXAMPLE D.2

Second quadrant (*a* negative, *b* positive); refer to Figure D.3.

$$\mathbf{P} = -3 + j4 \qquad \text{rectangular form}$$

$$r = \sqrt{(-3)^2 + 4^2} = \sqrt{9 + 16} = \sqrt{25} = 5$$

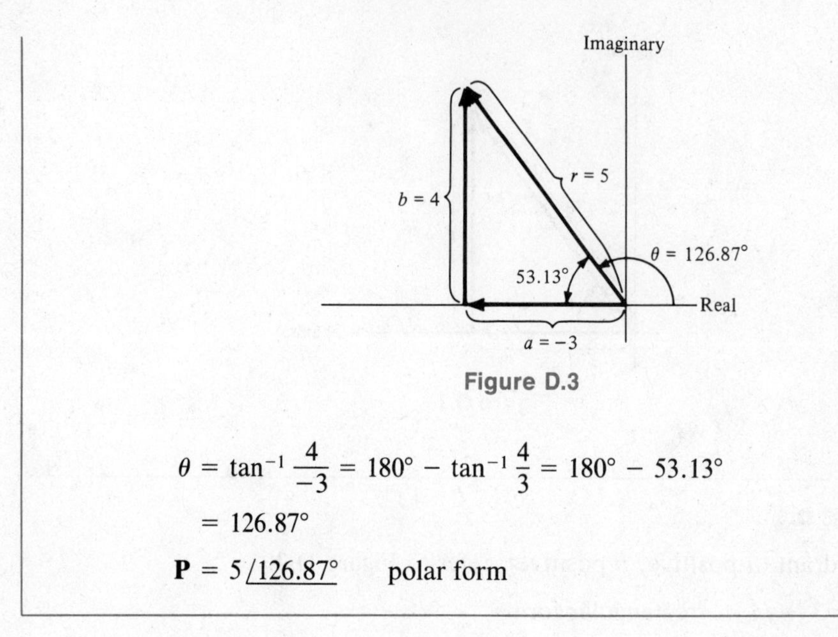

Figure D.3

$$\theta = \tan^{-1}\frac{4}{-3} = 180° - \tan^{-1}\frac{4}{3} = 180° - 53.13°$$
$$= 126.87°$$
$$\mathbf{P} = 5\underline{/126.87°} \quad \text{polar form}$$

EXAMPLE D.3

Third quadrant (a negative, b negative); refer to Figure D.4.

$$\mathbf{P} = -3 - j4 \quad \text{rectangular form}$$
$$r = \sqrt{(-3)^2 + (-4)^2} = \sqrt{9 + 16} = \sqrt{25} = 5$$
$$\theta = \tan^{-1}\frac{-4}{-3} = -180° + \tan^{-1}\frac{4}{3} = -180° + 53.13°$$
$$= -126.87°$$
$$\mathbf{P} = 5\underline{/-126.87°} \quad \text{polar form}$$

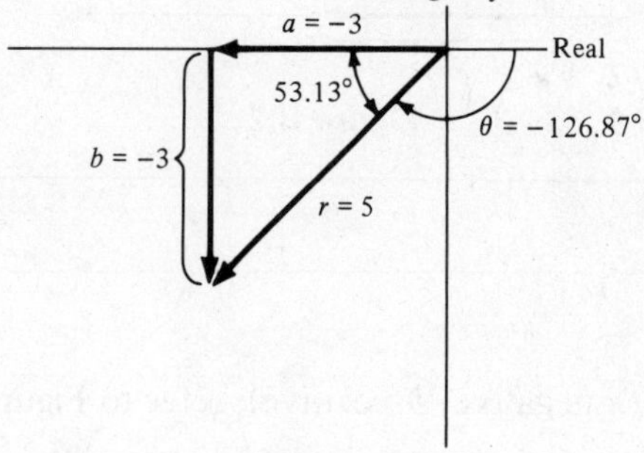

Figure D.4

EXAMPLE D.4

Fourth quadrant (a positive, b negative); refer to Figure D.5.

$$\mathbf{P} = 3 - j4 \qquad \text{rectangular form}$$

$$r = \sqrt{3^2 + (-4)^2} = \sqrt{9 + 16} = \sqrt{25} = 5$$

$$\theta = \tan^{-1}\frac{-4}{3} = -\tan^{-1}\frac{4}{3} = -53.13^\circ$$

$$\mathbf{P} = 5\,\underline{/-53.13^\circ} \qquad \text{polar form}$$

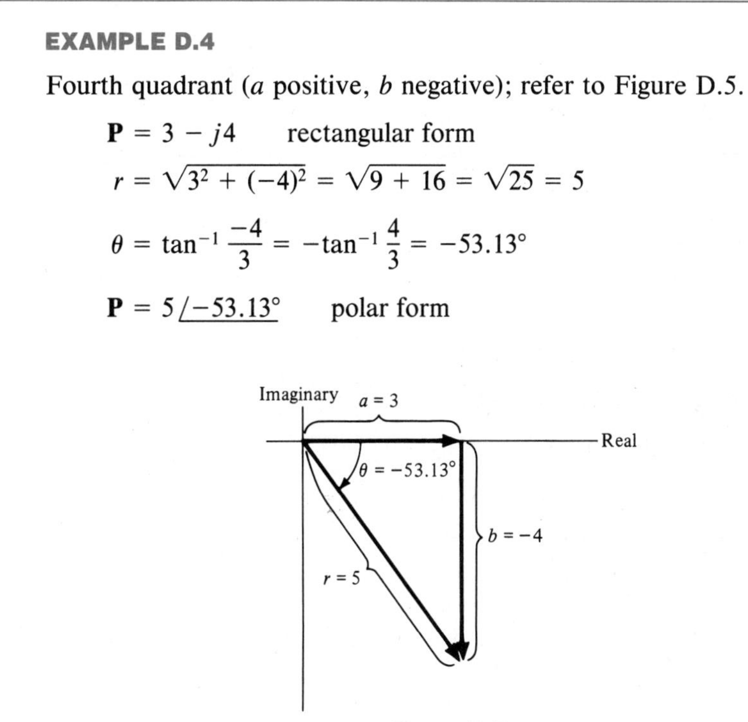

Figure D.5

Polar-to-Rectangular Conversion Using Trigonometry

EXAMPLE D.5

First quadrant ($0^\circ \leq \theta \leq 90^\circ$); refer to Figure D.2.

$$\mathbf{P} = 5\,\underline{/53.13^\circ} \qquad \text{polar form}$$

$$a = r\cos\theta = 5\cos(53.13^\circ) = 3$$

$$b = r\sin\theta = 5\sin(53.13^\circ) = 4$$

$$\mathbf{P} = 3 + j4 \qquad \text{rectangular form}$$

EXAMPLE D.6

Second quadrant ($90^\circ \leq \theta \leq 180^\circ$); refer to Figure D.3.

$$\mathbf{P} = 5\,\underline{/126.87^\circ} \qquad \text{polar form}$$

$a = 5 \cos(126.87°) = (5)(-0.6) = -3$

$b = 5 \sin(126.87°) = (5)(0.8) = 4$

$\mathbf{P} = -3 + j4$ rectangular form

EXAMPLE D.7

Third quadrant ($-90° \geq \theta \geq -180°$); refer to Figure D.4.

$\mathbf{P} = 5\underline{/-126.87°}$ polar form

$a = 5 \cos(-126.87°) = (5)(-0.6) = -3$

$b = 5 \sin(-126.87°) = (5)(-0.8) = -4$

$\mathbf{P} = -3 - j4$ rectangular form

EXAMPLE D.8

Fourth quadrant ($-90° \leq \theta \leq 0°$); refer to Figure D.5.

$\mathbf{P} = 5\underline{/-53.13°}$ polar form

$a = 5 \cos(-53.13°) = (5)(0.6) = 3$

$b = 5 \sin(-53.13°) = (5)(-0.8) = -4$

$\mathbf{P} = 3 - j4$ rectangular form

Conversion with Calculators

For all the following conversions, the calculator is assumed to be in the DEG (degree) mode. There are two popular conversion systems used in commercial calculators, namely, the *a and b system* and the *→POL and →REC system*. Each one is explained.

***a* AND *b* SYSTEM.** Two of the keys on some calculators, such as the Sharp scientific models (Sharp EL-506A, for example), are labeled *a* and *b*. These two keys are employed together with the →*xy* function and the →*r*θ function. The →*r*θ function converts the rectangular form of a complex number into its polar form. The →*xy* function converts the polar form of a complex number into its rectangular form.

For rectangular-to-polar conversion, the real part is stored in *a* and the imaginary part is stored in *b*. The →*r*θ function does the conversion and stores the magnitude (r) in a and the phase (θ) in b. In the following examples, please note the following:

3 means to press the calculator key labeled 3

[a] means to press the calculator key labeled a

EXAMPLE D.9

Convert to polar form:

$\mathbf{P} = 3 - j4$ rectangular form

Key in: 3 [a] 4 [b] [2^{nd}F] [$\rightarrow r\theta$] Answer: 5 (r)

Key in: [b] Answer: $-53.13°$ (θ)

$\mathbf{P} = 5\underline{/-53.13°}$ polar form

For the polar-to-rectangular conversion, the magnitude is stored in a and the angle is stored in b. The $\rightarrow xy$ function does the conversion and stores the real part answer in a and the imaginary part answer in b.

EXAMPLE D.10

Convert to rectangular form:

$\mathbf{P} = 5\underline{/-126.87°}$ polar form

Key in: 5 [b] 126.87 [+/−] [2^{nd}F] [$\rightarrow xy$] Answer: -3 (a)

Key in: [b] Answer: -4 (b)

$\mathbf{P} = -3 - j4$ rectangular form

→POL and →REC System. Two of the keys on some calculators (all the Casio calculators, the Texas Instruments calculators, and the Sharp EL-5200) are labeled →POL and →REC. These two keys convert into polar and into rectangular forms, respectively.

EXAMPLE D.11

Convert to polar:

$\mathbf{P} = 3 - j4$ rectangular form

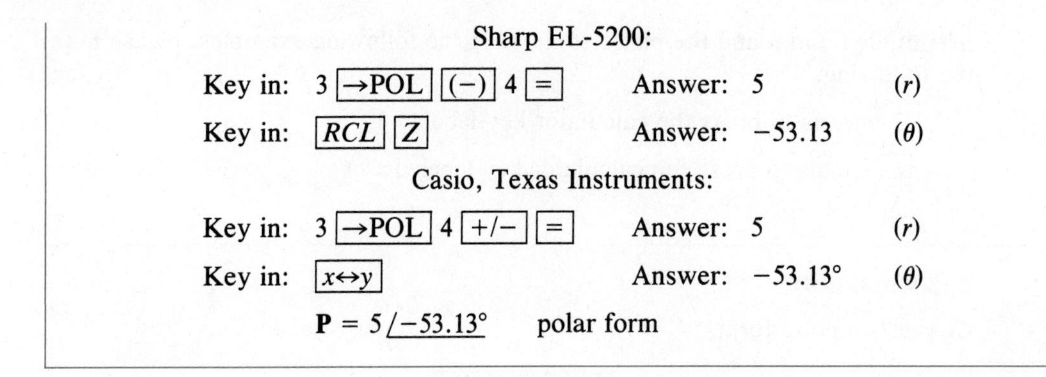

Sharp EL-5200:

Key in: 3 [→POL] [(−)] 4 [=] Answer: 5 (r)

Key in: [*RCL*] [*Z*] Answer: −53.13 (θ)

Casio, Texas Instruments:

Key in: 3 [→POL] 4 [+/−] [=] Answer: 5 (r)

Key in: [$x \leftrightarrow y$] Answer: −53.13° (θ)

$\mathbf{P} = 5\underline{/-53.13^\circ}$ polar form

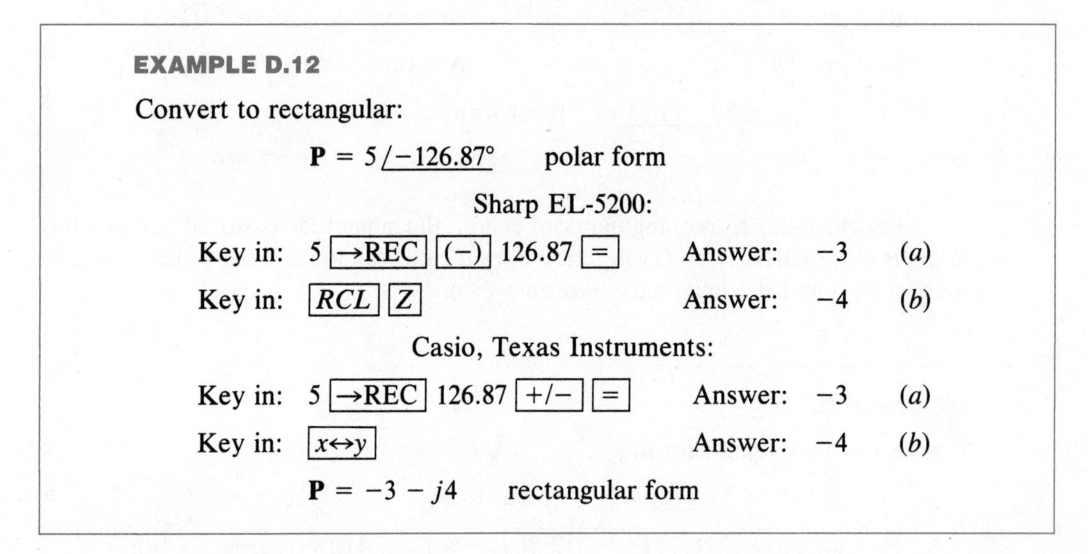

EXAMPLE D.12

Convert to rectangular:

$\mathbf{P} = 5\underline{/-126.87^\circ}$ polar form

Sharp EL-5200:

Key in: 5 [→REC] [(−)] 126.87 [=] Answer: −3 (a)

Key in: [*RCL*] [*Z*] Answer: −4 (b)

Casio, Texas Instruments:

Key in: 5 [→REC] 126.87 [+/−] [=] Answer: −3 (a)

Key in: [$x \leftrightarrow y$] Answer: −4 (b)

$\mathbf{P} = -3 - j4$ rectangular form

Complex Number Calculations

Some calculators (Sharp EL-506A, for example) have the capability of performing complex number calculations. Usually, all the numbers must be entered in rectangular form and all the answers are given in rectangular form. The following is extracted (by permission from Sharp Corporation) from the Sharp Model EL-506A Operation Manual.

COMPLEX NUMBER CALCULATION

The EL-506A can perform the addition, subtraction, multiplication, and division of complex numbers.

To carry out complex number calculations, the calculator must be first put in the Complex number (CPLX) mode, by pressing the [2ndF] key and the [CPLX] key.

The ''–'' indicator will appear at the top part of the display below the legend ''CPLX'' to show that the calculator is in the CPLX mode. To clear this mode, press the [2ndF] key and the [CPLX] again and the ''–'' indicator will disappear.

A complex number is represented in the form of a + bi, where ''a'' is the real part of the complex number and ''bi'' is the imaginary part. Press the [*a*] key to enter the value of the real part, and the [*b*] key, the value of the imaginary part. Press the [–] key to obtain the result of the calculation.

Notes:

1. In the CPLX mode, the four basic operations and coordinates conversion can be performed.
2. In CPLX mode, memory calculations, constant calculations, and calculations with parentheses cannot be performed.
3. If the value of either real part ''a'' or imaginary part ''b'' is 0, the calculation can be performed by entering the other part without value 0. In CPLX mode, [+], [−], [÷], [×], and [=] keys become inoperative unless the value of either of the two parts is input. The four basic arithmetic operations of complex numbers can be performed using the following formulas:

   Addition: $(a + bi) + (c + di) = (a + c) + (b + d)i$

   Subtraction: $(a + bi) - (c + di) = (a - c) + (b - d)i$

   Multiplication: $(a + bi) \times (c + di) = (ac - bd) + (ad + bc)i$

   Division: $(a + bi) \div (c + di) = \dfrac{ac + bd}{c^2 + d^2} + \dfrac{bc - ad}{c^2 + d^2}\,i$

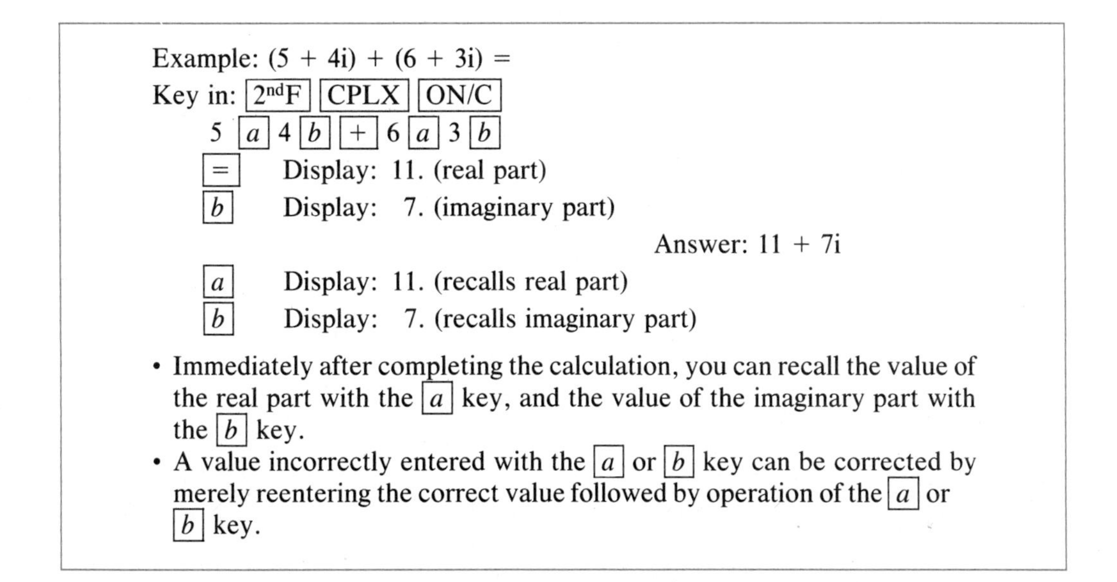

Example: (5 + 4i) + (6 + 3i) =

Key in: [2ndF] [CPLX] [ON/C]

5 [*a*] 4 [*b*] [+] 6 [*a*] 3 [*b*]

[=] Display: 11. (real part)

[*b*] Display: 7. (imaginary part)

Answer: 11 + 7i

[*a*] Display: 11. (recalls real part)

[*b*] Display: 7. (recalls imaginary part)

- Immediately after completing the calculation, you can recall the value of the real part with the [*a*] key, and the value of the imaginary part with the [*b*] key.
- A value incorrectly entered with the [*a*] or [*b*] key can be corrected by merely reentering the correct value followed by operation of the [*a*] or [*b*] key.

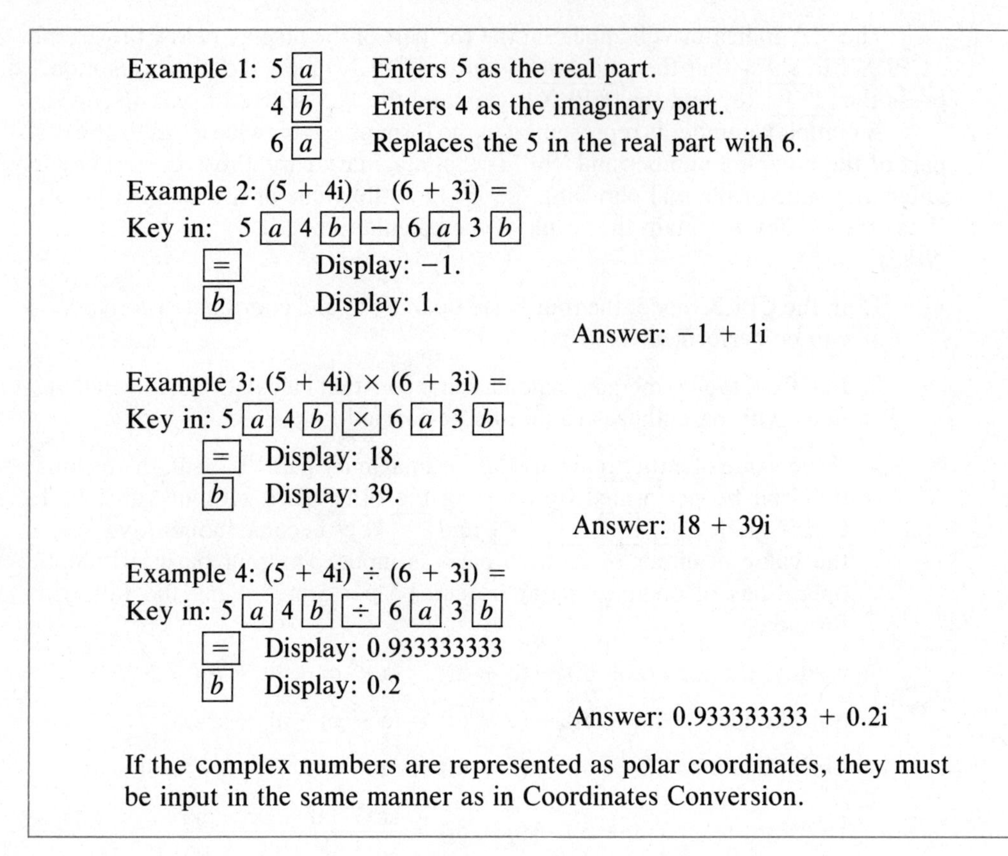

Example 1: 5 [a] Enters 5 as the real part.
4 [b] Enters 4 as the imaginary part.
6 [a] Replaces the 5 in the real part with 6.

Example 2: (5 + 4i) − (6 + 3i) =
Key in: 5 [a] 4 [b] [−] 6 [a] 3 [b]
[=] Display: −1.
[b] Display: 1.
Answer: −1 + 1i

Example 3: (5 + 4i) × (6 + 3i) =
Key in: 5 [a] 4 [b] [×] 6 [a] 3 [b]
[=] Display: 18.
[b] Display: 39.
Answer: 18 + 39i

Example 4: (5 + 4i) ÷ (6 + 3i) =
Key in: 5 [a] 4 [b] [÷] 6 [a] 3 [b]
[=] Display: 0.933333333
[b] Display: 0.2
Answer: 0.933333333 + 0.2i

If the complex numbers are represented as polar coordinates, they must be input in the same manner as in Coordinates Conversion.

APPENDIX E

Integral Calculus Derivation of Areas Under Some Functions

It is shown in Chapter 9 that the average value of a waveform is the area under the function divided by the period. Often calculus has to be used to find the area under a function. In electrical problems the function is usually either a voltage or a current as a function of time. Figure 9.13a shows a sinusoidal voltage that exists from 0 to π rad (half the period) and has a peak value of V_P. The area under half of the sinusoid is found as follows:

$$v = V_P \sin\theta \qquad 0 \leq \theta \geq \pi$$

$$\begin{aligned}
\text{Area} &= \int_0^{\pi} V_P \sin\theta \, d\theta \\
&= V_P \int_0^{\pi} \sin\theta \, d\theta \\
&= V_P[-\cos\theta]_0^{\pi} \\
&= V_P[-\cos\pi - (-\cos 0)] \\
&= V_P[-(-1) - (-1)] \\
&= V_P[1 + 1]
\end{aligned}$$

$$\boxed{\text{Area} = 2V_P}$$

Figure 9.13b shows a sinusoidal voltage squared. It exists from 0 to π rad ($\frac{1}{2}$ period) and has a peak value V_P^2.

The area under half the sinusoid squared is found as follows:

$$v = V_P \sin\theta \qquad 0 \leq \theta \geq \pi$$

$$v^2 = V_P^2 \sin^2\theta$$

$$\sin^2\theta = \tfrac{1}{2}(1 - \cos^2\theta)$$

$$v^2 = \frac{V_P^2}{2}(1 - \cos^2\theta)$$

$$\text{Area} = \int_0^\pi v^2\, d\theta$$

$$= \int_0^\pi \frac{V_P^2}{2}(1 - \cos^2\theta)\, d\theta$$

$$= \frac{V_P^2}{2}\int_0^\pi (1 - \cos^2\theta)\, d\theta$$

$$= \frac{V_P^2}{2}\left(\int_0^\pi d\theta - \int_0^\pi \cos^2\theta\, d\theta\right)$$

$$= \frac{V_P^2}{2}\left(\int_0^\pi d\theta - \frac{1}{2}\int_0^\pi \cos^2\theta\, 2d\theta\right)$$

$$= \frac{V_P^2}{2}\left(\theta\Big|_0^\pi - \frac{1}{2}\sin^2\theta\Big|_0^\pi\right)$$

$$= \frac{V_P^2}{2}\left((\pi - 0) - \frac{1}{2}(0 - 0)\right)$$

$$= \frac{V_P^2}{2}(\pi)$$

$$\boxed{\text{Area} = \tfrac{1}{2}\pi V_P^2}$$

Figure 9.13c shows a triangular voltage waveform with period T and peak value V_P. The area under the curve is found as follows:

$$v = \frac{V_P}{T}t \qquad 0 \leq t \geq T$$

$$\text{Area} = \int_0^T v\, dt$$

$$= \int_0^T \frac{V_P}{T}t\, dt$$

$$= \frac{V_P}{T} \int_0^T t \, dt$$

$$= \frac{V_P}{T} \left(\frac{t^2}{2} \Big|_0^T \right)$$

$$= \frac{V_P}{T} \left(\frac{T^2}{2} - \frac{0}{2} \right)$$

$$= \frac{V_P}{T} \frac{T^2}{2}$$

$$\boxed{\text{Area} = \frac{V_P T}{2}}$$

Figure 9.13d shows a waveform that varies as the square of the previously considered triangular voltage. The area under this squared waveform is found as follows:

$$v = \frac{V_P}{T} t \qquad 0 \leq t \geq T$$

$$v^2 = \left(\frac{V_P}{T} \right)^2 t^2$$

$$\text{Area} = \int_0^T \left(\frac{V_P}{T} \right)^2 t^2 \, dt$$

$$= \left(\frac{V_P}{T} \right)^2 \int_0^T t^2 \, dt$$

$$= \left(\frac{V_P}{T} \right)^2 \left(\frac{t^3}{3} \Big|_0^T \right)$$

$$= \left(\frac{V_P}{T} \right)^2 \left(\frac{T^3}{3} - \frac{0}{3} \right)$$

$$= \frac{V_P^2}{T^2} \frac{T^3}{3}$$

$$\boxed{\text{Area} = \tfrac{1}{3} V_P^2 T}$$

APPENDIX F

Calculus Derivation of Capacitive and Inductive Transients

In technological work the Greek letter Δ (uppercase delta) is used to show or denote change. A change in charge is denoted by ΔQ, whereas a change in time is denoted by Δt.

Recall Ampere's Law from Chapter 3:

$$I = \frac{Q}{t}$$

This relation is useful to compute the average, or dc, current.

Often the discussion involves currents that are not constant but are constantly changing, as in capacitor charging and discharging and in the building up and collapsing of inductor currents. For these cases Ampere's Law may be restated as:

$$I = \frac{\Delta Q}{\Delta t}$$

where: I => current (A)
Δt => change in time (s)
ΔQ => change in charge during the interval Δt (C)

An infinitesimally small change in time is denoted by the expression $\Delta t \to 0$ (read as: "delta t goes to zero"). If, in the preceding equation, the change in time is allowed to become infinitesimally small and therefore approach zero, Ampere's Law may be restated as:

$$i = \frac{d(q)}{dt}$$

where: i => instantaneous current (A)

$\frac{d}{dt}$ => instantaneous change with respect to time (1/s)

q => expression showing how the quantity of charge is changing with respect to time (C)

The symbol d/dt is referred to as the *derivative* with respect to time of whatever function follows it (in this case, q).

The process of finding a derivative is called *differentiation,* and to *differentiate* means to find a derivative. Tables of derivatives are found in any textbook on calculus.

From these considerations, it follows that the instantaneous change with respect to time of the expression for charge at time t is given by the value of the *derivative* of the charge expression evaluated at that time.

If the voltage across a capacitor is observed to vary with time, the amount of stored charge causing this voltage also varies. Therefore, both voltage and charge become time variables. Since variable quantities are usually represented by lower-case letters of the alphabet, Eq. (10.1a), after rearrangement, may be restated as:

$$q = Cv$$

Substituting this expression for q into Ampere's Law, the following is obtained:

$$i = \frac{d(q)}{dt} = \frac{d(Cv_C)}{dt}$$

Since the current refers to a capacitor, i is replaced by i_C. The capacitance symbol C may be removed from inside the derivative, since it is time independent. The following is obtained:

$$i_C = C\frac{d(v_C)}{dt} \qquad \textbf{(10.1b)}$$

where: i_C => instantaneous charging or discharging current (A)

C => capacitor value (F)

$\frac{d(v_C)}{dt}$ => derivative of the capacitor voltage expression (V/s)

This equation is stated in Chapter 10 as Eq. (10.1b).

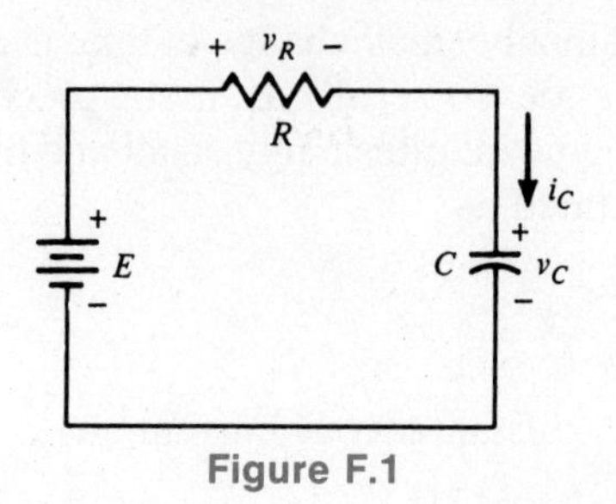

Figure F.1

Figure F.1 represents a circuit used to charge capacitor C. Apply KVL around the loop:

$$v_R + v_C = E$$

Note

$$v_R = i_C(R)$$

Recall

$$i_C = C\frac{d(v_C)}{dt}$$

Therefore,

$$v_R = RC\frac{d(v_C)}{dt}$$

Substitute:

$$RC\frac{d(v_C)}{dt} + v_C = E$$

Note that v_C is a function of its own rate of change. This type of equation is known as a differential equation. The solution for v_C by calculus methods follows.

$$RC\frac{d(v_C)}{dt} + v_C = E$$

$$RC\frac{d(v_C)}{dt} = E - v_C$$

$$RC\, d(v_C) = (E - v_C)\, dt$$

$$RC\frac{d(v_C)}{(E - v_C)} = dt$$

$$\frac{d(v_C)}{(E - v_C)} = \frac{1}{RC}\, dt$$

$$\int \frac{d(v_C)}{(E - v_C)} = \int \frac{1}{RC}\, dt$$

$$\int \frac{d(v_C)}{(E - v_C)} = \frac{1}{RC}\int dt$$

To integrate by parts, let $u = E - v_C$. Then $du = -d(v_C)$. Thus $d(v_C) = -du$. Substituting:

$$\int \frac{-du}{u} = \frac{1}{RC} \int dt$$

$$\int \frac{du}{u} = -\frac{1}{RC} \int dt$$

From integration tables:

$$\ln(u) = -\frac{t}{RC} + \text{constant}$$

Substituting:

$$\ln(E - v_C) = -\frac{t}{RC} + \text{constant}$$

Since at $t = 0$, $v_C = 0$, solve:

$$\ln(E - 0) = -\frac{0}{RC} + \text{constant}$$

$$\text{constant} = \ln(E)$$

Substituting back gives

$$\ln(E - v_C) = -\frac{t}{RC} + \ln(E)$$

$$\ln(E - v_C) - \ln(E) = -\frac{t}{RC}$$

$$\frac{E - v_C}{E} = e^{-t/RC}$$

$$E - v_C = Ee^{-t/RC}$$

$$v_C = E - Ee^{-t/RC}$$

$$\boxed{v_C = E(1 - e^{-t/RC})}$$

Since E is the final value of the capacitor voltage (V_F) and the term RC is given the name *time constant* and is represented by the Greek letter τ, the equation may be restated as:

$$\boxed{v_C = V_F(1 - e^{-t/\tau})} \qquad (10.12)$$

This equation represents the voltage buildup as the capacitor charges. It is stated in Chapter 10 as Eq. (10.12).

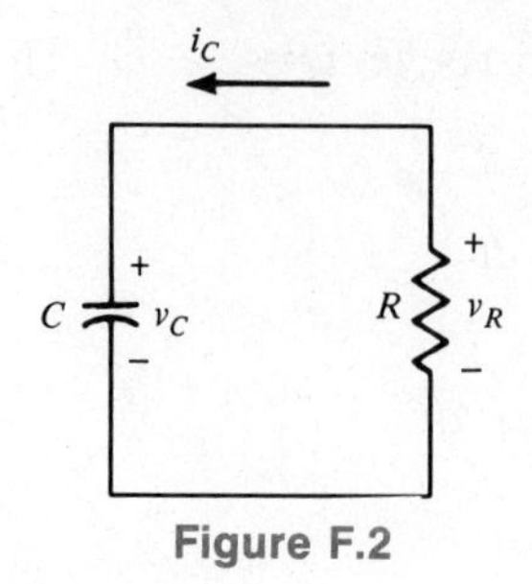

Figure F.2

Figure F.2 represents a circuit used to discharge capacitor C. The initial ($t = 0$) voltage across the capacitor is V_I.

By KVL,

$$v_C = -i_C(R)$$

Recall

$$i_C = C\,\frac{d(v_C)}{dt}$$

Therefore,

$$v_C = -RC\,\frac{d(v_C)}{dt}$$

Rearranging gives

$$\frac{d(v_C)}{v_C} = -\frac{1}{RC}\,dt$$

$$\int \frac{d(v_C)}{v_C} = \int -\frac{1}{RC}\,dt$$

$$\int \frac{d(v_C)}{v_C} = -\frac{1}{RC}\int dt$$

$$\ln(v_C) = -\frac{1}{RC}\,t + \text{constant}$$

Since at $t = 0$, $v_C = V_I$, solve for the constant:

$$\ln(V_I) = -\frac{0}{RC} + \text{constant}$$

$$\text{Constant} = \ln(V_I)$$

Substituting:

$$\ln(v_C) = -\frac{1}{RC}\,t + \ln(V_I)$$

$$\ln(v_C) - \ln(V_I) = -\frac{1}{RC}t$$

$$\ln\left(\frac{v_C}{V_I}\right) = -\frac{1}{RC}t$$

$$\frac{v_C}{V_I} = e^{-t/RC}$$

$$v_C = V_I e^{-t/RC}$$

Since RC is called τ, the equation may be restated as:

$$\boxed{v_C = V_I e^{-t/\tau}} \quad \textbf{(10.13)}$$

This equation represents the voltage collapse as a capacitor discharges and is stated in Chapter 10 as Eq. (10.13).

Equation (10.5b) may be restated in terms of the instantaneous voltage value, the derivative of the inductor current with respect to time, and the inductance, as follows:

$$v_L = L\frac{d(i_L)}{dt} \quad \textbf{(10.5b)}$$

This equation states that the induced countervoltage across the inductor is proportional to the time rate of change of the current through the inductor. This equation is listed in Chapter 10 as Eq. (10.5b).

Figure F.3 shows a circuit used to set up a constant current through an inductor.

By KVL,

$$v_R + v_L = E$$

Note that

$$v_R = i_L(R)$$

Recall

$$v_L = L\frac{d(i_L)}{dt}$$

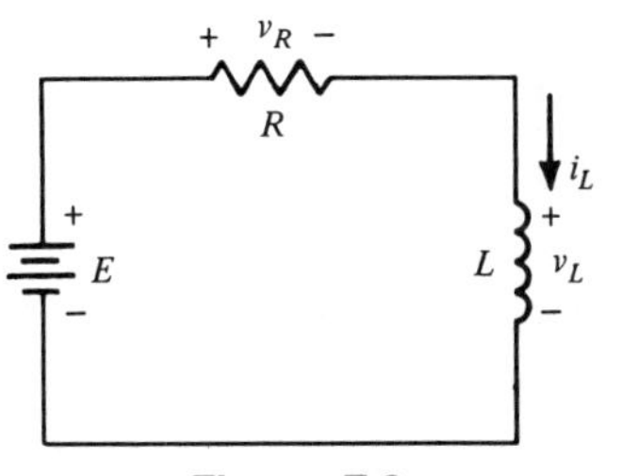

Figure F.3

Therefore,

$$i_L(R) + L\frac{d(i_L)}{dt} = E$$

$$L\frac{d(i_L)}{dt} = E - i_L(R)$$

$$L\,d(i_L) = (E - i_L(R))\,dt$$

$$L\frac{d(i_L)}{(E - i_L(R))} = dt$$

$$\int L\frac{d(i_L)}{(E - i_L(R))} = \int dt$$

To integrate by parts, let $u = E - i_L(R)$. Then $du = -R\,d(i_L)$; therefore,

$$d(i_L) = -\frac{1}{R}\,du$$

Substituting back gives

$$\int L\frac{\left(-\frac{1}{R}\,du\right)}{u} = \int dt$$

$$\int \left(-\frac{L}{R}\right)\frac{du}{u} = \int dt$$

$$-\frac{L}{R}\int \frac{du}{u} = \int dt$$

$$-\frac{L}{R}\ln(u) = t + \text{constant}$$

$$-\frac{L}{R}\ln(E - i_L(R)) = t + \text{constant}$$

Since at $t = 0$, $i_L = 0$, solve for the constant:

$$-\frac{L}{R}\ln(E - 0) = 0 + \text{constant}$$

$$\text{Constant} = -\frac{L}{R}\ln(E)$$

Substituting back gives

$$-\frac{L}{R}\ln(E - i_L(R)) = t - \frac{L}{R}\ln(E)$$

$$\frac{L}{R}\ln(E) - \frac{L}{R}\ln(E - i_L(R)) = t$$

$$\frac{L}{R}(\ln(E) - \ln(E - i_L(R))) = t$$

$$\frac{L}{R}\ln\left(\frac{E}{(E - i_L(R))}\right) = t$$

$$\ln\left(\frac{E}{(E - i_L(R))}\right) = \frac{t}{\left(\frac{L}{R}\right)}$$

$$\frac{E}{(E - i_L(R))} = e^{t/(L/R)}$$

$$\frac{(E - i_L(R))}{E} = e^{-t/(L/R)}$$

$$E - i_L(R) = Ee^{-t/(L/R)}$$

$$i_L(R) = E - Ee^{-t/(L/R)}$$

$$= E(1 - e^{-t/(L/R)})$$

$$i_L = \frac{E}{R}(1 - e^{-t/(L/R)})$$

Since E/R is the final inductor current (I_F) and L/R is called τ, the equation becomes

$$\boxed{i_L = I_F(1 - e^{-t/\tau})} \tag{10.14}$$

This equation represents the current buildup through an inductor and is stated in Chapter 10 as Eq. (10.14).

Figure F.4 shows a circuit used to collapse an existing inductor current. The initial current value (at $t = 0$) is I_I. Note

$$v_L = v_R$$

Recall

$$v_L = L\frac{d(i_L)}{dt}$$

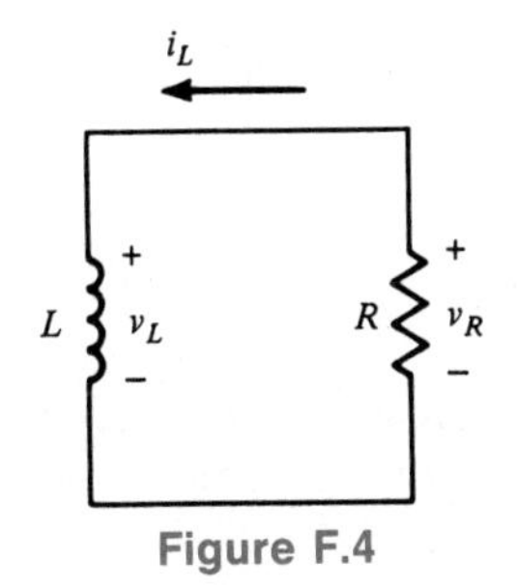

Figure F.4

Therefore,

$$L\frac{d(i_L)}{dt} = -i_L(R)$$

$$\frac{L}{R}\frac{d(i_L)}{i_L} = -dt$$

$$\int \frac{L}{R}\frac{d(i_L)}{i_L} = \int -dt$$

$$\frac{L}{R}\ln(i_L) = -t + \text{constant}$$

Since at $t = 0$, $i_L = I_I$, solve for the constant:

$$\frac{L}{R}\ln(I_I) = 0 + \text{constant}$$

$$\text{Constant} = \frac{L}{R}\ln(I_I)$$

Substituting back gives

$$\frac{L}{R}\ln(i_L) = -t + \frac{L}{R}\ln(I_I)$$

$$\frac{L}{R}\ln(i_L) - \frac{L}{R}\ln(I_I) = -t$$

$$\frac{L}{R}(\ln(i_L) - \ln(I_I)) = -t$$

$$\frac{L}{R}\ln\left(\frac{i_L}{I_I}\right) = -t$$

$$\ln\left(\frac{i_L}{I_I}\right) = -\frac{t}{(L/R)}$$

$$\frac{i_L}{I_I} = e^{-t/(L/R)}$$

$$i_L = I_I e^{-t/(L/R)}$$

Since (L/R) is τ, the equation may be rewritten as:

$$\boxed{i_L = I_I e^{-t/\tau}} \qquad \textbf{(10.15)}$$

This equation represents the current collapse through an inductor and is stated in Chapter 10 as Eq. (10.15).

APPENDIX G

Fourier Analysis of Waveforms

The French mathematician Baron Jean Baptiste Fourier, while studying heat-flow problems, showed in 1826 that arbitrary periodic functions can be represented as an infinite series of sinusoids, frequencies that are multiples of a basic frequency. The sinusoid at the basic frequency is called the *fundamental* component. The sinusoids at frequencies that are multiples of the fundamental frequency are called *harmonics*. The trigonometric Fourier series is given by

$$f(t) = A_0 + \sum_{n=1}^{\infty} (A_n \cos n\omega_0 t + B_n \sin n\omega_0 t) \tag{G.1a}$$

which is expressed as:

$$\begin{aligned} f(t) = A_0 &+ A_1 \cos\omega_0 t + A_2 \cos 2\omega_0 t + A_3 \cos 3\omega_0 t + \cdots \\ &+ B_1 \sin\omega_0 t + B_2 \sin 2\omega_0 t + B_3 \sin 3\omega_0 t + \cdots \end{aligned} \tag{G.1b}$$

where A_0 is the average (dc) value of the function, ω_0 is the fundamental radian frequency, and—for example—$3\omega_0$ is the third harmonic frequency.

The Fourier series can be expressed in an alternate form to Eqs. (G.1), recognizing from trigonometric relations that

$$A_n \cos n\omega_0 t + B_n \cos n\omega_0 t = C_n \cos(n\omega_0 t + \theta_n) \tag{G.2}$$

where

$$C_n = \sqrt{A_n^2 + B_n^2} \quad \text{and} \quad \theta_n = \tan^{-1}\frac{B_n}{A_n} \tag{G.3}$$

The alternate form of the Fourier series becomes

$$f(t) = C_0 + C_1 \cos(\omega_0 t + \theta_1) + C_2 \cos(2\omega_0 t + \theta_2) + \cdots + C_n \cos(n\omega_0 t + \theta_n) \quad \text{(G.4)}$$

Fourier analysis is used to express nonsinusoidal periodic waves as an infinite series of sinusoids. Since, in the study of electric and electronic circuits, sinusoidal excitations are often used and their response is analyzed, it is often more convenient to convert a nonsinusoidal waveform to its Fourier series and then use the sinusoidal components to analyze the response of the system.

The values of the coefficients (A's, B's, and C's) are determined by the following formulas:

$$A_0 = \frac{1}{T}\int_0^T f(t)\, dt \quad \text{(G.5)}$$

$$A_n = \frac{2}{T}\int_0^T f(t) \cos n\omega_0 t\, dt \quad \text{(G.6)}$$

$$B_n = \frac{2}{T}\int_0^T f(t) \sin n\omega_0 t\, dt \quad \text{(G.7)}$$

The following example illustrates the use of Eqs. (G.5) through (G.7) in evaluating the coefficients to determine the Fourier series of a square wave.

EXAMPLE G.1

Determine the Fourier series for the square wave shown in Figure G.1.

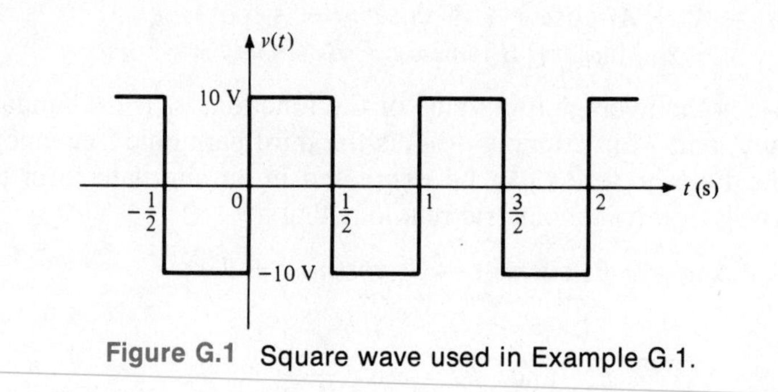

Figure G.1 Square wave used in Example G.1.

Solution

The square wave is a periodic waveform, mathematically described as:

$$v(t) = 10 \text{ V}, \quad 0 < t < 0.5 \text{ s}$$
$$= -10 \text{ V}, \quad 0.5 < t < 1 \text{ s}$$

Its period is $T = 1$ s, and since $T = 2\pi/\omega_0$,

$$\omega_0 = \frac{2\pi}{T} = 2 \text{ rad/s} \quad \text{or} \quad f_0 = 1 \text{ Hz}$$

Next, we evaluate the A_0 terms by Eq. (G.5):

$$A_0 = \frac{1}{1}\left[\int_0^{1/2} 10\, dt - \int_{1/2}^{1} 10\, dt\right]$$

$$= 10t\Big|_0^{1/2} - 10t\Big|_{1/2}^{1} = 5 - [10 - 5] = 0$$

This result is not surprising, since by observation it could also be concluded that the average (dB) value of the square wave is zero.

The A_n coefficients are determined by Eq. (G.6):

$$A_n = \frac{2}{1}\left[\int_0^{1/2} 10 \cos n2\pi t\, dt - \int_{1/2}^{1} 10 \cos n2\pi t\, dt\right]$$

$$= \frac{20}{2n\pi}\left[\sin n2\pi t\Big|_0^{1/2} - \sin n2\pi t\Big|_{1/2}^{1}\right]$$

$$= \frac{10}{n\pi}[\sin n\pi - (\sin n2\pi - \sin n\pi)]$$

$$= 0$$

To evaluate the B_n coefficients, apply Eq. (G.7):

$$B_n = \frac{2}{1}\left[\int_0^{1/2} 10 \sin n2\pi t\, dt - \int_{1/2}^{1} 10 \sin n2\pi t\, dt\right]$$

$$= \frac{20}{2n\pi}\left[-\cos n2\pi t\Big|_0^{1/2} + \cos n2\pi t\Big|_{1/2}^{1}\right]$$

$$= \frac{10}{n\pi}[-(\cos n\pi t - \cos 0) + (\cos n2\pi - \cos n\pi)]$$

For odd values of n, $n = 1, 3, 5, \ldots,$

$$B_n = \frac{10}{n\pi}[-(-1 - 1) + 1 + 1] = \frac{40}{n\pi}$$

For even values of n,

$$B_n = 0$$

Therefore, substituting the coefficients in Eq. (G.1b) yields

$$f(t) = \frac{40}{\pi}\left(\sin\omega_0 t + \frac{1}{3}\sin 3\omega_0 t + \frac{1}{5}\sin 5\omega_0 t + \cdots\right) \text{ V}$$

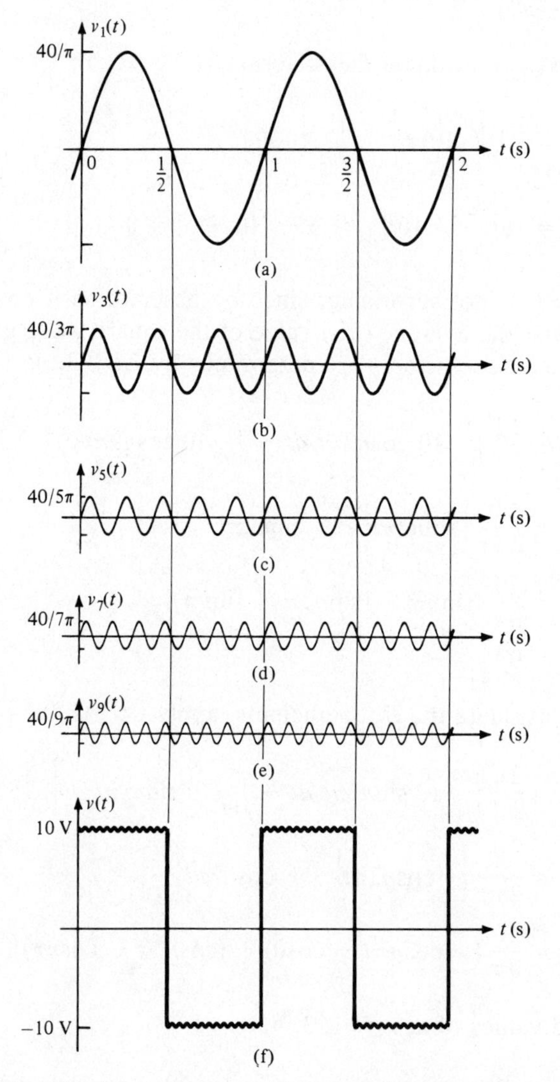

Figure G.2 Five significant terms of the square-wave Fourier series. **(a)** Fundamental. **(b)** 3rd Harmonic. **(c)** 5th Harmonic. **(d)** 7th Harmonic. **(e)** 9th Harmonic. **(f)** Result of summation of the first five terms.

The result of Example G.1 points out a number of interesting facts:

1. Some coefficients have a value of zero; hence, some terms can be missing from the Fourier series of a given waveform. In the case of Example G.1, only the odd harmonics are present.

2. The existing sinusoidal terms have coefficients that decrease in value for the higher-frequency harmonics. (This is generally true; however, there are some exceptions.)

3. The determination of the coefficients is tedious and can be very complicated, depending on the original waveform.

Five significant terms of the Fourier series for the square wave in Example G.1 are shown in Figure G.2. Notice that the summation of the five sinusoids results in a waveform that is very close to the square wave of Figure G.1. The discrepancy is due to a lack of higher harmonics, and the resultant wave does not have the sharp vertical edges and slight alternations on the top and bottom edges of the original square wave. Creating waveforms by addition of sinusoidal waves as ''prescribed'' by the Fourier series is called *synthesis*. Music synthesizers are electronic instruments that generate sinusoidal waves, which produce various sounds.

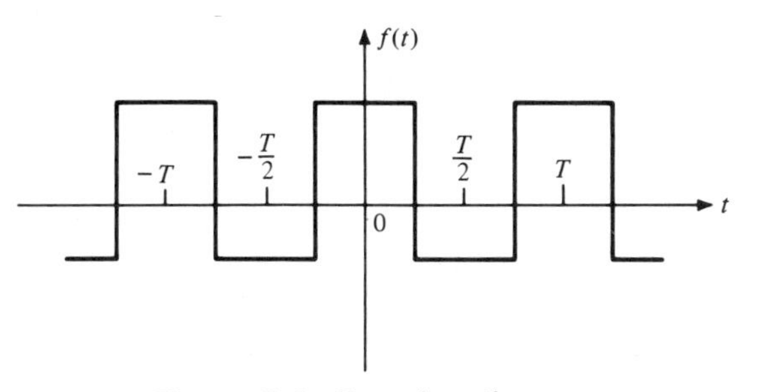

Figure G.3 Even function.

The absence of certain terms in the Fourier series of Example G.1 was predicted because of the *symmetry* of the wave. Depending on the type of symmetry that a wave possesses, it is possible to determine which terms will be present in the Fourier series and which will vanish.

A function that satisfies the condition

$$f(t) = f(-t) \tag{G.8}$$

is an *even function,* and a waveform that satisfies this condition is said to be symmetric about the *vertical axis*. The Fourier series will contain the *average* (*dc*) and *cosine* terms. The waveform shown in Figure G.3 is an even function.

An *odd function* is characterized by the condition

$$f(t) = -f(-t) \tag{G.9}$$

TABLE G.1
Waveforms and their Fourier series representations

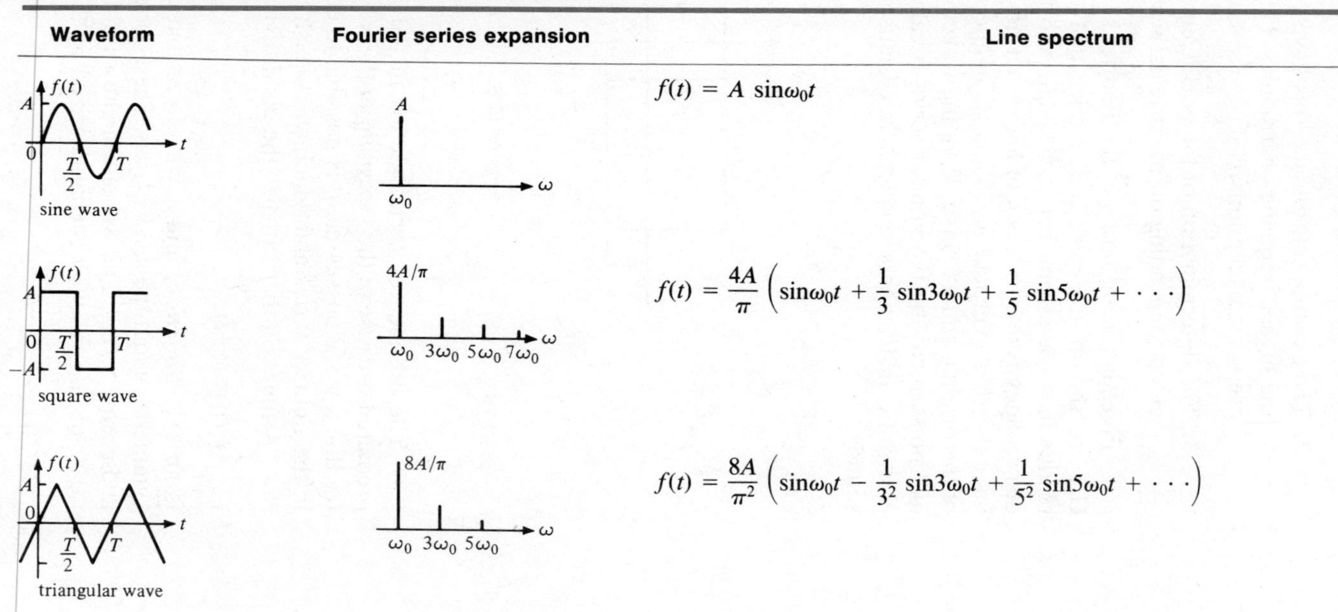

Waveform	Fourier series expansion	Line spectrum
sine wave		$f(t) = A \sin\omega_0 t$
square wave		$f(t) = \dfrac{4A}{\pi}\left(\sin\omega_0 t + \dfrac{1}{3}\sin 3\omega_0 t + \dfrac{1}{5}\sin 5\omega_0 t + \cdots\right)$
triangular wave		$f(t) = \dfrac{8A}{\pi^2}\left(\sin\omega_0 t - \dfrac{1}{3^2}\sin 3\omega_0 t + \dfrac{1}{5^2}\sin 5\omega_0 t + \cdots\right)$

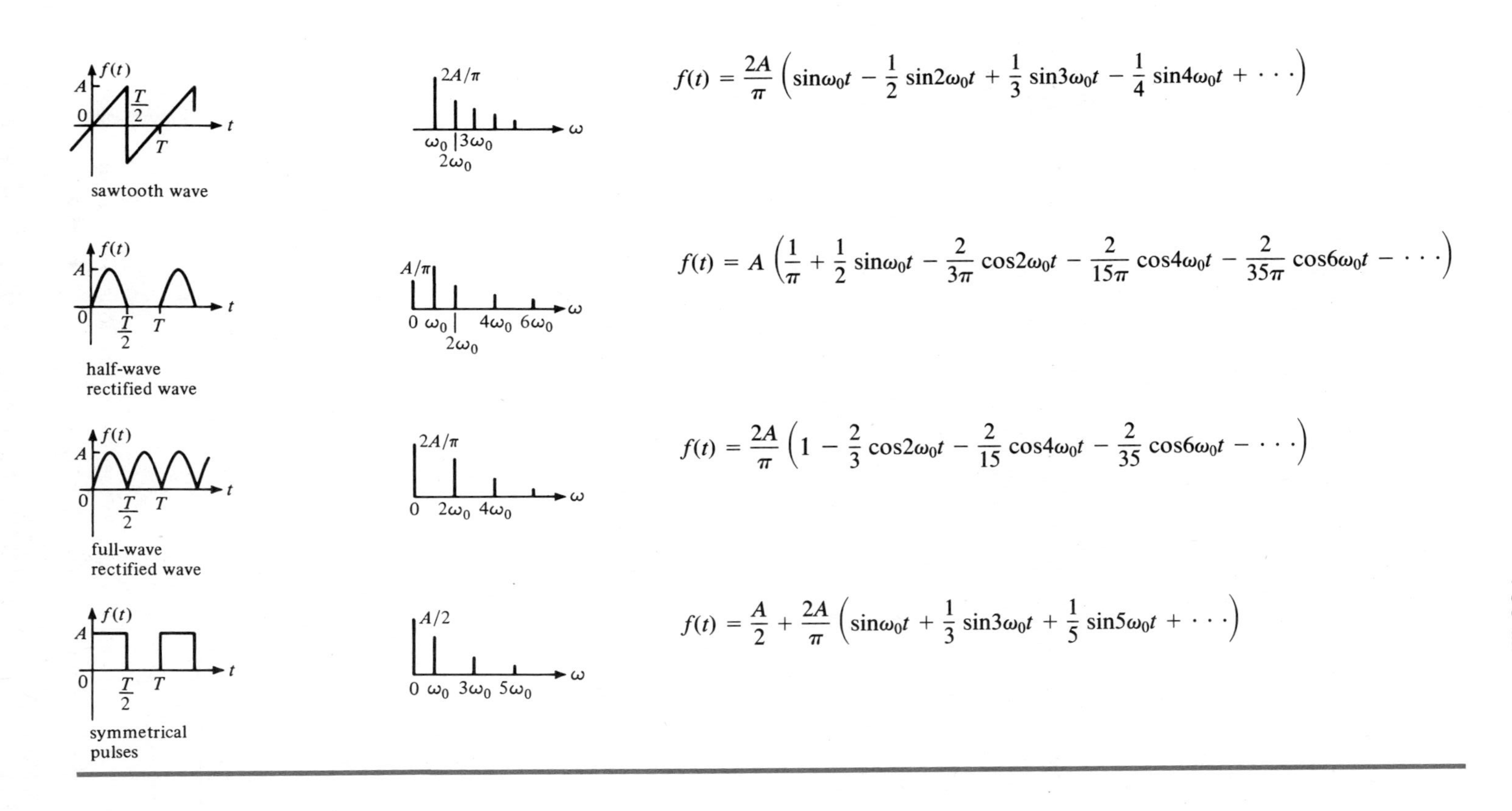
sawtooth wave
$f(t) = \frac{2A}{\pi}\left(\sin\omega_0 t - \frac{1}{2}\sin 2\omega_0 t + \frac{1}{3}\sin 3\omega_0 t - \frac{1}{4}\sin 4\omega_0 t + \cdots\right)$
half-wave rectified wave
$f(t) = A\left(\frac{1}{\pi} + \frac{1}{2}\sin\omega_0 t - \frac{2}{3\pi}\cos 2\omega_0 t - \frac{2}{15\pi}\cos 4\omega_0 t - \frac{2}{35\pi}\cos 6\omega_0 t - \cdots\right)$
full-wave rectified wave
$f(t) = \frac{2A}{\pi}\left(1 - \frac{2}{3}\cos 2\omega_0 t - \frac{2}{15}\cos 4\omega_0 t - \frac{2}{35}\cos 6\omega_0 t - \cdots\right)$
symmetrical pulses
$f(t) = \frac{A}{2} + \frac{2A}{\pi}\left(\sin\omega_0 t + \frac{1}{3}\sin 3\omega_0 t + \frac{1}{5}\sin 5\omega_0 t + \cdots\right)$

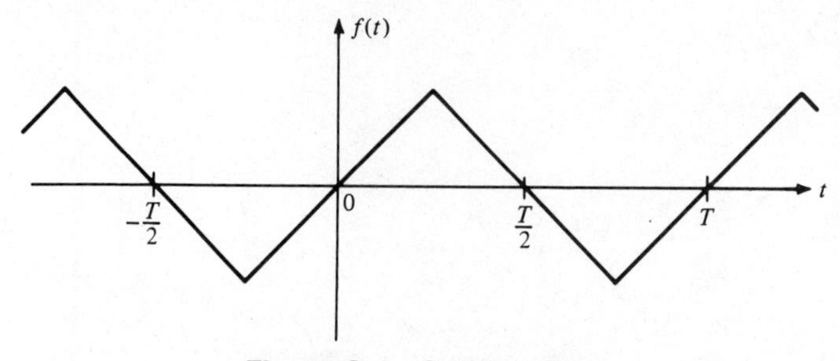

Figure G.4 Odd function.

The triangular wave of Figure G.4 is an odd function. Odd functions are symmetrical about a *point,* and their Fourier series expansions contain only *sine* terms.

Half-wave symmetry is characterized by the condition

$$f(t) = -f\left(t + \frac{T}{2}\right) \tag{G.10}$$

Half-wave symmetry is also known as *mirror symmetry,* and such a waveform is shown in Figure G.5. The Fourier series expansion of a wave with half-wave symmetry contains the *odd n terms* only.

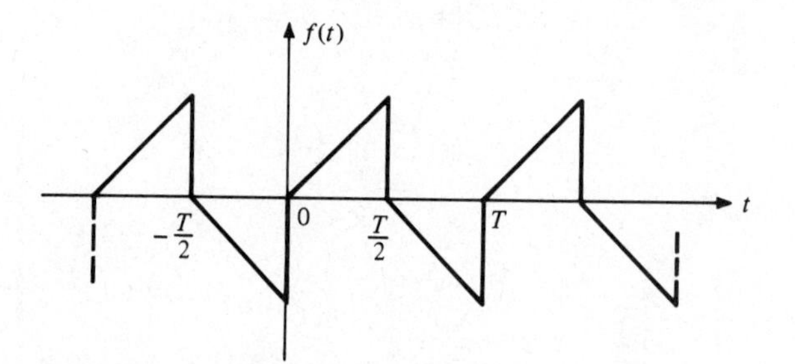

Figure G.5 Half-wave symmetry.

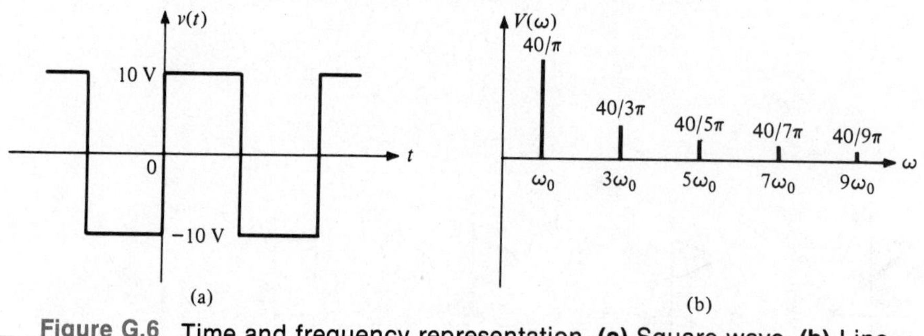

Figure G.6 Time and frequency representation. **(a)** Square wave. **(b)** Line spectrum of square wave.

The square wave used in Example G.1 is an odd function with half-wave symmetry, and the Fourier series expansion contains only odd harmonic sine terms.

Nonsinusoidal periodic waves are plotted as a function of time and can also be plotted from the Fourier series as a function of frequency. The resultant frequency plot is called a *line spectrum*. The line spectrum of the square wave from Example G.1 is shown in Figure G.6.

Several instruments are used to measure the harmonic content of nonsinusoidal waves. The most versatile of the instruments is the *spectrum analyzer* because it is capable of using a CRT (cathode-ray tube) to display the time spectrum of an arbitrary waveform in the form of an amplitude plot of the frequency components. A photograph of a spectrum analyzer is shown in Figure 15.27.

A number of waveforms, their Fourier series expansions, and their line spectrums are given in Table G.1.

APPENDIX H

Circuit Analysis with SPICE

SPICE

SPICE is an acronym for *S*imulation *P*rogram with *I*ntegrated *C*ircuit *E*mphasis. It was developed in the late 1960s at the University of California at Berkeley primarily by Lawrence Nagel, with significant contributions by Ellis Cohen. SPICE is a program that simulates analog circuits. Originally it was written for mainframe computers. With the proliferation and popularity of personal computers, several companies adapted the original SPICE program for desktop machines.

PSPICE and MICROSIM (*FREE PSPICE*)

MicroSim Corporation (20 Fairbanks, Irvine, CA 92718) produces PSPICE. The commercial version costs $950. An associated graphics program called Probe costs $450. MicroSim, however, offers a *classroom version* of these two programs at *no charge* to any professor of electronics in the United States who writes to them and asks for a copy. The classroom version comes on two diskettes, which may be reproduced at will and may be run on as many PCs as desired.

The authors strongly encourage course instructors to obtain and distribute this very versatile program.

MicroSim Corporation has generously given the authors permission to use the material from their PSPICE brochure in this book. All material from the MicroSim brochure is in *italics*.

WHAT IS PSPICE?

PSPICE is an analog circuit simulator. It will calculate the behavior of electrical circuits. It calculates a circuit's currents and voltages. Even more specifically, PSPICE contains a *circuit simulator language,* which allows a circuit to be described in terms of its interconnections and components. PSPICE also contains a program that processes the circuit description and, along with control statements, predicts circuit currents, voltages, and waveforms.

To circuit designers and engineers, PSPICE is an invaluable tool. With PSPICE they *can perform the same measurements that they would with an actual circuit and many others that would not be feasible with a breadboard.*

It must be emphasized however, that PSPICE is an answer tool and not a circuits learning tool. PSPICE will analyze the circuit but will not give insights on the analysis. Nevertheless, PSPICE used in conjunction with the methods and techniques presented in this text will serve the technologist well.

TYPES OF ANALYSES

PSPICE provides the following types of analyses:

- *Dc voltages and currents of the circuits.*
- *Ac response of the circuit. One or more inputs may be specified and swept over a range of frequencies. The voltages and currents of the circuit will be calculated. Both magnitude and phase are computed.*
- *Transient or time response of the circuit. One or more time-varying inputs may be specified. The voltages and currents of the circuit's response in time are calculated. Oscilloscope-type displays are one of the results of transient analysis.*

The free classroom version of PSPICE also performs noise analysis, temperature analysis, and Monte Carlo analysis. It further is capable of doing analog-to-digital (A-to-D) and digital-to-analog (D-to-A) conversions and analyses. PSPICE also offers interfaces to several PC-based schematic editors such as *Aptos Computer, Automated Images, Case Technology, FutureNet, Omation, OrCad, P-Cad, Phase Three Logic, and Viewlogic.* All these features and options are beyond the needs and goals of an introductory circuit analysis test such as this one and therefore are not examined here.

HOW TO USE PSPICE

There are three steps involved in getting your answers from PSPICE:

1. Use a word processor and the SPICE language to prepare an ASCII source file that describes your circuit.

2. Run the PSPICE program. PSPICE will process the source file and generate an object file that will contain the results.
3. Use DOS to print out the object file and, therefore, the results.

NAMING THE SOURCE AND OBJECT FILES

For reasons that will become apparent as you use SPICE programs, source files should have the extension .CIR and object files should have the extension .OUT. A circuit named SERIES should have its circuit description source file named SERIES.CIR, whereas its output or object file should be named SERIES.OUT. This method of naming the files allows you to refer to the circuit SERIES while clearly indicating whether you are referring to the source file (SERIES.CIR) or the output file (SERIES.OUT).

WORD PROCESSORS AND THE SOURCE FILE

The source file may be prepared using any available word processor. Some commercial SPICE programs provide their own word processor. PSPICE does not. Since all DOS packages contain a line editor called EDLIN, we will use EDLIN in its limited word processing capabilities. EDLIN may be used to create, display, and change files. Your DOS manual contains detailed EDLIN information. The examples provided here, however, will give enough information to allow you to use EDLIN to write and edit source files for PSPICE.

GETTING THE COMPUTER READY

After obtaining the classroom version of PSPICE from MicroSim, install PSPICE on your hard disk as follows:

1. Make sure you are running on DOS 3.0 or higher.
2. Create a directory called PSPICE.DIR as follows:

COMPUTER	YOU
C>	`CD\`
C>	`MD PSPICE.DIR`

3. Copy both diskettes containing PSPICE files to the newly created PSPICE.DIR directory as follows:

COMPUTER	YOU
C>	`CD\PSPICE.DIR`
C>	`COPY A:*.*`

4. Once both diskettes have been copied to your hard drive, every time you want to use PSPICE go to the PSPICE directory. This way all your PSPICE work will be contained in PSPICE.DIR.

GETTING THE CIRCUIT READY

The circuit to be analyzed must be prepared in the following manner:

1. All nodes must be numbered with positive integers.
2. The ground (reference) node must be numbered 0.
3. Each node must have at least two connections.
4. Each node must have a dc path to ground.
5. Each element must be labeled appropriately. Here we describe only elements used in this book. Electronic elements such as diodes, transistors, and ICs are not presented here.
6. Component names must start with a specific letter and may contain a combination of seven more letters or numbers. *Special characters, including spaces, may not be used.*
7. A very abbreviated list of circuit elements follows:

   C capacitor

   L inductor

   R resistor

   I independent current source

   V independent voltage source

   E linear voltage-controlled voltage source

   F linear current-controlled current source

   G linear voltage-controlled current source

   H linear current-controlled voltage source

CIRCUIT ELEMENT DESCRIPTION

An ELEMENT STATEMENT must have the following information:

1. The component NAME
2. The NODE NUMBERS to which it is connected
3. The VALUE of the component (units may be used or omitted)

The component value may contain the following factors:

T => 10^{12} G => 10^{9} MEG => 10^{6}

K => 10^{3} M => 10^{-3} U => 10^{-6}

N => 10^{-9} P => 10^{-12} F => 10^{-15}

EXAMPLES OF ELEMENT STATEMENTS

R1 1 2 10K	10-kΩ resistor connected between nodes 1 and 2
VIN 1 0 DC 15	15-V battery with the positive terminal at node 1 and negative terminal at node 0 (reference)
I2 3 4 AC 5 30	ac phasor current source of 5 A at 30° maintaining current from node 3 to node 4
VX 8 15	dummy voltage source used to measure the current from node 8 to node 15 (an ammeter)

CONTROL STATEMENTS

The TITLE line must be the first line in the source file. The contents of this line are printed on each page of the output.

A COMMENT is any line beginning with an asterisk (*). A comment is strictly for the writer's benefit and is ignored by the computer.

.END must be the last line in any SPICE source file. The period before the word END is required.

Among the many control statements offered by PSPICE, the ones that are needed for the circuits presented in this book are the following (note that each one starts with a period):

.AC	requests ac circuit analysis
.DC	allows current or voltage sweep analysis
.DISTO	requests distortion analysis
.END	must be the last line of file
.FOUR	requests Fourier analysis
.NODESET	sets initial guess for DC or TRANS
.PRINT	requests printout contents
.PLOT	requests plotted output
.TC	sets initial conditions for TRAN
.TRAN	requests transient analysis

HOW TO START EDLIN

To start EDLIN from DOS type:

```
edlin filename
```

At this point you will either want to create a new file or edit an old one. Each process is described next.

If this is a new file, the **filename** should be the name of the new file. When EDLIN does not find this file on the drive, EDLIN creates a new file with the name you specified. The following message and prompt are displayed:

```
New File
*_
```

The prompt for EDLIN is an asterisk (*).
The letter I (Insert) must be entered next.
EDLIN returns with the number 1, indicating it is awaiting the first line. After each line is entered, EDLIN returns with the next line number.

If an existing file is to be edited, **filename** should be the name of the file you wish to edit. When EDLIN finds the specified file, the file is loaded into memory, and it can be edited using EDLIN editing commands.

When the entering or editing session is completed, use Control C to return to the EDLIN prompt and then use the E (End) command to save the file on disk.

The EDLIN commands of interest to us are:

line #	Edits that line #
A	Appends lines
C	Copies lines
D	Deletes lines
E	Ends editing
I	Inserts lines
L	Lists text

The use of EDLIN will become more apparent as we do the following examples.

EXAMPLE H.1

Use PSPICE to solve Example 5.17, which asks to calculate the currents through each resistor in the circuit of Figure 5.23.

Figure 5.23 is reproduced as Figure H.1a.

Solution

Figure H.1b shows the circuit already prepared for PSPICE analysis. Note the following details:

1. A dummy voltage source has been inserted in each path where the current is to be determined. The dummy source polarity has been assigned so that the desired current direction through each source is + to − as the current goes through the source.

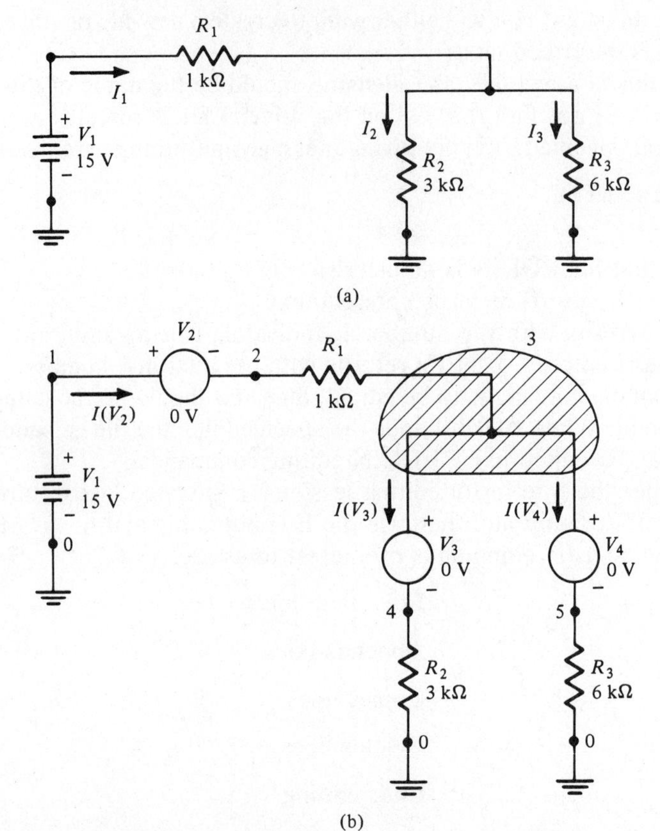

Figure H.1 Circuits for Example H.1. **(a)** Original circuit. **(b)** Circuit modified for PSPICE analysis.

2. Each voltage source has been labeled as V_1, V_2, V_3, and V_4.
3. Each passive component has been labeled R_1, R_2, and R_3.
4. The reference (ground) node has been assigned the number 0 (zero), and the other nodes have been assigned the numbers 1, 2, 3, 4, and 5.

Next write the circuit description using the PSPICE language, as follows:

```
EXAMPLE H.1, SPICE SOLUTION OF EXAMPLE 5.17
*LINE #1 ABOVE IS THE TITLE
V1 1 0 DC 15
V2 1 2
V3 3 4
V4 3 5
R1 2 3 1K
R2 4 0 3K
R3 5 0 6K
.END
```

To put this information into a computer file named EXH1.CIR proceed as follows (after each entry push the ENTER key):

```
COMPUTER      YOU
>C            CD\PSPICE.DIR       this puts you in the PSPICE directory
>C            EDLIN EXH1.CIR      this creates a file called EXH1.CIR (Example H.1 circuit
                                    description)
New File                          a new file is confirmed
*             I                   the EDLIN prompt appears; the I enables the Insert
                                    command
    1:*       EXAMPLE H.1, SPICE SOLUTION OF EXAMPLE 5.17
    2:*       *LINE #1 ABOVE IS THE TITLE
    3:*       V1 1 0 DC 15
    4:*       V2 1 2
    5:*       V3 3 4
    6:*       V4 3 5
    7:*       R1 2 3 1K
    8:*       R2 4 0 3K
    9:*       R3 5 0 6K
   10:*       .END
   11:*       ^C  (CONTROL C)     this ends the Insert mode
*             E                   this ends the EDLIN session and saves the file to disk under
                                    the name EXH1.CIR
C>            PSPICE              this calls for the PSPICE program
You may use <enter> alone to exit PSpice.
Input file name [.CIR]?
              EXH1.CIR
Output file name [EXH1.OUT]?
              push the <enter> key
```

PSPICE performs the requested solution. The computer then comes back with the usual DOS prompt. At this time, the solution may be either displayed on the screen with TYPE EXH1.OUT or a hard copy may be obtained from the printer with PRINT EXH1.OUT. During screen output, screen scrolling may be suspended with Control s and reinstated by pressing any other key. The commands are as follows:

```
COMPUTER      YOU
C>            TYPE EXH1.OUT       for screen output
C>            PRINT EXH1.OUT      for printer output
(If the computer asks for OUTPUT DEVICE, press ENTER to
 use the printer)
   C:PSPICE.DIR\EXH1.OUT is currently being printed out.
C>
```

In the printer output that follows, note that the program is reproduced first and then each node voltage is presented followed by the current through each voltage source. The current through V_2 and, therefore, through R_1 is 5 mA, the current through V_3 and, therefore, through R_2 is 3.33 mA, and the current through V_4 and, therefore, through R_3 is 1.667 mA.

```
******* 06/04/88 ******* Evaluation PSPICE (Jan. 1988) ******* 19:53:57 *******

EXH1, SPICE SOLUTION OF EXAMPLE 5.17

****     CIRCUIT DESCRIPTION

*******************************************************************************
```

```
*LINE #1 ABOVE IS THE TITLE
V1 1 0 DC 15
V2 1 2
V3 3 4
V4 3 5
R1 2 3 1K
R2 4 0 3K
R3 5 0 6K
.END

 ******* 06/04/88 ******* Evaluation PSPICE (Jan. 1988) ******* 19:53:57 *******

 EXH1, SPICE SOLUTION OF EXAMPLE 5.17

 ****     SMALL SIGNAL BIAS SOLUTION       TEMPERATURE =    27.000 DEG C

******************************************************************************

 NODE   VOLTAGE     NODE   VOLTAGE     NODE   VOLTAGE     NODE   VOLTAGE

(    1)   15.0000  (    2)   15.0000  (    3)   10.0000  (    4)   10.0000

(    5)   10.0000

    VOLTAGE SOURCE CURRENTS
    NAME         CURRENT

    V1          -5.000E-03
    V2           5.000E-03
    V3           3.333E-03
    V4           1.667E-03

    TOTAL POWER DISSIPATION   7.50E-02 WATTS

         JOB CONCLUDED

         TOTAL JOB TIME            4.67
```

EXAMPLE H.2

Use PSPICE to solve for the current through each resistor in the circuit for Examples 7.4, 7.8, and 7.19. The circuit is reproduced as Figure H.2a.

Solution

Figure H.2b shows the circuit already prepared for PSPICE analysis. Note that a dummy voltage source has been inserted in series with each resistor. Note also that each node has been labeled and that each passive component has been labeled.

The PSPICE circuit description file should be labeled EXH2.CIR and should be entered in the PSPICE.DIR directory using EDLIN as described in Example H.1. The PSPICE program for Example H.2 is:

```
EXAMPLE H.2, SPICE SOLUTION OF EXAMPLE 7.19
I1 2 3 DC 10
V1 1 2
V2 3 11
V3 8 0
V4 14 7
```

(*program continued on 596*)

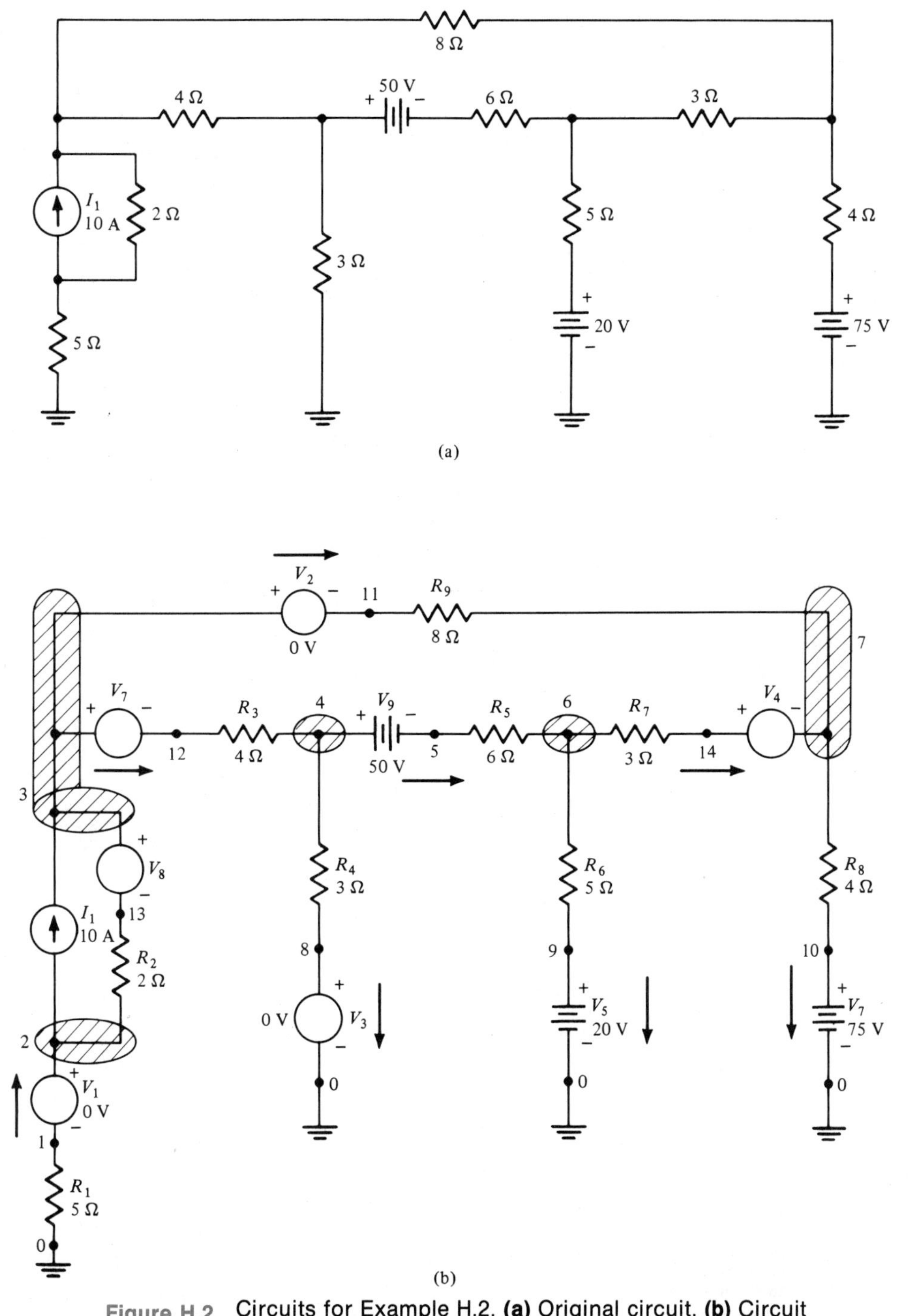

Figure H.2 Circuits for Example H.2. **(a)** Original circuit. **(b)** Circuit modified for PSPICE analysis. **(c)** Circuit showing resulting currents. **(d)** Circuit showing resulting node voltages.

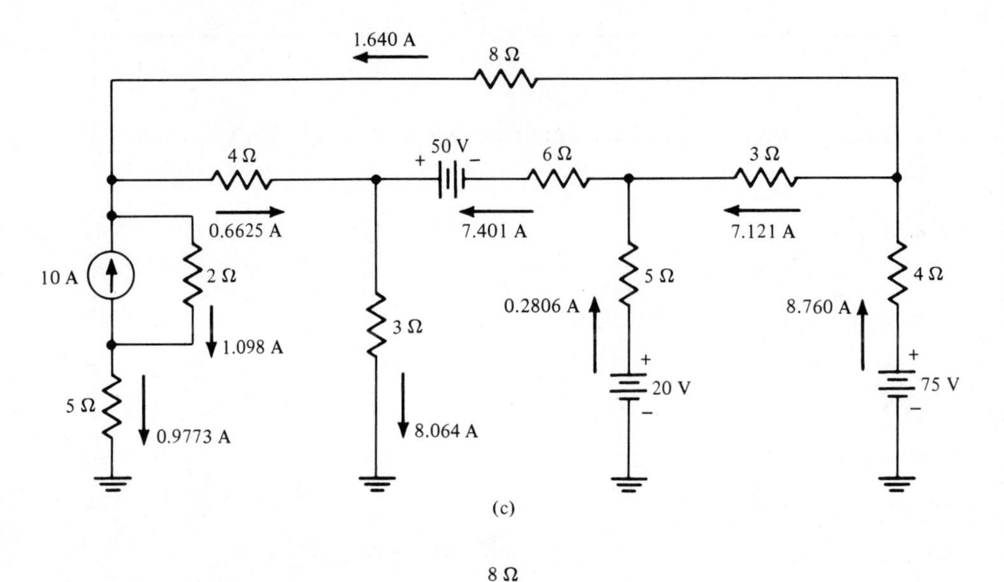

(c)

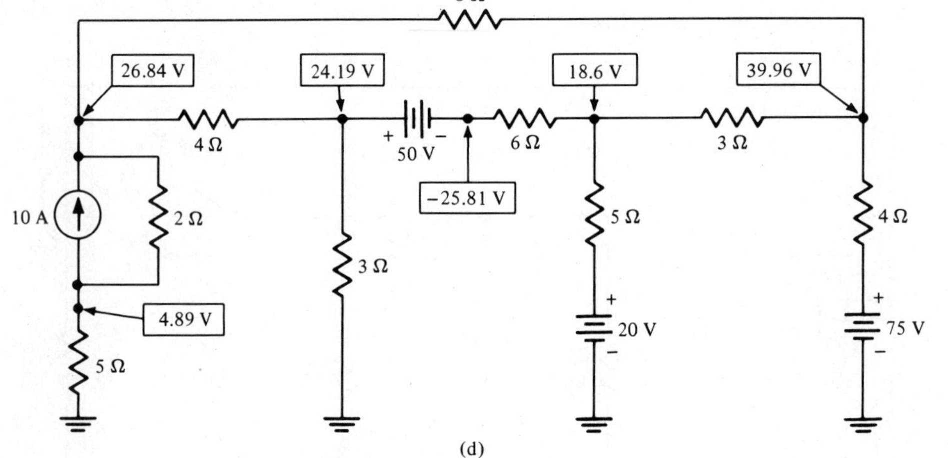

(d)

Figure H.2 (continued)

continued from program on 594

```
V5 9 0 DC 20
V6 10 0 DC 75
V7 3 12
V8 3 13
V9 4 5 DC 50
R1 1 0 5
R2 13 2 2
R3 12 4 4
R4 4 8 3
R5 5 6 6
R6 6 9 5
R7 6 14 3
R8 7 10 4
R9 11 7 8
.END
```

The printer output shows that the currents through the elements agree with the answers obtained in Example 7.19. Figure H.2c shows the circuit with the currents through the respective resistors. The negative signs obtained for some of the currents with PSPICE indicate that those currents have a direction opposite the ones assigned in Figure H.2b. Figure H.2c shows the currents with the correct directions. Compare Figure H.2c with Figure 7.11b to verify the results.

Figure H.2d shows the circuit with the node voltages that were obtained from the PSPICE analysis. These node voltages were rounded to two decimal places. Compare Figure H.2d with Figure 7.11c to verify the results.

The PSPICE printer output is:

```
******* 06/04/88 ******* Evaluation PSPICE (Jan. 1988) ******* 22:39:40 *******

EXAMPLE H.2, SPICE SOLUTION OF EXAMPLE 7.19

****     CIRCUIT DESCRIPTION

******************************************************************************

I1 2 3 DC 10
V1 1 2
V2 3 11
V3 8 0
V4 14 7
V5 9 0 DC 20
V6 10 0 DC 75
V7 3 12
V8 3 13
V9 4 5 DC 50
R1 1 0 5
R2 13 2 2
R3 12 4 4
R4 4 8 3
R5 5 6 6
R6 6 9 5
R7 6 14 3
R8 7 10 4
R9 11 7 8
.END

******* 06/04/88 ******* Evaluation PSPICE (Jan. 1988) ******* 22:39:40 *******

EXAMPLE H.2, SPICE SOLUTION OF EXAMPLE 7.19

****     SMALL SIGNAL BIAS SOLUTION       TEMPERATURE =   27.000 DEG C

******************************************************************************

 NODE   VOLTAGE     NODE   VOLTAGE     NODE   VOLTAGE     NODE   VOLTAGE

(    1)    4.8863  (    2)    4.8863  (    3)   26.8410  (    4)   24.1910

(    5)  -25.8090  (    6)   18.5970  (    7)   39.9590  (    8)    0.0000

(    9)   20.0000  (   10)   75.0000  (   11)   26.8410  (   12)   26.8410

(   13)   26.8410  (   14)   39.9590
```

```
VOLTAGE SOURCE CURRENTS
NAME          CURRENT

V1          -9.773E-01
V2          -1.640E+00
V3           8.064E+00
V4          -7.121E+00
V5          -2.806E-01
V6          -8.760E+00
V7           6.625E-01
V8           1.098E+01
V9          -7.401E+00

TOTAL POWER DISSIPATION  1.03E+03 WATTS

         JOB CONCLUDED

         TOTAL JOB TIME            9.33
```

EXAMPLE H.3

For the circuit shown in Figure H.3a (also Figure 11.4a for Example 11.3), use PSPICE to solve for the currents through the series elements.

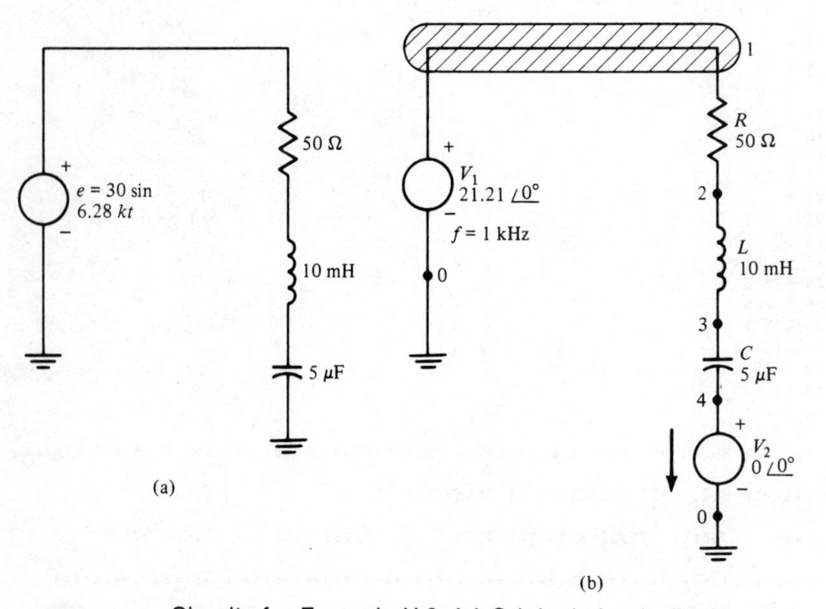

Figure H.3 Circuits for Example H.3. **(a)** Original circuit. **(b)** Circuit modified for PSPICE analysis.

Solution

Figure H.3b shows the circuit prepared for PSPICE analysis. The nodes are labeled 0, 1, 2, 3, and 4, the passive elements are labeled R, L, and C, the ac voltage source is labeled V_1, and the dummy source is labeled V_2.

As described in Example H.1, use EDLIN to create a file named EXH3.CIR. After inputting the following circuit description, run PSPICE and output the answer with DOS.

```
EXAMPLE H.3, PSPICE SOLUTION OF EXAMPLE 11.3
V1 1 0 AC 21.21 0
V2 4 0
R 1 2 50
L 2 3 10M
C 3 4 5U
.AC LIN 1 1K 1K
*THE ABOVE LINE ASKS FOR AN AC ANALYSIS AT 1 KHZ
.PRINT AC IM(V2) IP(V2) IR(V2) II(V2)
*THE ABOVE LINE ASKS TO OUTPUT THE FOLLOWING:
*IM(V2), THE MAGNITUDE OF THE CURRENT THROUGH V2
*IP(V2), THE PHASE OF THE CURRENT THROUGH V2
*IR(V2), THE REAL PART OF THE CURRENT THROUGH V2
*II(V2), THE IMAGINARY PART OF THE CURRENT THROUGH V2
.END
```

The PSPICE output that follows shows that the current through the series circuit is 360.5 mA at $-31.8°$ (or, in rectangular form, $306.4 - j190$ mA). This agrees with the answer to Example 11.3.

```
******* 06/05/88 ******* Evaluation PSPICE (Jan. 1988) ******* 11:00:10 *******

EXAMPLE H.3, PSPICE SOLUTION OF EXAMPLE 11.3

****     CIRCUIT DESCRIPTION

******************************************************************************

V1 1 0 AC 21.21 0
V2 4 0
R 1 2 50
L 2 3 10M
C 3 4 5U
.AC LIN 1 1K 1K
*THE ABOVE LINE ASKS FOR AN AC ANALYSIS AT 1 KHZ
.PRINT AC IM(V2) IP(V2) IR(V2) II(V2)
*THE ABOVE LINE ASKS TO OUTPUT THE FOLLOWING:
*IM(V2), THE MAGNITUDE OF THE CURRENT THROUGH V2
*IP(V2), THE PHASE OF THE CURRENT THROUGH V2
*IR(V2), THE REAL PART OF THE CURRENT THROUGH V2
*II(V2), THE IMAGINARY PART OF THE CURRENT THROUGH V2
.END

******* 06/05/88 ******* Evaluation PSPICE (Jan. 1988) ******* 11:00:10 *******

EXAMPLE H.3, PSPICE SOLUTION OF EXAMPLE 11.3

****     SMALL SIGNAL BIAS SOLUTION       TEMPERATURE =  27.000 DEG C

******************************************************************************

 NODE   VOLTAGE     NODE   VOLTAGE     NODE   VOLTAGE     NODE   VOLTAGE

(    1)     0.0000 (    2)     0.0000  (    3)     0.0000  (    4)     0.0000

   VOLTAGE SOURCE CURRENTS

    NAME          CURRENT

    V1            0.000E+00
    V2            0.000E+00

    TOTAL POWER DISSIPATION  0.00E+00 WATTS
```

```
******* 06/05/88 ******* Evaluation PSPICE (Jan. 1988) ******* 11:00:10 *******
EXAMPLE H.3, PSPICE SOLUTION OF EXAMPLE 11.3

****      AC ANALYSIS                          TEMPERATURE =  27.000 DEG C

******************************************************************************

 FREQ         IM(V2)        IP(V2)       IR(V2)       II(V2)

  1.000E+03   3.605E-01   -3.180E+01   3.064E-01   -1.900E-01

          JOB CONCLUDED

          TOTAL JOB TIME          5.00
```

EXAMPLE H.4

Obtain the magnitude (in decibels) and phase frequency response for the filter of Example 13.19 using PSPICE.

Solution

Figure H.4a shows the circuit. Figure H.4b shows the circuit already prepared for PSPICE analysis. Use the directions given in Example H.1 to set up, enter, and run PSPICE, and output the results for the circuit.

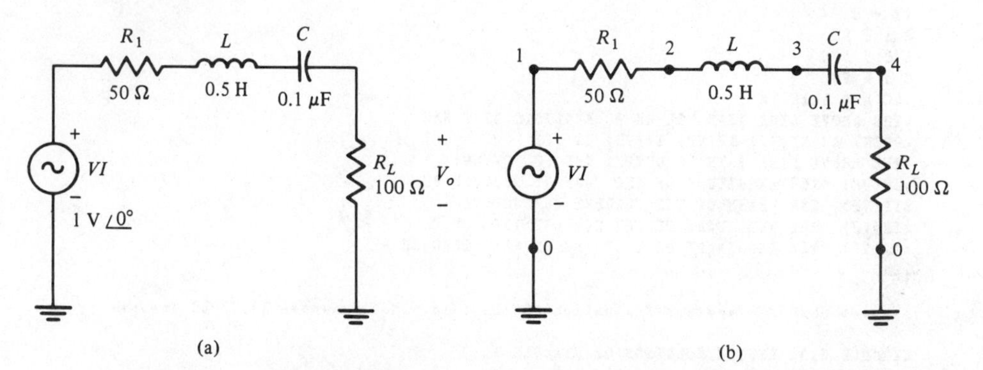

Figure H.4 Circuits for Example H.4. **(a)** Original circuit. **(b)** Circuit modified for PSPICE analysis.

The PSPICE source file should contain:

```
EXAMPLE H.4, PSPICE SOLUTION OF EXAMPLE 13.19
VI 1 0 AC 1 0
R1 1 2 50
L 2 3 0.5
C 3 4 0.1U
RL 4 0 100
.AC LIN 20 500 1000
*THE ABOVE LINE ASKS FOR AN ANALYSIS FROM 500 to 1KHZ
*WITH 20 STEPS
.PLOT AC VDB(4)
```

```
*THE ABOVE LINE ASKS FOR A PLOT OF
*THE VOLTAGE AT NODE 4 IN DB
.PLOT AC VP(4)
*THE ABOVE LINE ASKS FOR A PLOT OF
*THE PHASE OF THE VOLTAGE AT NODE 4
.WIDTH OUT=80
*THE ABOVE LINE ASKS FOR AN OUTPUT WIDTH OF 80 COLUMNS
.END
```

The PSPICE output is shown next. Notice that the output is 3 dB below maximum at approximately 680 Hz and 740 Hz. Follow the directions after the PSPICE output for modifying the EXH4.CIR file in order to obtain a frequency response between the upper and lower half-power frequencies.

```
******* 06/05/88 ******* Evaluation PSPICE (Jan. 1988) ******* 21:50:33 *******

EXAMPLE H.4, PSPICE SOLUTION OF EXAMPLE 13.19

****     CIRCUIT DESCRIPTION

******************************************************************************

V1 1 0 AC 1 0
R1 1 2 50
L 2 3 0.5
C 3 4 0.1U
RL 4 0 100
.AC LIN 20 500 1000
*THE ABOVE LINE ASKS FOR AN ANALYSIS FROM 500 TO 1KHZ
*WITH 20 STEPS
.PLOT AC VDB(4)
*THE ABOVE LINE ASKS FOR A PLOT OF
*THE VOLTAGE AT NODE 4 IN DB
.PLOT AC VP(4)
*THE ABOVE LINE ASKS FOR A PLOT OF
*THE PHASE OF THE VOLTAGE AT NODE 4
.WIDTH OUT=80
*THE ABOVE LINE ASKS FOR AN OUTPUT WIDTH OF 80 COLUMNS
.END

******* 06/05/88 ******* Evaluation PSPICE (Jan. 1988) ******* 21:50:33 *******

EXAMPLE H.4, PSPICE SOLUTION OF EXAMPLE 13.19

****     SMALL SIGNAL BIAS SOLUTION       TEMPERATURE =  27.000 DEG C

******************************************************************************

NODE   VOLTAGE     NODE   VOLTAGE     NODE   VOLTAGE     NODE   VOLTAGE

(    1)    0.0000 (    2)    0.0000 (    3)    0.0000 (    4)    0.0000

   VOLTAGE SOURCE CURRENTS

   NAME          CURRENT

   V1            0.000E+00

   TOTAL POWER DISSIPATION  0.00E+00 WATTS

******* 06/05/88 ******* Evaluation PSPICE (Jan. 1988) ******* 21:50:33 *******

EXAMPLE H.4, PSPICE SOLUTION OF EXAMPLE 13.19

****     AC ANALYSIS                      TEMPERATURE =  27.000 DEG C

******************************************************************************
```

```
 LEGEND:
*: VDB(4)
  FREQ        VDB(4)

(*)----------    -3.0000E+01  -2.0000E+01  -1.0000E+01   0.0000E+00   1.0000E+01
                        - - - - - - - - - - - - - - - - - - - - - - - - - -
  5.000E+02 -2.419E+01 .      *    .            .           .            .
  5.263E+02 -2.279E+01 .        *  .            .           .            .
  5.526E+02 -2.124E+01 .          *.            .           .            .
  5.789E+02 -1.948E+01 .           .*           .           .            .
  6.053E+02 -1.742E+01 .           .   *        .           .            .
  6.316E+02 -1.491E+01 .           .      *     .           .            .
  6.579E+02 -1.166E+01 .           .          * .           .            .
  6.842E+02 -7.299E+00 .           .            .  *        .            .
  7.105E+02 -3.533E+00 .           .            .       *   .            .
  7.368E+02 -6.674E+00 .           .            .   *       .            .
  7.632E+02 -1.079E+01 .           .           *.           .            .
  7.895E+02 -1.377E+01 .           .       *    .           .            .
  8.158E+02 -1.599E+01 .           .    *       .           .            .
  8.421E+02 -1.773E+01 .           .  *         .           .            .
  8.684E+02 -1.916E+01 .           .*           .           .            .
  8.947E+02 -2.037E+01 .           *            .           .            .
  9.211E+02 -2.140E+01 .          *.            .           .            .
  9.474E+02 -2.231E+01 .         * .            .           .            .
  9.737E+02 -2.312E+01 .        *  .            .           .            .
  1.000E+02 -2.385E+01 .       *   .            .           .            .
                        - - - - - - - - - - - - - - - - - - - - - - - - - -

 ******* 06/05/88 ******* Evaluation PSPICE (Jan. 1988) ******* 21:50:33 *******
 EXAMPLE H.4, PSPICE SOLUTION OF EXAMPLE 13.19
 ****     AC ANALYSIS                      TEMPERATURE =   27.000 DEG C
*****************************************************************************
  FREQ        VP(4)

(*)----------     -1.0000E+02  -5.0000E+01   0.0000E+00   5.0000E+01   1.0000E+02
                         - - - - - - - - - - - - - - - - - - - - - - - - - -
   5.000E+02  8.468E+01 .           .            .           .        *   .
   5.263E+02  8.375E+01 .           .            .           .        *   .
   5.526E+02  8.253E+01 .           .            .           .       *    .
   5.789E+02  8.084E+01 .           .            .           .       *    .
   6.053E+02  7.836E+01 .           .            .           .      *     .
   6.316E+02  7.436E+01 .           .            .           .     *      .
   6.579E+02  6.694E+01 .           .            .           .   *        .
   6.842E+02  4.966E+01 .           .            .           *            .
   7.105E+02  2.967E+00 .           .            .*          .            .
   7.368E+02 -4.592E+01 .           .*           .           .            .
   7.632E+02 -6.433E+01 .       *   .            .           .            .
   7.895E+02 -7.209E+01 .      *    .            .           .            .
   8.158E+02 -7.623E+01 .     *     .            .           .            .
   8.421E+02 -7.877E+01 .     *     .            .           .            .
   8.684E+02 -8.049E+01 .    *      .            .           .            .
   8.947E+02 -8.173E+01 .    *      .            .           .            .
   9.211E+02 -8.267E+01 .    *      .            .           .            .
   9.474E+02 -8.340E+01 .   *       .            .           .            .
   9.737E+02 -8.399E+01 .   *       .            .           .            .
                         - - - - - - - - - - - - - - - - - - - - - - - - - -
          JOB CONCLUDED

          TOTAL JOB TIME            16.75
```

In order to get a frequency response between 680 Hz and 740 Hz, follow these directions:

```
COMPUTER        YOU
C>              CD\PSPICE.DIR
C>              EDLIN EXH4.CIR
End of input file
```

```
*              L              this will list the file
*              7              this will make line 7
                                available for editing
7:*.AC LIN 20 500 1000

7:*            .AC LIN 20 680 740
*              8
8:*            *THE ABOVE LINE ASKS FOR AN ANALYSIS FROM 680
                TO 740 HZ
*              9
9:*            *WITH 20 STEPS
*              E              this ends the editing session
C>             PSPICE         this calls for PSPICE
                                after this, proceed as before
```

The PSPICE output for the modified file is shown next. Note that due to the narrower frequency range chosen, the half-power frequencies may now be read accurately as 689 Hz and 734 Hz, whereas the center frequency is observed at 712 Hz. Note also that the phase drops from +90° to −90°.

```
 ******* 06/05/88 ******* Evaluation PSPICE (Jan. 1988) ******* 22.11.18 *******

 EXAMPLE H.4, PSPICE SOLUTION OF EXAMPLE 13.19

 ****     CIRCUIT DESCRIPTION

*****************************************************************************

VI 1 0 AC 1 0
R1 1 2 50
L 2 3 0.5
C 3 4 0.1U
RL 4 0 100
.AC LIN 20 680 740
*THE ABOVE LINE ASKS FOR AN ANALYSIS FROM 680 to 740 HZ
*WITH 20 STEPS
.PLOT AC VDB(4)
*THE ABOVE LINE ASKS FOR A PLOT OF
*THE VOLTAGE AT NODE 4 IN DB
.PLOT AC VP(4)
*THE ABOVE LINE ASKS FOR A PLOT OF
*THE PHASE OF THE VOLTAGE AT NODE 4
.WIDTH OUT=80
*THE ABOVE LINE ASKS FOR AN OUTPUT WIDTH OF 80 COLUMNS
.END

 ******* 06/05/88 ******* Evaluation PSPICE (Jan. 1988) ******* 22:11:18 *******

 EXAMPLE H.4, PSPICE SOLUTION OF EXAMPLE 13.19

 ****     SMALL SIGNAL BIAS SOLUTION       TEMPERATURE =  27.000 DEG C

*****************************************************************************

 NODE   VOLTAGE     NODE   VOLTAGE     NODE   VOLTAGE     NODE   VOLTAGE

(    1)    0.0000  (    2)    0.0000  (    3)    0.0000  (    4)    0.0000

    VOLTAGE SOURCE CURRENTS
    NAME          CURRENT

    VI            0.000E+00

    TOTAL POWER DISSIPATION  0.00E+00 WATTS
```

```
******* 06/05/88 ******* Evaluation PSPICE (Jan. 1988) ******* 22:11:18 *******

 EXAMPLE H.4, PSPICE SOLUTION OF EXAMPLE 13.19

 ****     AC ANALYSIS                        TEMPERATURE =  27.000 DEG C

******************************************************************************

 LEGEND:
*: VDB(4)
  FREQ        VDB(4)

(*)----------   -1.0000E+01  -8.0000E+00  -6.0000E+00  -4.0000E+00  -2.0000E+00
                - - - - - - - - - - - - - - - - - - - - - - - - - - - - - - - -
  6.800E+02 -8.076E+00 .             *           .           .           .
  6.832E+02 -7.495E+00 .             .  *        .           .           .
  6.863E+02 -6.904E+00 .             .      *    .           .           .
  6.895E+02 -6.310E+00 .             .          *.           .           .
  6.926E+02 -5.723E+00 .             .           . *         .           .
  6.958E+02 -5.159E+00 .             .           .    *      .           .
  6.989E+02 -4.639E+00 .             .           .        *  .           .
  7.021E+02 -4.188E+00 .             .           .           *.          .
  7.053E+02 -3.835E+00 .             .           .           .*          .
  7.084E+02 -3.606E+00 .             .           .           .  *        .
  7.116E+02 -3.522E+00 .             .           .           .  *        .
  7.147E+02 -3.588E+00 .             .           .           .  *        .
  7.179E+02 -3.797E+00 .             .           .           .*          .
  7.211E+02 -4.127E+00 .             .           .           *.          .
  7.242E+02 -4.550E+00 .             .           .        *  .           .
  7.274E+02 -5.039E+00 .             .           .     *     .           .
  7.305E+02 -5.568E+00 .             .           . *         .           .
  7.337E+02 -6.118E+00 .             .          *.           .           .
  7.368E+02 -6.674E+00 .             .       *   .           .           .
  7.400E+02 -7.225E+00 .             .   *       .           .           .
                - - - - - - - - - - - - - - - - - - - - - - - - - - - - - - - -
```

```
 ******* 06/05/88 ******* Evaluation PSPICE (Jan. 1988) ******* 22:11:18 *******

 EXAMPLE H.4, PSPICE SOLUTION OF EXAMPLE 13.19

 ****     AC ANALYSIS                        TEMPERATURE =  27.000 DEG C

******************************************************************************

  FREQ        VP(4)

(*)----------   -5.0000E+01   0.0000E+00   5.0000E+01   1.0000E+02   1.5000E+02
                - - - - - - - - - - - - - - - - - - - - - - - - - - - - - - - -
  6.800E+02  5.370E+01 .             .           .*          .           .
  6.832E+02  5.073E+01 .             .           *           .           .
  6.863E+02  4.735E+01 .             .          *.           .           .
  6.895E+02  4.349E+01 .             .         * .           .           .
  6.926E+02  3.909E+01 .             .        *  .           .           .
  6.958E+02  3.409E+01 .             .       *   .           .           .
  6.989E+02  2.845E+01 .             .     *     .           .           .
  7.021E+02  2.216E+01 .             .    *      .           .           .
  7.053E+02  1.530E+01 .             .  *        .           .           .
  7.084E+02  7.986E+00 .             . *         .           .           .
  7.116E+02  4.407E-01 .             *           .           .           .
  7.147E+02 -7.087E+00 .           * .           .           .           .
  7.179E+02 -1.435E+01 .         *   .           .           .           .
  7.211E+02 -2.114E+01 .        *    .           .           .           .
  7.242E+02 -2.734E+01 .      *      .           .           .           .
  7.274E+02 -3.289E+01 .    *        .           .           .           .
  7.305E+02 -3.781E+01 .  *          .           .           .           .
  7.337E+02 -4.213E+01 . *           .           .           .           .
  7.368E+02 -4.592E+01 .*            .           .           .           .
  7.400E+02 -4.924E+01 *             .           .           .           .
                - - - - - - - - - - - - - - - - - - - - - - - - - - - - - - - -
```

```
JOB CONCLUDED

TOTAL JOB TIME          16.70
```

The preceding four examples were solved with PSPICE and should serve as an introduction to this very popular and versatile circuit analysis program.

If you want to get better at PSPICE, there are many circuit analysis textbooks that have more examples. The authors also recommend the *PSPICE User's Guide,* which may be purchased from MicroSim for $40. Another book that is also very helpful is *SPICE: A Guide to Circuit Simulation and Analysis Using PSPICE.* The book was written by Paul Tuinenga and is available from Prentice-Hall.

Answers to Odd-Numbered Problems

Chapter 2

2.1 50 m
2.3 75 N
2.5 250 J
2.7 4700; 0.000354; 820,000; 0.0000019
2.9 100,000 microunits; 37,000 units; 2200 kilounits; 0.0000047 microunits; 3300 nanounits; 4.5 kilounits; 220,000 microunits
2.11 500 milliamps; 0.7 volts; 1.2 milliamps; 0.01 microfarads; 0.01 henrys
2.13 3.70×10^{-5} N
2.15 6 mm
2.17 36 miles/hour
2.19 1.25×10^{9} electrons

Chapter 3

3.1 4 A
3.3 6 mC
3.5 60 s
3.7 200 mJ
3.9 1.472×10^{1} A; 1.2×10^{3} V; 1.72×10^{-7} C; 7.02×10^{-8} A
3.11 4 min
3.13 −6 V
3.15 6000 J
3.17 60 μJ/s
3.19 17,314,560 J
3.21 30 V
3.23 1.67 min

Chapter 4

4.1 12.5 V
4.3 6 A; 12 Ω; 500 V
4.5 $87.60
4.7 24 h
4.9 0.28 Ω
4.11 6.88×10^{6} Ω
4.13 31.62 V
4.15 1.94×10^{-2} Ω
4.17 0.332 Ω
4.19 0.436 Ω
4.21 Red, Red, Red, Silver
4.23 22 kΩ ± 5%; 68 kΩ ± 10%

Chapter 5

5.1 100 Ω; 9 kΩ; 7.02 kΩ; 10.2 kΩ
5.3 2.77 mA; 81.1 mA; 2.5 μA; 1.79 mA

5.5 35 V; 25 V; 4.5 V
5.7 3.4 V
5.9 10 mA
5.11 5 mA
5.13 5 mA
5.15 1.33 mA; 2 mA; 0.67 mA
5.17 0.2 mS
5.19 30 Ω; 33.3 mS
5.21 4 kΩ; 3 kΩ
5.23 30 kΩ
5.25 27 mA; 18 mA; 9 mA; 9 mA; 3 mA; 6 mA; 2 mA; 2 mA; 2 mA; 9 mA

Chapter 6

6.1 1.96 A; 11.76 V; 5.88 V; 23.52 V
6.3 0.5 mA
6.5 −2.5 V
6.7 118.81 V
6.9 1.42 A
6.11 12 A
6.13 64.2 mA
6.15 5.15 Ω; −10.14 V
6.17 2.4 Ω
6.19 ≈73%; 3.07 A

Chapter 7

7.1 −2; −3; −12; 24
7.3 4, −0.5; 4.235, −0.353
7.5 0.323 A, −1.129 A, 0.323 A↑, 1.129 A↑, 1.452 A↓
7.7 5 A↓
7.9 1.274 A↓
7.11 8.721 V, 5.778 V, 8.439 V
7.13 1.552, −2.460, −1.802

Chapter 8

8.1 2.5 kΩ; 3.33 mA
8.3 −0.5 V
8.5 2.66 Ω; −6.68 V
8.7 4 kΩ; 17.45 V
8.9 0.23 A
8.11 −0.5 V
8.15 14.5 kΩ
8.17 −3.11 A
8.19 4.8 V
8.21 0.8 mA
8.23 0.55 mA
8.25 1.6 mA
8.27 −0.424 V
8.29 96.45 Ω

Chapter 9

9.1 60 Hz, 1 kHz, 250 Hz, 20 kHz
9.3 60 V, 21.21 V
9.5 0.52 rad, 0.79 rad, 1.57 rad
9.7 45°, 120°, 270°
9.9 50 sin62.8t
9.13 $127.26\angle 0°$ V, $30\angle -30°$ mA
9.15 10 V
9.17 70.7 sin(125.6t − 30°) mV, 106.05 sin(125.6t + 60°) mA
9.19 4.41 V
9.21 9 V
9.23 12.73 V

Chapter 10

10.1 15 μF, 12 V, 60 μC, 120 μC, 720 μJ
10.3 18 V
10.5 6 mH
10.7 10 ms, 50 ms, 2 ms
10.9 39 mA
10.11 3 mA
10.13 0.223, 0.135, 0.223
10.17 10 s
10.19 1.59 kΩ, 634 Ω, 531 Ω, 1.59 Ω
10.21 $1.59\angle -90°$ kΩ, $634\angle -90°$ Ω, $531\angle -90°$ Ω, $1.59\angle -90°$ Ω
10.23 $629\angle 90°$ μS, $1.58\angle 90°$ mS, $1.88\angle 90°$ mS, $629\angle 90°$ mS
10.25 1.256 Ω, 3.14 Ω, 3.768 Ω, 1.256 kΩ
10.27 $1.256\angle 90°$ Ω, $3.14\angle 90°$ Ω, $3.768\angle 90°$ Ω, $1.256\angle 90°$ kΩ
10.29 $796\angle -90°$ mS, $318\angle -90°$ mS, $265\angle -90°$ mS, $796\angle -90°$ mS

Chapter 11

11.1 17.7 sin(2000t + 115°) A
11.3 4.98 μF
11.5 $20\angle 60°$ Ω
11.7 $41.09\angle 13.25°$ Ω, $61.79\angle 50°$ Ω, $102.34\angle 67°$ Ω, $942.85\angle 87.6°$ Ω
11.9 5.59 A
11.11 473 Ω
11.13 $10.61\angle 0°$ V, 400 Hz, 199 Ω, −j199 Ω, $538.15\angle -21.7°$ Ω, $19.72\angle 21.7°$ mA, $3.92\angle -68.3°$ V
11.15 $12\angle 0°$ V, 2 kHz, 15.92 kΩ, 20 kΩ, j37.68 kΩ, −j15.92 kΩ, 50 mS, −j26.54 mS, j62.81 mS, $61.77\angle 35.96°$ mS, $741.24\angle 35.96°$ mA, $600\angle 0°$ mA, $318.48\angle -90°$ mA, $753.72\angle 90°$ mA
11.17 $2\angle -20.74°$ mA
11.19 $4.32\angle 30.74°$ V

Chapter 12

12.1 $42.38\angle 13.5^\circ$ V, $3.53\angle 13.5^\circ$ kΩ
12.3 $4.7\angle 27.86^\circ$ V
12.5 $48.62\angle -24.08^\circ$ V, $2.76\angle -24.08^\circ$ kΩ
12.7 $142.2\angle -28.43^\circ$ V, $7.11\angle -28.43^\circ$ kΩ
12.9 $218.07\angle 23.58^\circ$ V, $41.42\angle 13.7^\circ$ kΩ, $4.26\angle 12.53^\circ$ mA
12.11 $4.26\angle 12.6^\circ$ mA
12.13 $4.34\angle 25.8^\circ$ kΩ, $1.27\angle 18.86^\circ$ mA, $0.43\angle 58.56^\circ$ mA
12.15 $5.8\angle -15^\circ$ V, 1.5 Ω, $3.87\angle -15^\circ$ V
12.17 $3.86\angle -15.01^\circ$ V
12.19 $5.71 + 5.94 \sin(6280t - 39.97^\circ)$ V
12.21 $0.5\angle 28.72^\circ$ mS, $-0.31\angle -22.78^\circ$ mS, $-0.31\angle -22.8^\circ$ mS, $0.36\angle 9.39^\circ$ mS
12.23 $2.53\angle 37.78^\circ$ kΩ, $1.59\angle -90^\circ$ kΩ, $1.59\angle -90^\circ$ kΩ, $1.88\angle -57.83^\circ$ kΩ
12.25 $3.83\angle 44.68^\circ$ kΩ, $0.85\angle -32.17^\circ$, $-0.85\angle -32.17^\circ$, $0.53\angle 57.83^\circ$ mS
12.27 $2\angle -28.72^\circ$ kΩ, $0.62\angle -51.52^\circ$, $-0.62\angle -51.52^\circ$, $0.38\angle 40^\circ$ mS
12.29 $3.85\angle 44.68^\circ$ kΩ, $0.85\angle -32.23^\circ$, $-0.85\angle -32.23^\circ$, $0.5\angle 57.67^\circ$ mS

Chapter 13

13.1 225 kHz; 2 A
13.3 $2828\angle 90^\circ$
13.5 12.6 mH
13.7 15.9 MHz; 1.59 MHz
13.9 127 Hz; 424 Hz
13.11 493 kHz; 503 kHz
13.13 1.58 MHz
13.15 0.00656 μF
13.17 6 kHz
13.19 0.159×10^{-3} H; 636 pF
13.21 −6.99 dB
13.23 0.792 W
13.25 5.44 dBW; 34 dBm
13.27 50.97 dB
13.29 1.59 kHz
13.31 125.6 Hz
13.33 10,366 Hz; 10,405 Hz
13.35 643.7 kHz; 654.3 kHz
13.37 1.59 H; 0.0318 μF

Chapter 14

14.1 160 VA; 138.72 W; 80 VAR
14.3 34.16 VA; 33.93 W
14.5 1.37 W; 0 VAR
14.7 0.97 lagging
14.9 677.7 μF
14.11 $60.62\angle 15^\circ$ V
14.13 220 V
14.15 4.473 kW
14.17 444.44 V
14.19 7.81 kVAR (inductive)
14.21 13
14.23 $\frac{1}{15.8}$

Chapter 15

15.1 8 mΩ
15.3 49.6 kΩ
15.5 19.9 kΩ; 20 kΩ; 160 kΩ
15.7 1 MΩ; 10 kΩ
15.9 3.7 V; 3.53 V
15.11 6.3 V; 6.03 V
15.13 0.53 V; 25 kHz
15.15 2.5%

INDEX